STUDENT'S SOLUTIONS MANUAL

JUDITH A. PENNA
Indiana University Purdue University Indianapolis

INTRODUCTORY AND INTERMEDIATE ALGEBRA

SECOND EDITION

Marvin L. Bittinger
Indiana University Purdue University Indianapolis

Judith A. Beecher
Indiana University Purdue University Indianapolis

PEARSON
Addison Wesley

Boston   San Francisco   New York
London   Toronto   Sydney   Tokyo   Singapore   Madrid
Mexico City   Munich   Paris   Cape Town   Hong Kong   Montreal

Reproduced by Pearson Addison-Wesley from electronic files supplied by the author.

Copyright © 2003 Pearson Education, Inc.
Publishing as Pearson Addison-Wesley, 75 Arlington Street, Boston, MA 02116

All rights reserved. No part of this publication may be reproduced, stored in a retrieval system, or transmitted, in any form or by any means, electronic, mechanical, photocopying, recording, or otherwise, without the prior written permission of the publisher. Printed in the United States of America.

ISBN   0-201-79668-6

1 2 3 4 5 6 7 CRS 06 05 04 03

Contents

Chapter 1	1
Chapter 2	13
Chapter 3	43
Chapter 4	61
Chapter 5	85
Chapter 6	123
Chapter 7	161
Chapter 8	183
Chapter 9	211
Chapter 10	231
Chapter 11	253
Chapter 12	289

Chapter 1

Introduction to Real Numbers and Algebraic Expressions

Exercise Set 1.1

1. Substitute 34 for n: $600(34) = 20,400$, so $20,400 is collected if 34 students enroll.

   Substitute 78 for n: $600(78) = 46,800$, so $46,800 is collected if 78 students enroll.

   Substitute 250 for n: $600(250) = 150,000$, so $150,000 is collected if 250 students enroll.

3. Substitute 45 m for b and 86 m for h, and carry out the multiplication:
$$A = \frac{1}{2}bh = \frac{1}{2}(45 \text{ m})(86 \text{ m})$$
$$= \frac{1}{2}(45)(86)(\text{m})(\text{m})$$
$$= 1935 \text{ m}^2$$

5. Substitute 65 for r and 4 for t, and carry out the multiplication:
$$d = rt = 65 \cdot 4 = 260 \text{ mi}$$

7. We substitute 6 ft for l and 4 ft for w in the formula for the area of a rectangle.
$$A = lw = (6 \text{ ft})(4 \text{ ft})$$
$$= (6)(4)(\text{ft})(\text{ft})$$
$$= 24 \text{ ft}^2$$

9. $8x = 8 \cdot 7 = 56$

11. $\dfrac{a}{b} = \dfrac{24}{3} = 8$

13. $\dfrac{3p}{q} = \dfrac{3 \cdot 2}{6} = \dfrac{6}{6} = 1$

15. $\dfrac{x+y}{5} = \dfrac{10+20}{5} = \dfrac{30}{5} = 6$

17. $\dfrac{x-y}{8} = \dfrac{20-4}{8} = \dfrac{16}{8} = 2$

19. $b + 7$, or $7 + b$

21. $c - 12$

23. $4 + q$, or $q + 4$

25. $a + b$, or $b + a$

27. $x \div y$, or $\dfrac{x}{y}$, or x/y, or $x \cdot \dfrac{1}{y}$

29. $x + w$, or $w + x$

31. $n - m$

33. $x + y$, or $y + x$

35. $2z$

37. $3m$

39. Let s represent your salary. Then we have $89\%s$, or $0.89s$.

41. The distance traveled is the product of the speed and the time. Thus, Danielle traveled $65t$ miles.

43. $50 - x$

45. Discussion and Writing Exercise

47. $x + 3y$

49. $2x - 3$

Exercise Set 1.2

1. The integer -1286 corresponds to 1286 ft below sea level; the integer 14,410 corresponds to 14,410 ft above sea level.

3. The integer 24 corresponds to 24° above zero; the integer -2 corresponds to 2° below zero.

5. The integer $-5,600,000,000,000$ corresponds to the total public debt of $5,600,000,000,000.

7. The integer -34 describes the situation from the Alley Cats' point of view. The integer 34 describes the situation from the Strikers' point of view.

9. The number $\dfrac{10}{3}$ can be named $3\dfrac{1}{3}$, or $3.3\overline{3}$. The graph is $\dfrac{1}{3}$ of the way from 3 to 4.

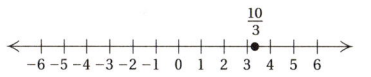

11. The graph of -5.2 is $\dfrac{2}{10}$ of the way from -5 to -6.

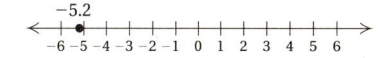

13. We first find decimal notation for $\frac{7}{8}$. Since $\frac{7}{8}$ means $7 \div 8$, we divide.

    $$\begin{array}{r} 0.875 \\ 8\overline{\smash{)}7.000} \\ \underline{64} \\ 60 \\ \underline{56} \\ 40 \\ \underline{40} \\ 0 \end{array}$$

    Thus $\frac{7}{8} = 0.875$, so $-\frac{7}{8} = -0.875$.

15. $\frac{5}{6}$ means $5 \div 6$, so we divide.

    $$\begin{array}{r} 0.833\ldots \\ 6\overline{\smash{)}5.000} \\ \underline{48} \\ 20 \\ \underline{18} \\ 20 \\ \underline{18} \\ 2 \end{array}$$

    We have $\frac{5}{6} = 0.8\overline{3}$.

17. $\frac{7}{6}$ means $7 \div 6$, so we divide.

    $$\begin{array}{r} 1.166\ldots \\ 6\overline{\smash{)}7.000} \\ \underline{6} \\ 10 \\ \underline{6} \\ 40 \\ \underline{36} \\ 40 \\ \underline{36} \\ 4 \end{array}$$

    We have $\frac{7}{6} = 1.1\overline{6}$.

19. $\frac{2}{3}$ means $2 \div 3$, so we divide.

    $$\begin{array}{r} 0.666\ldots \\ 3\overline{\smash{)}2.000} \\ \underline{18} \\ 20 \\ \underline{18} \\ 20 \\ \underline{18} \\ 2 \end{array}$$

    We have $\frac{2}{3} = 0.\overline{6}$.

21. We first find decimal notation for $\frac{1}{2}$. Since $\frac{1}{2}$ means $1 \div 2$, we divide.

    $$\begin{array}{r} 0.5 \\ 2\overline{\smash{)}1.0} \\ \underline{10} \\ 0 \end{array}$$

    Thus $\frac{1}{2} = 0.5$, so $-\frac{1}{2} = -0.5$

23. $\frac{1}{10}$ means $1 \div 10$, so we divide.

    $$\begin{array}{r} 0.1 \\ 10\overline{\smash{)}1.0} \\ \underline{10} \\ 0 \end{array}$$

    We have $\frac{1}{10} = 0.1$

25. Since 8 is to the right of 0, we have $8 > 0$.

27. Since -8 is to the left of 3, we have $-8 < 3$.

29. Since -8 is to the left of 8, we have $-8 < 8$.

31. Since -8 is to the left of -5, we have $-8 < -5$.

33. Since -5 is to the right of -11, we have $-5 > -11$.

35. Since -6 is to the left of -5, we have $-6 < -5$.

37. Since 2.14 is to the right of 1.24, we have $2.14 > 1.24$.

39. Since -14.5 is to the left of 0.011, we have $-14.5 < 0.011$.

41. Since -12.88 is to the left of -6.45, we have $-12.88 < -6.45$.

43. Convert to decimal notation $\frac{5}{12} = 0.4166\ldots$ and $\frac{11}{25} = 0.44$. Since $0.4166\ldots$ is to the left of 0.44, $\frac{5}{12} < \frac{11}{25}$.

45. $-3 \geq -11$ is true since $-3 > -11$ is true.

47. $0 \geq 8$ is false since neither $0 > 8$ nor $0 = 8$ is true.

49. $x < -6$ has the same meaning as $-6 > x$.

51. $y \geq -10$ has the same meaning as $-10 \leq y$.

53. The distance of -3 from 0 is 3, so $|-3| = 3$.

55. The distance of 10 from 0 is 10, so $|10| = 10$.

57. The distance of 0 from 0 is 0, so $|0| = 0$.

59. The distance of -24 from 0 is 24, so $|-24| = 24$.

61. The distance of $-\frac{2}{3}$ from 0 is $\frac{2}{3}$, so $\left|-\frac{2}{3}\right| = \frac{2}{3}$.

63. The distance of $\frac{0}{4}$ from 0 is $\frac{0}{4}$, or 0, so $\left|\frac{0}{4}\right| = 0$.

65. The distance of $-3\frac{5}{8}$ from 0 is $3\frac{5}{8}$, so $\left|-3\frac{5}{8}\right| = 3\frac{5}{8}$.

67. Discussion and Writing Exercise

69. $\dfrac{5c}{d} = \dfrac{5 \cdot 15}{25} = \dfrac{75}{25} = 3$

71. $\dfrac{q-r}{8} = \dfrac{30-6}{8} = \dfrac{24}{8} = 3$

73. $-\dfrac{2}{3}, \dfrac{1}{2}, -\dfrac{3}{4}, -\dfrac{5}{6}, \dfrac{3}{8}, \dfrac{1}{6}$ can be written in decimal notation as $-0.\overline{6}, 0.5, -0.75, -0.8\overline{3}, 0.375, 0.1\overline{6}$, respectively. Listing from least to greatest, we have
$$-\dfrac{5}{6}, -\dfrac{3}{4}, -\dfrac{2}{3}, \dfrac{1}{6}, \dfrac{3}{8}, \dfrac{1}{2}.$$

75. $0.\overline{1} = \dfrac{0.\overline{3}}{3} = \dfrac{\frac{1}{3}}{3} = \dfrac{1}{3} \cdot \dfrac{1}{3} = \dfrac{1}{9}$

77. First consider $0.\overline{5}$.
$$0.\overline{5} = 0.\overline{3} \cdot \dfrac{5}{3} = \dfrac{1}{3} \cdot \dfrac{5}{3} = \dfrac{5}{9}$$
Then, $5.\overline{5} = 5 + 0.\overline{5} = 5 + \dfrac{5}{9} = 5\dfrac{5}{9}$, or $\dfrac{50}{9}$.

Exercise Set 1.3

1. $2 + (-9)$ The absolute values are 2 and 9. The difference is $9 - 2$, or 7. The negative number has the larger absolute value, so the answer is negative. $2 + (-9) = -7$

3. $-11 + 5$ The absolute values are 11 and 5. The difference is $11 - 5$, or 6. The negative number has the larger absolute value, so the answer is negative. $-11 + 5 = -6$

5. $-8 + 8$ A negative and a positive number. The numbers have the same absolute value. The sum is 0. $-8 + 8 = 0$

7. $-3 + (-5)$ Two negatives. Add the absolute values, getting 8. Make the answer negative. $-3 + (-5) = -8$

9. $-7 + 0$ One number is 0. The answer is the other number. $-7 + 0 = -7$

11. $0 + (-27)$ One number is 0. The answer is the other number. $0 + (-27) = -27$

13. $17 + (-17)$ A negative and a positive number. The numbers have the same absolute value. The sum is 0. $17 + (-17) = 0$

15. $-17 + (-25)$ Two negatives. Add the absolute values, getting 42. Make the answer negative. $-17 + (-25) = -42$

17. $18 + (-18)$ A positive and a negative number. The numbers have the same absolute value. The sum is 0. $18 + (-18) = 0$

19. $-28 + 28$ A negative and a positive number. The numbers have the same absolute value. The sum is 0. $-28 + 28 = 0$

21. $8 + (-5)$ The absolute values are 8 and 5. The difference is $8 - 5$, or 3. The positive number has the larger absolute value, so the answer is positive. $8 + (-5) = 3$

23. $-4 + (-5)$ Two negatives. Add the absolute values, getting 9. Make the answer negative. $-4 + (-5) = -9$

25. $13 + (-6)$ The absolute values are 13 and 6. The difference is $13 - 6$, or 7. The positive number has the larger absolute value, so the answer is positive. $13 + (-6) = 7$

27. $-25 + 25$ A negative and a positive number. The numbers have the same absolute value. The sum is 0. $-25 + 25 = 0$

29. $53 + (-18)$ The absolute values are 53 and 18. The difference is $53 - 18$, or 35. The positive number has the larger absolute value, so the answer is positive. $53 + (-18) = 35$

31. $-8.5 + 4.7$ The absolute values are 8.5 and 4.7. The difference is $8.5 - 4.7$, or 3.8. The negative number has the larger absolute value, so the answer is negative. $-8.5 + 4.7 = -3.8$

33. $-2.8 + (-5.3)$ Two negatives. Add the absolute values, getting 8.1. Make the answer negative. $-2.8 + (-5.3) = -8.1$

35. $-\dfrac{3}{5} + \dfrac{2}{5}$ The absolute values are $\dfrac{3}{5}$ and $\dfrac{2}{5}$. The difference is $\dfrac{3}{5} - \dfrac{2}{5}$, or $\dfrac{1}{5}$. The negative number has the larger absolute value, so the answer is negative. $-\dfrac{3}{5} + \dfrac{2}{5} = -\dfrac{1}{5}$

37. $-\dfrac{2}{9} + \left(-\dfrac{5}{9}\right)$ Two negatives. Add the absolute values, getting $\dfrac{7}{9}$. Make the answer negative. $-\dfrac{2}{9} + \left(-\dfrac{5}{9}\right) = -\dfrac{7}{9}$

39. $-\dfrac{5}{8} + \dfrac{1}{4}$ The absolute values are $\dfrac{5}{8}$ and $\dfrac{1}{4}$. The difference is $\dfrac{5}{8} - \dfrac{2}{8}$, or $\dfrac{3}{8}$. The negative number has the larger absolute value, so the answer is negative. $-\dfrac{5}{8} + \dfrac{1}{4} = -\dfrac{3}{8}$

41. $-\dfrac{5}{8} + \left(-\dfrac{1}{6}\right)$ Two negatives. Add the absolute values, getting $\dfrac{15}{24} + \dfrac{4}{24}$, or $\dfrac{19}{24}$. Make the answer negative. $-\dfrac{5}{8} + \left(-\dfrac{1}{6}\right) = -\dfrac{19}{24}$

43. $-\dfrac{3}{8} + \dfrac{5}{12}$ The absolute values are $\dfrac{3}{8}$ and $\dfrac{5}{12}$. The difference is $\dfrac{10}{24} - \dfrac{9}{24}$, or $\dfrac{1}{24}$. The positive number has the larger absolute value, so the answer is positive. $-\dfrac{3}{8} + \dfrac{5}{12} = \dfrac{1}{24}$

45. $76 + (-15) + (-18) + (-6)$
a) Add the negative numbers: $-15 + (-18) + (-6) = -39$
b) Add the results: $76 + (-39) = 37$

47. $-44 + \left(-\dfrac{3}{8}\right) + 95 + \left(-\dfrac{5}{8}\right)$
a) Add the negative numbers: $-44 + \left(-\dfrac{3}{8}\right) + \left(-\dfrac{5}{8}\right) = -45$
b) Add the results: $-45 + 95 = 50$

49. We add from left to right.
$$\begin{aligned}&98+(-54)+113+(-998)+44+(-612)\\=\ &44+113+(-998)+44+(-612)\\=\ &157+(-998)+44+(-612)\\=\ &-841+44+(-612)\\=\ &-797+(-612)\\=\ &-1409\end{aligned}$$

51. The additive inverse of 24 is -24 because $24+(-24)=0$.

53. The additive inverse of -26.9 is 26.9 because $-26.9+26.9=0$.

55. If $x=8$, then $-x=-8$. (The opposite of 8 is -8.)

57. If $x=-\frac{13}{8}$ then $-x=-\left(-\frac{13}{8}\right)=\frac{13}{8}$. (The opposite of $-\frac{13}{8}$ is $\frac{13}{8}$.)

59. If $x=-43$ then $-(-x)=-(-(-43))=-43$. (The opposite of the opposite of -43 is -43.)

61. If $x=\frac{4}{3}$ then $-(-x)=-\left(-\frac{4}{3}\right)=\frac{4}{3}$. (The opposite of the opposite of $\frac{4}{3}$ is $\frac{4}{3}$.)

63. $-(-24)=24$ (The opposite of -24 is 24.)

65. $-\left(-\frac{3}{8}\right)=\frac{3}{8}$ (The opposite of $-\frac{3}{8}$ is $\frac{3}{8}$.)

67. Let $E=$ the elevation of Mauna Kea above sea level.

$$\underbrace{\text{Elevation above sea level}}_{\downarrow}\ \text{is}\ \underbrace{\text{total height}}_{\downarrow}\ \text{plus}\ \underbrace{\text{elevation below sea level.}}_{\downarrow}$$
$$E\ \ =\ \ 33{,}480\ \ +\ \ (-19{,}684)$$

We carry out the addition.
$$E=33{,}480+(-19{,}684)=13{,}796$$
The elevation of Mauna Kea is 13,796 ft above sea level.

69. Let $T=$ the final temperature. We will express the rise in temperature as a positive number and a decrease in the temperature as a negative number.

$$\underbrace{\text{Final temperature}}_{\downarrow}\ \text{is}\ \underbrace{\text{original temperature}}_{\downarrow}\ \text{plus}$$
$$T\ \ =\ \ 32\ \ +$$
$$\underbrace{\text{rise in temperature}}_{\downarrow}\ \text{plus}\ \underbrace{\text{decrease in temperature.}}_{\downarrow}$$
$$15\ \ +\ \ (-50)$$

We add from left to right.
$$\begin{aligned}T&=32+15+(-50)\\&=47+(-50)\\&=-3\end{aligned}$$
The final temperature was $-3°$F.

71. Let $S=$ the sum of the profits and losses. We add the five numbers in the bar graph to find S.
$$S=\$10{,}500+(-\$16{,}600)+(-\$12{,}800)+(-\$9600)+\$8200=-\$20{,}300$$
The sum of the profits and losses is $-\$20{,}300$.

73. Let $B=$ the new balance in the account. We will express the deposit as a positive number and the amounts of the checks as negative numbers.

$$\underbrace{\text{New balance}}_{\downarrow}\ \text{is}\ \underbrace{\text{original balance}}_{\downarrow}\ \text{plus}\ \underbrace{\text{amount of first check}}_{\downarrow}\ \text{plus}$$
$$B\ \ =\ \ 460\ \ +\ \ (-530)\ \ +$$
$$\underbrace{\text{amount of deposit}}_{\downarrow}\ \text{plus}\ \underbrace{\text{amount of second check.}}_{\downarrow}$$
$$75\ \ +\ \ (-90)$$

We add from left to right.
$$\begin{aligned}B&=460+(-530)+75+(-90)\\&=-70+75+(-90)\\&=5+(-90)\\&=-85\end{aligned}$$
The balance in the account is $-\$85$.

75. Discussion and Writing Exercise

77. We first find decimal notation for $\frac{5}{8}$. Since $\frac{5}{8}$ means $5\div8$, we divide.

$$\begin{array}{r}0.625\\8\overline{)5.000}\\\underline{48}\\20\\\underline{16}\\40\\\underline{40}\\0\end{array}$$

Thus $\frac{5}{8}=0.625$, so $-\frac{5}{8}=-0.625$.

79. First we find decimal notation for $\frac{1}{12}$. Since $\frac{1}{12}$ means $1\div12$, we divide.

$$\begin{array}{r}0.0833\ldots\\12\overline{)1.0000}\\\underline{96}\\40\\\underline{36}\\40\\\underline{36}\\4\end{array}$$

Thus $\frac{1}{12}=0.08\overline{3}$, so $-\frac{1}{12}=-0.08\overline{3}$.

81. The distance of 2.3 from 0 is 2.3, so $|2.3|=2.3$.

83. The distance of $-\frac{4}{5}$ from 0 is $\frac{4}{5}$, so $\left|-\frac{4}{5}\right|=\frac{4}{5}$.

Exercise Set 1.4

85. When x is positive, the opposite of x, $-x$, is negative, so $-x$ is negative for all positive numbers x.

87. If a is positive, $-a$ is negative. Thus $-a + b$, the sum of two negative numbers, is negative. The correct answer is (b).

Exercise Set 1.4

1. $2 - 9 = 2 + (-9) = -7$
3. $0 - 4 = 0 + (-4) = -4$
5. $-8 - (-2) = -8 + 2 = -6$
7. $-11 - (-11) = -11 + 11 = 0$
9. $12 - 16 = 12 + (-16) = -4$
11. $20 - 27 = 20 + (-27) = -7$
13. $-9 - (-3) = -9 + 3 = -6$
15. $-40 - (-40) = -40 + 40 = 0$
17. $7 - 7 = 7 + (-7) = 0$
19. $7 - (-7) = 7 + 7 = 14$
21. $8 - (-3) = 8 + 3 = 11$
23. $-6 - 8 = -6 + (-8) = -14$
25. $-4 - (-9) = -4 + 9 = 5$
27. $1 - 8 = 1 + (-8) = -7$
29. $-6 - (-5) = -6 + 5 = -1$
31. $8 - (-10) = 8 + 10 = 18$
33. $0 - 10 = 0 + (-10) = -10$
35. $-5 - (-2) = -5 + 2 = -3$
37. $-7 - 14 = -7 + (-14) = -21$
39. $0 - (-5) = 0 + 5 = 5$
41. $-8 - 0 = -8 + 0 = -8$
43. $7 - (-5) = 7 + 5 = 12$
45. $2 - 25 = 2 + (-25) = -23$
47. $-42 - 26 = -42 + (-26) = -68$
49. $-71 - 2 = -71 + (-2) = -73$
51. $24 - (-92) = 24 + 92 = 116$
53. $-50 - (-50) = -50 + 50 = 0$
55. $-\frac{3}{8} - \frac{5}{8} = -\frac{3}{8} + \left(-\frac{5}{8}\right) = -\frac{8}{8} = -1$
57. $\frac{3}{4} - \frac{2}{3} = \frac{3}{4} + \left(-\frac{2}{3}\right) = \frac{9}{12} + \left(-\frac{8}{12}\right) = \frac{1}{12}$
59. $-\frac{3}{4} - \frac{2}{3} = -\frac{3}{4} + \left(-\frac{2}{3}\right) = -\frac{9}{12} + \left(-\frac{8}{12}\right) = -\frac{17}{12}$
61. $-\frac{5}{8} - \left(-\frac{3}{4}\right) = -\frac{5}{8} + \frac{3}{4} = -\frac{5}{8} + \frac{6}{8} = \frac{1}{8}$
63. $6.1 - (-13.8) = 6.1 + 13.8 = 19.9$
65. $-2.7 - 5.9 = -2.7 + (-5.9) = -8.6$
67. $0.99 - 1 = 0.99 + (-1) = -0.01$
69. $-79 - 114 = -79 + (-114) = -193$
71. $0 - (-500) = 0 + 500 = 500$
73. $-2.8 - 0 = -2.8 + 0 = -2.8$
75. $7 - 10.53 = 7 + (-10.53) = -3.53$
77. $\frac{1}{6} - \frac{2}{3} = \frac{1}{6} + \left(-\frac{2}{3}\right) = \frac{1}{6} + \left(-\frac{4}{6}\right) = -\frac{3}{6}$, or $-\frac{1}{2}$
79. $-\frac{4}{7} - \left(-\frac{10}{7}\right) = -\frac{4}{7} + \frac{10}{7} = \frac{6}{7}$
81. $-\frac{7}{10} - \frac{10}{15} = -\frac{7}{10} + \left(-\frac{10}{15}\right) = -\frac{21}{30} + \left(-\frac{20}{30}\right) = -\frac{41}{30}$
83. $\frac{1}{5} - \frac{1}{3} = \frac{1}{5} + \left(-\frac{1}{3}\right) = \frac{3}{15} + \left(-\frac{5}{15}\right) = -\frac{2}{15}$
85. $18 - (-15) - 3 - (-5) + 2 = 18 + 15 + (-3) + 5 + 2 = 37$
87. $-31 + (-28) - (-14) - 17 = (-31) + (-28) + 14 + (-17) = -62$
89. $-34 - 28 + (-33) - 44 = (-34) + (-28) + (-33) + (-44) = -139$
91. $-93 - (-84) - 41 - (-56) = (-93) + 84 + (-41) + 56 = 6$
93. $-5 - (-30) + 30 + 40 - (-12) = (-5) + 30 + 30 + 40 + 12 = 107$
95. $132 - (-21) + 45 - (-21) = 132 + 21 + 45 + 21 = 219$

97. Let D = the difference in elevation.

$$\underbrace{\text{Difference in elevation}}_{D} \underbrace{\text{is}}_{=} \underbrace{\text{larger depth}}_{11{,}033} \underbrace{\text{minus}}_{-} \underbrace{\text{smaller depth.}}_{8648}$$

We carry out the subtraction.
$$D = 11{,}033 - 8648 = 2385$$

The difference in elevation is 2385 m.

99. Let A = the amount owed.

$$\underbrace{\text{Amount owed}}_{A} \underbrace{\text{is}}_{=} \underbrace{\text{amount of charge}}_{476.89} \underbrace{\text{minus}}_{-} \underbrace{\text{amount of return.}}_{128.95}$$

We subtract.
$$A = 476.89 - 128.95 = 347.94$$

Laura owes $347.94.

101. a) We subtract the number of home runs allowed from the number of home runs hit.
Home run differential = 197 − 120 = 77

b) We subtract the number of home runs allowed from the number of home runs hit.
Home run differential = 153 − 194 = −41

103. Let D = the difference in elevation.

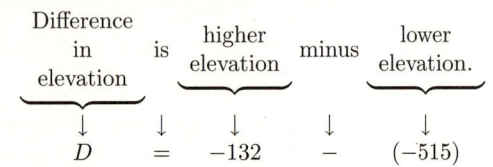

$$\underbrace{\text{Difference in elevation}}_{D} \text{ is } \underbrace{\text{higher elevation}}_{-132} \text{ minus } \underbrace{\text{lower elevation.}}_{(-515)}$$

We carry out the subtraction.
$D = -132 - (-515) = -132 + 515 = 383$

Lake Assal is 383 ft lower than the Valdes Peninsula.

105. Discussion and Writing Exercise

107. $y + 7$, or $7 + y$

109. $a - h$

111. $s + r$, or $r + s$

113. Use a calculator.
$123,907 - 433,789 = -309,882$

115. False. $3 - 0 = 3, 0 - 3 = -3, 3 - 0 \neq 0 - 3$

117. True

119. True by definition of opposites.

121. a) We add the values assigned to the cards.
$-1 + (-1) + 1 + 1 + 1 + (-1) + (-1) + 0 + (-1) + (-1) + 1 = -2$

b) Yes, because the final count is negative.

Exercise Set 1.5

1. -8

3. -48

5. -24

7. -72

9. 16

11. 42

13. -120

15. -238

17. 1200

19. 98

21. -72

23. -12.4

25. 30

27. 21.7

29. $\frac{2}{3} \cdot \left(-\frac{3}{5}\right) = -\left(\frac{2 \cdot 3}{3 \cdot 5}\right) = -\left(\frac{2}{5} \cdot \frac{3}{3}\right) = -\frac{2}{5}$

31. $-\frac{3}{8} \cdot \left(-\frac{2}{9}\right) = \frac{3 \cdot 2}{8 \cdot 9} = \frac{3 \cdot 2 \cdot 1}{4 \cdot 2 \cdot 3 \cdot 3} = \frac{3 \cdot 2}{3 \cdot 2} \cdot \frac{1}{4 \cdot 3} = \frac{1}{12}$

33. -17.01

35. $-\frac{5}{9} \cdot \frac{3}{4} = -\left(\frac{5 \cdot 3}{9 \cdot 4}\right) = -\frac{5 \cdot 3}{3 \cdot 3 \cdot 4} = -\frac{5}{3 \cdot 4} \cdot \frac{3}{3} = -\frac{5}{12}$

37. $7 \cdot (-4) \cdot (-3) \cdot 5 = 7 \cdot 12 \cdot 5 = 7 \cdot 60 = 420$

39. $-\frac{2}{3} \cdot \frac{1}{2} \cdot \left(-\frac{6}{7}\right) = -\frac{2}{6} \cdot \left(-\frac{6}{7}\right) = \frac{2 \cdot 6}{7 \cdot 6} = \frac{2}{7} \cdot \frac{6}{6} = \frac{2}{7}$

41. $-3 \cdot (-4) \cdot (-5) = 12 \cdot (-5) = -60$

43. $-2 \cdot (-5) \cdot (-3) \cdot (-5) = 10 \cdot 15 = 150$

45. $-\frac{2}{45}$

47. $-7 \cdot (-21) \cdot 13 = 147 \cdot 13 = 1911$

49. $-4 \cdot (-1.8) \cdot 7 = (7.2) \cdot 7 = 50.4$

51. $-\frac{1}{9} \cdot \left(-\frac{2}{3}\right) \cdot \left(\frac{5}{7}\right) = \frac{2}{27} \cdot \frac{5}{7} = \frac{10}{189}$

53. $4 \cdot (-4) \cdot (-5) \cdot (-12) = -16 \cdot (60) = -960$

55. $0.07 \cdot (-7) \cdot 6 \cdot (-6) = 0.07 \cdot 6 \cdot (-7) \cdot (-6) = 0.42 \cdot (42) = 17.64$

57. $\left(-\frac{5}{6}\right)\left(\frac{1}{8}\right)\left(-\frac{3}{7}\right)\left(-\frac{1}{7}\right) = \left(-\frac{5}{48}\right)\left(\frac{3}{49}\right) = -\frac{5 \cdot 3}{16 \cdot 3 \cdot 49} = -\frac{5}{16 \cdot 49} \cdot \frac{3}{3} = -\frac{5}{784}$

59. 0, The product of 0 and any real number is 0.

61. $(-8)(-9)(-10) = 72(-10) = -720$

63. $(-6)(-7)(-8)(-9)(-10) = 42 \cdot 72 \cdot (-10) = 3024 \cdot (-10) = -30,240$

65. $(-3x)^2 = (-3 \cdot 7)^2$ Substituting
$= (-21)^2$ Multiplying inside the parentheses
$= (-21)(-21)$ Evaluating the power
$= 441$

$-3x^2 = -3(7)^2$ Substituting
$= -3 \cdot 49$ Evaluating the power
$= -147$

67. When $x = 2$: $5x^2 = 5(2)^2$ Substituting
$= 5 \cdot 4$ Evaluating the power
$= 20$

When $x = -2$: $5x^2 = 5(-2)^2$ Substituting
$= 5 \cdot 4$ Evaluating the power
$= 20$

69. Let w = the total weight change. Since Dave's weight decreases 2 lb each week for 10 weeks we have
$$w = 10 \cdot (-2) = -20.$$
Thus, the total weight change is -20 lb.

71. This is a multistep problem. First we find the number of degrees the temperature dropped. Since it dropped $3°C$ each minute for 18 minutes we have a drop d given by
$$d = 18 \cdot (-3) = -54.$$
Now let T = the temperature at 10:18 AM.
$$T = 0 + (-54) = -54$$
The temperature was $-54°C$ at 10:18 AM.

73. This is a multistep problem. First we find the total decrease in price. Since it decreased $1.38 each hour for 8 hours we have a decrease in price d given by
$$d = 8(-\$1.38) = -\$11.04.$$
Now let P = the price of the stock after 8 hours.
$$P = \$23.75 + (-\$11.04) = \$12.71$$
After 8 hours the price of the stock was $12.71.

75. This is a multistep problem. First we find the total distance the diver rises. Since the diver rises 7 meters each minute for 9 minutes, the total distance d the diver rises is given by
$$d = 9 \cdot 7 = 63.$$
Now let E = the diver's elevation after 9 minutes.
$$E = -95 + 63 = -32$$
The diver's elevation is -32 m, or 32 m below the surface.

77. Discussion and Writing Exercise

79.
$$\frac{x - 2y}{3} = \frac{20 - 2 \cdot 7}{3}$$
$$= \frac{20 - 14}{3}$$
$$= \frac{6}{3}$$
$$= 2$$

81. Since -10 is to the right of -12, it is true that $-10 > -12$.

83. Since 4 is to the right of -8, it is false that $4 < -8$.

85. If a is positive and b is negative, then ab is negative and thus $-ab$ is positive. The correct answer is (a).

87. To locate $2x$, start at 0 and measure off two adjacent lengths of x to the right of 0.

To locate $3x$, start at 0 and measure off three adjacent lengths of x to the right of 0.

To locate $2y$, start at 0 and measure off two adjacent lengths of y to the right of 0.

To locate $-x$, start at 0 and measure off the length x to the left of 0.

To locate $-y$, start at 0 and measure off the length y to the left of 0.

To locate $x + y$, start at 0 and measure off the length x to the right of 0 followed by the length y immediately to the right of x. (We could also measure off y followed by x.)

To locate $x - y$, start at 0 and measure off the length x to the right of 0. Then, from that point, measure off the length y going to the left.

To locate $x - 2y$, first locate $x - y$ as described above. Then, from that point, measure off another length y going to the left.

Exercise Set 1.6

1. $48 \div (-6) = -8$ Check: $-8(-6) = 48$

3. $\dfrac{28}{-2} = -14$ Check: $-14(-2) = 28$

5. $\dfrac{-24}{8} = -3$ Check: $-3 \cdot 8 = -24$

7. $\dfrac{-36}{-12} = 3$ Check: $3(-12) = -36$

9. $\dfrac{-72}{9} = -8$ Check: $-8 \cdot 9 = -72$

11. $-100 \div (-50) = 2$ Check: $2(-50) = -100$

13. $-108 \div 9 = -12$ Check: $9(-12) = -108$

15. $\dfrac{200}{-25} = -8$ Check: $-8(-25) = 200$

17. Not defined

19. $\dfrac{-23}{-2} = \dfrac{23}{2}$ Check: $\dfrac{23}{2}(-2) = -23$

21. The reciprocal of $\dfrac{15}{7}$ is $\dfrac{7}{15}$ because $\dfrac{15}{7} \cdot \dfrac{7}{15} = 1$.

23. The reciprocal of $-\dfrac{47}{13}$ is $-\dfrac{13}{47}$ because $\left(-\dfrac{47}{13}\right) \cdot \left(-\dfrac{13}{47}\right) = 1$.

25. The reciprocal of 13 is $\dfrac{1}{13}$ because $13 \cdot \dfrac{1}{13} = 1$.

27. The reciprocal of 4.3 is $\dfrac{1}{4.3}$ because $4.3 \cdot \dfrac{1}{4.3} = 1$.

29. The reciprocal of $-\dfrac{1}{7.1}$ is -7.1 because $\left(-\dfrac{1}{7.1}\right)(-7.1) = 1$.

31. The reciprocal of $\dfrac{p}{q}$ is $\dfrac{q}{p}$ because $\dfrac{p}{q} \cdot \dfrac{q}{p} = 1$.

33. The reciprocal of $\dfrac{1}{4y}$ is $4y$ because $\dfrac{1}{4y} \cdot 4y = 1$.

35. The reciprocal of $\dfrac{2a}{3b}$ is $\dfrac{3b}{2a}$ because $\dfrac{2a}{3b} \cdot \dfrac{3b}{2a} = 1$.

37. $4 \cdot \dfrac{1}{17}$

39. $8 \cdot \left(-\dfrac{1}{13}\right)$

41. $13.9 \cdot \left(-\dfrac{1}{1.5}\right)$

43. $x \cdot y$

45. $(3x+4)\left(\dfrac{1}{5}\right)$

47. $(5a-b)\left(\dfrac{1}{5a+b}\right)$

49. $\dfrac{3}{4} \div \left(-\dfrac{2}{3}\right) = \dfrac{3}{4} \cdot \left(-\dfrac{3}{2}\right) = -\dfrac{9}{8}$

51. $-\dfrac{5}{4} \div \left(-\dfrac{3}{4}\right) = -\dfrac{5}{4} \cdot \left(-\dfrac{4}{3}\right) = \dfrac{20}{12} = \dfrac{5 \cdot 4}{3 \cdot 4} = \dfrac{5}{3}$

53. $-\dfrac{2}{7} \div \left(-\dfrac{4}{9}\right) = -\dfrac{2}{7} \cdot \left(-\dfrac{9}{4}\right) = \dfrac{18}{28} = \dfrac{9 \cdot 2}{14 \cdot 2} = \dfrac{9}{14}$

55. $-\dfrac{3}{8} \div \left(-\dfrac{8}{3}\right) = -\dfrac{3}{8} \cdot \left(-\dfrac{3}{8}\right) = \dfrac{9}{64}$

57. $-6.6 \div 3.3 = -2$ Do the long division. Make the answer negative.

59. $\dfrac{-11}{-13} = \dfrac{11}{13}$ The opposite of a number divided by the opposite of another number is the quotient of the two numbers.

61. $\dfrac{48.6}{-3} = -16.2$ Do the long division. Make the answer negative.

63. $\dfrac{-9}{17-17} = \dfrac{-9}{0}$ Division by 0 is not defined.

65. $\dfrac{-4}{54} \approx -0.074 \approx -7.4\%$

67. $\dfrac{97}{229} \approx 0.424 \approx 42.4\%$

69. Discussion and Writing Exercise

71. $\dfrac{1}{4} - \dfrac{1}{2} = \dfrac{1}{4} + \left(-\dfrac{1}{2}\right) = \dfrac{1}{4} + \left(-\dfrac{2}{4}\right) = -\dfrac{1}{4}$

73. $35 \cdot (-1.2) = -42$

75. $13.4 + (-4.9)$ The absolute values are 13.4 and 4.9. The difference is $13.4 - 4.9$, or 8.5. The positive number has the larger absolute value, so the answer is positive.
$13.4 + (-4.9) = 8.5$

77. First we find decimal notation for $\dfrac{1}{11}$. Since $\dfrac{1}{11}$ means $1 \div 11$, we divide.

$\quad\quad 0.0\,9\,0\,9\ldots$
$11\overline{\smash{)}1.0\,0\,0\,0}$
$\quad\quad\underline{9\,9}$
$\quad\quad1\,0\,0$
$\quad\quad\underline{9\,9}$
$\quad\quad1$

Thus $\dfrac{1}{11} = 0.\overline{09}$, so $-\dfrac{1}{11} = -0.\overline{09}$.

79. We find $\dfrac{1}{-10.5}$ using a calculator. If a reciprocal key is available, enter -10.5 and then use the reciprocal key. If a reciprocal key is not available, find $\dfrac{1}{-10.5}$, or $1 \div (-10.5)$. In either case, the result is $-0.\overline{095238}$.

Use the reciprocal key again or enter $\dfrac{1}{\frac{1}{-10.5}}$ to find the reciprocal of the first result. The calculator returns -10.5, the original number.

81. $-a$ is positive and b is negative, so $\dfrac{-a}{b}$ is the quotient of a positive and a negative number and, thus, is negative.

83. a is negative and $-b$ is positive, so $\dfrac{a}{-b}$ is the quotient of a negative number and a positive number and, thus, is negative. Then $-\left(\dfrac{a}{-b}\right)$ is the opposite of a negative number and, thus, is positive.

85. $-a$ and $-b$ are both positive, so $\dfrac{-a}{-b}$ is the quotient of two positive numbers and, thus, is positive. Then $-\left(\dfrac{-a}{-b}\right)$ is the opposite of a positive number and, thus, is negative.

Exercise Set 1.7

1. Note that $5y = 5 \cdot y$. We multiply by 1, using y/y as an equivalent expression for 1:
$$\dfrac{3}{5} = \dfrac{3}{5} \cdot 1 = \dfrac{3}{5} \cdot \dfrac{y}{y} = \dfrac{3y}{5y}$$

3. Note that $15x = 3 \cdot 5x$. We multiply by 1, using $5x/5x$ as an equivalent expression for 1:
$$\dfrac{2}{3} = \dfrac{2}{3} \cdot 1 = \dfrac{2}{3} \cdot \dfrac{5x}{5x} = \dfrac{10x}{15x}$$

5. $-\dfrac{24a}{16a} = -\dfrac{3 \cdot 8a}{2 \cdot 8a}$
$\phantom{-\dfrac{24a}{16a}} = -\dfrac{3}{2} \cdot \dfrac{8a}{8a}$
$\phantom{-\dfrac{24a}{16a}} = -\dfrac{3}{2} \cdot 1 \quad \left(\dfrac{8a}{8a} = 1\right)$
$\phantom{-\dfrac{24a}{16a}} = -\dfrac{3}{2} \quad\quad$ Identity property of 1

7. $-\dfrac{42ab}{36ab} = -\dfrac{7 \cdot 6ab}{6 \cdot 6ab}$
$\phantom{-\dfrac{42ab}{36ab}} = -\dfrac{7}{6} \cdot \dfrac{6ab}{6ab}$
$\phantom{-\dfrac{42ab}{36ab}} = -\dfrac{7}{6} \cdot 1 \quad \left(\dfrac{6ab}{6ab} = 1\right)$
$\phantom{-\dfrac{42ab}{36ab}} = -\dfrac{7}{6} \quad\quad$ Identity property of 1

9. $8 + y$, commutative law of addition

11. nm, commutative law of multiplication

13. $xy + 9$, commutative law of addition
$9 + yx$, commutative law of multiplication

Exercise Set 1.7

15. $c + ab$, commutative law of addition
$ba + c$, commutative law of multiplication

17. $(a + b) + 2$, associative law of addition

19. $8(xy)$, associative law of multiplication

21. $a + (b + 3)$, associative law of addition

23. $(3a)b$, associative law of multiplication

25. a) $(a + b) + 2 = a + (b + 2)$, associative law of addition
b) $(a + b) + 2 = (b + a) + 2$, commutative law of addition
c) $(a + b) + 2 = (b + a) + 2$ Using the commutative law first,
$= b + (a + 2)$ then the associative law
There are other correct answers.

27. a) $5 + (v + w) = (5 + v) + w$, associative law of addition
b) $5 + (v + w) = 5 + (w + v)$, commutative law of addition
c) $5 + (v + w) = 5 + (w + v)$ Using the commutative law first,
$= (5 + w) + v$ then the associative law
There are other correct answers.

29. a) $(xy)3 = x(y3)$, associative law of multiplication
b) $(xy)3 = (yx)3$, commutative law of multiplication
c) $(xy)3 = (yx)3$ Using the commutative law first,
$= y(x3)$ then the associative law
There are other correct answers.

31. a) $7(ab) = (7a)b$
b) $7(ab) = (7a)b = b(7a)$
c) $7(ab) = 7(ba) = (7b)a$
There are other correct answers.

33. $2(b + 5) = 2 \cdot b + 2 \cdot 5 = 2b + 10$

35. $7(1 + t) = 7 \cdot 1 + 7 \cdot t = 7 + 7t$

37. $6(5x + 2) = 6 \cdot 5x + 6 \cdot 2 = 30x + 12$

39. $7(x + 4 + 6y) = 7 \cdot x + 7 \cdot 4 + 7 \cdot 6y = 7x + 28 + 42y$

41. $7(x - 3) = 7 \cdot x - 7 \cdot 3 = 7x - 21$

43. $-3(x - 7) = -3 \cdot x - (-3) \cdot 7 = -3x - (-21) = -3x + 21$

45. $\frac{2}{3}(b - 6) = \frac{2}{3} \cdot b - \frac{2}{3} \cdot 6 = \frac{2}{3}b - 4$

47. $7.3(x - 2) = 7.3 \cdot x - 7.3 \cdot 2 = 7.3x - 14.6$

49. $-\frac{3}{5}(x - y + 10) = -\frac{3}{5} \cdot x - \left(-\frac{3}{5}\right) \cdot y + \left(-\frac{3}{5}\right) \cdot 10 =$
$-\frac{3}{5}x - \left(-\frac{3}{5}y\right) + (-6) = -\frac{3}{5}x + \frac{3}{5}y - 6$

51. $-9(-5x - 6y + 8) = -9(-5x) - (-9)6y + (-9)8$
$= 45x - (-54y) + (-72) = 45x + 54y - 72$

53. $-4(x - 3y - 2z) = -4 \cdot x - (-4)3y - (-4)2z$
$= -4x - (-12y) - (-8z) = -4x + 12y + 8z$

55. $3.1(-1.2x + 3.2y - 1.1) = 3.1(-1.2x) + (3.1)3.2y - 3.1(1.1)$
$= -3.72x + 9.92y - 3.41$

57. $4x + 3z$ Parts are separated by plus signs. The terms are $4x$ and $3z$.

59. $7x + 8y - 9z = 7x + 8y + (-9z)$ Separating parts with plus signs
The terms are $7x$, $8y$, and $-9z$.

61. $2x + 4 = 2 \cdot x + 2 \cdot 2 = 2(x + 2)$

63. $30 + 5y = 5 \cdot 6 + 5 \cdot y = 5(6 + y)$

65. $14x + 21y = 7 \cdot 2x + 7 \cdot 3y = 7(2x + 3y)$

67. $5x + 10 + 15y = 5 \cdot x + 5 \cdot 2 + 5 \cdot 3y = 5(x + 2 + 3y)$

69. $8x - 24 = 8 \cdot x - 8 \cdot 3 = 8(x - 3)$

71. $32 - 4y = 4 \cdot 8 - 4 \cdot y = 4(8 - y)$

73. $8x + 10y - 22 = 2 \cdot 4x + 2 \cdot 5y - 2 \cdot 11 = 2(4x + 5y - 11)$

75. $ax - a = a \cdot x - a \cdot 1 = a(x - 1)$

77. $ax - ay - az = a \cdot x - a \cdot y - a \cdot z = a(x - y - z)$

79. $18x - 12y + 6 = 6 \cdot 3x - 6 \cdot 2y + 6 \cdot 1 = 6(3x - 2y + 1)$

81. $\frac{2}{3}x - \frac{5}{3}y + \frac{1}{3} = \frac{1}{3} \cdot 2x - \frac{1}{3} \cdot 5y + \frac{1}{3} \cdot 1 =$
$\frac{1}{3}(2x - 5y + 1)$

83. $9a + 10a = (9 + 10)a = 19a$

85. $10a - a = 10a - 1 \cdot a = (10 - 1)a = 9a$

87. $2x + 9z + 6x = 2x + 6x + 9z = (2 + 6)x + 9z = 8x + 9z$

89. $7x + 6y^2 + 9y^2 = 7x + (6 + 9)y^2 = 7x + 15y^2$

91. $41a + 90 - 60a - 2 = 41a - 60a + 90 - 2$
$= (41 - 60)a + (90 - 2)$
$= -19a + 88$

93. $23 + 5t + 7y - t - y - 27$
$= 23 - 27 + 5t - 1 \cdot t + 7y - 1 \cdot y$
$= (23 - 27) + (5 - 1)t + (7 - 1)y$
$= -4 + 4t + 6y$, or $4t + 6y - 4$

95. $\frac{1}{2}b + \frac{1}{2}b = \left(\frac{1}{2} + \frac{1}{2}\right)b = 1b = b$

97. $2y + \frac{1}{4}y + y = 2y + \frac{1}{4}y + 1 \cdot y = \left(2 + \frac{1}{4} + 1\right)y = 3\frac{1}{4}y$, or $\frac{13}{4}y$

99. $11x - 3x = (11 - 3)x = 8x$

101. $6n - n = (6 - 1)n = 5n$

103. $y - 17y = (1 - 17)y = -16y$

105.
$-8 + 11a - 5b + 6a - 7b + 7$
$= 11a + 6a - 5b - 7b - 8 + 7$
$= (11 + 6)a + (-5 - 7)b + (-8 + 7)$
$= 17a - 12b - 1$

107. $9x + 2y - 5x = (9 - 5)x + 2y = 4x + 2y$

109. $11x + 2y - 4x - y = (11 - 4)x + (2 - 1)y = 7x + y$

111. $2.7x + 2.3y - 1.9x - 1.8y = (2.7 - 1.9)x + (2.3 - 1.8)y = 0.8x + 0.5y$

113.
$\frac{13}{2}a + \frac{9}{5}b - \frac{2}{3}a - \frac{3}{10}b - 42$
$= \left(\frac{13}{2} - \frac{2}{3}\right)a + \left(\frac{9}{5} - \frac{3}{10}\right)b - 42$
$= \left(\frac{39}{6} - \frac{4}{6}\right)a + \left(\frac{18}{10} - \frac{3}{10}\right)b - 42$
$= \frac{35}{6}a + \frac{15}{10}b - 42$
$= \frac{35}{6}a + \frac{3}{2}b - 42$

115. Discussion and Writing Exercise

117. $9w = 9 \cdot 20 = 180$

119. Since -43 is to the left of -40, it is true that $-43 < -40$.

121. $-6 \leq 6$ is true since $-6 = -6$ is true.

123. No; for any replacement other than 5 the two expressions do not have the same value. For example, let $t = 2$. Then $3 \cdot 2 + 5 = 6 + 5 = 11$, but $3 \cdot 5 + 2 = 15 + 2 = 17$.

125. Yes; commutative law of addition

127.
$q + qr + qrs + qrst$ There are no like terms.
$= q \cdot 1 + q \cdot r + q \cdot rs + q \cdot rst$
$= q(1 + r + rs + rst)$ Factoring

Exercise Set 1.8

1. $-(2x + 7) = -2x - 7$ Changing the sign of each term

3. $-(8 - x) = -8 + x$ Changing the sign of each term

5. $-4a + 3b - 7c$

7. $-6x + 8y - 5$

9. $-3x + 5y + 6$

11. $8x + 6y + 43$

13. $9x - (4x + 3) = 9x - 4x - 3$ Removing parentheses by changing the sign of every term
$= 5x - 3$ Collecting like terms

15. $2a - (5a - 9) = 2a - 5a + 9 = -3a + 9$

17. $2x + 7x - (4x + 6) = 2x + 7x - 4x - 6 = 5x - 6$

19. $2x - 4y - 3(7x - 2y) = 2x - 4y - 21x + 6y = -19x + 2y$

21.
$15x - y - 5(3x - 2y + 5z)$
$= 15x - y - 15x + 10y - 25z$ Multiplying each term in parentheses by -5
$= 9y - 25z$

23. $(3x + 2y) - 2(5x - 4y) = 3x + 2y - 10x + 8y = -7x + 10y$

25.
$(12a - 3b + 5c) - 5(-5a + 4b - 6c)$
$= 12a - 3b + 5c + 25a - 20b + 30c$
$= 37a - 23b + 35c$

27. $[9 - 2(5 - 4)] = [9 - 2 \cdot 1]$ Computing $5 - 4$
$= [9 - 2]$ Computing $2 \cdot 1$
$= 7$

29. $8[7 - 6(4 - 2)] = 8[7 - 6(2)] = 8[7 - 12] = 8[-5] = -40$

31.
$[4(9 - 6) + 11] - [14 - (6 + 4)]$
$= [4(3) + 11] - [14 - 10]$
$= [12 + 11] - [14 - 10]$
$= 23 - 4$
$= 19$

33.
$[10(x + 3) - 4] + [2(x - 1) + 6]$
$= [10x + 30 - 4] + [2x - 2 + 6]$
$= [10x + 26] + [2x + 4]$
$= 10x + 26 + 2x + 4$
$= 12x + 30$

35.
$[7(x + 5) - 19] - [4(x - 6) + 10]$
$= [7x + 35 - 19] - [4x - 24 + 10]$
$= [7x + 16] - [4x - 14]$
$= 7x + 16 - 4x + 14$
$= 3x + 30$

37.
$3\{[7(x - 2) + 4] - [2(2x - 5) + 6]\}$
$= 3\{[7x - 14 + 4] - [4x - 10 + 6]\}$
$= 3\{[7x - 10] - [4x - 4]\}$
$= 3\{7x - 10 - 4x + 4\}$
$= 3\{3x - 6\}$
$= 9x - 18$

39.
$4\{[5(x - 3) + 2] - 3[2(x + 5) - 9]\}$
$= 4\{[5x - 15 + 2] - 3[2x + 10 - 9]\}$
$= 4\{[5x - 13] - 3[2x + 1]\}$
$= 4\{5x - 13 - 6x - 3\}$
$= 4\{-x - 16\}$
$= -4x - 64$

41. $8 - 2 \cdot 3 - 9 = 8 - 6 - 9$ Multiplying
$= 2 - 9$ Doing all additions and subtractions in order from left to right
$= -7$

43. $(8 - 2 \cdot 3) - 9 = (8 - 6) - 9$ Multiplying inside the parentheses
$= 2 - 9$ Subtracting inside the parentheses
$= -7$

45. $[(-24) \div (-3)] \div \left(-\frac{1}{2}\right) = 8 \div \left(-\frac{1}{2}\right) = 8 \cdot (-2) = -16$

47. $16 \cdot (-24) + 50 = -384 + 50 = -334$

49. $2^4 + 2^3 - 10 = 16 + 8 - 10 = 24 - 10 = 14$

Exercise Set 1.8 11

51. $5^3 + 26 \cdot 71 - (16 + 25 \cdot 3) = 5^3 + 26 \cdot 71 - (16 + 75) =$
$5^3 + 26 \cdot 71 - 91 = 125 + 26 \cdot 71 - 91 = 125 + 1846 - 91 =$
$1971 - 91 = 1880$

53. $4 \cdot 5 - 2 \cdot 6 + 4 = 20 - 12 + 4 = 8 + 4 = 12$

55. $4^3/8 = 64/8 = 8$

57. $8(-7) + 6(-5) = -56 - 30 = -86$

59. $19 - 5(-3) + 3 = 19 + 15 + 3 = 34 + 3 = 37$

61. $9 \div (-3) + 16 \div 8 = -3 + 2 = -1$

63. $-4^2 + 6 = -16 + 6 = -10$

65. $-8^2 - 3 = -64 - 3 = -67$

67. $12 - 20^3 = 12 - 8000 = -7988$

69. $2 \cdot 10^3 - 5000 = 2 \cdot 1000 - 5000 = 2000 - 5000 = -3000$

71. $6[9 - (3 - 4)] = 6[9 - (-1)] = 6[9 + 1] = 6[10] = 60$

73. $-1000 \div (-100) \div 10 = 10 \div 10 = 1$

75. $8 - (7 - 9) = 8 - (-2) = 8 + 2 = 10$

77. $\dfrac{10 - 6^2}{9^2 + 3^2} = \dfrac{10 - 36}{81 + 9} = \dfrac{-26}{90} = -\dfrac{13}{45}$

79. $\dfrac{3(6-7) - 5 \cdot 4}{6 \cdot 7 - 8(4-1)} = \dfrac{3(-1) - 5 \cdot 4}{42 - 8 \cdot 3} = \dfrac{-3 - 20}{42 - 24} = -\dfrac{23}{18}$

81. $\dfrac{2^3 - 3^2 + 12 \cdot 5}{-32 \div (-16) \div (-4)} = \dfrac{8 - 9 + 12 \cdot 5}{-32 \div (-16) \div (-4)} =$
$\dfrac{8 - 9 + 60}{2 \div (-4)} = \dfrac{8 - 9 + 60}{-\frac{1}{2}} = \dfrac{-1 + 60}{-\frac{1}{2}} = \dfrac{59}{-\frac{1}{2}} =$
$59(-2) = -118$

83. Discussion and Writing Exercise

85. $8(2x - 3) = 8 \cdot 2x - 8 \cdot 3 = 16x - 24$

87. $\dfrac{2}{3}(y - 9) = \dfrac{2}{3} \cdot y - \dfrac{2}{3} \cdot 9 = \dfrac{2}{3}y - \dfrac{18}{3} = \dfrac{2}{3}y - 6$

89. $35 - 5b = 5 \cdot 7 - 5 \cdot b = 5(7 - b)$

91. $15x - 60y = 15 \cdot x - 15 \cdot 4y = 15(x - 4y)$

93. $6y + 2x - 3a + c = 6y - (-2x) - 3a - (-c) = 6y - (-2x + 3a - c)$

95. $6m + 3n - 5m + 4b = 6m - (-3n) - 5m - (-4b) =$
$6m - (-3n + 5m - 4b)$

97. $\{x - [f - (f - x)] + [x - f]\} - 3x$
$= \{x - [f - f + x] + [x - f]\} - 3x$
$= \{x - [x] + [x - f]\} - 3x$
$= \{x - x + x - f\} - 3x = x - f - 3x = -2x - f$

99. a) $x^2 + 3 = 7^2 + 3 = 49 + 3 = 52;$
$x^2 + 3 = (-7)^2 + 3 = 49 + 3 = 52;$
$x^2 + 3 = (-5.013)^2 + 3 = 25.130169 + 3 = 28.130169$

b) $1 - x^2 = 1 - 5^2 = 1 - 25 = -24;$
$1 - x^2 = 1 - (-5)^2 = 1 - 25 = -24;$
$1 - x^2 = 1 - (-10.455)^2 = 1 - 109.307025 =$
-108.307025

101. $\dfrac{-15 + 20 + 50 + (-82) + (-7) + (-2)}{6} = \dfrac{-36}{6} = -6$

Chapter 2

Solving Equations and Inequalities

Exercise Set 2.1

1. $x + 17 = 32$ Writing the equation
   $15 + 17 \ ? \ 32$ Substituting 15 for x
   $32 \ | $ TRUE

   Since the left-hand and right-hand sides are the same, 15 is a solution of the equation.

3. $x - 7 = 12$ Writing the equation
   $21 - 7 \ ? \ 12$ Substituting 21 for x
   $14 \ | $ FALSE

   Since the left-hand and right-hand sides are not the same, 21 is not a solution of the equation.

5. $6x = 54$ Writing the equation
   $6(-7) \ ? \ 54$ Substituting
   $-42 \ | $ FALSE

   -7 is not a solution of the equation.

7. $\dfrac{x}{6} = 5$ Writing the equation
   $\dfrac{30}{6} \ ? \ 5$ Substituting
   $5 \ | $ TRUE

   5 is a solution of the equation.

9. $5x + 7 = 107$
   $5 \cdot 19 + 7 \ ? \ 107$ Substituting
   $95 + 7 \ | $
   $102 \ | $ FALSE

   19 is not a solution of the equation.

11. $7(y - 1) = 63$
    $7(-11 - 1) \ ? \ 63$ Substituting
    $7(-12) \ | $
    $-84 \ | $ FALSE

    -11 is not a solution of the equation.

13. $x + 2 = 6$
    $x + 2 - 2 = 6 - 2$ Subtracting 2 on both sides
    $x = 4$ Simplifying
    Check: $x + 2 = 6$
    $4 + 2 \ ? \ 6$
    $6 \ | $ TRUE

    The solution is 4.

15. $x + 15 = -5$
    $x + 15 - 15 = -5 - 15$ Subtracting 15 on both sides
    $x = -20$
    Check: $x + 15 = -5$
    $-20 + 15 \ ? \ -5$
    $-5 \ | $ TRUE

    The solution is -20.

17. $x + 6 = -8$
    $x + 6 - 6 = -8 - 6$
    $x = -14$
    Check: $x + 6 = -8$
    $-14 + 6 \ ? \ -8$
    $-8 \ | $ TRUE

    The solution is -14.

19. $x + 16 = -2$
    $x + 16 - 16 = -2 - 16$
    $x = -18$
    Check: $x + 16 = -2$
    $-18 + 16 \ ? \ -2$
    $-2 \ | $ TRUE

    The solution is -18.

21. $x - 9 = 6$
    $x - 9 + 9 = 6 + 9$
    $x = 15$
    Check: $x - 9 = 6$
    $15 - 9 \ ? \ 6$
    $6 \ | $ TRUE

    The solution is 15.

23. $x - 7 = -21$
    $x - 7 + 7 = -21 + 7$
    $x = -14$
    Check: $x - 7 = -21$
    $-14 - 7 \ ? \ -21$
    $-21 \ | $ TRUE

    The solution is -14.

25. $5 + t = 7$
    $-5 + 5 + t = -5 + 7$
    $t = 2$
    Check: $5 + t = 7$
    $5 + 2 \ ? \ 7$
    $7 \ | $ TRUE

    The solution is 2.

27. $-7 + y = 13$
$7 + (-7) + y = 7 + 13$
$y = 20$

Check: $\dfrac{-7 + y = 13}{-7 + 20 \; ? \; 13}$
$ 13 \; | \quad$ TRUE

The solution is 20.

29. $-3 + t = -9$
$3 + (-3) + t = 3 + (-9)$
$t = -6$

Check: $\dfrac{-3 + t = -9}{-3 + (-6) \; ? \; -9}$
$ -9 \; | \quad$ TRUE

The solution is -6.

31. $x + \dfrac{1}{2} = 7$
$x + \dfrac{1}{2} - \dfrac{1}{2} = 7 - \dfrac{1}{2}$
$x = 6\dfrac{1}{2}$

Check: $\dfrac{x + \dfrac{1}{2} = 7}{6\dfrac{1}{2} + \dfrac{1}{2} \; ? \; 7}$
$\phantom{6\dfrac{1}{2} + \dfrac{1}{2} \; ? \;} 7 \; | \quad$ TRUE

The solution is $6\dfrac{1}{2}$.

33. $12 = a - 7.9$
$12 + 7.9 = a - 7.9 + 7.9$
$19.9 = a$

Check: $\dfrac{12 = a - 7.9}{12 \; ? \; 19.9 - 7.9}$
$ | \; 12 \quad$ TRUE

The solution is 19.9.

35. $r + \dfrac{1}{3} = \dfrac{8}{3}$
$r + \dfrac{1}{3} - \dfrac{1}{3} = \dfrac{8}{3} - \dfrac{1}{3}$
$r = \dfrac{7}{3}$

Check: $\dfrac{r + \dfrac{1}{3} = \dfrac{8}{3}}{\dfrac{7}{3} + \dfrac{1}{3} \; ? \; \dfrac{8}{3}}$
$\phantom{\dfrac{7}{3} + \dfrac{1}{3} \; ? \;} \dfrac{8}{3} \; | \quad$ TRUE

The solution is $\dfrac{7}{3}$.

37. $m + \dfrac{5}{6} = -\dfrac{11}{12}$
$m + \dfrac{5}{6} - \dfrac{5}{6} = -\dfrac{11}{12} - \dfrac{5}{6}$
$m = -\dfrac{11}{12} - \dfrac{5}{6} \cdot \dfrac{2}{2}$
$m = -\dfrac{11}{12} - \dfrac{10}{12}$
$m = -\dfrac{21}{12} = -\dfrac{3 \cdot 7}{3 \cdot 4}$
$m = -\dfrac{7}{4}$

Check: $\dfrac{m + \dfrac{5}{6} = -\dfrac{11}{12}}{-\dfrac{7}{4} + \dfrac{5}{6} \; ? \; -\dfrac{11}{12}}$
$ -\dfrac{21}{12} + \dfrac{10}{12}$
$ -\dfrac{11}{12} \; | \quad$ TRUE

The solution is $-\dfrac{7}{4}$.

39. $x - \dfrac{5}{6} = \dfrac{7}{8}$
$x - \dfrac{5}{6} + \dfrac{5}{6} = \dfrac{7}{8} + \dfrac{5}{6}$
$x = \dfrac{7}{8} \cdot \dfrac{3}{3} + \dfrac{5}{6} \cdot \dfrac{4}{4}$
$x = \dfrac{21}{24} + \dfrac{20}{24}$
$x = \dfrac{41}{24}$

Check: $\dfrac{x - \dfrac{5}{6} = \dfrac{7}{8}}{\dfrac{41}{24} - \dfrac{5}{6} \; ? \; \dfrac{7}{8}}$
$ \dfrac{41}{24} - \dfrac{20}{24} \; | \; \dfrac{21}{24}$
$ \dfrac{21}{24} \; | \quad$ TRUE

The solution is $\dfrac{41}{24}$.

41. $-\dfrac{1}{5} + z = -\dfrac{1}{4}$
$\dfrac{1}{5} - \dfrac{1}{5} + z = \dfrac{1}{5} - \dfrac{1}{4}$
$z = \dfrac{1}{5} \cdot \dfrac{4}{4} - \dfrac{1}{4} \cdot \dfrac{5}{5}$
$z = \dfrac{4}{20} - \dfrac{5}{20}$
$z = -\dfrac{1}{20}$

Exercise Set 2.1   15

Check: $\dfrac{-\dfrac{1}{5} + z = -\dfrac{1}{4}}{\begin{array}{c|c}-\dfrac{1}{5} + \left(-\dfrac{1}{20}\right) \ ? \ -\dfrac{1}{4} \\ -\dfrac{4}{20} + \left(-\dfrac{1}{20}\right) & -\dfrac{5}{20} \\ -\dfrac{5}{20} & \end{array}}$ TRUE

The solution is $-\dfrac{1}{20}$.

43. $\quad x + 2.3 = 7.4$
$\quad x + 2.3 - 2.3 = 7.4 - 2.3$
$\quad x = 5.1$

Check: $\dfrac{x + 2.3 = 7.4}{\begin{array}{c|c} 5.1 + 2.3 \ ? \ 7.4 \\ 7.4 & \end{array}}$ TRUE

The solution is 5.1.

45. $\quad 7.6 = x - 4.8$
$\quad 7.6 + 4.8 = x - 4.8 + 4.8$
$\quad 12.4 = x$

Check: $\dfrac{7.6 = x - 4.8}{\begin{array}{c|c} 7.6 \ ? \ 12.4 - 4.8 \\ & 7.6 \end{array}}$ TRUE

The solution is 12.4.

47. $\quad -9.7 = -4.7 + y$
$\quad 4.7 + (-9.7) = 4.7 + (-4.7) + y$
$\quad -5 = y$

Check: $\dfrac{-9.7 = -4.7 + y}{\begin{array}{c|c} -9.7 \ ? \ -4.7 + (-5) \\ & -9.7 \end{array}}$ TRUE

The solution is -5.

49. $\quad 5\dfrac{1}{6} + x = 7$
$\quad -5\dfrac{1}{6} + 5\dfrac{1}{6} + x = -5\dfrac{1}{6} + 7$
$\quad x = -\dfrac{31}{6} + \dfrac{42}{6}$
$\quad x = \dfrac{11}{6}, \text{ or } 1\dfrac{5}{6}$

Check: $\dfrac{5\dfrac{1}{6} + x = 7}{\begin{array}{c|c} 5\dfrac{1}{6} + 1\dfrac{5}{6} \ ? \ 7 \\ 7 & \end{array}}$ TRUE

The solution is $\dfrac{11}{6}$, or $1\dfrac{5}{6}$.

51. $\quad q + \dfrac{1}{3} = -\dfrac{1}{7}$
$\quad q + \dfrac{1}{3} - \dfrac{1}{3} = -\dfrac{1}{7} - \dfrac{1}{3}$
$\quad q = -\dfrac{1}{7} \cdot \dfrac{3}{3} - \dfrac{1}{3} \cdot \dfrac{7}{7}$
$\quad q = -\dfrac{3}{21} - \dfrac{7}{21}$
$\quad q = -\dfrac{10}{21}$

Check: $\dfrac{q + \dfrac{1}{3} = -\dfrac{1}{7}}{\begin{array}{c|c} -\dfrac{10}{21} + \dfrac{1}{3} \ ? \ -\dfrac{1}{7} \\ -\dfrac{10}{21} + \dfrac{7}{21} & -\dfrac{3}{21} \\ -\dfrac{3}{21} & \end{array}}$ TRUE

The solution is $-\dfrac{10}{21}$.

53. Discussion and Writing Exercise

55. $-3 + (-8)$ Two negative numbers. We add the absolute values, getting 11, and make the answer negative.
$-3 + (-8) = -11$

57. $-\dfrac{2}{3} \cdot \dfrac{5}{8} = -\dfrac{2 \cdot 5}{3 \cdot 8} = -\dfrac{\not{2} \cdot 5}{3 \cdot \not{2} \cdot 4} = -\dfrac{5}{12}$

59. $\dfrac{2}{3} \div \left(-\dfrac{4}{9}\right) = \dfrac{2}{3} \cdot \left(-\dfrac{9}{4}\right) = -\dfrac{2 \cdot 9}{3 \cdot 4} = -\dfrac{\not{2} \cdot \not{3} \cdot 3}{\not{3} \cdot \not{2} \cdot 2} = -\dfrac{3}{2}$

61. $-\dfrac{2}{3} - \left(-\dfrac{5}{8}\right) = -\dfrac{2}{3} + \dfrac{5}{8}$
$\qquad = -\dfrac{2}{3} \cdot \dfrac{8}{8} + \dfrac{5}{8} \cdot \dfrac{3}{3}$
$\qquad = -\dfrac{16}{24} + \dfrac{15}{24}$
$\qquad = -\dfrac{1}{24}$

63. The translation is $83 - x$.

65. $\quad -356.788 = -699.034 + t$
$\quad 699.034 + (-356.788) = 699.034 + (-699.034) + t$
$\quad 342.246 = t$

The solution is 342.246.

16 Chapter 2: Solving Equations and Inequalities

67. $$x + \frac{4}{5} = -\frac{2}{3} - \frac{4}{15}$$
$$x + \frac{4}{5} = -\frac{2}{3} \cdot \frac{5}{5} - \frac{4}{15} \quad \text{Adding on the right side}$$
$$x + \frac{4}{5} = -\frac{10}{15} - \frac{4}{15}$$
$$x + \frac{4}{5} = -\frac{14}{15}$$
$$x + \frac{4}{5} - \frac{4}{5} = -\frac{14}{15} - \frac{4}{5}$$
$$x = -\frac{14}{15} - \frac{4}{5} \cdot \frac{3}{3}$$
$$x = -\frac{14}{15} - \frac{12}{15}$$
$$x = -\frac{26}{15}$$
The solution is $-\frac{26}{15}$.

69. $$16 + x - 22 = -16$$
$$x - 6 = -16 \quad \text{Adding on the left side}$$
$$x - 6 + 6 = -16 + 6$$
$$x = -10$$
The solution is -10.

71. $$x + 3 = 3 + x$$
$$x + 3 - 3 = 3 + x - 3$$
$$x = x$$
$x = x$ is true for all real numbers. Thus the solution is all real numbers.

73. $$-\frac{3}{2} + x = -\frac{5}{17} - \frac{3}{2}$$
$$\frac{3}{2} - \frac{3}{2} + x = \frac{3}{2} - \frac{5}{17} - \frac{3}{2}$$
$$x = \left(\frac{3}{2} - \frac{3}{2}\right) - \frac{5}{17}$$
$$x = -\frac{5}{17}$$
The solution is $-\frac{5}{17}$.

75. $$|x| + 6 = 19$$
$$|x| + 6 - 6 = 19 - 6$$
$$|x| = 13$$
x represents a number whose distance from 0 is 13. Thus $x = -13$ or $x = 13$.
The solutions are -13 and 13.

Exercise Set 2.2

1. $$6x = 36$$
$$\frac{6x}{6} = \frac{36}{6} \quad \text{Dividing by 6 on both sides}$$
$$1 \cdot x = 6 \quad \text{Simplifying}$$
$$x = 6 \quad \text{Identity property of 1}$$
Check: $\dfrac{6x = 36}{6 \cdot 6 \; ? \; 36}$
$\quad\quad\quad\quad 36 \;|\quad\quad$ TRUE
The solution is 6.

3. $$5x = 45$$
$$\frac{5x}{5} = \frac{45}{5} \quad \text{Dividing by 5 on both sides}$$
$$1 \cdot x = 9 \quad \text{Simplifying}$$
$$x = 9 \quad \text{Identity property of 1}$$
Check: $\dfrac{5x = 45}{5 \cdot 9 \; ? \; 45}$
$\quad\quad\quad\quad 45 \;|\quad\quad$ TRUE
The solution is 9.

5. $$84 = 7x$$
$$\frac{84}{7} = \frac{7x}{7} \quad \text{Dividing by 7 on both sides}$$
$$12 = 1 \cdot x$$
$$12 = x$$
Check: $\dfrac{84 = 7x}{84 \; ? \; 7 \cdot 12}$
$\quad\quad\quad\quad |\; 84\quad\quad$ TRUE
The solution is 12.

7. $$-x = 40$$
$$-1 \cdot x = 40$$
$$\frac{-1 \cdot x}{-1} = \frac{40}{-1}$$
$$1 \cdot x = -40$$
$$x = -40$$
Check: $\dfrac{-x = 40}{-(-40) \; ? \; 40}$
$\quad\quad\quad\quad 40 \;|\quad\quad$ TRUE
The solution is -40.

9. $$-x = -1$$
$$-1 \cdot x = -1$$
$$\frac{-1 \cdot x}{-1} = \frac{-1}{-1}$$
$$1 \cdot x = 1$$
$$x = 1$$
Check: $\dfrac{-x = -1}{-(1) \; ? \; -1}$
$\quad\quad\quad\quad -1 \;|\quad\quad$ TRUE
The solution is 1.

11. $$7x = -49$$
$$\frac{7x}{7} = \frac{-49}{7}$$
$$1 \cdot x = -7$$
$$x = -7$$
Check: $\dfrac{7x = -49}{7(-7) \; ? \; -49}$
$\quad\quad\quad\quad -49 \;|\quad\quad$ TRUE
The solution is -7.

Exercise Set 2.2

13. $-12x = 72$

$\dfrac{-12x}{-12} = \dfrac{72}{-12}$

$1 \cdot x = -6$

$x = -6$

Check: $\begin{array}{c|c} -12x = 72 \\ \hline -12(-6) \; ? \; 72 \\ 72 \; | \end{array}$ TRUE

The solution is -6.

15. $-21x = -126$

$\dfrac{-21x}{-21} = \dfrac{-126}{-21}$

$1 \cdot x = 6$

$x = 6$

Check: $\begin{array}{c|c} -21x = -126 \\ \hline -21 \cdot 6 \; ? \; -126 \\ -126 \; | \end{array}$ TRUE

The solution is 6.

17. $\dfrac{t}{7} = -9$

$7 \cdot \dfrac{1}{7}t = 7 \cdot (-9)$

$1 \cdot t = -63$

$t = -63$

Check: $\begin{array}{c|c} \dfrac{t}{7} = -9 \\ \hline \dfrac{-63}{7} \; ? \; -9 \\ -9 \; | \end{array}$ TRUE

The solution is -63.

19. $\dfrac{3}{4}x = 27$

$\dfrac{4}{3} \cdot \dfrac{3}{4}x = \dfrac{4}{3} \cdot 27$

$1 \cdot x = \dfrac{4 \cdot \cancel{3} \cdot 3 \cdot 3}{\cancel{3} \cdot 1}$

$x = 36$

Check: $\begin{array}{c|c} \dfrac{3}{4}x = 27 \\ \hline \dfrac{3}{4} \cdot 36 \; ? \; 27 \\ 27 \; | \end{array}$ TRUE

The solution is 36.

21. $\dfrac{-t}{3} = 7$

$3 \cdot \dfrac{1}{3} \cdot (-t) = 3 \cdot 7$

$-t = 21$

$-1 \cdot (-1 \cdot t) = -1 \cdot 21$

$1 \cdot t = -21$

$t = -21$

Check: $\begin{array}{c|c} \dfrac{-t}{3} = 7 \\ \hline \dfrac{-(-21)}{3} \; ? \; 7 \\ \dfrac{21}{3} \\ 7 \; | \end{array}$ TRUE

The solution is -21.

23. $-\dfrac{m}{3} = \dfrac{1}{5}$

$-\dfrac{1}{3} \cdot m = \dfrac{1}{5}$

$-3 \cdot \left(-\dfrac{1}{3} \cdot m\right) = -3 \cdot \dfrac{1}{5}$

$m = -\dfrac{3}{5}$

Check: $\begin{array}{c|c} -\dfrac{m}{3} = \dfrac{1}{5} \\ \hline -\dfrac{-\dfrac{3}{5}}{3} \; ? \; \dfrac{1}{5} \\ -\left(-\dfrac{3}{5} \div 3\right) \\ -\left(-\dfrac{3}{5} \cdot \dfrac{1}{3}\right) \\ -\left(-\dfrac{1}{5}\right) \\ \dfrac{1}{5} \; | \end{array}$ TRUE

The solution is $-\dfrac{3}{5}$.

25. $-\dfrac{3}{5}r = \dfrac{9}{10}$

$-\dfrac{5}{3} \cdot \left(-\dfrac{3}{5}r\right) = -\dfrac{5}{3} \cdot \dfrac{9}{10}$

$1 \cdot r = -\dfrac{\cancel{5} \cdot \cancel{3} \cdot 3}{\cancel{3} \cdot \cancel{5} \cdot 2}$

$r = -\dfrac{3}{2}$

Check: $\begin{array}{c|c} -\dfrac{3}{5}r = \dfrac{9}{10} \\ \hline -\dfrac{3}{5} \cdot \left(-\dfrac{3}{2}\right) \; ? \; \dfrac{9}{10} \\ \dfrac{9}{10} \; | \end{array}$ TRUE

The solution is $-\dfrac{3}{2}$.

27.
$$-\frac{3}{2}r = -\frac{27}{4}$$
$$-\frac{2}{3} \cdot \left(-\frac{3}{2}r\right) = -\frac{2}{3} \cdot \left(-\frac{27}{4}\right)$$
$$1 \cdot r = \frac{2 \cdot 3 \cdot 3 \cdot 3}{3 \cdot 2 \cdot 2}$$
$$r = \frac{9}{2}$$

Check:
$$\begin{array}{c|c} -\frac{3}{2}r = -\frac{27}{4} \\ \hline -\frac{3}{2} \cdot \frac{9}{2} \;?\; -\frac{27}{4} \\ -\frac{27}{4} \;\bigg|\; \end{array} \text{TRUE}$$

The solution is $\frac{9}{2}$.

29.
$$6.3x = 44.1$$
$$\frac{6.3x}{6.3} = \frac{44.1}{6.3}$$
$$1 \cdot x = 7$$
$$x = 7$$

Check:
$$\begin{array}{c|c} 6.3x = 44.1 \\ \hline 6.3 \cdot 7 \;?\; 44.1 \\ 44.1 \;\bigg|\; \end{array} \text{TRUE}$$

The solution is 7.

31.
$$-3.1y = 21.7$$
$$\frac{-3.1y}{-3.1} = \frac{21.7}{-3.1}$$
$$1 \cdot y = -7$$
$$y = -7$$

Check:
$$\begin{array}{c|c} 3.1y = 21.7 \\ \hline -3.1(-7) \;?\; 21.7 \\ 21.7 \;\bigg|\; \end{array} \text{TRUE}$$

The solution is -7.

33.
$$38.7m = 309.6$$
$$\frac{38.7m}{38.7} = \frac{309.6}{38.7}$$
$$1 \cdot m = 8$$
$$m = 8$$

Check:
$$\begin{array}{c|c} 38.7m = 309.6 \\ \hline 38.7 \cdot 8 \;?\; 309.6 \\ 309.6 \;\bigg|\; \end{array} \text{TRUE}$$

The solution is 8.

35.
$$-\frac{2}{3}y = -10.6$$
$$-\frac{3}{2} \cdot \left(-\frac{2}{3}y\right) = -\frac{3}{2} \cdot (-10.6)$$
$$1 \cdot y = \frac{31.8}{2}$$
$$y = 15.9$$

Check:
$$\begin{array}{c|c} -\frac{2}{3}y = -10.6 \\ \hline -\frac{2}{3} \cdot (15.9) \;?\; -10.6 \\ -\frac{31.8}{3} \\ -10.6 \;\bigg|\; \end{array} \text{TRUE}$$

The solution is 15.9.

37.
$$\frac{-x}{5} = 10$$
$$5 \cdot \frac{-x}{5} = 5 \cdot 10$$
$$-x = 50$$
$$-1 \cdot (-x) = -1 \cdot 50$$
$$x = -50$$

Check:
$$\begin{array}{c|c} \frac{-x}{5} = 10 \\ \hline \frac{-(-50)}{5} \;?\; 10 \\ \frac{50}{5} \\ 10 \;\bigg|\; \end{array} \text{TRUE}$$

The solution is -50.

39.
$$-\frac{t}{2} = 7$$
$$2 \cdot \left(-\frac{t}{2}\right) = 2 \cdot 7$$
$$-t = 14$$
$$-1 \cdot (-t) = -1 \cdot 14$$
$$t = -14$$

Check:
$$\begin{array}{c|c} -\frac{t}{2} = 7 \\ \hline -\frac{-14}{2} \;?\; 7 \\ -(-7) \\ 7 \;\bigg|\; \end{array} \text{TRUE}$$

The solution is -14.

41. Discussion and Writing Exercise

43. $3x + 4x = (3+4)x = 7x$

45. $-4x + 11 - 6x + 18x = (-4 - 6 + 18)x + 11 = 8x + 11$

47. $3x - (4 + 2x) = 3x - 4 - 2x = x - 4$

49. $8y - 6(3y + 7) = 8y - 18y - 42 = -10y - 42$

51. The translation is $8r$ miles.

53.
$$-0.2344m = 2028.732$$
$$\frac{-0.2344m}{-0.2344} = \frac{2028.732}{-0.2344}$$
$$1 \cdot m = -8655$$
$$m = -8655$$

The solution is -8655.

Exercise Set 2.3

55. For all x, $0 \cdot x = 0$. There is no solution to $0 \cdot x = 9$.

57. $\quad 2|x| = -12$
$$\frac{2|x|}{2} = \frac{-12}{2}$$
$$1 \cdot |x| = -6$$
$$|x| = -6$$

Absolute value cannot be negative. The equation has no solution.

59. $\quad 3x = \dfrac{b}{a}$
$$\frac{1}{3} \cdot 3x = \frac{1}{3} \cdot \frac{b}{a}$$
$$x = \frac{b}{3a}$$

The solution is $\dfrac{b}{3a}$.

61. $\quad \dfrac{a}{b} x = 4$
$$\frac{b}{a} \cdot \frac{a}{b} x = \frac{b}{a} \cdot 4$$
$$x = \frac{4b}{a}$$

The solution is $\dfrac{4b}{a}$.

Exercise Set 2.3

1. $\quad 5x + 6 = 31$
$\quad 5x + 6 - 6 = 31 - 6 \quad$ Subtracting 6 on both sides
$\quad\quad\quad 5x = 25 \quad$ Simplifying
$\quad\quad\quad \dfrac{5x}{5} = \dfrac{25}{5} \quad$ Dividing by 5 on both sides
$\quad\quad\quad\quad x = 5 \quad$ Simplifying

Check: $\quad \underline{5x + 6 = 31}$
$\quad\quad\quad 5 \cdot 5 + 6 \;?\; 31$
$\quad\quad\quad\quad 25 + 6 \;\Big|$
$\quad\quad\quad\quad\quad 31 \;\Big|\quad$ TRUE

The solution is 5.

3. $\quad 8x + 4 = 68$
$\quad 8x + 4 - 4 = 68 - 4 \quad$ Subtracting 4 on both sides
$\quad\quad\quad 8x = 64 \quad$ Simplifying
$\quad\quad\quad \dfrac{8x}{8} = \dfrac{64}{8} \quad$ Dividing by 8 on both sides
$\quad\quad\quad\quad x = 8 \quad$ Simplifying

Check: $\quad \underline{8x + 4 = 68}$
$\quad\quad\quad 8 \cdot 8 + 4 \;?\; 68$
$\quad\quad\quad\quad 64 + 4 \;\Big|$
$\quad\quad\quad\quad\quad 68 \;\Big|\quad$ TRUE

The solution is 8.

5. $\quad 4x - 6 = 34$
$\quad 4x - 6 + 6 = 34 + 6 \quad$ Adding 6 on both sides
$\quad\quad\quad 4x = 40$
$\quad\quad\quad \dfrac{4x}{4} = \dfrac{40}{4} \quad$ Dividing by 4 on both sides
$\quad\quad\quad\quad x = 10$

Check: $\quad \underline{4x - 6 = 34}$
$\quad\quad\quad 4 \cdot 10 - 6 \;?\; 34$
$\quad\quad\quad\quad 40 - 6 \;\Big|$
$\quad\quad\quad\quad\quad 34 \;\Big|\quad$ TRUE

The solution is 10.

7. $\quad 3x - 9 = 33$
$\quad 3x - 9 + 9 = 33 + 9$
$\quad\quad\quad 3x = 42$
$\quad\quad\quad \dfrac{3x}{3} = \dfrac{42}{3}$
$\quad\quad\quad\quad x = 14$

Check: $\quad \underline{3x - 9 = 33}$
$\quad\quad\quad 3 \cdot 14 - 9 \;?\; 33$
$\quad\quad\quad\quad 42 - 9 \;\Big|$
$\quad\quad\quad\quad\quad 33 \;\Big|\quad$ TRUE

The solution is 14.

9. $\quad 7x + 2 = -54$
$\quad 7x + 2 - 2 = -54 - 2$
$\quad\quad\quad 7x = -56$
$\quad\quad\quad \dfrac{7x}{7} = \dfrac{-56}{7}$
$\quad\quad\quad\quad x = -8$

Check: $\quad \underline{7x + 2 = -54}$
$\quad\quad\quad 7(-8) + 2 \;?\; -54$
$\quad\quad\quad\quad -56 + 2 \;\Big|$
$\quad\quad\quad\quad\quad -54 \;\Big|\quad$ TRUE

The solution is -8.

11. $\quad -45 = 6y + 3$
$\quad -45 - 3 = 6y + 3 - 3$
$\quad\quad\quad -48 = 6y$
$\quad\quad\quad \dfrac{-48}{6} = \dfrac{6y}{6}$
$\quad\quad\quad\quad -8 = y$

Check: $\quad \underline{-45 = 6y + 3}$
$\quad\quad\quad -45 \;?\; 6(-8) + 3$
$\quad\quad\quad\quad \Big|\; -48 + 3$
$\quad\quad\quad\quad \Big|\; -45 \quad$ TRUE

The solution is -8.

13. $\quad -4x + 7 = 35$
$\quad -4x + 7 - 7 = 35 - 7$
$\quad\quad\quad -4x = 28$
$\quad\quad\quad \dfrac{-4x}{-4} = \dfrac{28}{-4}$
$\quad\quad\quad\quad x = -7$

Check: $\dfrac{-4x+7=35}{\begin{array}{r}-4(-7)+7\ ?\ 35\\28+7\\35\end{array}\bigg|\ \text{TRUE}}$

The solution is -7.

15. $-7x-24=-129$
$-7x-24+24=-129+24$
$-7x=-105$
$\dfrac{-7x}{-7}=\dfrac{-105}{-7}$
$x=15$

Check: $\dfrac{-7x-24=-129}{\begin{array}{r}-7\cdot 15-24\ ?\ -129\\-105-24\\-129\end{array}\bigg|\ \text{TRUE}}$

The solution is 15.

17. $5x+7x=72$
$12x=72$ Collecting like terms
$\dfrac{12x}{12}=\dfrac{72}{12}$ Dividing by 12 on both sides
$x=6$

Check: $\dfrac{5x+7x=72}{\begin{array}{r}5\cdot 6+7\cdot 6\ ?\ 72\\30+42\\72\end{array}\bigg|\ \text{TRUE}}$

The solution is 6.

19. $8x+7x=60$
$15x=60$ Collecting like terms
$\dfrac{15x}{15}=\dfrac{60}{15}$ Dividing by 15 on both sides
$x=4$

Check: $\dfrac{8x+7x=60}{\begin{array}{r}8\cdot 4+7\cdot 4\ ?\ 60\\32+28\\60\end{array}\bigg|\ \text{TRUE}}$

The solution is 4.

21. $4x+3x=42$
$7x=42$
$\dfrac{7x}{7}=\dfrac{42}{7}$
$x=6$

Check: $\dfrac{4x+3x=42}{\begin{array}{r}4\cdot 6+3\cdot 6\ ?\ 42\\24+18\\42\end{array}\bigg|\ \text{TRUE}}$

The solution is 6.

23. $-6y-3y=27$
$-9y=27$
$\dfrac{-9y}{-9}=\dfrac{27}{-9}$
$y=-3$

Check: $\dfrac{-6y-3y=27}{\begin{array}{r}-6(-3)-3(-3)\ ?\ 27\\18+9\\27\end{array}\bigg|\ \text{TRUE}}$

The solution is -3.

25. $-7y-8y=-15$
$-15y=-15$
$\dfrac{-15y}{-15}=\dfrac{-15}{-15}$
$y=1$

Check: $\dfrac{-7y-8y=-15}{\begin{array}{r}-7\cdot 1-8\cdot 1\ ?\ -15\\-7-8\\-15\end{array}\bigg|\ \text{TRUE}}$

The solution is 1.

27. $x+\dfrac{1}{3}x=8$
$\left(1+\dfrac{1}{3}\right)x=8$
$\dfrac{4}{3}x=8$
$\dfrac{3}{4}\cdot\dfrac{4}{3}x=\dfrac{3}{4}\cdot 8$
$x=6$

Check: $\dfrac{x+\dfrac{1}{3}x=8}{\begin{array}{r}6+\dfrac{1}{3}\cdot 6\ ?\ 8\\6+2\\8\end{array}\bigg|\ \text{TRUE}}$

The solution is 6.

29. $10.2y-7.3y=-58$
$2.9y=-58$
$\dfrac{2.9y}{2.9}=\dfrac{-58}{2.9}$
$y=-20$

Check: $\dfrac{10.2y-7.3y=-58}{\begin{array}{r}10.2(-20)-7.3(-20)\ ?\ -58\\-204+146\\-58\end{array}\bigg|\ \text{TRUE}}$

The solution is -20.

31. $8y - 35 = 3y$

$8y = 3y + 35$ Adding 35 and simplifying

$8y - 3y = 35$ Subtracting $3y$ and simplifying

$5y = 35$ Collecting like terms

$\dfrac{5y}{5} = \dfrac{35}{5}$ Dividing by 5

$y = 7$

Check: $\begin{array}{c|c} 8y - 35 = 3y \\ \hline 8 \cdot 7 - 35 \;?\; 3 \cdot 7 \\ 56 - 35 & 21 \\ 21 & \end{array}$ TRUE

The solution is 7.

33. $8x - 1 = 23 - 4x$

$8x + 4x = 23 + 1$ Adding 1 and $4x$ and simplifying

$12x = 24$ Collecting like terms

$\dfrac{12x}{12} = \dfrac{24}{12}$ Dividing by 12

$x = 2$

Check: $\begin{array}{c|c} 8x - 1 = 23 - 4x \\ \hline 8 \cdot 2 - 1 \;?\; 23 - 4 \cdot 2 \\ 16 - 1 & 23 - 8 \\ 15 & 15 \end{array}$ TRUE

The solution is 2.

35. $2x - 1 = 4 + x$

$2x - x = 4 + 1$ Adding 1 and $-x$

$x = 5$ Collecting like terms

Check: $\begin{array}{c|c} 2x - 1 = 4 + x \\ \hline 2 \cdot 5 - 1 \;?\; 4 + 5 \\ 10 - 1 & 9 \\ 9 & \end{array}$ TRUE

The solution is 5.

37. $6x + 3 = 2x + 11$

$6x - 2x = 11 - 3$

$4x = 8$

$\dfrac{4x}{4} = \dfrac{8}{4}$

$x = 2$

Check: $\begin{array}{c|c} 6x + 3 = 2x + 11 \\ \hline 6 \cdot 2 + 3 \;?\; 2 \cdot 2 + 11 \\ 12 + 3 & 4 + 11 \\ 15 & 15 \end{array}$ TRUE

The solution is 2.

39. $5 - 2x = 3x - 7x + 25$

$5 - 2x = -4x + 25$

$4x - 2x = 25 - 5$

$2x = 20$

$\dfrac{2x}{2} = \dfrac{20}{2}$

$x = 10$

Check: $\begin{array}{c|c} 5 - 2x = 3x - 7x + 25 \\ \hline 5 - 2 \cdot 10 \;?\; 3 \cdot 10 - 7 \cdot 10 + 25 \\ 5 - 20 & 30 - 70 + 25 \\ -15 & -40 + 25 \\ & -15 \end{array}$ TRUE

The solution is 10.

41. $4 + 3x - 6 = 3x + 2 - x$

$3x - 2 = 2x + 2$ Collecting like terms on each side

$3x - 2x = 2 + 2$

$x = 4$

Check: $\begin{array}{c|c} 4 + 3x - 6 = 3x + 2 - x \\ \hline 4 + 3 \cdot 4 - 6 \;?\; 3 \cdot 4 + 2 - 4 \\ 4 + 12 - 6 & 12 + 2 - 4 \\ 16 - 6 & 14 - 4 \\ 10 & 10 \end{array}$ TRUE

The solution is 4.

43. $4y - 4 + y + 24 = 6y + 20 - 4y$

$5y + 20 = 2y + 20$

$5y - 2y = 20 - 20$

$3y = 0$

$y = 0$

Check: $\begin{array}{c|c} 4y - 4 + y + 24 = 6y + 20 - 4y \\ \hline 4 \cdot 0 - 4 + 0 + 24 \;?\; 6 \cdot 0 + 20 - 4 \cdot 0 \\ 0 - 4 + 0 + 24 & 0 + 20 - 0 \\ 20 & 20 \end{array}$ TRUE

The solution is 0.

45. $\dfrac{7}{2}x + \dfrac{1}{2}x = 3x + \dfrac{3}{2} + \dfrac{5}{2}x$

The least common multiple of all the denominators is 2. We multiply by 2 on both sides.

$$2\left(\dfrac{7}{2}x + \dfrac{1}{2}x\right) = 2\left(3x + \dfrac{3}{2} + \dfrac{5}{2}x\right)$$

$$2 \cdot \dfrac{7}{2}x + 2 \cdot \dfrac{1}{2}x = 2 \cdot 3x + 2 \cdot \dfrac{3}{2} + 2 \cdot \dfrac{5}{2}x$$

$$7x + x = 6x + 3 + 5x$$

$$8x = 11x + 3$$

$$8x - 11x = 3$$

$$-3x = 3$$

$$\dfrac{-3x}{-3} = \dfrac{3}{-3}$$

$$x = -1$$

22   Chapter 2: Solving Equations and Inequalities

Check:
$$\frac{7}{2}x + \frac{1}{2}x = 3x + \frac{3}{2} + \frac{5}{2}x$$

$$\begin{array}{c|c} \frac{7}{2}(-1) + \frac{1}{2}(-1) \;?\; 3(-1) + \frac{3}{2} + \frac{5}{2}(-1) \\ -\frac{7}{2} - \frac{1}{2} & -3 + \frac{3}{2} - \frac{5}{2} \\ -4 & -\frac{8}{2} \\ & -4 \quad \text{TRUE} \end{array}$$

The solution is -1.

47. $\dfrac{2}{3} + \dfrac{1}{4}t = \dfrac{1}{3}$

The least common multiple of all the denominators is 12. We multiply by 12 on both sides.

$$12\left(\frac{2}{3} + \frac{1}{4}t\right) = 12 \cdot \frac{1}{3}$$
$$12 \cdot \frac{2}{3} + 12 \cdot \frac{1}{4}t = 12 \cdot \frac{1}{3}$$
$$8 + 3t = 4$$
$$3t = 4 - 8$$
$$3t = -4$$
$$\frac{3t}{3} = \frac{-4}{3}$$
$$t = -\frac{4}{3}$$

Check:
$$\frac{2}{3} + \frac{1}{4}t = \frac{1}{3}$$

$$\begin{array}{c|c} \frac{2}{3} + \frac{1}{4}\left(-\frac{4}{3}\right) \;?\; \frac{1}{3} \\ \frac{2}{3} - \frac{1}{3} & \\ \frac{1}{3} & \text{TRUE} \end{array}$$

The solution is $-\dfrac{4}{3}$.

49. $\dfrac{2}{3} + 3y = 5y - \dfrac{2}{15}$,   LCM is 15

$$15\left(\frac{2}{3} + 3y\right) = 15\left(5y - \frac{2}{15}\right)$$
$$15 \cdot \frac{2}{3} + 15 \cdot 3y = 15 \cdot 5y - 15 \cdot \frac{2}{15}$$
$$10 + 45y = 75y - 2$$
$$10 + 2 = 75y - 45y$$
$$12 = 30y$$
$$\frac{12}{30} = \frac{30y}{30}$$
$$\frac{2}{5} = y$$

Check:
$$\frac{2}{3} + 3y = 5y - \frac{2}{15}$$

$$\begin{array}{c|c} \frac{2}{3} + 3 \cdot \frac{2}{5} \;?\; 5 \cdot \frac{2}{5} - \frac{2}{15} \\ \frac{2}{3} + \frac{6}{5} & 2 - \frac{2}{15} \\ \frac{10}{15} + \frac{18}{15} & \frac{30}{15} - \frac{2}{15} \\ \frac{28}{15} & \frac{28}{15} \quad \text{TRUE} \end{array}$$

The solution is $\dfrac{2}{5}$.

51. $\dfrac{5}{3} + \dfrac{2}{3}x = \dfrac{25}{12} + \dfrac{5}{4}x + \dfrac{3}{4}$,   LCM is 12

$$12\left(\frac{5}{3} + \frac{2}{3}x\right) = 12\left(\frac{25}{12} + \frac{5}{4}x + \frac{3}{4}\right)$$
$$12 \cdot \frac{5}{3} + 12 \cdot \frac{2}{3}x = 12 \cdot \frac{25}{12} + 12 \cdot \frac{5}{4}x + 12 \cdot \frac{3}{4}$$
$$20 + 8x = 25 + 15x + 9$$
$$20 + 8x = 15x + 34$$
$$20 - 34 = 15x - 8x$$
$$-14x = 7x$$
$$\frac{-14}{7} = \frac{7x}{7}$$
$$-2 = x$$

Check:
$$\frac{5}{3} + \frac{2}{3}x = \frac{25}{12} + \frac{5}{4}x + \frac{3}{4}$$

$$\begin{array}{c|c} \frac{5}{3} + \frac{2}{3}(-2) \;?\; \frac{25}{12} + \frac{5}{4}(-2) + \frac{3}{4} \\ \frac{5}{3} - \frac{4}{3} & \frac{25}{12} - \frac{5}{2} + \frac{3}{4} \\ \frac{1}{3} & \frac{25}{12} - \frac{30}{12} + \frac{9}{12} \\ & \frac{4}{12} \\ & \frac{1}{3} \quad \text{TRUE} \end{array}$$

The solution is -2.

53. $2.1x + 45.2 = 3.2 - 8.4x$

     Greatest number of decimal places is 1

$$10(2.1x + 45.2) = 10(3.2 - 8.4x)$$

     Multiplying by 10 to clear decimals

$$10(2.1x) + 10(45.2) = 10(3.2) - 10(8.4x)$$
$$21x + 452 = 32 - 84x$$
$$21x + 84x = 32 - 452$$
$$105x = -420$$
$$\frac{105x}{105} = \frac{-420}{105}$$
$$x = -4$$

Exercise Set 2.3

Check:
$$\frac{2.1x + 45.2 = 3.2 - 8.4x}{\begin{array}{c|c} 2.1(-4) + 45.2 \;?\; 3.2 - 8.4(-4) \\ -8.4 + 45.2 & 3.2 + 33.6 \\ 36.8 & 36.8 \quad \text{TRUE} \end{array}}$$

The solution is -4.

55.
$$1.03 - 0.62x = 0.71 - 0.22x$$
Greatest number of decimal places is 2
$$100(1.03 - 0.62x) = 100(0.71 - 0.22x)$$
Multiplying by 100 to clear decimals
$$100(1.03) - 100(0.62x) = 100(0.71) - 100(0.22x)$$
$$103 - 62x = 71 - 22x$$
$$32 = 40x$$
$$\frac{32}{40} = \frac{40x}{40}$$
$$\frac{4}{5} = x, \text{ or}$$
$$0.8 = x$$

Check:
$$\frac{1.03 - 0.62x = 0.71 - 0.22x}{\begin{array}{c|c} 1.03 - 0.62(0.8) \;?\; 0.71 - 0.22(0.8) \\ 1.03 - 0.496 & 0.71 - 0.176 \\ 0.534 & 0.534 \quad \text{TRUE} \end{array}}$$

The solution is $\frac{4}{5}$, or 0.8.

57.
$$\frac{2}{7}x - \frac{1}{2}x = \frac{3}{4}x + 1, \text{ LCM is 28}$$
$$28\left(\frac{2}{7}x - \frac{1}{2}x\right) = 28\left(\frac{3}{4}x + 1\right)$$
$$28 \cdot \frac{2}{7}x - 28 \cdot \frac{1}{2}x = 28 \cdot \frac{3}{4}x + 28 \cdot 1$$
$$8x - 14x = 21x + 28$$
$$-6x = 21x + 28$$
$$-6x - 21x = 28$$
$$-27x = 28$$
$$x = -\frac{28}{27}$$

Check:
$$\frac{\frac{2}{7}x - \frac{1}{2}x = \frac{3}{4}x + 1}{\begin{array}{c|c} \frac{2}{7}\left(-\frac{28}{27}\right) - \frac{1}{2}\left(-\frac{28}{27}\right) \;?\; \frac{3}{4}\left(-\frac{28}{27}\right) + 1 \\ -\frac{8}{27} + \frac{14}{27} & -\frac{21}{27} + 1 \\ \frac{6}{27} & \frac{6}{27} \quad \text{TRUE} \end{array}}$$

The solution is $-\frac{28}{27}$.

59.
$$3(2y - 3) = 27$$
$$6y - 9 = 27 \qquad \text{Using a distributive law}$$
$$6y = 27 + 9 \qquad \text{Adding 9}$$
$$6y = 36$$
$$y = 6 \qquad \text{Dividing by 6}$$

Check:
$$\frac{3(2y - 3) = 27}{\begin{array}{c|c} 3(2 \cdot 6 - 3) \;?\; 27 \\ 3(12 - 3) \\ 3 \cdot 9 \\ 27 & \text{TRUE} \end{array}}$$

The solution is 6.

61.
$$40 = 5(3x + 2)$$
$$40 = 15x + 10 \qquad \text{Using a distributive law}$$
$$40 - 10 = 15x$$
$$30 = 15x$$
$$2 = x$$

Check:
$$\frac{40 = 5(3x + 2)}{\begin{array}{c|c} 40 \;?\; 5(3 \cdot 2 + 2) \\ & 5(6 + 2) \\ & 5 \cdot 8 \\ & 40 \quad \text{TRUE} \end{array}}$$

The solution is 2.

63.
$$-23 + y = y + 25$$
$$-y - 23 + y = -y + y + 25$$
$$-23 = 25 \qquad \text{FALSE}$$

The equation has no solution.

65.
$$-23 + x = x - 23$$
$$-x - 23 + x = -x + x - 23$$
$$-23 = -23 \qquad \text{TRUE}$$

All real numbers are solutions.

67.
$$2(3 + 4m) - 9 = 45$$
$$6 + 8m - 9 = 45 \qquad \text{Collecting like terms}$$
$$8m - 3 = 45$$
$$8m = 45 + 3$$
$$8m = 48$$
$$m = 6$$

Check:
$$\frac{2(3 + 4m) - 9 = 45}{\begin{array}{c|c} 2(3 + 4 \cdot 6) - 9 \;?\; 45 \\ 2(3 + 24) - 9 \\ 2 \cdot 27 - 9 \\ 54 - 9 \\ 45 & \text{TRUE} \end{array}}$$

The solution is 6.

69. $5r - (2r + 8) = 16$
$5r - 2r - 8 = 16$
$3r - 8 = 16$ Collecting like terms
$3r = 16 + 8$
$3r = 24$
$r = 8$

Check: $\dfrac{5r - (2r + 8) = 16}{\begin{array}{c|c} 5\cdot 8 - (2\cdot 8 + 8) \; ? \; 16 \\ 40 - (16 + 8) \\ 40 - 24 \\ 16 \end{array}}$ TRUE

The solution is 8.

71. $6 - 2(3x - 1) = 2$
$6 - 6x + 2 = 2$
$8 - 6x = 2$
$8 - 2 = 6x$
$6 = 6x$
$1 = x$

Check: $\dfrac{6 - 2(3x - 1) = 2}{\begin{array}{c} 6 - 2(3 \cdot 1 - 1) \; ? \; 2 \\ 6 - 2(3 - 1) \\ 6 - 2 \cdot 2 \\ 6 - 4 \\ 2 \end{array}}$ TRUE

The solution is 1.

73. $5x + 5 - 7x = 15 - 12x + 10x - 10$
$-2x + 5 = 5 - 2x$ Collecting like terms
$2x - 2x + 5 = 2x + 5 - 2x$ Adding $2x$
$5 = 5$ TRUE

All real numbers are solutions.

75. $22x - 5 - 15x + 3 = 10x - 4 - 3x + 11$
$7x - 2 = 7x + 7$ Collecting like terms
$-7x + 7x - 2 = -7x + 7x + 7$
$-2 = 7$ FALSE

The equation has no solution.

77. $5(d + 4) = 7(d - 2)$
$5d + 20 = 7d - 14$
$20 + 14 = 7d - 5d$
$34 = 2d$
$17 = d$

Check: $\dfrac{5(d+4) = 7(d-2)}{\begin{array}{c|c} 5(17 + 4) \; ? \; 7(17 - 2) \\ 5 \cdot 21 & 7 \cdot 15 \\ 105 & 105 \end{array}}$ TRUE

The solution is 17.

79. $8(2t + 1) = 4(7t + 7)$
$16t + 8 = 28t + 28$
$16t - 28t = 28 - 8$
$-12t = 20$
$t = -\dfrac{20}{12}$
$t = -\dfrac{5}{3}$

Check: $\dfrac{8(2t+1) = 4(7t+7)}{\begin{array}{c|c} 8\left(2\left(-\dfrac{5}{3}\right) + 1\right) \; ? \; 4\left(7\left(-\dfrac{5}{3}\right) + 7\right) \\ 8\left(-\dfrac{10}{3} + 1\right) & 4\left(-\dfrac{35}{3} + 7\right) \\ 8\left(-\dfrac{7}{3}\right) & 4\left(-\dfrac{14}{3}\right) \\ -\dfrac{56}{3} & -\dfrac{56}{3} \end{array}}$ TRUE

The solution is $-\dfrac{5}{3}$.

81. $3(r - 6) + 2 = 4(r + 2) - 21$
$3r - 18 + 2 = 4r + 8 - 21$
$3r - 16 = 4r - 13$
$13 - 16 = 4r - 3r$
$-3 = r$

Check: $\dfrac{3(r-6) + 2 = 4(r+2) - 21}{\begin{array}{c|c} 3(-3 - 6) + 2 \; ? \; 4(-3 + 2) - 21 \\ 3(-9) + 2 & 4(-1) - 21 \\ -27 + 2 & -4 - 21 \\ -25 & -25 \end{array}}$ TRUE

The solution is -3.

83. $19 - (2x + 3) = 2(x + 3) + x$
$19 - 2x - 3 = 2x + 6 + x$
$16 - 2x = 3x + 6$
$16 - 6 = 3x + 2x$
$10 = 5x$
$2 = x$

Check: $\dfrac{19 - (2x + 3) = 2(x + 3) + x}{\begin{array}{c|c} 19 - (2 \cdot 2 + 3) \; ? \; 2(2 + 3) + 2 \\ 19 - (4 + 3) & 2 \cdot 5 + 2 \\ 19 - 7 & 10 + 2 \\ 12 & 12 \end{array}}$ TRUE

The solution is 2.

85.
$$2[4-2(3-x)]-1 = 4[2(4x-3)+7]-25$$
$$2[4-6+2x]-1 = 4[8x-6+7]-25$$
$$2[-2+2x]-1 = 4[8x+1]-25$$
$$-4+4x-1 = 32x+4-25$$
$$4x-5 = 32x-21$$
$$-5+21 = 32x-4x$$
$$16 = 28x$$
$$\frac{16}{28} = x$$
$$\frac{4}{7} = x$$

The check is left to the student.
The solution is $\frac{4}{7}$.

87.
$$11-4(x+1)-3 = 11+2(4-2x)-16$$
$$11-4x-4-3 = 11+8-4x-16$$
$$4-4x = 3-4x$$
$$4x+4-4x = 4x+3-4x$$
$$4 = 3 \quad \text{FALSE}$$

The equation has no solution.

89.
$$22x-1-12x = 5(2x-1)+4$$
$$22x-1-12x = 10x-5+4$$
$$10x-1 = 10x-1$$
$$-10x+10x-1 = -10x+10x-1$$
$$-1 = -1 \quad \text{TRUE}$$

All real numbers are solutions.

91.
$$0.7(3x+6) = 1.1-(x+2)$$
$$2.1x+4.2 = 1.1-x-2$$
$$10(2.1x+4.2) = 10(1.1-x-2) \quad \text{Clearing decimals}$$
$$21x+42 = 11-10x-20$$
$$21x+42 = -10x-9$$
$$21x+10x = -9-42$$
$$31x = -51$$
$$x = -\frac{51}{31}$$

The check is left to the student.
The solution is $-\frac{51}{31}$.

93. Discussion and Writing Exercise

95. Do the long division. The answer is negative.

$$\begin{array}{r} 6.5 \\ 3.4_\wedge\overline{)2\,2.1_\wedge0} \\ \underline{2\,0\,4} \\ 1\,7\,0 \\ \underline{1\,7\,0} \\ 0 \end{array}$$

$-22.1 \div 3.4 = -6.5$

97. $7x-21-14y = 7 \cdot x - 7 \cdot 3 - 7 \cdot 2y = 7(x-3-2y)$

99. Since we are using a calculator we will not clear the decimals.
$$0.008+9.62x-42.8 = 0.944x+0.0083-x$$
$$9.62x-42.792 = -0.056x+0.0083$$
$$9.62x+0.056x = 0.0083+42.792$$
$$9.676x = 42.8003$$
$$x = \frac{42.8003}{9.676}$$
$$x \approx 4.4233464$$

The solution is approximately 4.4233464.

101. First we multiply to remove the parentheses.
$$\frac{2}{3}\left(\frac{7}{8}-4x\right)-\frac{5}{8} = \frac{3}{8}$$
$$\frac{7}{12}-\frac{8}{3}x-\frac{5}{8} = \frac{3}{8}, \text{ LCM is 24}$$
$$24\left(\frac{7}{12}-\frac{8}{3}x-\frac{5}{8}\right) = 24 \cdot \frac{3}{8}$$
$$24 \cdot \frac{7}{12} - 24 \cdot \frac{8}{3}x - 24 \cdot \frac{5}{8} = 9$$
$$14-64x-15 = 9$$
$$-1-64x = 9$$
$$-64x = 10$$
$$x = -\frac{10}{64}$$
$$x = -\frac{5}{32}$$

The solution is $-\frac{5}{32}$.

Exercise Set 2.4

1. a) We substitute 1900 for a and calculate B.
$$B = 30a = 30 \cdot 1900 = 57,000$$
The minimum furnace output is 57,000 Btu's.

b) $B = 30a$
$\frac{B}{30} = \frac{30a}{30}$ Dividing by 30
$\frac{B}{30} = a$

3. a) We substitute 8 for t and calculate M.
$$M = \frac{1}{5} \cdot 8 = \frac{8}{5}, \text{ or } 1\frac{3}{5}$$
The storm is $1\frac{3}{5}$ miles away.

b) $M = \frac{1}{5}t$
$5 \cdot M = 5 \cdot \frac{1}{5}t$
$5M = t$

5. a) We substitute 21,345 for n and calculate f.
$$f = \frac{21,345}{15} = 1423$$
There are 1423 full-time equivalent students.

b) $f = \dfrac{n}{15}$

$15 \cdot f = 15 \cdot \dfrac{n}{15}$

$15f = n$

7. We substitute 84 for c and 8 for w and calculate D.

$D = \dfrac{c}{w} = \dfrac{84}{8} = 10.5$

The calorie density is 10.5 calories per oz.

9. We substitute 7 for n and calculate N.

$N = n^2 - n = 7^2 - 7 = 49 - 7 = 42$

42 games are played.

11. $y = 5x$

$\dfrac{y}{5} = \dfrac{5x}{5}$

$\dfrac{y}{5} = x$

13. $a = bc$

$\dfrac{a}{b} = \dfrac{bc}{b}$

$\dfrac{a}{b} = c$

15. $y = 13 + x$

$y - 13 = 13 + x - 13$

$y - 13 = x$

17. $y = x + b$

$y - b = x + b - b$

$y - b = x$

19. $y = 5 - x$

$y - 5 = 5 - x - 5$

$y - 5 = -x$

$-1 \cdot (y - 5) = -1 \cdot (-x)$

$-y + 5 = x$, or

$5 - y = x$

21. $y = a - x$

$y - a = a - x - a$

$y - a = -x$

$-1 \cdot (y - a) = -1 \cdot (-x)$

$-y + a = x$, or

$a - y = x$

23. $8y = 5x$

$\dfrac{8y}{8} = \dfrac{5x}{8}$

$y = \dfrac{5x}{8}$, or $\dfrac{5}{8}x$

25. $By = Ax$

$\dfrac{By}{A} = \dfrac{Ax}{A}$

$\dfrac{By}{A} = x$

27. $W = mt + b$

$W - b = mt + b - b$

$W - b = mt$

$\dfrac{W - b}{m} = \dfrac{mt}{m}$

$\dfrac{W - b}{m} = t$

29. $y = bx + c$

$y - c = bx + c - c$

$y - c = bx$

$\dfrac{y - c}{b} = \dfrac{bx}{b}$

$\dfrac{y - c}{b} = x$

31. $A = \dfrac{a + b + c}{3}$

$3A = a + b + c$ Multiplying by 3

$3A - a - c = b$ Subtracting a and c

33. $A = at + b$

$A - b = at$ Subtracting b

$\dfrac{A - b}{a} = t$ Dividing by a

35. $A = bh$

$\dfrac{A}{b} = \dfrac{bh}{b}$ Dividing by b

$\dfrac{A}{b} = h$

37. $P = 2l + 2w$

$P - 2l = 2l + 2w - 2l$ Subtracting $2l$

$P - 2l = 2w$

$\dfrac{P - 2l}{2} = \dfrac{2w}{2}$ Dividing by 2

$\dfrac{P - 2l}{2} = w$, or

$\dfrac{1}{2}P - l = w$

39. $A = \dfrac{a + b}{2}$

$2A = a + b$ Multiplying by 2

$2A - b = a$ Subtracting b

41. $F = ma$

$\dfrac{F}{m} = \dfrac{ma}{m}$ Dividing by m

$\dfrac{F}{m} = a$

43. $E = mc^2$

$\dfrac{E}{m} = \dfrac{mc^2}{m}$ Dividing by m

$\dfrac{E}{m} = c^2$

45. $Ax + By = c$

$\qquad Ax = c - By \quad$ Subtracting By

$\qquad \dfrac{Ax}{A} = \dfrac{c - By}{A} \quad$ Dividing by A

$\qquad x = \dfrac{c - By}{A}$

47. $v = \dfrac{3k}{t}$

$\qquad tv = t \cdot \dfrac{3k}{t} \quad$ Multiplying by t

$\qquad tv = 3k$

$\qquad \dfrac{tv}{v} = \dfrac{3k}{v} \quad$ Dividing by v

$\qquad t = \dfrac{3k}{v}$

49. Discussion and Writing Exercise

51. $2a - b = 2 \cdot 2 - 3 = 4 - 3 = 1$

53. $0.082 + (-9.407) = -9.325$

55. $-45.8 - (-32.6) = -45.8 + 32.6 = -13.2$

57. $-\dfrac{2}{3} + \dfrac{5}{6} = -\dfrac{2}{3} \cdot \dfrac{2}{2} + \dfrac{5}{6}$

$\qquad = -\dfrac{4}{6} + \dfrac{5}{6}$

$\qquad = \dfrac{1}{6}$

59. a) We substitute 120 for w, 67 for h, and 23 for a and calculate K.

$\qquad K = 917 + 6(w + h - a)$

$\qquad K = 917 + 6(120 + 67 - 23)$

$\qquad K = 917 + 6(164)$

$\qquad K = 917 + 984$

$\qquad K = 1901$ calories

b) Solve for a:

$\qquad K = 917 + 6(w + h - a)$

$\qquad K = 917 + 6w + 6h - 6a$

$\qquad K + 6a = 917 + 6w + 6h$

$\qquad 6a = 917 + 6w + 6h - K$

$\qquad a = \dfrac{917 + 6w + 6h - K}{6}$

Solve for h:

$\qquad K = 917 + 6(w + h - a)$

$\qquad K = 917 + 6w + 6h - 6a$

$\qquad K - 917 - 6w + 6a = 6h$

$\qquad \dfrac{K - 917 - 6w + 6a}{6} = h$

Solve for w:

$\qquad K = 917 + 6(w + h - a)$

$\qquad K = 917 + 6w + 6h - 6a$

$\qquad K - 917 - 6h + 6a = 6w$

$\qquad \dfrac{K - 917 - 6h + 6a}{6} = w$

61. $A = \dfrac{1}{2}ah + \dfrac{1}{2}bh$

$\qquad 2A = 2\left(\dfrac{1}{2}ah + \dfrac{1}{2}bh\right) \quad$ Clearing the fractions

$\qquad 2A = ah + bh$

$\qquad 2A - ah = bh \quad$ Subtracting ah

$\qquad \dfrac{2A - ah}{h} = b \quad$ Dividing by h

$\qquad A = \dfrac{1}{2}ah + \dfrac{1}{2}bh$

$\qquad 2A = ah + bh \quad$ Clearing fractions as above

$\qquad 2A = h(a + b) \quad$ Factoring

$\qquad \dfrac{2A}{a + b} = h \quad$ Dividing by $a + b$

63. $A = lw$

When l and w both double, we have

$\qquad 2l \cdot 2w = 4lw = 4A$,

so A quadruples.

65. $A = \dfrac{1}{2}bh$

When b increases by 4 units we have

$\qquad \dfrac{1}{2}(b + 4)h = \dfrac{1}{2}bh + 2h = A + 2h$,

so A increases by $2h$ units.

Exercise Set 2.5

1. *Translate*.

What percent of 180 is 36?

$\qquad p \cdot 180 = 36$

Solve. We divide by 36 on both sides and convert the answer to percent notation.

$\qquad p \cdot 180 = 36$

$\qquad \dfrac{p \cdot 180}{180} = \dfrac{36}{180}$

$\qquad p = 0.2$

$\qquad p = 20\%$

Thus, 36 is 20% of 180. The answer is 20%.

3. *Translate*.

45 is 30% of what?

$\qquad 45 = 30\% \cdot b$

Solve. We solve the equation.

$\qquad 45 = 30\% \cdot b$

$\qquad 45 = 0.3b \quad$ Converting to decimal notation

$\qquad \dfrac{45}{0.3} = \dfrac{b}{0.3}$

$\qquad 150 = b$

Thus, 45 is 30% of 150. The answer is 150.

5. Translate.

What is 65% of 840?

$a = 65\% \cdot 840$

Solve. We convert 65% to decimal notation and multiply.

$$a = 65\% \cdot 840$$
$$a = 0.65 \times 840$$
$$a = 546$$

Thus, 546 is 65% of 840. The answer is 546.

7. Translate.

30 is what percent of 125?

$30 = p \cdot 125$

Solve. We solve the equation.

$$30 = p \cdot 125$$
$$\frac{30}{125} = \frac{p \cdot 125}{125}$$
$$0.24 = p$$
$$24\% = p$$

Thus, 30 is 24% of 125. The answer is 24%.

9. Translate.

12% of what number is 0.3?

$12\% \cdot b = 0.3$

Solve. We solve the equation.

$$12\% \cdot b = 0.3$$
$$0.12b = 0.3 \quad \text{Converting to decimal notation}$$
$$\frac{b}{0.12} = \frac{0.3}{0.12}$$
$$b = 2.5$$

Thus, 12% of 2.5 is 0.3. The answer is 2.5.

11. Translate.

2 is what percent of 40?

$2 = p \cdot 40$

Solve. We divide by 40 on both sides and convert the answer to percent notation.

$$2 = p \cdot 40$$
$$\frac{2}{40} = \frac{p \cdot 40}{40}$$
$$0.05 = p$$
$$5\% = p$$

Thus, 2 is 5% of 40. The answer is 5%.

13. Translate.

What percent of 68 is 17?

$p \cdot 68 = 17$

Solve. We divide by 68 on both sides and then convert to percent notation.

$$p \cdot 68 = 17$$
$$p = \frac{17}{68}$$
$$p = 0.25 = 25\%$$

The answer is 25%.

15. Translate.

What is 35% of 240?

$a = 35\% \cdot 240$

Solve. We convert 35% to decimal notation and multiply.

$$a = 35\% \cdot 240$$
$$a = 0.35 \cdot 240$$
$$a = 84$$

The answer is 84.

17. Translate.

What percent of 125 is 30?

$p \cdot 125 = 30$

Solve. We divide by 125 on both sides and then convert to percent notation.

$$p \cdot 125 = 30$$
$$p = \frac{30}{125}$$
$$p = 0.24 = 24\%$$

The answer is 24%.

19. Translate.

What percent of 300 is 48?

$p \cdot 300 = 48$

Solve. We divide by 300 on both sides and then convert to percent notation.

$$p \cdot 300 = 48$$
$$p = \frac{48}{300}$$
$$p = 0.16 = 16\%$$

The answer is 16%.

21. Translate.

14 is 30% of what number?

$14 = 30\% \cdot b$

Solve. We solve the equation.

$$14 = 0.3b \quad (30\% = 0.3)$$
$$\frac{14}{0.3} = b$$
$$46.\overline{6} = b$$

The answer is $46.\overline{6}$, or $46\frac{2}{3}$, or $\frac{140}{3}$.

Exercise Set 2.5

23. *Translate*.

What is 2% of 40?

$a = 2\% \cdot 40$

Solve. We convert 2% to decimal notation and multiply.

$a = 2\% \cdot 40$
$a = 0.02 \cdot 40$
$a = 0.8$

The answer is 0.8.

25. *Translate*.

0.8 is 16% of what number?

$0.8 = 16\% \cdot b$

Solve. We solve the equation.

$0.8 = 0.16b \quad (16\% = 0.16)$
$\dfrac{0.8}{0.16} = b$
$5 = b$

The answer is 5.

27. *Translate*.

54 is 135% of what number?

$54 = 135\% \cdot b$

Solve. We solve the equation.

$54 = 1.35b \quad (135\% = 1.35)$
$\dfrac{54}{1.35} = b$
$40 = b$

The answer is 40.

29. First we reword and translate.

What is 3% of $6600?

$a = 3\% \cdot 6600$

Solve. We convert 3% to decimal notation and multiply.

$a = 3\% \cdot 6600 = 0.03 \cdot 6600 = 198$

The price of the dog is $198.

31. First we reword and translate.

What is 24% of $6600?

$a = 24\% \cdot 6600$

Solve. We convert 24% to decimal notation and multiply.

$a = 24\% \cdot 6600 = 0.24 \cdot 6600 = 1584$

Veterinarian expenses are $1584.

33. First we reword and translate.

What is 8% of $6600?

$a = 8\% \cdot 6600$

Solve. We convert 8% to decimal notation and multiply.

$a = 8\% \cdot 6600 = 0.08 \cdot 6600 = 528$

The cost of supplies is $528.

35. To find the percent of cars manufactured in the U.S., we first reword and translate.

11.9 million is what percent of 17.4 million?

$11.9 = p \cdot 17.4$

Solve. We divide by 17.4 on both sides and convert to percent notation.

$11.9 = p \cdot 17.4$
$\dfrac{11.9}{17.4} = p$
$0.684 \approx p$
$68.4\% \approx p$

About 68.4% of the cars were manufactured in the U.S.

To find the percent of cars manufactured in Asia, we first reword and translate.

4.5 million is what percent of 17.4 million?

$4.5 = p \cdot 17.4$

Solve. We divide by 17.4 on both sides and convert to percent notation.

$4.5 = p \cdot 17.4$
$\dfrac{4.5}{17.4} = p$
$0.259 \approx p$
$25.9\% \approx p$

About 25.9% of the cars were manufactured in Asia.

To find the percent of cars manufactured in Europe, we subtract the two percents found above from 100%.

$100\% - 68.4\% - 25.9\% = 5.7\%$

About 5.7% of the cars were manufactured in Europe.

37. First we reword and translate.

193 is 32% of what number?

$193 = 32\% \cdot b$

Solve. We solve the equation.

$193 = 0.32 \cdot b \quad (32\% = 0.32)$
$\dfrac{193}{0.32} = b$
$603 \approx b$

Sammy Sosa had 603 at-bats.

39. First we reword and translate.

What is 8% of $3500?

$a = 8\% \cdot 3500$

Solve. We convert 8% to decimal notation and multiply.

$a = 8\% \cdot 3500 = 0.08 \cdot 3500 = 280$

Sarah will pay $280 in interest.

41. a) First we reword and translate.

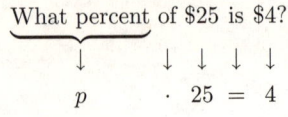

What percent of $25 is $4?

$p \cdot 25 = 4$

Solve. We divide by 25 on both sides and convert to percent notation.

$$p \cdot 25 = 4$$
$$\frac{p \cdot 25}{25} = \frac{4}{25}$$
$$p = 0.16$$
$$p = 16\%$$

The tip was 16% of the cost of the meal.

b) We add to find the total cost of the meal, including tip:

$25 + $4 = $29

43. a) First we reword and translate.

What is 15% of $25?

$a = 15\% \cdot 25$

Solve. We convert 15% to decimal notation and multiply.

$$a = 15\% \cdot 25$$
$$a = 0.15 \times 25$$
$$a = 3.75$$

The tip was $3.75.

b) We add to find the total cost of the meal, including tip:

$25 + $3.75 = $28.75

45. a) First we reword and translate.

15% of what is $4.32?

$15\% \cdot b = 4.32$

Solve. We solve the equation.

$$15\% \cdot b = 4.32$$
$$0.15 \cdot b = 4.32$$
$$\frac{0.15 \cdot b}{0.15} = \frac{4.32}{0.15}$$
$$b = 28.8$$

The cost of the meal before the tip was $28.80.

b) We add to find the total cost of the meal, including tip:

$28.80 + $4.32 = $33.12

47. First we reword and translate.

8% of what is 16?

$8\% \cdot b = 16$

Solve. We solve the equation.

$$8\% \cdot b = 16$$
$$0.08 \cdot b = 16$$
$$\frac{0.08 \cdot b}{0.08} = \frac{16}{0.08}$$
$$b = 200$$

There were 200 women in the original study.

49. First we reword and translate.

What is 16.5% of 191?

$a = 16.5\% \cdot 191$

Solve. We convert 16.5% to decimal notation and multiply.

$$a = 16.5\% \cdot 191$$
$$a = 0.165 \cdot 191$$
$$a = 31.515 \approx 31.5$$

About 31.5 lb of the author's body weight is fat.

51. We subtract to find the increase.

$735 − $430 = $305

The increase is $305.

Now we find the percent increase.

$305 is what percent of $430?

$305 = p \cdot 430$

We divide by 430 on both sides and then convert to percent notation.

$$305 = p \cdot 430$$
$$\frac{305}{430} = p$$
$$0.71 \approx p$$
$$71\% \approx p$$

The percent increase is about 71%.

53. First we find the increase in the rate for smokers.

Rate increase is 100% of $780.

$a = 100\% \cdot 780$

We convert 100% to decimal notation and multiply.

$a = 100\% \cdot \$780 = 1 \cdot 780 = 780$

The rate increase is $780.

Now we add the rate increase to the rate for nonsmokers to find the rate for smokers.

$780 + $780 = $1560

55. We subtract to find the increase.

$2955 − $1645 = $1310

The increase is $1310.

Now we find the percent increase.

$1310 is what percent of $1645?

$1310 = p \cdot 1645$

Exercise Set 2.6

We divide by 1645 on both sides and then convert to percent notation.
$$1310 = p \cdot 1645$$
$$\frac{1310}{1645} = p$$
$$0.80 \approx p$$
$$80\% \approx p$$

The percent increase is about 80%.

57. Discussion and Writing Exercise

59. $-3 - 8 = -3 + (-8) = -11$

61. $-\frac{3}{5} + \frac{1}{5}$ The absolute values are $\frac{3}{5}$ and $\frac{1}{5}$. The difference is $\frac{3}{5} - \frac{1}{5}$, or $\frac{2}{5}$. The negative number has the larger absolute value, so the answer is negative. $-\frac{3}{5} + \frac{1}{5} = -\frac{2}{5}$

63.
$$-5a + 3c - 2(c - 3a)$$
$$= -5a + 3c - 2 \cdot c - 2(-3a)$$
$$= -5a + 3c - 2c + 6a$$
$$= (-5 + 6)a + (3 - 2)c$$
$$= 1 \cdot a + 1 \cdot c$$
$$= a + c$$

65. $-6.5 + 2.6 = -3.9$ The absolute values are 6.5 and 2.6. The difference is 3.9. The negative number has the larger absolute value, so the answer is negative, -3.9.

67. Since 6 ft = 6×1 ft = 6×12 in. = 72 in., we can express 6 ft, 4 in. as 72 in. + 4 in., or 76 in.

Translate. We reword the problem.

96.1% of what is 76 in.?

$96.1\% \cdot b = 76$

Solve. We solve the equation.
$$96.1\% \cdot b = 76$$
$$0.961 \cdot b = 76$$
$$\frac{0.961 \cdot b}{0.961} = \frac{76}{0.961}$$
$$b \approx 79$$

Note that 79 in. = 72 in. + 7 in. = 6 ft, 7 in.

Jaraan's final adult height will be about 6 ft, 7 in.

Exercise Set 2.6

1. *Familiarize.* Using the labels on the drawing in the text, we let $x =$ the length of the shorter piece, in inches, and $3x =$ the length of the longer piece, in inches.

Translate. We reword the problem.

The length of the shorter piece plus the length of the longer piece is 240 ft.

$x + 3x = 240$

Solve. We solve the equation.
$$x + 3x = 240$$
$$4x = 240 \quad \text{Collecting like terms}$$
$$\frac{4x}{4} = \frac{240}{4}$$
$$x = 60$$

If x is 60, then $3x = 3 \cdot 60$, or 180.

Check. 180 is three times 60, and $60 + 180 = 240$. The answer checks.

State. The lengths of the pieces are 60 in. and 180 in.

3. *Familiarize.* Let $c =$ the cost of one box of Wheaties.

Translate.

Total cost is Number of boxes times Price of one box

$14.68 = 4 \cdot c$

Solve. We solve the equation.
$$14.68 = 4 \cdot c$$
$$\frac{14.68}{4} = c \quad \text{Dividing by 4}$$
$$3.67 = c$$

Check. If one box of Wheaties costs \$3.67, then 4 boxes cost 4(\$3.67), or \$14.68. The answer checks.

State. One box of Wheaties costs \$3.67.

5. *Familiarize.* Let $d =$ the amount spent on women's dresses, in billions of dollars.

Translate.

Amount spent on blouses was \$0.2 billion more than amount spent on dresses

$6.5 = 0.2 + d$

Solve. We solve the equation.
$$6.5 = 0.2 + d$$
$$6.5 - 0.2 = 0.2 + d - 0.2 \quad \text{Subtracting 0.2}$$
$$6.3 = d$$

Check. If we add \$0.2 billion to \$6.3 billion, we get \$6.5 billion. The answer checks.

State. \$6.3 billion was spent on dresses.

7. *Familiarize.* Let $d =$ the musher's distance from Nome, in miles. Then $2d =$ the distance from Anchorage, in miles. This is the number of miles the musher has completed. The sum of the two distances is the length of the race, 1049 miles.

Translate.

Distance from Nome plus distance from Anchorage is 1049 mi.

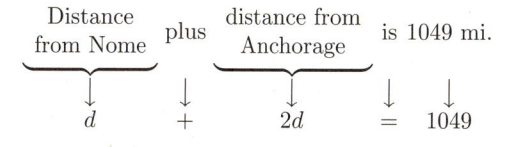

$d + 2d = 1049$

Carry out. We solve the equation.
$$d + 2d = 1049$$
$$3d = 1049 \quad \text{Collecting like terms}$$
$$\frac{3d}{3} = \frac{1049}{3}$$
$$d = \frac{1049}{3}$$

If $d = \frac{1049}{3}$, then $2d = 2 \cdot \frac{1049}{3} = \frac{2098}{3} = 699\frac{1}{3}$.

Check. $\frac{2098}{3}$ is twice $\frac{1049}{3}$, and $\frac{1049}{3} + \frac{2098}{3} = \frac{3147}{3} = 1049$. The result checks.

State. The musher has traveled $699\frac{1}{3}$ miles.

9. **Familiarize.** Using the labels on the drawing in the text, we let $x =$ the smaller number and $x + 1 =$ the larger number.

   Translate. We reword the problem.

   First number + second number is 547
   x + $(x+1)$ = 547

   Solve. We solve the equation.
   $$x + (x+1) = 547$$
   $$2x + 1 = 547 \quad \text{Collecting like terms}$$
   $$2x + 1 - 1 = 547 - 1 \quad \text{Subtracting 1}$$
   $$2x = 546$$
   $$\frac{2x}{2} = \frac{546}{2} \quad \text{Dividing by 2}$$
   $$x = 273$$

   If x is 273, then $x + 1$ is 274.

   Check. 273 and 274 are consecutive integers, and their sum is 547. The answer checks.

   State. The page numbers are 273 and 274.

11. **Familiarize.** Let $a =$ the first number. Then $a + 1 =$ the second number, and $a + 2 =$ the third number.

    Translate. We reword the problem.

    First number + second number + third number is 126
    a + $(a+1)$ + $(a+2)$ = 114

    Solve. We solve the equation.
    $$a + (a+1) + (a+2) = 126$$
    $$3a + 3 = 126 \quad \text{Collecting like terms}$$
    $$3a + 3 - 3 = 126 - 3$$
    $$3a = 123$$
    $$\frac{3a}{3} = \frac{123}{3}$$
    $$a = 41$$

    If a is 41, then $a + 1$ is 42 and $a + 2$ is 43.

    Check. 41, 42, and 43 are consecutive integers, and their sum is 126. The answer checks.

    State. The numbers are 41, 42, and 43.

13. **Familiarize.** Let $x =$ the first odd integer. Then $x + 2 =$ the next odd integer and $(x+2) + 2$, or $x + 4 =$ the third odd integer.

    Translate. We reword the problem.

    First odd integer + second odd integer + third odd integer is 189
    x + $(x+2)$ + $(x+4)$ = 189

    Solve. We solve the equation.
    $$x + (x+2) + (x+4) = 189$$
    $$3x + 6 = 189 \quad \text{Collecting like terms}$$
    $$3x + 6 - 6 = 189 - 6$$
    $$3x = 183$$
    $$\frac{3x}{3} = \frac{183}{3}$$
    $$x = 61$$

    If x is 61, then $x + 2$ is 63 and $x + 4$ is 65.

    Check. 61, 63, and 65 are consecutive odd integers, and their sum is 189. The answer checks.

    State. The integers are 61, 63, and 65.

15. **Familiarize.** Using the labels on the drawing in the text, we let $w =$ the width and $3w + 6 =$ the length. The perimeter P of a rectangle is given by the formula $2l + 2w = P$, where $l =$ the length and $w =$ the width.

    Translate. Substitute $3w + 6$ for l and 124 for P:
    $$2l + 2w = P$$
    $$2(3w+6) + 2w = 124$$

    Solve. We solve the equation.
    $$2(3w+6) + 2w = 124$$
    $$6w + 12 + 2w = 124$$
    $$8w + 12 = 124$$
    $$8w + 12 - 12 = 124 - 12$$
    $$8w = 112$$
    $$\frac{8w}{8} = \frac{112}{8}$$
    $$w = 14$$

    The possible dimensions are $w = 14$ ft and $l = 3w + 6 = 3(14) + 6$, or 48 ft.

    Check. The length, 48 ft, is 6 ft more than three times the width, 14 ft. The perimeter is $2(48 \text{ ft}) + 2(14 \text{ ft}) = 96 \text{ ft} + 28 \text{ ft} = 124 \text{ ft}$. The answer checks.

    State. The width is 14 ft, and the length is 48 ft.

17. **Familiarize.** Let $p =$ the regular price of the shoes. At 15% off, Amy paid 85% of the regular price.

    Translate.
    $63.75 is 85% of the regular price.
    63.75 = 0.85 · p

Exercise Set 2.6

Solve. We solve the equation.
$$63.75 = 0.85p$$
$$\frac{63.75}{0.08} = p \quad \text{Dividing both sides by 0.85}$$
$$75 = p$$

Check. 85% of $75, or 0.85($75), is $63.75. The answer checks.

State. The regular price was $75.

19. *Familiarize*. Let $b =$ the price of the book itself. When the sales tax rate is 5%, the tax paid on the book is 5% of b, or $0.05b$.

    Translate.

    $\underbrace{\text{Price of book}}_{b} \text{ plus } \underbrace{\text{sales tax}}_{0.05b} \text{ is } \$89.25.$

    $b + 0.05b = 89.25$

    Solve. We solve the equation.
    $$b + 0.05b = 89.25$$
    $$1.05b = 89.25$$
    $$b = \frac{89.25}{1.05}$$
    $$b = 85$$

    Check. 5% of $85, or 0.05($85), is $4.25 and $85 + $4.25 is $89.25, the total cost. The answer checks.

    State. The book itself cost $85.

21. *Familiarize*. Let $n =$ the number of visits required for a total parking cost of $27.00. The parking cost for each $1\frac{1}{2}$ hour visit is $1.50 for the first hour plus $1.00 for part of a second hour, or $2.50. Then the total parking cost for n visits is $2.50n$ dollars.

    Translate. We reword the problem.

    $\underbrace{\text{Total parking cost}}_{2.50n} \text{ is } \$27.00.$

    $2.50n = 27.00$

    Solve. We solve the equation.
    $$2.5n = 27$$
    $$10(2.5n) = 10(27) \quad \text{Clearing the decimal}$$
    $$25n = 270$$
    $$\frac{25n}{25} = \frac{270}{25}$$
    $$n = 10.8$$

    If the total parking cost is $27.00 for 10.8 visits, then the cost will be more than $27.00 for 11 or more visits.

    Check. The parking cost for 10 visits is $2.50(10), or $25, and the parking cost for 11 visits is $2.50(11), or $27.50. Since 11 is the smallest number for which the parking cost exceeds $27.00, the answer checks.

    State. The minimum number of weekly visits for which it is worthwhile to buy a parking pass is 11.

23. *Familiarize*. Let $x =$ the measure of the first angle. Then $3x =$ the measure of the second angle, and $x + 40 =$ the measure of the third angle. Recall that the sum of measures of the angles of a triangle is 180°.

    Translate.

    $\underbrace{\text{Measure of first angle}}_{x} + \underbrace{\text{measure of second angle}}_{3x} + \underbrace{\text{measure of third angle}}_{(x+40)} \text{ is } 180.$

    $x + 3x + (x + 40) = 180$

    Solve. We solve the equation.
    $$x + 3x + (x + 40) = 180$$
    $$5x + 40 = 180$$
    $$5x + 40 - 40 = 180 - 40$$
    $$5x = 140$$
    $$\frac{5x}{5} = \frac{140}{5}$$
    $$x = 28$$

    Possible answers for the angle measures are as follows:

    First angle: $x = 28°$

    Second angle: $3x = 3(28) = 84°$

    Third angle: $x + 40 = 28 + 40 = 68°$

    Check. Consider 28°, 84°, and 68°. The second angle is three times the first, and the third is 40° more than the first. The sum, 28° + 84° + 68°, is 180°. These numbers check.

    State. The measures of the angles are 28°, 84°, and 68°.

25. *Familiarize*. Using the labels on the drawing in the text, we let $x =$ the measure of the first angle, $x + 5 =$ the measure of the second angle, and $3x + 10 =$ the measure of the third angle. Recall that the sum of measures of the angles of a triangle is 180°.

    Translate.

    $\underbrace{\text{Measure of first angle}}_{x} + \underbrace{\text{measure of second angle}}_{(x+5)} + \underbrace{\text{measure of third angle}}_{(3x+10)} \text{ is } 180.$

    $x + (x+5) + (3x+10) = 180$

    Solve. We solve the equation.
    $$x + (x + 5) + (3x + 10) = 180$$
    $$5x + 15 = 180$$
    $$5x + 15 - 15 = 180 - 15$$
    $$5x = 165$$
    $$\frac{5x}{5} = \frac{165}{5}$$
    $$x = 33$$

    Possible answers for the angle measures are as follows:

    First angle: $x = 33°$

    Second angle: $x + 5 = 33 + 5 = 38°$

    Third angle: $3x + 10 = 3(33) + 10 = 109°$

    Check. The second angle is 5° more than the first, and the third is 10° more than 3 times the first. The sum, 33° + 38° + 109°, is 180°. The numbers check.

    State. The measures of the angles are 33°, 38°, and 109°.

27. **Familiarize.** Let a = the amount Sarah invested. The investment grew by 28% of a, or $0.28a$.

Translate.

$$\underbrace{\text{Amount invested}}_{\downarrow} \text{ plus } \underbrace{\text{amount of growth}}_{\downarrow} \text{ is } \$448.$$
$$a \quad + \quad 0.28a \quad = \quad 448$$

Solve. We solve the equation.
$$a + 0.28a = 448$$
$$1.28a = 448$$
$$a = 350$$

Check. 28% of $350 is $0.28(\$350)$, or $98, and $350 + $98 = $448. The answer checks.

State. Sarah invested $350.

29. **Familiarize.** Let b = the balance in the account at the beginning of the month. The balance grew by 2% of b, or $0.02b$.

Translate.

$$\underbrace{\text{Original balance}}_{\downarrow} \text{ plus } \underbrace{\text{amount of growth}}_{\downarrow} \text{ is } \$870.$$
$$b \quad + \quad 0.02b \quad = \quad 870$$

Solve. We solve the equation.
$$b + 0.02b = 870$$
$$1.02b = 870$$
$$b \approx \$852.94$$

Check. 2% of $852.94 is $0.02(\$852.94)$, or $17.06, and $852.94 + $17.06 = $870. The answer checks.

State. The balance at the beginning of the month was $852.94.

31. **Familiarize.** The total cost is the initial charge plus the mileage charge. Let d = the distance, in miles, that Courtney can travel for $12. The mileage charge is the cost per mile times the number of miles traveled or $0.75d$.

Translate.

$$\underbrace{\text{Initial charge}}_{\downarrow} \text{ plus } \underbrace{\text{mileage charge}}_{\downarrow} \text{ is } \$12.$$
$$3 \quad + \quad 0.75d \quad = \quad 12$$

Solve. We solve the equation.
$$3 + 0.75d = 12$$
$$0.75d = 9$$
$$d = 12$$

Check. A 12-mi taxi ride from the airport would cost $3 + 12(\$0.75)$, or $3 + $9, or $12. The answer checks.

State. Courtney can travel 12 mi from the airport for $12.

33. **Familiarize.** Let c = the cost of the meal before the tip. We know that the cost of the meal before the tip plus the tip, 15% of the cost, is the total cost, $41.40.

Translate.

$$\underbrace{\text{Cost of meal}}_{\downarrow} \text{ plus } \underbrace{\text{tip}}_{\downarrow} \text{ is } \$41.40$$
$$c \quad + \quad 15\%c \quad = \quad 41.40$$

Solve. We solve the equation.
$$c + 15\%c = 41.40$$
$$c + 0.15c = 41.40$$
$$1c + 0.15c = 41.40$$
$$1.15c = 41.40$$
$$\frac{1.15c}{1.15} = \frac{41.40}{1.15}$$
$$c = 36$$

Check. We find 15% of $36 and add it to $36:
$15\% \times \$36 = 0.15 \times \$36 = \$5.40$ and $\$36 + \$5.40 = \$41.40$.
The answer checks.

State. The cost of the meal before the tip was added was $36.

35. Discussion and Writing Exercise

37. $$-\frac{4}{5} - \frac{3}{8} = -\frac{4}{5} + \left(-\frac{3}{8}\right)$$
$$= -\frac{32}{40} + \left(-\frac{15}{40}\right)$$
$$= -\frac{47}{40}$$

39. $$-\frac{4}{5} \cdot \frac{3}{8} = -\frac{4 \cdot 3}{5 \cdot 8}$$
$$= -\frac{4 \cdot 3}{5 \cdot 2 \cdot 4}$$
$$= -\frac{\cancel{4} \cdot 3}{5 \cdot 2 \cdot \cancel{4}}$$
$$= -\frac{3}{10}$$

41. $$\frac{1}{10} \div \left(-\frac{1}{100}\right) = \frac{1}{10} \cdot \left(-\frac{100}{1}\right) = -\frac{1 \cdot 100}{10 \cdot 1} =$$
$$-\frac{\cancel{1} \cdot \cancel{10} \cdot 10}{\cancel{10} \cdot \cancel{1} \cdot 1} = -\frac{10}{1} = -10$$

43. $-25.6(-16) = 409.6$

45. $-25.6 + (-16) = -41.6$

47. **Familiarize.** Let a = the original number of apples. Then $\frac{1}{3}a$, $\frac{1}{4}a$, $\frac{1}{8}a$, and $\frac{1}{5}a$ are given to four people, respectively. The fifth and sixth people get 10 apples and 1 apple, respectively.

Translate. We reword the problem.

$$\underbrace{\text{The total number of apples}}_{\downarrow} \text{ is } \underbrace{a}_{\downarrow}$$
$$\frac{1}{3}a + \frac{1}{4}a + \frac{1}{8}a + \frac{1}{5}a + 10 + 1 = a$$

Solve. We solve the equation.
$$\frac{1}{3}a + \frac{1}{4}a + \frac{1}{8}a + \frac{1}{5}a + 10 + 1 = a, \text{ LCD is } 120$$
$$120\left(\frac{1}{3}a + \frac{1}{4}a + \frac{1}{8}a + \frac{1}{5}a + 11\right) = 120 \cdot a$$
$$40a + 30a + 15a + 24a + 1320 = 120a$$
$$109a + 1320 = 120a$$
$$1320 = 11a$$
$$120 = a$$

Check. If the original number of apples was 120, then the first four people got $\frac{1}{3} \cdot 120$, $\frac{1}{4} \cdot 120$, $\frac{1}{8} \cdot 120$, and $\frac{1}{5} \cdot 120$, or 40, 30, 15, and 24 apples, respectively. Adding all the apples we get $40 + 30 + 15 + 24 + 10 + 1$, or 120. The result checks.

State. There were originally 120 apples in the basket.

49. Divide the largest triangle into three triangles, each with a vertex at the center of the circle and with height x as shown.

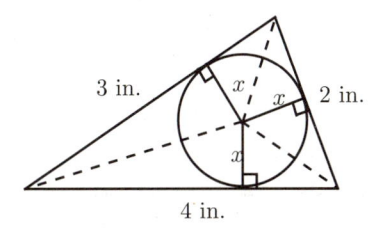

Then the sum of the areas of the three smaller triangles is the area of the original triangle. We have:
$$\frac{1}{2} \cdot 3x + \frac{1}{2} \cdot 2x + \frac{1}{2} \cdot 4x = 2.9047$$
$$2\left(\frac{1}{2} \cdot 3x + \frac{1}{2} \cdot 2x + \frac{1}{2} \cdot 4x\right) = 2(2.9047)$$
$$3x + 2x + 4x = 5.8094$$
$$9x = 5.8094$$
$$x \approx 0.65$$

Thus, x is about 0.65 in.

51. **Familiarize.** Let $p =$ the price of the gasoline as registered on the pump. Then the sales tax will be $9\% p$.

Translate. We reword the problem.

Price on pump plus sales tax is $10
$$p + 9\%p = 10$$

Solve. We solve the equation.
$$p + 9\%p = 10$$
$$1p + 0.09p = 10$$
$$1.09p = 10$$
$$\frac{1.09p}{1.09} = \frac{10}{1.09}$$
$$p \approx 9.17$$

Check. We find 9% of $9.17 and add it to $9.17:
$$9\% \times \$9.17 = 0.09 \times \$9.17 \approx \$0.83$$
Then $\$9.17 + \$0.83 = \$10$, so $9.17 checks.

State. The attendant should have filled the tank until the pump read $9.17, not $9.10.

Exercise Set 2.7

1. $x > -4$
   a) Since $4 > -4$ is true, 4 is a solution.
   b) Since $0 > -4$ is true, 0 is a solution.
   c) Since $-4 > -4$ is false, -4 is not a solution.
   d) Since $6 > -4$ is true, 6 is a solution.
   e) Since $5.6 > -4$ is true, 5.6 is a solution.

3. $x \geq 6.8$
   a) Since $-6 \geq 6.8$ is false, -6 is not a solution.
   b) Since $0 \geq 6.8$ is false, 0 is not a solution.
   c) Since $6 \geq 6.8$ is false, 6 is not a solution.
   d) Since $8 \geq 6.8$ is true, 8 is a solution.
   e) Since $-3\frac{1}{2} \geq 6.8$ is false, $-3\frac{1}{2}$ is not a solution.

5. The solutions of $x > 4$ are those numbers greater than 4. They are shown on the graph by shading all points to the right of 4. The open circle at 4 indicates that 4 is not part of the graph.

7. The solutions of $t < -3$ are those numbers less than -3. They are shown on the graph by shading all points to the left of -3. The open circle at -3 indicates that -3 is not part of the graph.

9. The solutions of $m \geq -1$ are are shown by shading the point for -1 and all points to the right of -1. The closed circle at -1 indicates that -1 is part of the graph.

11. In order to be a solution of the inequality $-3 < x \leq 4$, a number must be a solution of both $-3 < x$ and $x \leq 4$. The solution set is graphed as follows:

The open circle at -3 means that -3 is not part of the graph. The closed circle at 4 means that 4 is part of the graph.

13. In order to be a solution of the inequality $0 < x < 3$, a number must be a solution of both $0 < x$ and $x < 3$. The solution set is graphed as follows:

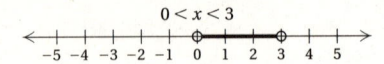

$$0 < x < 3$$
$$\xleftarrow{\quad}\overset{\circ}{\underset{-5\ -4\ -3\ -2\ -1\ \ 0\ \ 1\ \ 2\ \ 3\ \ 4\ \ 5}{\quad}}\overset{\circ}{\xrightarrow{\quad}}$$

The open circles at 0 and at 3 mean that 0 and 3 are not part of the graph.

15.
$$x + 7 > 2$$
$$x + 7 - 7 > 2 - 7 \quad \text{Subtracting 7}$$
$$x > -5 \quad \text{Simplifying}$$

The solution set is $\{x | x > -5\}$.

The graph is as follows:

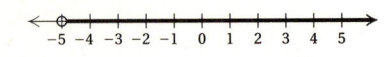

17.
$$x + 8 \leq -10$$
$$x + 8 - 8 \leq -10 - 8 \quad \text{Subtracting 8}$$
$$x \leq -18 \quad \text{Simplifying}$$

The solution set is $\{x | x \leq -18\}$.

The graph is as follows:

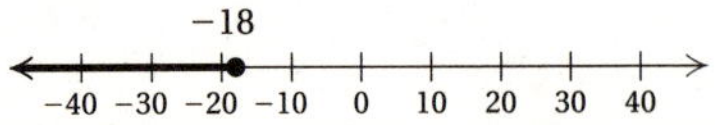

19.
$$y - 7 > -12$$
$$y - 7 + 7 > -12 + 7 \quad \text{Adding 7}$$
$$y > -5 \quad \text{Simplifying}$$

The solution set is $\{y | y > -5\}$.

21.
$$2x + 3 > x + 5$$
$$2x + 3 - 3 > x + 5 - 3 \quad \text{Subtracting 3}$$
$$2x > x + 2 \quad \text{Simplifying}$$
$$2x - x > x + 2 - x \quad \text{Subtracting } x$$
$$x > 2 \quad \text{Simplifying}$$

The solution set is $\{x | x > 2\}$.

23.
$$3x + 9 \leq 2x + 6$$
$$3x + 9 - 9 \leq 2x + 6 - 9 \quad \text{Subtracting 9}$$
$$3x \leq 2x - 3 \quad \text{Simplifying}$$
$$3x - 2x \leq 2x - 3 - 2x \quad \text{Subtracting } 2x$$
$$x \leq -3 \quad \text{Simplifying}$$

The solution set is $\{x | x \leq -3\}$.

25.
$$5x - 6 < 4x - 2$$
$$5x - 6 + 6 < 4x - 2 + 6$$
$$5x < 4x + 4$$
$$5x - 4x < 4x + 4 - 4x$$
$$x < 4$$

The solution set is $\{x | x < 4\}$.

27.
$$-9 + t > 5$$
$$-9 + t + 9 > 5 + 9$$
$$t > 14$$

The solution set is $\{t | t > 14\}$.

29.
$$y + \frac{1}{4} \leq \frac{1}{2}$$
$$y + \frac{1}{4} - \frac{1}{4} \leq \frac{1}{2} - \frac{1}{4}$$
$$y \leq \frac{2}{4} - \frac{1}{4} \quad \text{Obtaining a common denominator}$$
$$y \leq \frac{1}{4}$$

The solution set is $\left\{y \middle| y \leq \frac{1}{4}\right\}$.

31.
$$x - \frac{1}{3} > \frac{1}{4}$$
$$x - \frac{1}{3} + \frac{1}{3} > \frac{1}{4} + \frac{1}{3}$$
$$x > \frac{3}{12} + \frac{4}{12} \quad \text{Obtaining a common denominator}$$
$$x > \frac{7}{12}$$

The solution set is $\left\{x \middle| x > \frac{7}{12}\right\}$.

33.
$$5x < 35$$
$$\frac{5x}{5} < \frac{35}{5} \quad \text{Dividing by 5}$$
$$x < 7$$

The solution set is $\{x | x < 7\}$. The graph is as follows:

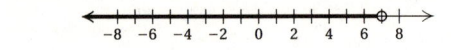

35.
$$-12x > -36$$
$$\frac{-12x}{-12} < \frac{-36}{-12} \quad \text{Dividing by } -12$$
$$\quad\quad\quad\quad \text{The symbol has to be reversed.}$$
$$x < 3 \quad \text{Simplifying}$$

The solution set is $\{x | x < 3\}$. The graph is as follows:

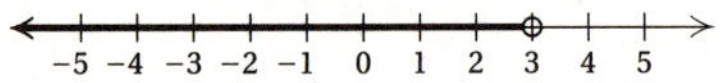

37.
$$5y \geq -2$$
$$\frac{5y}{5} \geq \frac{-2}{5} \quad \text{Dividing by 5}$$
$$y \geq -\frac{2}{5}$$

The solution set is $\left\{y \middle| y \geq -\frac{2}{5}\right\}$.

39.
$$-2x \leq 12$$
$$\frac{-2x}{-2} \geq \frac{12}{-2} \quad \text{Dividing by } -2$$
$$\quad\quad\quad\quad \text{The symbol has to be reversed.}$$
$$x \geq -6 \quad \text{Simplifying}$$

The solution set is $\{x | x \geq -6\}$.

41.
$$-4y \geq -16$$
$$\frac{-4y}{-4} \leq \frac{-16}{-4} \quad \text{Dividing by } -4$$
$$\quad\quad\quad\quad \text{The symbol has to be reversed.}$$
$$y \leq 4 \quad \text{Simplifying}$$

The solution set is $\{y | y \leq 4\}$.

Exercise Set 2.7

43. $-3x < -17$
$\dfrac{-3x}{-3} > \dfrac{-17}{-3}$ Dividing by -3
↑ The symbol has to be reversed.
$x > \dfrac{17}{3}$ Simplifying

The solution set is $\left\{x \middle| x > \dfrac{17}{3}\right\}$.

45. $-2y > \dfrac{1}{7}$
$-\dfrac{1}{2} \cdot (-2y) < -\dfrac{1}{2} \cdot \dfrac{1}{7}$
↑ The symbol has to be reversed.
$y < -\dfrac{1}{14}$

The solution set is $\left\{y \middle| y < -\dfrac{1}{14}\right\}$.

47. $-\dfrac{6}{5} \le -4x$
$-\dfrac{1}{4} \cdot \left(-\dfrac{6}{5}\right) \ge -\dfrac{1}{4} \cdot (-4x)$
$\dfrac{6}{20} \ge x$
$\dfrac{3}{10} \ge x$, or $x \le \dfrac{3}{10}$

The solution set is $\left\{x \middle| \dfrac{3}{10} \ge x\right\}$, or $\left\{x \middle| x \le \dfrac{3}{10}\right\}$.

49. $4 + 3x < 28$
$-4 + 4 + 3x < -4 + 28$ Adding -4
$3x < 24$ Simplifying
$\dfrac{3x}{3} < \dfrac{24}{3}$ Dividing by 3
$x < 8$

The solution set is $\{x | x < 8\}$.

51. $3x - 5 \le 13$
$3x - 5 + 5 \le 13 + 5$ Adding 5
$3x \le 18$
$\dfrac{3x}{3} \le \dfrac{18}{3}$ Dividing by 3
$x \le 6$

The solution set is $\{x | x \le 6\}$.

53. $13x - 7 < -46$
$13x - 7 + 7 < -46 + 7$
$13x < -39$
$\dfrac{13x}{13} < \dfrac{-39}{13}$
$x < -3$

The solution set is $\{x | x < -3\}$.

55. $30 > 3 - 9x$
$30 - 3 > 3 - 9x - 3$ Subtracting 3
$27 > -9x$
$\dfrac{27}{-9} < \dfrac{-9x}{-9}$ Dividing by -9
↑ The symbol has to be reversed.
$-3 < x$

The solution set is $\{x | -3 < x\}$, or $\{x | x > -3\}$.

57. $4x + 2 - 3x \le 9$
$x + 2 \le 9$ Collecting like terms
$x + 2 - 2 \le 9 - 2$
$x \le 7$

The solution set is $\{x | x \le 7\}$.

59. $-3 < 8x + 7 - 7x$
$-3 < x + 7$ Collecting like terms
$-3 - 7 < x + 7 - 7$
$-10 < x$

The solution set is $\{x | -10 < x\}$, or $\{x | x > -10\}$.

61. $6 - 4y > 4 - 3y$
$6 - 4y + 4y > 4 - 3y + 4y$ Adding $4y$
$6 > 4 + y$
$-4 + 6 > -4 + 4 + y$ Adding -4
$2 > y$, or $y < 2$

The solution set is $\{y | 2 > y\}$, or $\{y | y < 2\}$.

63. $5 - 9y \le 2 - 8y$
$5 - 9y + 9y \le 2 - 8y + 9y$
$5 \le 2 + y$
$-2 + 5 \le -2 + 2 + y$
$3 \le y$, or $y \ge 3$

The solution set is $\{y | 3 \le y\}$, or $\{y | y \ge 3\}$.

65. $19 - 7y - 3y < 39$
$19 - 10y < 39$ Collecting like terms
$-19 + 19 - 10y < -19 + 39$
$-10y < 20$
$\dfrac{-10y}{-10} > \dfrac{20}{-10}$
↑ The symbol has to be reversed.
$y > -2$

The solution set is $\{y | y > -2\}$.

67. $2.1x + 45.2 > 3.2 - 8.4x$
$10(2.1x + 45.2) > 10(3.2 - 8.4x)$ Multiplying by 10 to clear decimals
$21x + 452 > 32 - 84x$
$21x + 84x > 32 - 452$ Adding $84x$ and subtracting 452
$105x > -420$
$x > -4$ Dividing by 105

The solution set is $\{x | x > -4\}$.

69. $\dfrac{x}{3} - 2 \le 1$
$3\left(\dfrac{x}{3} - 2\right) \le 3 \cdot 1$ Multiplying by 3 to clear the fraction
$x - 6 \le 3$ Simplifying
$x \le 9$ Adding 6

The solution set is $\{x | x \le 9\}$.

71. $\dfrac{y}{5} + 1 \le \dfrac{2}{5}$
$5\left(\dfrac{y}{5} + 1\right) \le 5 \cdot \dfrac{2}{5}$ Clearing fractions
$y + 5 \le 2$
$y \le -3$ Subtracting 5

The solution set is $\{y | y \le -3\}$.

73. $3(2y - 3) < 27$
$6y - 9 < 27$ Removing parentheses
$6y < 36$ Adding 9
$y < 6$ Dividing by 6

The solution set is $\{y|y < 6\}$.

75. $2(3 + 4m) - 9 \geq 45$
$6 + 8m - 9 \geq 45$ Removing parentheses
$8m - 3 \geq 45$ Collecting like terms
$8m \geq 48$ Adding 3
$m \geq 6$ Dividing by 8

The solution set is $\{m|m \geq 6\}$.

77. $8(2t + 1) > 4(7t + 7)$
$16t + 8 > 28t + 28$
$16t - 28t > 28 - 8$
$-12t > 20$
$t < -\dfrac{20}{12}$ Dividing by -12 and reversing the symbol
$t < -\dfrac{5}{3}$

The solution set is $\left\{t \middle| t < -\dfrac{5}{3}\right\}$.

79. $3(r - 6) + 2 < 4(r + 2) - 21$
$3r - 18 + 2 < 4r + 8 - 21$
$3r - 16 < 4r - 13$
$-16 + 13 < 4r - 3r$
$-3 < r$, or $r > -3$

The solution set is $\{r|r > -3\}$.

81. $0.8(3x + 6) \geq 1.1 - (x + 2)$
$2.4x + 4.8 \geq 1.1 - x - 2$
$10(2.4x + 4.8) \geq 10(1.1 - x - 2)$ Clearing decimals
$24x + 48 \geq 11 - 10x - 20$
$24x + 48 \geq -10x - 9$ Collecting like terms
$24x + 10x \geq -9 - 48$
$34x \geq -57$
$x \geq -\dfrac{57}{34}$

The solution set is $\left\{x \middle| x \geq -\dfrac{57}{34}\right\}$.

83. $\dfrac{5}{3} + \dfrac{2}{3}x < \dfrac{25}{12} + \dfrac{5}{4}x + \dfrac{3}{4}$

The number 12 is the least common multiple of all the denominators. We multiply by 12 on both sides.

$12\left(\dfrac{5}{3} + \dfrac{2}{3}x\right) < 12\left(\dfrac{25}{12} + \dfrac{5}{4}x + \dfrac{3}{4}\right)$

$12 \cdot \dfrac{5}{3} + 12 \cdot \dfrac{2}{3}x < 12 \cdot \dfrac{25}{12} + 12 \cdot \dfrac{5}{4}x + 12 \cdot \dfrac{3}{4}$

$20 + 8x < 25 + 15x + 9$
$20 + 8x < 34 + 15x$
$20 - 34 < 15x - 8x$
$-14 < 7x$
$-2 < x$, or $x > -2$

The solution set is $\{x|x > -2\}$.

85. Discussion and Writing Exercise

87. $-56 + (-18)$ Two negative numbers. Add the absolute values and make the answer negative.
$-56 + (-18) = -74$

89. $-\dfrac{3}{4} + \dfrac{1}{8}$ One negative and one positive number. Find the difference of the absolute values. Then make the answer negative, since the negative number has the larger absolute value.
$-\dfrac{3}{4} + \dfrac{1}{8} = -\dfrac{6}{8} + \dfrac{1}{8} = -\dfrac{5}{8}$

91. $-56 - (-18) = -56 + 18 = -38$

93. $-2.3 - 7.1 = -2.3 + (-7.1) = -9.4$

95. $5 - 3^2 + (8 - 2)^2 \cdot 4 = 5 - 3^2 + 6^2 \cdot 4$
$= 5 - 9 + 36 \cdot 4$
$= 5 - 9 + 144$
$= -4 + 144$
$= 140$

97. $5(2x - 4) - 3(4x + 1) = 10x - 20 - 12x - 3 = -2x - 23$

99. $|x| < 3$

a) Since $|0| = 0$ and $0 < 3$ is true, 0 is a solution.
b) Since $|-2| = 2$ and $2 < 3$ is true, -2 is a solution.
c) Since $|-3| = 3$ and $3 < 3$ is false, -3 is not a solution.
d) Since $|4| = 4$ and $4 < 3$ is false, 4 is not a solution.
e) Since $|3| = 3$ and $3 < 3$ is false, 3 is not a solution.
f) Since $|1.7| = 1.7$ and $1.7 < 3$ is true, 1.7 is a solution.
g) Since $|-2.8| = 2.8$ and $2.8 < 3$ is true, -2.8 is a solution.

101. $x + 3 \leq 3 + x$
$x - x \leq 3 - 3$ Subtracting x and 3
$0 \leq 0$

We get an inequality that is true for all values of x, so the inequality is true for all real numbers.

Exercise Set 2.8

1. $n \geq 7$

3. $w > 2$ kg

5. 90 mph $< s <$ 110 mph

7. $a \leq 1,200,000$

9. $c \leq \$1.50$

11. $x > 8$

13. $y \leq -4$

15. $n \geq 1300$

17. $a \leq 500$ L

19. $3x + 2 < 13$, or $2 + 3x < 13$

Exercise Set 2.8   39

21. *Familiarize*. Let s represent the score on the fourth test.

Translate.

$$\underbrace{\frac{82+76+78+s}{4}}_{\text{The average score}} \underbrace{\geq}_{\text{is at least}} \underbrace{80}_{80}.$$

Solve.

$$\frac{82+76+78+s}{4} \geq 80$$
$$4\left(\frac{82+76+78+s}{4}\right) \geq 4 \cdot 80$$
$$82+76+78+s \geq 320$$
$$236+s \geq 320$$
$$s \geq 84$$

Check. As a partial check we show that the average is at least 80 when the fourth test score is 84.

$$\frac{82+76+78+84}{4} = \frac{320}{4} = 80$$

State. The student will get at least a B if the score on the fourth test is at least 84. The solution set is $\{s|x \geq 84\}$.

23. *Familiarize*. We use the formula for converting Celsius temperatures to Fahrenheit temperatures, $F = \frac{9}{5}C + 32$.

Translate.

$$\underbrace{\frac{9}{5}C + 32}_{\text{Fahrenheit temperature}} \underbrace{\leq}_{\text{is less than}} \underbrace{1945.4}_{1945.4}.$$

Solve.

$$\frac{9}{5}C + 32 < 1945.4$$
$$\frac{9}{5}C < 1913.4$$
$$\frac{5}{9} \cdot \frac{9}{5}C < \frac{5}{9}(1913.4)$$
$$C < 1063$$

Check. As a partial check we can show that the Fahrenheit temperature is less than 1945.4° for a Celsius temperature less than 1063° and is greater than 1945.4° for a Celsius temperature greater than 1063°.

$$F = \frac{9}{5} \cdot 1062 + 32 = 1943.6 < 1945.4$$
$$F = \frac{9}{5} \cdot 1064 + 32 = 1947.2 > 1945.4$$

State. Gold stays solid for temperatures less than 1063°C. The solution set is $\{C|C < 1063°\}$.

25. *Familiarize*. $R = -0.075t + 3.85$

In the formula R represents the world record and t represents the years since 1930. When $t = 0$ (1930), the record was $-0.075 \cdot 0 + 3.85$, or 3.85 minutes. When $t = 2$ (1932), the record was $-0.075(2) + 3.85$, or 3.7 minutes. For what values of t will $-0.075t + 3.85$ be less than 3.5?

Translate. The record is to be less than 3.5. We have the inequality

$$R < 3.5.$$

To find the t values which satisfy this condition we substitute $-0.075t + 3.85$ for R.

$$-0.075t + 3.85 < 3.5$$

Solve.

$$-0.075t + 3.85 < 3.5$$
$$-0.075t < 3.5 - 3.85$$
$$-0.075t < -0.35$$
$$t > \frac{-0.35}{-0.075}$$
$$t > 4\frac{2}{3}$$

Check. With inequalities it is impossible to check each solution. But we can check to see if the solution set we obtained seems reasonable.

When $t = 4\frac{1}{2}$, $R = -0.075(4.5) + 3.85$, or 3.5125.

When $t = 4\frac{2}{3}$, $R = -0.075\left(\frac{14}{3}\right) + 3.85$, or 3.5.

When $t = 4\frac{3}{4}$, $R = -0.075(4.75) + 3.85$, or 3.49375.

Since $r = 3.5$ when $t = 4\frac{2}{3}$ and R decreases as t increases, R will be less than 3.5 when t is greater than $4\frac{2}{3}$.

State. The world record will be less than 3.5 minutes more than $4\frac{2}{3}$ years after 1930. If we let $Y =$ the year, then the solution set is $\{Y|Y \geq 1935\}$.

27. *Familiarize*. As in the drawing in the text, we let $L =$ the length of the envelope. Recall that the area of a rectangle is the product of the length and the width.

Translate.

$$\underbrace{L}_{\text{Length}} \underbrace{\cdot}_{\text{times}} \underbrace{3\frac{1}{2}}_{\text{width}} \underbrace{\geq}_{\text{is at least}} \underbrace{17\frac{1}{2}}_{17\frac{1}{2} \text{ in}^2}$$

Solve.

$$L \cdot 3\frac{1}{2} \geq 17\frac{1}{2}$$
$$L \cdot \frac{7}{2} \geq \frac{35}{2}$$
$$L \cdot \frac{7}{2} \cdot \frac{2}{7} \geq \frac{35}{2} \cdot \frac{2}{7}$$
$$L \geq 5$$

The solution set is $\{L|L \geq 5\}$.

Check. We can obtain a partial check by substituting a number greater than or equal to 5 in the inequality. For example, when $L = 6$:

$$L \cdot 3\frac{1}{2} = 6 \cdot 3\frac{1}{2} = 6 \cdot \frac{7}{2} = 21 \geq 17\frac{1}{2}$$

The result appears to be correct.

State. Lengths of 5 in. or more will satisfy the constraints. The solution set is $\{L|L \geq 5 \text{ in.}\}$.

29. *Familiarize*. Let $c =$ the number of copies Myra has made. The total cost of the copies is the setup fee of \$5 plus \$4 times the number of copies, or $\$4 \cdot c$.

Translate.

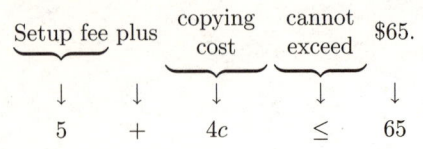

Setup fee	plus	copying cost	cannot exceed	$65.
↓	↓	↓	↓	↓
5	+	4c	≤	65

Solve. We solve the inequality.

$$5 + 4c \leq 65$$
$$4c \leq 60$$
$$c \leq 15$$

Check. As a partial check, we show that Myra can have 15 copies made and not exceed her $65 budget.

$$\$5 + \$4 \cdot 15 = 5 + 60 = \$65$$

State. Myra can have 15 or fewer copies made and stay within her budget.

31. ***Familiarize.*** Let m represent the length of a telephone call, in minutes.

   Translate.

$0.75 charge	plus	charge for time used	is at least	$3.00.
↓	↓	↓	↓	↓
0.75	+	0.45m	≥	3

   Solve. We solve the inequality.

   $$0.75 + 0.45m \geq 3$$
   $$0.45m \geq 2.25$$
   $$m \geq 5$$

   Check. As a partial check, we can show that if a call lasts 5 minutes it costs at least $3.00:

   $$\$0.75 + \$0.45(5) = \$0.75 + \$2.25 = \$3.00.$$

   State. Simon's calls last at least 5 minutes each.

33. ***Familiarize.*** Let c = the number of courses for which Angelica registers. Her total tuition is the $35 registration fee plus $375 times the number of courses for which she registers, or $375 \cdot c$.

   Translate.

Registration fee	plus	fee for courses	cannot exceed	$1000.
↓	↓	↓	↓	↓
35	+	375 · c	≤	1000

   Solve. We solve the inequality.

   $$35 + 375c \leq 1000$$
   $$375c \leq 965$$
   $$c \leq 2.57\overline{3}$$

   Check. Although the solution set of the inequality is all numbers less than or equal to $2.57\overline{3}$, since c represents the number of courses for which Angelica registers, we round down to 2. If she registers for 2 courses, her tuition is $35 + \$375 \cdot 2$, or $785 which does not exceed $1000. If she registers for 3 courses, her tuition is $35 + \$375 \cdot 3$, or $1160 which exceeds $1000.

   State. Angelica can register for at most 2 courses.

35. ***Familiarize.*** Let s = the number of servings of fruits or vegetables Dale eats on Saturday.

   Translate.

Average number of fruit or vegetable servings	is at least	5.
↓	↓	↓
$\dfrac{4+6+7+4+6+4+s}{7}$	≥	5

   Solve. We first multiply by 7 to clear the fraction.

   $$7\left(\frac{4+6+7+4+6+4+s}{7}\right) \geq 7 \cdot 5$$
   $$4+6+7+4+6+4+s \geq 35$$
   $$31 + s \geq 35$$
   $$s \geq 4$$

   Check. As a partial check, we show that Dale can eat 4 servings of fruits or vegetables on Saturday and average at least 5 servings per day for the week:

   $$\frac{4+6+7+4+6+4+4}{7} = \frac{35}{7} = 5$$

   State. Dale should eat at least 4 servings of fruits or vegetables on Saturday.

37. ***Familiarize.*** We first make a drawing. We let l represent the length, in feet.

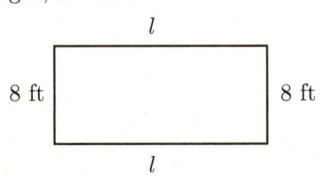

   [Rectangle with length l on top and bottom, width 8 ft on left and right sides]

   The perimeter is $P = 2l + 2w$, or $2l + 2 \cdot 8$, or $2l + 16$.

   Translate. We translate to 2 inequalities.

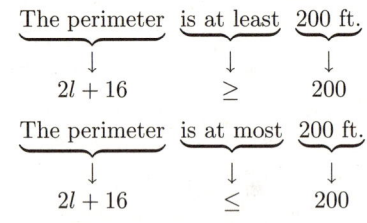

The perimeter	is at least	200 ft.
↓	↓	↓
2l + 16	≥	200

The perimeter	is at most	200 ft.
↓	↓	↓
2l + 16	≤	200

   Solve. We solve each inequality.

   $$2l + 16 \geq 200 \qquad 2l + 16 \leq 200$$
   $$2l \geq 184 \qquad\qquad 2l \leq 184$$
   $$l \geq 92 \qquad\qquad\quad l \leq 92$$

   Check. We check to see if the solutions seem reasonable.

   When $l = 91$ ft, $P = 2 \cdot 91 + 16$, or 198 ft.
   When $l = 92$ ft, $P = 2 \cdot 92 + 16$, or 200 ft.
   When $l = 93$ ft, $P = 2 \cdot 93 + 16$, or 202 ft.

   From these calculations, it appears that the solutions are correct.

   State. Lengths greater than or equal to 92 ft will make the perimeter at least 200 ft. Lengths less than or equal to 92 ft will make the perimeter at most 200 ft.

Exercise Set 2.8

39. ***Familiarize.*** Using the label on the drawing in the text, we let L represent the length.

The area is the length times the width, or $4L$.

Translate.

$$\underbrace{\text{Area}}_{\downarrow \atop 4L} \quad \underbrace{\text{is less than}}_{\downarrow \atop <} \quad \underbrace{86 \text{ cm}^2.}_{\downarrow \atop 86}$$

Solve.
$$4L < 86$$
$$L < 21.5$$

Check. We check to see if the solution seems reasonable.

When $L = 22$, the area is $22 \cdot 4$, or 88 cm^2.

When $L = 21.5$, the area is $21.5(4)$, or 86 cm^2.

When $L = 21$, the area is $21 \cdot 4$, or 84 cm^2.

From these calculations, it would appear that the solution is correct.

State. The area will be less than 86 cm^2 for lengths less than 21.5 cm.

41. ***Familiarize.*** Let v = the blue book value of the car. Since the car was repaired, we know that \$8500 does not exceed $0.8v$ or, in other words, $0.8v$ is at least \$8500.

Translate.

$$\underbrace{\text{80\% of the} \atop \text{blue book value}}_{\downarrow \atop 0.8v} \quad \underbrace{\text{is at least}}_{\downarrow \atop \geq} \quad \underbrace{\$8500.}_{\downarrow \atop 8500}$$

Solve.
$$0.8v \geq 8500$$
$$v \geq \frac{8500}{0.8}$$
$$v \geq 10,625$$

Check. As a partial check, we show that 80% of \$10,625 is at least \$8500:
$$0.8(\$10,625) = \$8500$$

State. The blue book value of the car was at least \$10,625.

43. ***Familiarize.*** Let r = the amount of fat in a serving of the regular peanut butter, in grams. If reduced fat peanut butter has at least 25% less fat than regular peanut butter, then it has at most 75% as much fat as the regular peanut butter.

Translate.

$$\underbrace{12 \text{ g of fat}}_{\downarrow \atop 12} \quad \underbrace{\text{is at most}}_{\downarrow \atop \leq} \quad \underbrace{75\%}_{\downarrow \atop 0.75} \text{ of } \underbrace{\text{the amount of} \atop \text{fat in regular} \atop \text{peanut butter.}}_{\downarrow \atop r}$$

Solve.
$$12 \leq 0.75r$$
$$16 \leq r$$

Check. As a partial check, we show that 12 g of fat does not exceed 75% of 16 g of fat:
$$0.75(16) = 12$$

State. Regular peanut butter contains at least 16 g of fat per serving.

45. ***Familiarize.*** Let w = the number of weeks after July 1. After w weeks the water level has dropped $\frac{2}{3}w$ ft.

Translate.

$$\underbrace{\text{Original} \atop \text{depth}}_{\downarrow \atop 25} \quad \underbrace{\text{minus}}_{\downarrow \atop -} \quad \underbrace{\text{drop in} \atop \text{water level}}_{\downarrow \atop \frac{2}{3}w} \quad \underbrace{\text{does not} \atop \text{exceed}}_{\downarrow \atop \leq} \quad \underbrace{21 \text{ ft.}}_{\downarrow \atop 21}$$

Solve. We solve the inequality.
$$25 - \frac{2}{3}w \leq 21$$
$$-\frac{2}{3}w \leq -4$$
$$w \geq -\frac{3}{2}(-4)$$
$$w \geq 6$$

Check. As a partial check we show that the water level is 21 ft 6 weeks after July 1.
$$25 - \frac{2}{3} \cdot 6 = 25 - 4 = 21 \text{ ft}$$

Since the water level continues to drop during the weeks after July 1, the answer seems reasonable.

State. The water level will not exceed 21 ft for dates at least 6 weeks after July 1.

47. ***Familiarize.*** Let h = the height of the triangle, in ft. Recall that the formula for the area of a triangle with base b and height h is $A = \frac{1}{2}bh$.

Translate.

$$\underbrace{\text{Area}}_{\downarrow \atop \frac{1}{2}\left(1\frac{1}{2}\right)h} \quad \underbrace{\text{is at least}}_{\downarrow \atop \geq} \quad \underbrace{3 \text{ ft}^2.}_{\downarrow \atop 3}$$

Solve. We solve the inequality.
$$\frac{1}{2}\left(1\frac{1}{2}\right)h \geq 3$$
$$\frac{1}{2} \cdot \frac{3}{2} \cdot h \geq 3$$
$$\frac{3}{4}h \geq 3$$
$$h \geq \frac{4}{3} \cdot 3$$
$$h \geq 3$$

Check. As a partial check, we show that the area of the triangle is 3 ft^2 when the height is 4 ft.
$$\frac{1}{2}\left(1\frac{1}{2}\right)(4) = \frac{1}{2} \cdot \frac{3}{2} \cdot \frac{4}{1} = 3$$

State. The height should be at least 4 ft.

49. Familiarize. The average number of calls per week is the sum of the calls for the three weeks divided by the number of weeks, 3. We let c represent the number of calls made during the third week.

Translate. The average of the three weeks is given by
$$\frac{17 + 22 + c}{3}.$$
Since the average must be at least 20, this means that it must be greater than or equal to 20. Thus, we can translate the problem to the inequality
$$\frac{17 + 22 + c}{3} \geq 20.$$
Solve. We first multiply by 3 to clear the fraction.
$$3\left(\frac{17 + 22 + c}{3}\right) \geq 3 \cdot 20$$
$$17 + 22 + c \geq 60$$
$$39 + c \geq 60$$
$$c \geq 21$$

Check. Suppose c is a number greater than or equal to 21. Then by adding 17 and 22 on both sides of the inequality we get
$$17 + 22 + c \geq 17 + 22 + 21$$
$$17 + 22 + c \geq 60$$
so
$$\frac{17 + 22 + c}{3} \geq \frac{60}{3}, \text{ or } 20.$$
State. 21 calls or more will maintain an average of at least 20 for the three-week period.

51. Discussion and Writing Exercise.

53. $-3 + 2(-5)^2(-3) - 7 = -3 + 2(25)(-3) - 7$
$= -3 + 50(-3) - 7$
$= -3 - 150 - 7$
$= -153 - 7$
$= -160$

55. $23(2x - 4) - 15(10 - 3x) = 46x - 92 - 150 + 45x = 91x - 242$

57. Familiarize. We use the formula $F = \frac{9}{5}C + 32$.

Translate. We are interested in temperatures such that $5° < F < 15°$. Substituting for F, we have:
$$5 < \frac{9}{5}C + 32 < 15$$
Solve.
$$5 < \frac{9}{5}C + 32 < 15$$
$$5 \cdot 5 < 5\left(\frac{9}{5}C + 32\right) < 5 \cdot 15$$
$$25 < 9C + 160 < 75$$
$$-135 < 9C < -85$$
$$-15 < C < -9\frac{4}{9}$$
Check. The check is left to the student.

State. Green ski wax works best for temperatures between $-15°C$ and $-9\frac{4}{9}°C$.

59. Familiarize. Let f = the fat content of a serving of regular tortilla chips, in grams. A product that contains 60% less fat than another product has 40% of the fat content of that product. If Reduced Fat Tortilla Pops cannot be labeled lowfat, then they contain at least 3 g of fat.

Translate.

40% of	the fat content of regular tortilla chips	is at least	3 grams of fat
↓ ↓	↓	↓	↓
0.4 ·	f	$\geq$	3

Solve.
$$0.4f \geq 3$$
$$f \geq 7.5$$

Check. As a partial check, we show that 40% of 7.5 g is not less than 3 g.
$$0.4(7.5) = 3$$
State. A serving of regular tortilla chips contains at least 7.5 g of fat.

Chapter 3

Graphs of Linear Equations

Exercise Set 3.1

1. The section of the circle graph for women representing heart disease shows that 47.6% of women 65 and older die of heart disease.

3. The section of the circle graph for men representing heart disease shows that 44.6% of men 65 and older die of heart disease.

5. *Familiarize*. The circle graph shows that 33.0% of men 65 and older die of cancer. Let c = the number of men who would be expected to die of cancer.

   Translate. We reword and translate.

   What is 33.0% of 150,000?
   $c\ =\ 33.0\%\ \cdot\ 150{,}000$

   Solve. We carry out the computation.
   $c = 33.0\% \cdot 150{,}000 = 0.330 \cdot 150{,}000 = 49{,}500$

   Check. We repeat the calculation. The answer checks.

   State. In a group of 150,000 men age 65 and older, 49,500 of them would be expected to die of cancer.

7. *Familiarize*. The circle graph shows that 12.6% of women 65 and older die of a stroke. Let s = the number of women who would be expected to die of a stroke.

   Translate. We reword and translate.

   What is 12.6% of 150,000?
   $s\ =\ 12.6\%\ \cdot\ 150{,}000$

   Solve. We carry out the computation.
   $s = 12.6\% \cdot 150{,}000 = 0.126 \cdot 150{,}000 = 18{,}900$

   Check. We repeat the calculation. The answer checks.

   State. In a group of 150,000 women age 65 and older, 18,900 of them would be expected to die of a stroke.

9. We go to the top of the bar that is above the body weight 200 lb. Then we move horizontally from the top of the bar to the vertical scale listing numbers of drinks. It appears approximately 6 drinks will give a 200-lb person a blood-alcohol level of 0.10%.

11. We see that the bars for weights above 200 lb extend beyond the 6 drink level. Thus, the weight of someone who can consume 6 drinks without reaching a blood-alcohol level of 0.10% is greater than 200 lb.

13. From $3\frac{1}{2}$ on the vertical scale we move horizontally until we reach a bar whose top is above the horizontal line on which we are moving. The first such bar corresponds to a body weight of 120 lb. Thus, we can conclude an individual weighs more than 120 lb if $3\frac{1}{2}$ drinks are consumed without reaching a blood-alcohol level of 0.10%.

15. First locate 1995 on the horizontal axis and then move up to the line. Now move across to the vertical scale and read that there were approximately 17,000 alcohol-related deaths in 1995.

17. The lowest point on the graph occurs above 1998. Thus, the lowest number of deaths occurred in 1998.

19. In Exercise 15 we found that there were approximately 17,000 alcohol-related deaths in 1995. To find the number of alcohol-related deaths in 1998, first locate 1998 on the horizontal scale and then move up to the line. Now move across to the vertical scale and read that there were approximately 16,000 alcohol-related deaths in 1998. We subtract to find the decrease:
    $$17{,}000 - 16{,}000 = 1000$$
    Thus, alcohol-related deaths decreased by about 1000 from 1995 to 1998.

21. $(2, 5)$ is 2 units right and 5 units up.
    $(-1, 3)$ is 1 unit left and 3 units up.
    $(3, -2)$ is 3 units right and 2 units down.
    $(-2, -4)$ is 2 units left and 4 units down.
    $(0, 4)$ is 0 units left or right and 4 units up.
    $(0, -5)$ is 0 units left or right and 5 units down.
    $(5, 0)$ is 5 units right and 0 units up or down.
    $(-5, 0)$ is 5 units left and 0 units up or down.

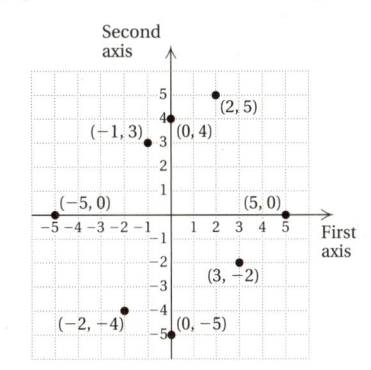

23. Since the first coordinate is negative and the second coordinate positive, the point $(-5, 3)$ is located in quadrant II.

25. Since the first coordinate is positive and the second coordinate negative, the point $(100, -1)$ is in quadrant IV.

27. Since both coordinates are negative, the point $(-6, -29)$ is in quadrant III.

29. Since both coordinates are positive, the point $(3.8, 9.2)$ is in quadrant I.

31. Since the first coordinate is negative and the second coordinate is positive, the point $\left(-\dfrac{1}{3}, \dfrac{15}{7}\right)$ is in quadrant II.

33. Since the first coordinate is positive and the second coordinate is negative, the point $\left(12\dfrac{7}{8}, -1\dfrac{1}{2}\right)$ is in quadrant IV.

35. A point with both coordinates positive is in quadrant I.

37. In quadrant II, the first coordinate is negative and the second coordinate is positive.

39.

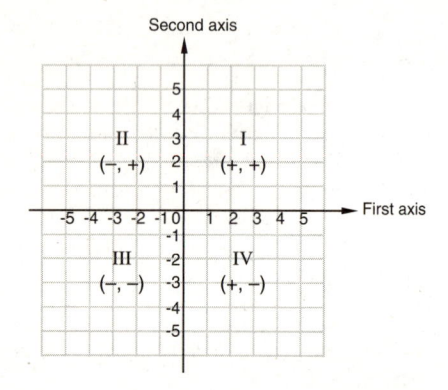

If the first coordinate is positive, then the point must be in either quadrant I or quadrant IV.

41. If the first and second coordinates are equal, they must either be both positive or both negative. The point must be in either quadrant I (both positive) or quadrant III (both negative).

43.

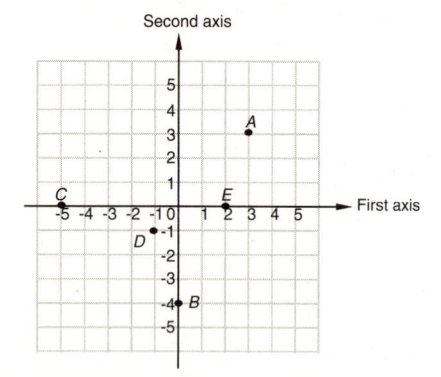

Point A is 3 units right and 3 units up. The coordinates of A are $(3, 3)$.

Point B is 0 units left or right and 4 units down. The coordinates of B are $(0, -4)$.

Point C is 5 units left and 0 units up or down. The coordinates of C are $(-5, 0)$.

Point D is 1 unit left and 1 unit down. The coordinates of D are $(-1, -1)$.

Point E is 2 units right and 0 units up or down. The coordinates of E are $(2, 0)$.

45. Discussion and Writing Exercise

47. The distance of -12 from 0 is 12, so $|-12| = 12$.

49. The distance of 0 from 0 is 0, so $|0| = 0$.

51. The distance of -3.4 from 0 is 3.4, so $|-3.4| = 3.4$.

53. The distance of $\dfrac{2}{3}$ from 0 is $\dfrac{2}{3}$, so $\left|\dfrac{2}{3}\right| = \dfrac{2}{3}$.

55. *Familiarize.* Let $p =$ the average price of a ticket to a Boston Red Sox game in 2000. Then the price in 2001 was $p + 27.4\%p$, or $p + 0.274p$, or $1.274p$.

Translate.

$$\underbrace{\text{The price in 2001}}_{1.274p} \underbrace{\text{was}}_{=} \underbrace{\$36.08.}_{36.08}$$

Solve. We solve the equation.

$$1.274p = 36.08$$
$$p = \dfrac{36.08}{1.274}$$
$$p \approx 28.32$$

Check. 27.4% of $28.32 is about $7.76 and $28.32 + $7.76 = $36.08. The answer checks.

State. The average price of a ticket to a Boston Red Sox game was about $28.32 in 2000.

57.

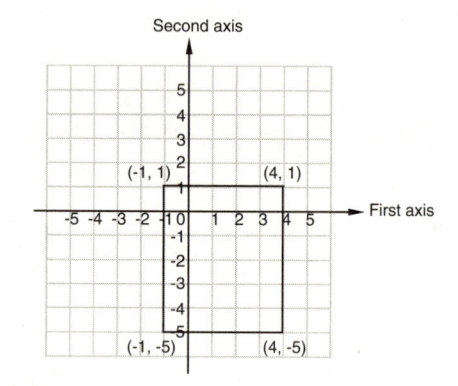

The coordinates of the fourth vertex are $(-1, -5)$.

59. Answers may vary.

We select eight points such that the sum of the coordinates for each point is 6.

$$\begin{array}{ll}
(-1, 7) & -1 + 7 = 6 \\
(0, 6) & 0 + 6 = 6 \\
(1, 5) & 1 + 5 = 6 \\
(2, 4) & 2 + 4 = 6 \\
(3, 3) & 3 + 3 = 6 \\
(4, 2) & 4 + 2 = 6 \\
(5, 1) & 5 + 1 = 6 \\
(6, 0) & 6 + 0 = 6
\end{array}$$

Exercise Set 3.2   45

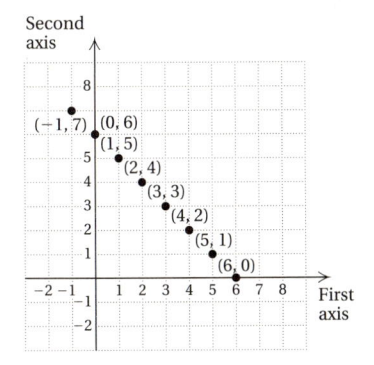

61.

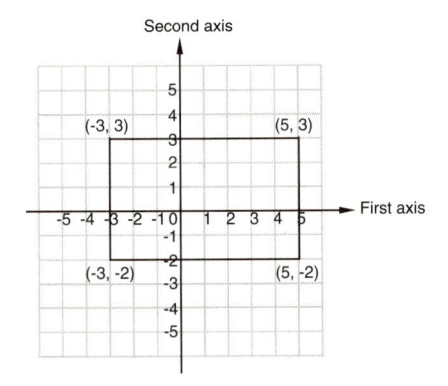

The length is 8, and the width is 5.
$P = 2l + 2w$
$P = 2 \cdot 8 + 2 \cdot 5 = 16 + 10 = 26$

Exercise Set 3.2

1. We substitute 2 for x and 9 for y (alphabetical order of variables).
$$\begin{array}{c|c} y = 3x - 1 \\ \hline 9 \;?\; 3 \cdot 2 - 1 \\ 6 - 1 \\ 5 & \text{FALSE} \end{array}$$
Since $9 = 5$ is false, the pair $(2, 9)$ is not a solution.

3. We substitute 4 for x and 2 for y.
$$\begin{array}{c|c} 2x + 3y = 12 \\ \hline 2 \cdot 4 + 3 \cdot 2 \;?\; 12 \\ 8 + 6 \\ 14 & \text{FALSE} \end{array}$$
Since $14 = 12$ is false, the pair $(4, 2)$ is not a solution.

5. We substitute 3 for a and -1 for b.
$$\begin{array}{c|c} 3a - 4b = 13 \\ \hline 3 \cdot 3 - 4(-1) \;?\; 13 \\ 9 + 4 \\ 13 & \text{TRUE} \end{array}$$
Since $13 = 13$ is true, the pair $(3, -1)$ is a solution.

7. To show that a pair is a solution, we substitute, replacing x with the first coordinate and y with the second coordinate in each pair.
$$\begin{array}{c|c} y = x - 5 \\ \hline -1 \;?\; 4 - 5 \\ -1 & \text{TRUE} \end{array} \qquad \begin{array}{c|c} y = x - 5 \\ \hline -4 \;?\; 1 - 5 \\ -4 & \text{TRUE} \end{array}$$
In each case the substitution results in a true equation. Thus, $(4, -1)$ and $(1, -4)$ are both solutions of $y = x - 5$. We graph these points and sketch the line passing through them.

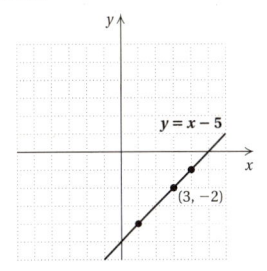

The line appears to pass through $(3, -2)$ also. We check to determine if $(3, -2)$ is a solution of $y = x - 5$.
$$\begin{array}{c|c} y = x - 5 \\ \hline -2 \;?\; 3 - 5 \\ -2 & \text{TRUE} \end{array}$$
Thus, $(3, -2)$ is another solution. There are other correct answers, including $(-1, -6)$, $(2, -3)$, $(0, -5)$, $(5, 0)$, and $(6, 1)$.

9. To show that a pair is a solution, we substitute, replacing x with the first coordinate and y with the second coordinate in each pair.
$$\begin{array}{c|c} y = \frac{1}{2}x + 3 \\ \hline 5 \;?\; \frac{1}{2} \cdot 4 + 3 \\ 2 + 3 \\ 5 & \text{TRUE} \end{array} \qquad \begin{array}{c|c} y = \frac{1}{2}x + 3 \\ \hline 2 \;?\; \frac{1}{2}(-2) + 3 \\ -1 + 3 \\ 2 & \text{TRUE} \end{array}$$
In each case the substitution results in a true equation. Thus, $(4, 5)$ and $(-2, 2)$ are both solutions of $y = \frac{1}{2}x + 3$. We graph these points and sketch the line passing through them.

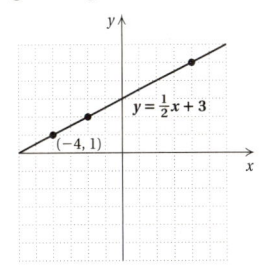

The line appears to pass through $(-4, 1)$ also. We check to determine if $(-4, 1)$ is a solution of $y = \frac{1}{2}x + 3$.

$$y = \frac{1}{2}x + 3$$

$$\begin{array}{c|l} \hline 1 \ ? \ \dfrac{1}{2}(-4) + 3 \\ & -2 + 3 \\ & 1 \quad \text{TRUE} \end{array}$$

Thus, $(-4, 1)$ is another solution. There are other correct answers, including $(-6, 0)$, $(0, 3)$, $(2, 4)$, and $(6, 6)$.

11. To show that a pair is a solution, we substitute, replacing x with the first coordinate and y with the second coordinate in each pair.

$$\begin{array}{c|l} 4x - 2y = 10 \\ \hline 4 \cdot 0 - 2(-5) \ ? \ 10 \\ 10 \ | \ \text{TRUE} \end{array}$$

$$\begin{array}{c|l} 4x - 2y = 10 \\ \hline 4 \cdot 4 - 2 \cdot 3 \ ? \ 10 \\ 16 - 6 \\ 10 \ | \ \text{TRUE} \end{array}$$

In each case the substitution results in a true equation. Thus, $(0, -5)$ and $(4, 3)$ are both solutions of $4x - 2y = 10$. We graph these points and sketch the line passing through them.

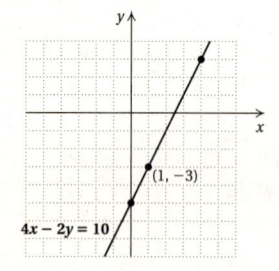

The line appears to pass through $(1, -3)$ also. We check to determine if $(1, -3)$ is a solution of $4x - 2y = 10$.

$$\begin{array}{c|l} 4x - 2y = 10 \\ \hline 4 \cdot 1 - 2(-3) \ ? \ 10 \\ 4 + 6 \\ 10 \ | \ \text{TRUE} \end{array}$$

Thus, $(1, -3)$ is another solution. There are other correct answers, including $(2, -1)$, $(3, 1)$, and $(5, 5)$.

13. $y = x + 1$

The equation is in the form $y = mx + b$. The y-intercept is $(0, 1)$. We find five other pairs.

When $x = -2$, $y = -2 + 1 = -1$.
When $x = -1$, $y = -1 + 1 = 0$.
When $x = 1$, $y = 1 + 1 = 2$.
When $x = 2$, $y = 2 + 1 = 3$.
When $x = 3$, $y = 3 + 1 = 4$.

x	y
-2	-1
-1	0
0	1
1	2
2	3
3	4

Plot these points, draw the line they determine, and label the graph $y = x + 1$.

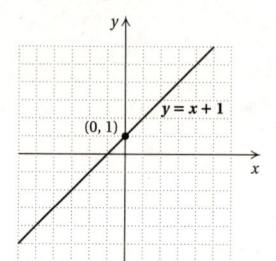

15. $y = x$

The equation is equivalent to $y = x + 0$. The y-intercept is $(0, 0)$. We find five other points.

When $x = -2$, $y = -2$.
When $x = -1$, $y = -1$.
When $x = 1$, $y = 1$.
When $x = 2$, $y = 2$.
When $x = 3$, $y = 3$.

x	y
-2	-2
-1	-1
0	0
1	1
2	2
3	3

Plot these points, draw the line they determine, and label the graph $y = x$.

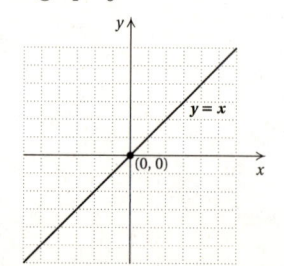

17. $y = \dfrac{1}{2}x$

The equation is equivalent to $y = \dfrac{1}{2}x + 0$. The y-intercept is $(0, 0)$. We find two other points.

Exercise Set 3.2

When $x = -2$, $y = \frac{1}{2}(-2) = -1$.

When $x = 4$, $y = \frac{1}{2} \cdot 4 = 2$.

x	y
-2	-1
0	0
4	2

Plot these points, draw the line they determine, and label the graph $y = \frac{1}{2}x$.

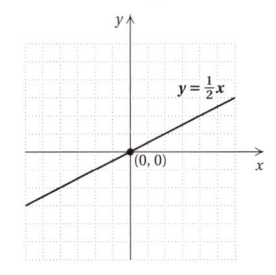

19. $y = x - 3$

The equation is equivalent to $y = x + (-3)$. The y-intercept is $(0, -3)$. We find two other points.

When $x = -2$, $y = -2 - 3 = -5$.

When $x = 4$, $y = 4 - 3 = 1$.

x	y
-2	-5
0	-3
4	1

Plot these points, draw the line they determine, and label the graph $y = x - 3$.

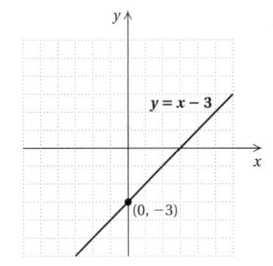

21. $y = 3x - 2 = 3x + (-2)$

The y-intercept is $(0, -2)$. We find two other points.

When $x = -2$, $y = 3(-2) + 2 = -6 + 2 = -4$.

When $x = 1$, $y = 3 \cdot 1 + 2 = 3 + 2 = 5$.

x	y
-2	-4
0	-2
1	5

Plot these points, draw the line they determine, and label the graph $y = 3x + 2$.

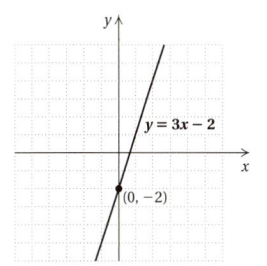

23. $y = \frac{1}{2}x + 1$

The y-intercept is $(0, 1)$. We find two other points using multiples of 2 for x to avoid fractions.

When $x = -4$, $y = \frac{1}{2}(-4) + 1 = -2 + 1 = -1$.

When $x = 4$, $y = \frac{1}{2} \cdot 4 + 1 = 2 + 1 = 3$.

x	y
-4	-1
0	1
4	3

Plot these points, draw the line they determine, and label the graph $y = \frac{1}{2}x + 1$.

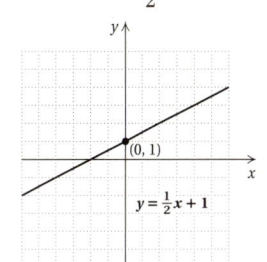

25. $x + y = -5$

$y = -x - 5$

$y = -x + (-5)$

The y-intercept is $(0, -5)$. We find two other points.

When $x = -4$, $y = -(-4) - 5 = 4 - 5 = -1$.

When $x = -1$, $y = -(-1) - 5 = 1 - 5 = -4$.

x	y
-4	-1
0	-5
-1	-4

Plot these points, draw the line they determine, and label the graph $x + y = -5$.

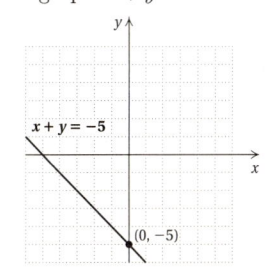

27. $y = \dfrac{5}{3}x - 2 = \dfrac{5}{3}x + (-2)$

The y-intercept is $(0, -2)$. We find two other points using multiples of 3 for x to avoid fractions.

When $x = -3$, $y = \dfrac{5}{3}(-3) - 2 = -5 - 2 = -7$.

When $x = 3$, $y = \dfrac{5}{3} \cdot 3 - 2 = 5 - 2 = 3$.

x	y
-3	-7
0	-2
3	3

Plot these points, draw the line they determine, and label the graph $y = \dfrac{5}{3}x - 2$.

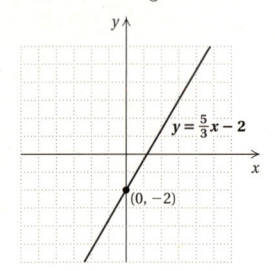

29. $x + 2y = 8$
$2y = -x + 8$
$y = -\dfrac{1}{2}x + 4$

The y-intercept is $(0, 4)$. We find two other points using multiples of 2 for x to avoid fractions.

When $x = -2$, $y = -\dfrac{1}{2}(-2) + 4 = 1 + 4 = 5$.

When $x = 4$, $y = -\dfrac{1}{2} \cdot 4 + 4 = -2 + 4 = 2$.

x	y
-2	5
0	4
4	2

Plot these points, draw the line they determine, and label the graph $x + 2y = 8$.

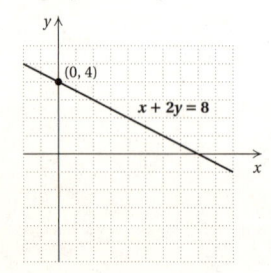

31. $y = \dfrac{3}{2}x + 1$

The y-intercept is $(0, 1)$. We find two other points using multiples of 2 for x to avoid fractions.

When $x = -4$, $y = \dfrac{3}{2}(-4) + 1 = -6 + 1 = -5$.

When $x = 2$, $y = \dfrac{3}{2} \cdot 2 + 1 = 3 + 1 = 4$.

x	y
-4	-5
0	1
2	4

Plot these points, draw the line they determine, and label the graph $y = \dfrac{3}{2}x + 1$.

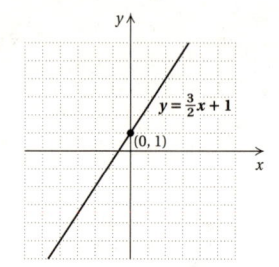

33. $8x - 2y = -10$
$-2y = -8x - 10$
$y = 4x + 5$

The y-intercept is $(0, 5)$. We find two other points.

When $x = -2$, $y = 4(-2) + 5 = -8 + 5 = -3$.

When $x = -1$, $y = 4(-1) + 5 = -4 + 5 = 1$.

x	y
-2	-3
-1	1
0	5

Plot these points, draw the line they determine, and label the graph $8x - 2y = -10$.

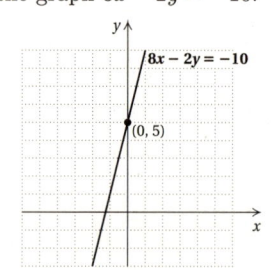

35. $8y + 2x = -4$
$8y = -2x - 4$
$y = -\dfrac{1}{4}x - \dfrac{1}{2}$
$y = -\dfrac{1}{4}x + \left(-\dfrac{1}{2}\right)$

The y-intercept is $\left(0, -\dfrac{1}{2}\right)$. We find two other points.

When $x = -2$, $y = -\dfrac{1}{4}(-2) - \dfrac{1}{2} = \dfrac{1}{2} - \dfrac{1}{2} = 0$.

When $x = 2$, $y = -\dfrac{1}{4} \cdot 2 - \dfrac{1}{2} = -\dfrac{1}{2} - \dfrac{1}{2} = -1$.

Exercise Set 3.2

x	y
-2	0
0	$-\dfrac{1}{2}$
2	-1

Plot these points, draw the line they determine, and label the graph $8y + 2x = -4$.

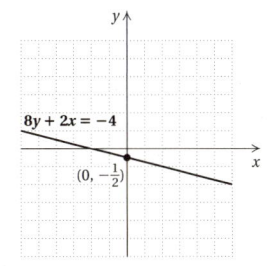

37. a) We substitute 0, 4, and 6 for t and then calculate V.

If $t = 0$, then $V = -50 \cdot 0 + 300 = \$300$.

If $t = 4$, then $V = -50 \cdot 4 + 300 = -200 + 300 = \$100$.

If $t = 6$, then $V = -50 \cdot 6 + 300 = -300 + 300 = \$0$.

b) We plot the three ordered pairs we found in part (a). Note the negative t- and V-values have no meaning in this problem.

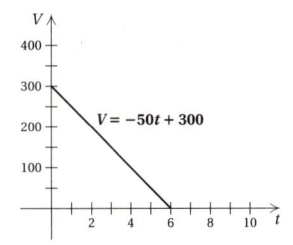

To use the graph to estimate the value of the software after 5 years we must determine which V-value is paired with $t = 5$. We locate 5 on the t-axis, go up to the graph, and then find the value on the V-axis that corresponds to that point. It appears that after 5 years the value of the software is \$50.

c) Substitute 150 for V and then solve for t.
$$V = -50t + 300$$
$$150 = -50t + 300$$
$$-150 = -50t$$
$$3 = t$$

The value of the software is \$150 after 3 years.

39. a) When $d = 1$, $N = 0.1(1) + 7 = 0.1 + 7 = 7.1$ gal.

In 1996, $d = 1996 - 1991 = 5$. When $d = 5$, $N = 0.1(5) + 7 = 0.5 + 7 = 7.5$ gal.

In 2001, $d = 2001 - 1991 = 10$. When $d = 10$, $N = 0.1(10) + 7 = 1 + 7 = 8$ gal.

In 2011, $d = 2011 - 1991 = 20$. When $d = 20$, $N = 0.1(20) + 7 = 2 + 7 = 9$ gal.

b) Plot the four ordered pairs we found in part (a). Note that negative d- and N-values have no meaning in this problem.

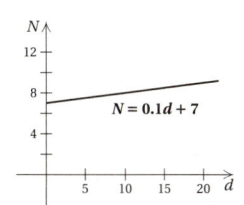

To use the graph to estimate what tea consumption was in 1997 we must determine which N-value is paired with 1997, or with $d = 6$. We locate 6 on the d-axis, go up to the graph, and then find the value on the N-axis that corresponds to that point. It appears that tea consumption was about 7.6 gallons in 1997.

c) Substitute 8.5 for N and then solve for d.
$$N = 0.1d + 7$$
$$8.5 = 0.1d + 7$$
$$1.5 = 0.1d$$
$$15 = d$$

Tea consumption will be about 8.5 gallons 15 years after 1991, or in 2006.

41. Discussion and Writing Exercise

43. $63 = 9x$

$\dfrac{63}{9} = \dfrac{9x}{9}$ Dividing by 9 on both sides

$7 = 1 \cdot x$ Simplifying

$7 = x$ Identity property of 1

The solution is 7.

45. $13x = -52$

$\dfrac{13x}{13} = \dfrac{-52}{13}$ Dividing by 13 on both sides

$1 \cdot x = -4$ Simplifying

$x = -4$ Identity property of 1

The solution is -4.

47. $\dfrac{1}{10}x = \dfrac{2}{5}$

$10 \cdot \dfrac{1}{10}x = 10 \cdot \dfrac{2}{5}$ Multiplying by 10 to clear fractions

$1 \cdot x = 4$ Simplifying

$x = 4$ Identity property of 1

The solution is 4.

49. First we find decimal notation for $\frac{7}{8}$.

$$
\begin{array}{r}
0.875 \\
8\overline{)7.000} \\
\underline{64} \\
60 \\
\underline{56} \\
40 \\
\underline{40} \\
0
\end{array}
$$

Since $\frac{7}{8} = 0.875$, then $-\frac{7}{8} = -0.875$.

51.
$$
\begin{array}{r}
1.828125 \\
64\overline{)117.000000} \\
\underline{64} \\
530 \\
\underline{512} \\
180 \\
\underline{128} \\
520 \\
\underline{512} \\
80 \\
\underline{64} \\
160 \\
\underline{128} \\
320 \\
\underline{320} \\
0
\end{array}
$$

$\frac{117}{64} = 1.828125$

53. Note that the sum of the coordinates of each point on the graph is 5. Thus, we have $x + y = 5$, or $y = -x + 5$.

55. Note that each y-coordinate is 2 more than the corresponding x-coordinate. Thus, we have $y = x + 2$.

Exercise Set 3.3

1. (a) The graph crosses the y-axis at $(0, 5)$, so the y-intercept is $(0, 5)$.

(b) The graph crosses the x-axis at $(2, 0)$, so the x-intercept is $(2, 0)$.

3. (a) The graph crosses the y-axis at $(0, -4)$, so the y-intercept is $(0, -4)$.

(b) The graph crosses the x-axis at $(3, 0)$, so the x-intercept is $(3, 0)$.

5. $3x + 5y = 15$

(a) To find the y-intercept, let $x = 0$. This is the same as covering up the x-term and then solving.

$$5y = 15$$
$$y = 3$$

The y-intercept is $(0, 3)$.

(b) To find the x-intercept, let $y = 0$. This is the same as covering up the y-term and then solving.

$$3x = 15$$
$$x = 5$$

The x-intercept is $(5, 0)$.

7. $7x - 2y = 28$

(a) To find the y-intercept, let $x = 0$. This is the same as covering up the x-term and then solving.

$$-2y = 28$$
$$y = -14$$

The y-intercept is $(0, -14)$.

(b) To find the x-intercept, let $y = 0$. This is the same as covering up the y-term and then solving.

$$7x = 28$$
$$x = 4$$

The x-intercept is $(4, 0)$.

9. $-4x + 3y = 10$

(a) To find the y-intercept, let $x = 0$. This is the same as covering up the x-term and then solving.

$$3y = 10$$
$$y = \frac{10}{3}$$

The y-intercept is $\left(0, \frac{10}{3}\right)$.

(b) To find the x-intercept, let $y = 0$. This is the same as covering up the y-term and then solving.

$$-4x = 10$$
$$x = -\frac{5}{2}$$

The x-intercept is $\left(-\frac{5}{2}, 0\right)$.

11. $6x - 3 = 9y$

$6x - 9y = 3 \quad$ Writing the equation in the form $Ax + By = C$

(a) To find the y-intercept, let $x = 0$. This is the same as covering up the x-term and then solving.

$$-9y = 3$$
$$y = -\frac{1}{3}$$

The y-intercept is $\left(0, -\frac{1}{3}\right)$.

Exercise Set 3.3 51

(b) To find the x-intercept, let $y = 0$. This is the same as covering up the y-term and then solving.
$$6x = 3$$
$$x = \frac{1}{2}$$
The x-intercept is $\left(\frac{1}{2}, 0\right)$.

13. $x + 3y = 6$

To find the x-intercept, let $y = 0$. Then solve for x.
$$x + 3y = 6$$
$$x + 3 \cdot 0 = 6$$
$$x = 6$$
Thus, $(6, 0)$ is the x-intercept.

To find the y-intercept, let $x = 0$. Then solve for y.
$$x + 3y = 6$$
$$0 + 3y = 6$$
$$3y = 6$$
$$y = 2$$
Thus, $(0, 2)$ is the y-intercept.

Plot these points and draw the line.

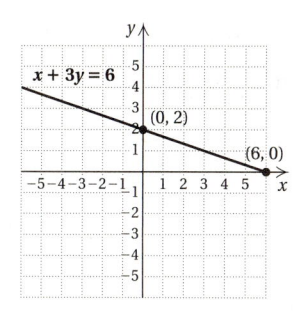

A third point should be used as a check. We substitute any value for x and solve for y.

We let $x = 3$. Then
$$x + 3y = 6$$
$$3 + 3y = 6$$
$$3y = 3$$
$$y = 1$$
The point $(3, 1)$ is on the graph, so the graph is probably correct.

15. $-x + 2y = 4$

To find the x-intercept, let $y = 0$. Then solve for x.
$$-x + 2y = 4$$
$$-x + 2 \cdot 0 = 4$$
$$-x = 4$$
$$x = -4$$
Thus, $(-4, 0)$ is the x-intercept.

To find the y-intercept, let $x = 0$. Then solve for y.
$$-x + 2y = 4$$
$$-0 + 2y = 4$$
$$2y = 4$$
$$y = 2$$
Thus, $(0, 2)$ is the y-intercept.

Plot these points and draw the line.

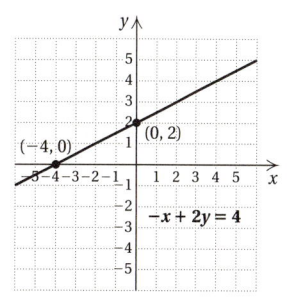

A third point should be used as a check. We substitute any value for x and solve for y.

We let $x = 4$. Then
$$-x + 2y = 4$$
$$-4 + 2y = 4$$
$$2y = 8$$
$$y = 4$$
The point $(4, 4)$ is on the graph, so the graph is probably correct.

17. $3x + y = 6$

To find the x-intercept, let $y = 0$. Then solve for x.
$$3x + y = 6$$
$$3x + 0 = 6$$
$$3x = 6$$
$$x = 2$$
Thus, $(2, 0)$ is the x-intercept.

To find the y-intercept, let $x = 0$. Then solve for y.
$$3x + y = 6$$
$$3 \cdot 0 + y = 6$$
$$y = 6$$
Thus, $(0, 6)$ is the y-intercept.

Plot these points and draw the line.

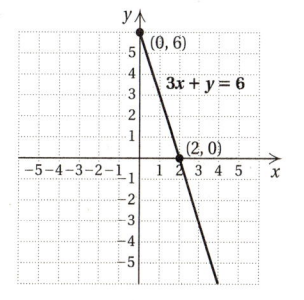

A third point should be used as a check. We substitute any value for x and solve for y.

We let $x = 1$. Then
$$3x + y = 6$$
$$3 \cdot 1 + y = 6$$
$$3 + y = 6$$
$$y = 3$$

The point $(1, 3)$ is on the graph, so the graph is probably correct.

19. $2y - 2 = 6x$

To find the x-intercept, let $y = 0$. Then solve for x.
$$2y - 2 = 6x$$
$$2 \cdot 0 - 2 = 6x$$
$$-2 = 6x$$
$$-\frac{1}{3} = x$$

Thus, $\left(-\frac{1}{3}, 0\right)$ is the x-intercept.

To find the y-intercept, let $x = 0$. Then solve for y.
$$2y - 2 = 6x$$
$$2y - 2 = 6 \cdot 0$$
$$2y - 2 = 0$$
$$2y = 2$$
$$y = 1$$

Thus, $(0, 1)$ is the y-intercept.

It is helpful to plot another point since the intercepts are so close together. This point can also serve as a check.

We let $x = 1$. Then
$$2y - 2 = 6x$$
$$2y - 2 = 6 \cdot 1$$
$$2y - 2 = 6$$
$$2y = 8$$
$$y = 4$$

Plot the point $(1, 4)$ and the intercepts and draw the line.

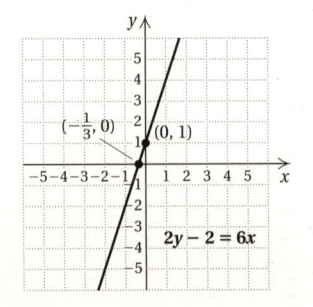

21. $3x - 9 = 3y$

To find the x-intercept, let $y = 0$. Then solve for x.
$$3x - 9 = 3y$$
$$3x - 9 = 3 \cdot 0$$
$$3x - 9 = 0$$
$$3x = 9$$
$$x = 3$$

Thus, $(3, 0)$ is the x-intercept.

To find the y-intercept, let $x = 0$. Then solve for y.
$$3x - 9 = 3y$$
$$3 \cdot 0 - 9 = 3y$$
$$-9 = 3y$$
$$-3 = y$$

Thus, $(0, -3)$ is the y-intercept.

Plot these points and draw the line.

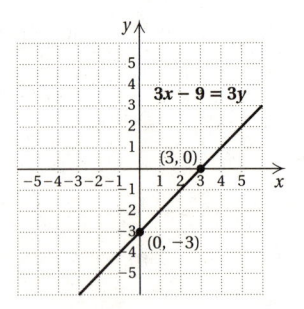

A third point should be used as a check. We substitute any value for x and solve for y.

We let $x = 1$. Then
$$3x - 9 = 3y$$
$$3 \cdot 1 - 9 = 3y$$
$$3 - 9 = 3y$$
$$-6 = 3y$$
$$-2 = y$$

The point $(1, -2)$ is on the graph, so the graph is probably correct.

23. $2x - 3y = 6$

To find the x-intercept, let $y = 0$. Then solve for x.
$$2x - 3y = 6$$
$$2x - 3 \cdot 0 = 6$$
$$2x = 6$$
$$x = 3$$

Thus, $(3, 0)$ is the x-intercept.

To find the y-intercept, let $x = 0$. Then solve for y.
$$2x - 3y = 6$$
$$2 \cdot 0 - 3y = 6$$
$$-3y = 6$$
$$y = -2$$

Thus, $(0, -2)$ is the y-intercept.

Exercise Set 3.3

Plot these points and draw the line.

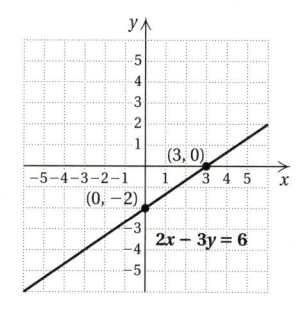

A third point should be used as a check. We substitute any value for x and solve for y.

We let $x = -3$.
$$2x - 3y = 6$$
$$2(-3) - 3y = 6$$
$$-6 - 3y = 6$$
$$-3y = 12$$
$$y = -4$$

The point $(-3, -4)$ is on the graph, so the graph is probably correct.

25. $4x + 5y = 20$

To find the x-intercept, let $y = 0$. Then solve for x.
$$4x + 5y = 20$$
$$4x + 5 \cdot 0 = 20$$
$$4x = 20$$
$$x = 5$$

Thus, $(5, 0)$ is the x-intercept.

To find the y-intercept, let $x = 0$. Then solve for y.
$$4x + 5y = 20$$
$$4 \cdot 0 + 5y = 20$$
$$5y = 20$$
$$y = 4$$

Thus, $(0, 4)$ is the y-intercept.

Plot these points and draw the graph.

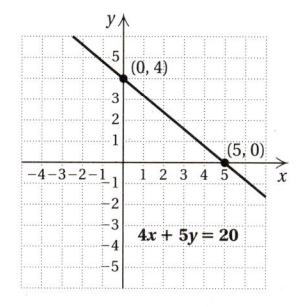

A third point should be used as a check. We substitute any value for x and solve for y.

We let $x = 4$. Then
$$4x + 5y = 20$$
$$4 \cdot 4 + 5y = 20$$
$$16 + 5y = 20$$
$$5y = 4$$
$$y = \frac{4}{5}$$

The point $\left(4, \frac{4}{5}\right)$ is on the graph, so the graph is probably correct.

27. $2x + 3y = 8$

To find the x-intercept, let $y = 0$. Then solve for x.
$$2x + 3y = 8$$
$$2x + 3 \cdot 0 = 8$$
$$2x = 8$$
$$x = 4$$

Thus, $(4, 0)$ is the x-intercept.

To find the y-intercept, let $x = 0$. Then solve for y.
$$2x + 3y = 8$$
$$2 \cdot 0 + 3y = 8$$
$$3y = 8$$
$$y = \frac{8}{3}$$

Thus, $\left(0, \frac{8}{3}\right)$ is the y-intercept.

Plot these points and draw the graph.

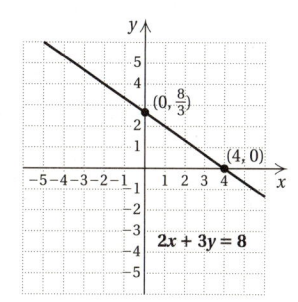

A third point should be used as a check.

We let $x = 1$. Then
$$2x + 3y = 8$$
$$2 \cdot 1 + 3y = 8$$
$$2 + 3y = 8$$
$$3y = 6$$
$$y = 2$$

The point $(1, 2)$ is on the graph, so the graph is probably correct.

29. $x - 3 = y$

To find the x-intercept, let $y = 0$. Then solve for x.
$$x - 3 = y$$
$$x - 3 = 0$$
$$x = 3$$

Thus, $(3, 0)$ is the x-intercept.

To find the y-intercept, let $x = 0$. Then solve for y.
$$x - 3 = y$$
$$0 - 3 = y$$
$$-3 = y$$
Thus, $(0, -3)$ is the y-intercept.

Plot these points and draw the line.

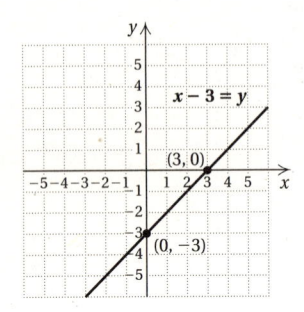

[Graph of $x - 3 = y$ showing points $(3, 0)$ and $(0, -3)$]

A third point should be used as a check.

We let $x = -2$. Then
$$x - 3 = y$$
$$-2 - 3 = y$$
$$-5 = y$$
The point $(-2, -5)$ is on the graph, so the graph is probably correct.

31. $3x - 2 = y$

To find the x-intercept, let $y = 0$. Then solve for x.
$$3x - 2 = y$$
$$3x - 2 = 0$$
$$3x = 2$$
$$x = \frac{2}{3}$$
Thus, $\left(\frac{2}{3}, 0\right)$ is the x-intercept.

To find the y-intercept, let $x = 0$. Then solve for y.
$$3x - 2 = y$$
$$3 \cdot 0 - 2 = y$$
$$-2 = y$$
Thus, $(0, -2)$ is the y-intercept.

Plot these points and draw the line.

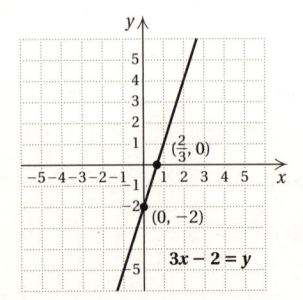

[Graph of $3x - 2 = y$ showing points $\left(\frac{2}{3}, 0\right)$ and $(0, -2)$]

A third point should be used as a check.

We let $x = 2$. Then
$$3x - 2 = y$$
$$3 \cdot 2 - 2 = y$$
$$6 - 2 = y$$
$$4 = y$$
The point $(2, 4)$ is on the graph, so the graph is probably correct.

33. $6x - 2y = 12$

To find the x-intercept, let $y = 0$. Then solve for x.
$$6x - 2y = 12$$
$$6x - 2 \cdot 0 = 12$$
$$6x = 12$$
$$x = 2$$
Thus, $(2, 0)$ is the x-intercept.

To find the y-intercept, let $x = 0$. Then solve for y.
$$6x - 2y = 12$$
$$6 \cdot 0 - 2y = 12$$
$$-2y = 12$$
$$y = -6$$
Thus, $(0, -6)$ is the y-intercept.

Plot these points and draw the line.

[Graph of $6x - 2y = 12$ showing points $(2, 0)$ and $(0, -6)$]

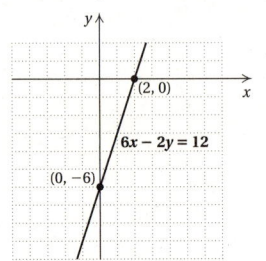

We use a third point as a check.

We let $x = 1$. Then
$$6x - 2y = 12$$
$$6 \cdot 1 - 2y = 12$$
$$6 - 2y = 12$$
$$-2y = 6$$
$$y = -3$$
The point $(1, -3)$ is on the graph, so the graph is probably correct.

35. $3x + 4y = 5$

To find the x-intercept, let $y = 0$. Then solve for x.
$$3x + 4y = 5$$
$$3x + 4 \cdot 0 = 5$$
$$3x = 5$$
$$x = \frac{5}{3}$$
Thus, $\left(\frac{5}{3}, 0\right)$ is the x-intercept.

Exercise Set 3.3

To find the y-intercept, let $x = 0$. Then solve for y.
$$3x + 4y = 5$$
$$3 \cdot 0 + 4y = 5$$
$$4y = 5$$
$$y = \frac{5}{4}$$

Thus, $\left(0, \frac{5}{4}\right)$ is the y-intercept.

It is helpful to plot another point since the intercepts are so close together. This point can also serve as a check.

We let $x = 3$. Then
$$3x + 4y = 5$$
$$3 \cdot 3 + 4y = 5$$
$$9 + 4y = 5$$
$$4y = -4$$
$$y = -1$$

Plot the point $(3, -1)$ and the intercepts and draw the line.

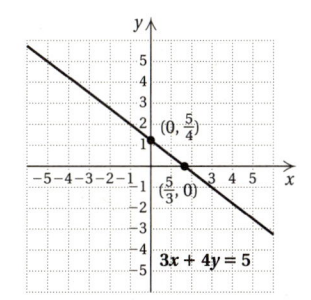

37. $y = -3 - 3x$

To find the x-intercept, let $y = 0$. Then solve for x.
$$y = -3 - 3x$$
$$0 = -3 - 3x$$
$$3x = -3$$
$$x = -1$$

Thus, $(-1, 0)$ is the x-intercept.

To find the y-intercept, let $x = 0$. Then solve for y.
$$y = -3 - 3x$$
$$y = -3 - 3 \cdot 0$$
$$y = -3$$

Thus, $(0, -3)$ is the y-intercept.

Plot these points and draw the graph.

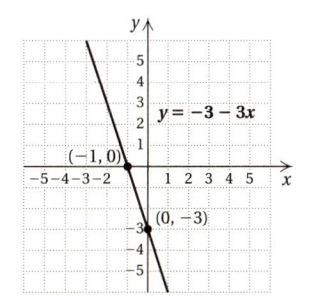

We use a third point as a check.

We let $x = -2$. Then
$$y = -3 - 3x$$
$$y = -3 - 3 \cdot (-2)$$
$$y = -3 + 6$$
$$y = 3$$

The point $(-2, 3)$ is on the graph, so the graph is probably correct.

39. $y - 3x = 0$

To find the x-intercept, let $y = 0$. Then solve for x.
$$0 - 3x = 0$$
$$-3x = 0$$
$$x = 0$$

Thus, $(0, 0)$ is the x-intercept. Note that this is also the y-intercept.

In order to graph the line, we will find a second point.

When $x = 1$, $y - 3 \cdot 1 = 0$
$$y - 3 = 0$$
$$y = 3$$

Plot the points and draw the graph.

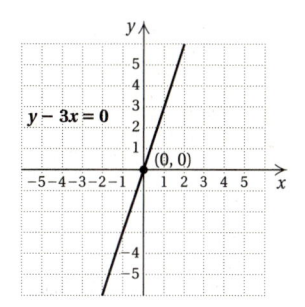

We use a third point as a check.

We let $x = -1$. Then
$$y - 3(-1) = 0$$
$$y + 3 = 0$$
$$y = -3$$

The point $(-1, -3)$ is on the graph, so the graph is probably correct.

41. $x = -2$

Any ordered pair $(-2, y)$ is a solution. The variable x must be -2, but y can be any number we choose. A few solutions are listed below. Plot these points and draw the line.

x	y
-2	-2
-2	0
-2	4

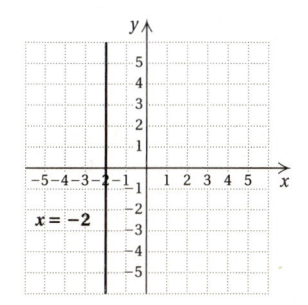

43. $y = 2$

Any ordered pair $(x, 2)$ is a solution. The variable y must be 2, but x can be any number we choose. A few solutions are listed below. Plot these points and draw the line.

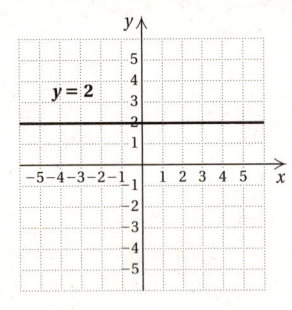

x	y
-3	2
0	2
2	2

45. $x = 2$

Any ordered pair $(2, y)$ is a solution. The variable x must be 2, but y can be any number we choose. A few solutions are listed below. Plot these points and draw the line.

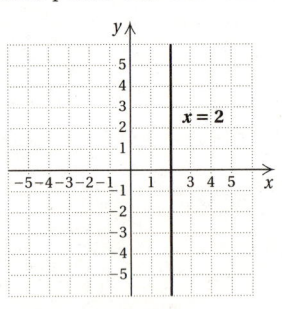

x	y
2	-1
2	4
2	5

47. $y = 0$

Any ordered pair $(x, 0)$ is a solution. The variable y must be 0, but x can be any number we choose. A few solutions are listed below. Plot these points and draw the line.

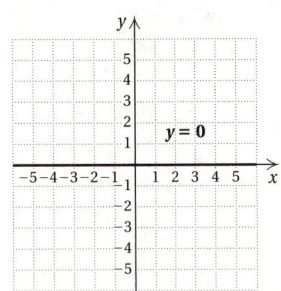

x	y
-5	0
-1	0
3	0

49. $x = \dfrac{3}{2}$

Any ordered pair $\left(\dfrac{3}{2}, y\right)$ is a solution. The variable x must be $\dfrac{3}{2}$, but y can be any number we choose. A few solutions are listed below. Plot these points and draw the line.

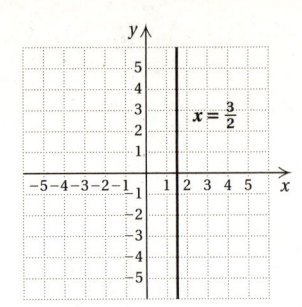

x	y
$\dfrac{3}{2}$	-2
$\dfrac{3}{2}$	0
$\dfrac{3}{2}$	4

51. $3y = -5$

$y = -\dfrac{5}{3}$ Solving for y

Any ordered pair $\left(x, -\dfrac{5}{3}\right)$ is a solution. A few solutions are listed below. Plot these points and draw the line.

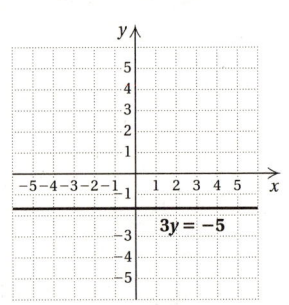

x	y
-3	$-\dfrac{5}{3}$
0	$-\dfrac{5}{3}$
2	$-\dfrac{5}{3}$

53. $4x + 3 = 0$

$4x = -3$

$x = -\dfrac{3}{4}$ Solving for x

Any ordered pair $\left(-\dfrac{3}{4}, y\right)$ is a solution. A few solutions are listed below. Plot these points and draw the line.

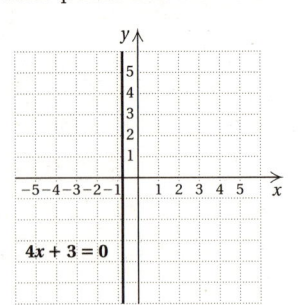

x	y
$-\dfrac{3}{4}$	-2
$-\dfrac{3}{4}$	0
$-\dfrac{3}{4}$	3

55. $48 - 3y = 0$

$-3y = -48$

$y = 16$ Solving for y

Any ordered pair $(x, 16)$ is a solution. A few solutions are listed below. Plot these points and draw the line.

x	y
-4	16
0	16
2	16

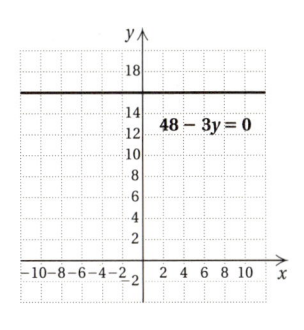

[Graph showing horizontal line $48 - 3y = 0$]

57. Note that every point on the horizontal line passing through $(0, -1)$ has -1 as the y-coordinate. Thus, the equation of the line is $y = -1$.

59. Note that every point on the vertical line passing through $(4, 0)$ has 4 as the x-coordinate. Thus, the equation of the line is $x = 4$.

61. Discussion and Writing Exercise

63. *Familiarize.* Let $p = $ the percent of desserts sold that will be pie.

Translate. We reword the problem.

40 is what percent of 250?

$40 = p \cdot 250$

Solve. We solve the equation.
$$40 = p \cdot 250$$
$$\frac{40}{250} = \frac{p \cdot 250}{250}$$
$$0.16 = p$$
$$16\% = p$$

Check. We can find 16% of 250:
$$16\% \cdot 250 = 0.16 \cdot 250 = 40$$
The answer checks.

State. 16% of the desserts sold will be pie.

65. $-1.6x < 64$

$\dfrac{-1.6x}{-1.6} > \dfrac{64}{-1.6}$ Dividing by -1.6 and reversing the inequality symbol

$x > -40$

The solution set is $\{x | x > -40\}$.

67. $x + (x - 1) < (x + 2) - (x + 1)$
$2x - 1 < x + 2 - x - 1$
$2x - 1 < 1$
$2x < 2$
$x < 1$

The solution set is $\{x | x < 1\}$.

69. A line parallel to the x-axis has an equation of the form $y = b$. Since the y-coordinate of one point on the line is -4, then $b = -4$ and the equation is $y = -4$.

71. Substitute -4 for x and 0 for y.
$$3(-4) + k = 5 \cdot 0$$
$$-12 + k = 0$$
$$k = 12$$

Exercise Set 3.4

1. We consider (x_1, y_1) to be $(-3, 5)$ and (x_2, y_2) to be $(4, 2)$.
$$m = \frac{y_2 - y_1}{x_2 - x_1} = \frac{2 - 5}{4 - (-3)} = \frac{-3}{7} = -\frac{3}{7}$$

3. We can choose any two points. We consider (x_1, y_1) to be $(-3, -1)$ and (x_2, y_2) to be $(0, 1)$.
$$m = \frac{y_2 - y_1}{x_2 - x_1} = \frac{1 - (-1)}{0 - (-3)} = \frac{2}{3}$$

5. We can choose any two points. We consider (x_1, y_1) to be $(-4, -2)$ and (x_2, y_2) to be $(4, 4)$.
$$m = \frac{y_2 - y_1}{x_2 - x_1} = \frac{4 - (-2)}{4 - (-4)} = \frac{6}{8} = \frac{3}{4}$$

7. We consider (x_1, y_1) to be $(-4, -2)$ and (x_2, y_2) to be $(3, -2)$.
$$m = \frac{y_2 - y_1}{x_2 - x_1} = \frac{-2 - (-2)}{3 - (-4)} = \frac{0}{7} = 0$$

9. We plot $(-2, 4)$ and $(3, 0)$ and draw the line containing these points.

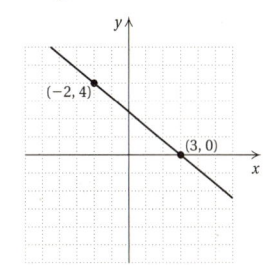

[Graph showing line through $(-2, 4)$ and $(3, 0)$]

To find the slope, consider (x_1, y_1) to be $(-2, 4)$ and (x_2, y_2) to be $(3, 0)$.
$$m = \frac{y_2 - y_1}{x_2 - x_1} = \frac{0 - 4}{3 - (-2)} = \frac{-4}{5} = -\frac{4}{5}$$

11. We plot $(-4, 0)$ and $(-5, -3)$ and draw the line containing these points.

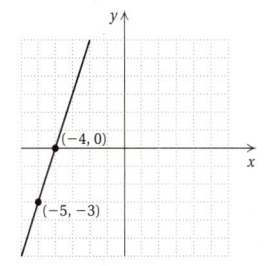

[Graph showing line through $(-4, 0)$ and $(-5, -3)$]

To find the slope, consider (x_1, y_1) to be $(-4, 0)$ and (x_2, y_2) to be $(-5, -3)$.
$$m = \frac{y_2 - y_1}{x_2 - x_1} = \frac{-3 - 0}{-5 - (-4)} = \frac{-3}{-1} = 3$$

13. We plot $(-4, 2)$ and $(2, -3)$ and draw the line containing these points.

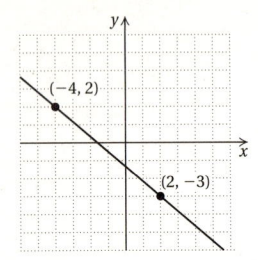

To find the slope, consider (x_1, y_1) to be $(-4, 2)$ and (x_2, y_2) to be $(2, -3)$.
$$m = \frac{y_2 - y_1}{x_2 - x_1} = \frac{-3 - 2}{2 - (-4)} = \frac{-5}{6} = -\frac{5}{6}$$

15. We plot $(5, 3)$ and $(-3, -4)$ and draw the line containing these points.

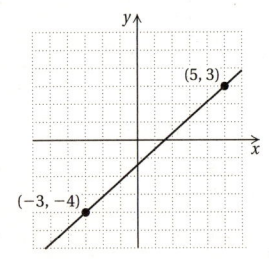

To find the slope, consider (x_1, y_1) to be $(5, 3)$ and (x_2, y_2) to be $(-3, -4)$.
$$m = \frac{y_2 - y_1}{x_2 - x_1} = \frac{-4 - 3}{-3 - 5} = \frac{-7}{-8} = \frac{7}{8}$$

17. $m = \dfrac{-\frac{1}{2} - \frac{3}{2}}{2 - 5} = \dfrac{-2}{-3} = \dfrac{2}{3}$

19. $m = \dfrac{-2 - 3}{4 - 4} = \dfrac{-5}{0}$

Since division by 0 is not defined, the slope is not defined.

21. $m = \dfrac{\text{rise}}{\text{run}} = \dfrac{2.4}{8.2} = \dfrac{2.4}{8.2} \cdot \dfrac{10}{10} = \dfrac{24}{82}$
$= \dfrac{\cancel{2} \cdot 12}{\cancel{2} \cdot 41} = \dfrac{12}{41}$

23. $m = \dfrac{\text{rise}}{\text{run}} = \dfrac{56}{258} = \dfrac{\cancel{2} \cdot 28}{\cancel{2} \cdot 129} = \dfrac{28}{129}$

25. Long's Peak rises $14{,}255$ ft $- 9600$ ft $= 4655$ ft.
Grade $= \dfrac{4655}{15{,}840} \approx 0.294 \approx 29.4\%$

27. The rate of change is the slope of the line. We can use any two ordered pairs to find the slope. We choose $(2, 50)$ and $(8, 200)$.
Rate of change $= \dfrac{200 \text{ mi} - 50 \text{ mi}}{8 \text{ gal} - 2 \text{ gal}} = \dfrac{150 \text{ mi}}{6 \text{ gal}} =$
25 miles per gallon

29. The rate of change is the slope of the line. We can use any two ordered pairs to find the slope. We choose $(2, 2000)$ and $(4, 1000)$. (Note that units on the vertical axis are given in thousands.)
Rate of change $= \dfrac{\$1000 - \$2000}{4 \text{ yr} - 2 \text{ yr}} = \dfrac{-\$1000}{2 \text{ yr}} = -\$500$ per year

31. The rate of change is the slope of the line. We can use any two ordered pairs to find the slope. We choose $(1990, 550{,}043)$ and $(2000, 626{,}932)$.
Rate of change $= \dfrac{626{,}932 - 550{,}043}{2000 - 1990} = \dfrac{76{,}889}{10} \approx$ 7689 people per year

33. $y = -10x + 7$

The equation is in the form $y = mx + b$, where $m = -10$. Thus, the slope is -10.

35. $y = 3.78x - 4$

The equation is in the form $y = mx + b$, where $m = 3.78$. Thus, the slope is 3.78.

37. We solve for y, obtaining an equation of the form $y = mx + b$.
$$3x - y = 4$$
$$-y = -3x + 4$$
$$-1(-y) = -1(-3x + 4)$$
$$y = 3x - 4$$
The slope is 3.

39. We solve for y, obtaining an equation of the form $y = mx + b$.
$$x + 5y = 10$$
$$5y = -x + 10$$
$$y = \frac{1}{5}(-x + 10)$$
$$y = -\frac{1}{5}x + 2$$
The slope is $-\dfrac{1}{5}$.

41. We solve for y, obtaining an equation of the form $y = mx + b$.
$$3x + 2y = 6$$
$$2y = -3x + 6$$
$$y = \frac{1}{2}(-3x + 6)$$
$$y = -\frac{3}{2}x + 3$$
The slope is $-\dfrac{3}{2}$.

Exercise Set 3.4

43. We solve for y, obtaining an equation of the form $y = mx + b$.

$$5x - 7y = 14$$
$$-7y = -5x + 14$$
$$y = -\frac{1}{7}(-5x + 14)$$
$$y = \frac{5}{7}x - 2$$

The slope is $\frac{5}{7}$.

45. $y = -2.74x$

The equation is in the form $y = mx + b$, where $m = -2.74$. Thus, the slope is -2.74.

47. We solve for y, obtaining an equation of the form $y = mx + b$.

$$9x = 3y + 5$$
$$9x - 5 = 3y$$
$$\frac{1}{3}(9x - 5) = y$$
$$3x - \frac{5}{3} = y$$

The slope is 3.

49. We solve for y, obtaining an equation of the form $y = mx + b$.

$$5x - 4y + 12 = 0$$
$$5x + 12 = 4y$$
$$\frac{1}{4}(5x + 12) = y$$
$$\frac{5}{4}x + 3 = y$$

The slope is $\frac{5}{4}$.

51. $y = 4$

The equation can be thought of as $y = 0 \cdot x + 4$, so the slope is 0.

53. Discussion and Writing Exercise

55.
$$15x = -60$$
$$\frac{15x}{15} = \frac{-60}{15} \quad \text{Dividing by 15 on both sides}$$
$$1 \cdot x = -4 \quad \text{Simplifying}$$
$$x = -4 \quad \text{Identity property of 1}$$

The solution is -4.

57.
$$-x = 37$$
$$-1 \cdot x = 37$$
$$\frac{-1 \cdot x}{-1} = \frac{37}{-1} \quad \text{Dividing by } -1 \text{ on both sides}$$
$$1 \cdot x = -37 \quad \text{Simplifying}$$
$$x = -37 \quad \text{Identity property of 1}$$

The solution is -37.

59. *Translate*.

What is 15% of $23.80?
$a = 15\% \cdot 23.80$

Solve. We convert to decimal notation and multiply.

$a = 15\% \cdot 23.80 = 0.15 \cdot 23.80 = 3.57$

The answer is $3.57.

61. *Familiarize*. Let $p =$ the percent of the cost of the meal represented by the tip.

Translate. We reword the problem.

$8.50 is what percent of $42.50?
$8.50 = p \cdot 42.50$

Solve. We solve the equation.

$$8.50 = p \cdot 42.50$$
$$0.2 = p$$
$$20\% = p$$

Check. We can find 20% of 42.50.

$20\% \cdot 42.50 = 0.2 \cdot 42.50 = 8.50$

The answer checks.

State. The tip was 20% of the cost of the meal.

63. *Familiarize*. Let $c =$ the cost of the meal before the tip was added. Then the tip is $15\% \cdot c$.

Translate. We reword the problem.

Cost of meal plus tip is total cost
$c + 15\% \cdot c = 51.92$

Solve. We solve the equation.

$$c + 15\% \cdot c = 51.92$$
$$1 \cdot c + 0.15c = 51.92$$
$$1.15c = 51.92$$
$$c \approx 45.15$$

Check. We can find 15% of 45.15 and then add this to 45.15.

$15\% \cdot 45.15 = 0.15 \cdot 45.15 \approx 6.77$ and $45.15 + 6.77 = 51.92$

The answer checks.

State. Before the tip the meal cost $45.15.

65. $y = 0.35x - 7$

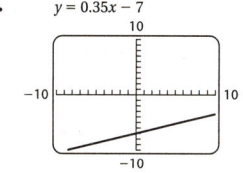

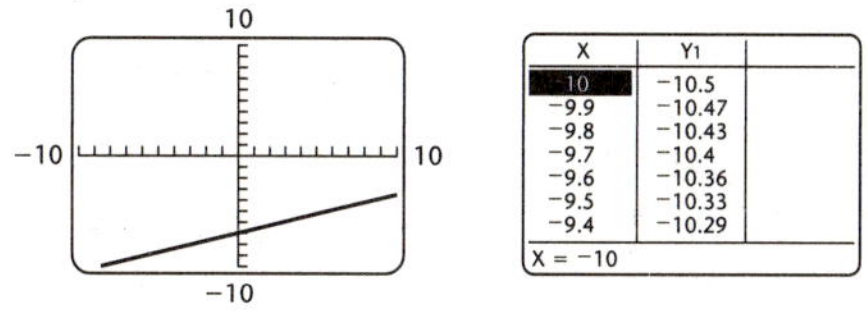

67. $y = x^3 - 5$

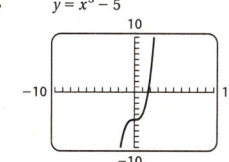

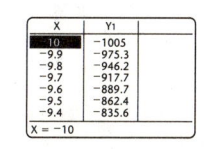

Chapter 4
Polynomials: Operations

Exercise Set 4.1

1. 3^4 means $3 \cdot 3 \cdot 3 \cdot 3$.
3. $(-1.1)^5$ means $(-1.1)(-1.1)(-1.1)(-1.1)(-1.1)$.
5. $\left(\frac{2}{3}\right)^4$ means $\left(\frac{2}{3}\right)\left(\frac{2}{3}\right)\left(\frac{2}{3}\right)\left(\frac{2}{3}\right)$.
7. $(7p)^2$ means $(7p)(7p)$.
9. $8k^3$ means $8 \cdot k \cdot k \cdot k$.
11. $a^0 = 1, a \neq 0$
13. $b^1 = b$
15. $\left(\frac{2}{3}\right)^0 = 1$
17. $8.38^0 = 1$
19. $(ab)^1 = ab$
21. $ab^1 = a \cdot b^1 = ab$
23. $m^3 = 3^3 = 3 \cdot 3 \cdot 3 = 27$
25. $p^1 = 19^1 = 19$
27. $x^4 = 4^4 = 4 \cdot 4 \cdot 4 \cdot 4 = 256$
29. $y^2 - 7 = 10^2 - 7$
    $= 100 - 7$ Evaluating the power
    $= 93$ Subtracting
31. $x^1 + 3 = 7^1 + 3$
    $= 7 + 3 \quad (7^1 = 7)$
    $= 10$

    $x^0 + 3 = 7^0 + 3$
    $= 1 + 3 \quad (7^0 = 1)$
    $= 4$
33. $A = \pi r^2 \approx 3.14 \times (34 \text{ ft})^2$
    $\approx 3.14 \times 1156 \text{ ft}^2$ Evaluating the power
    $\approx 3629.84 \text{ ft}^2$
35. $3^{-2} = \frac{1}{3^2} = \frac{1}{9}$
37. $10^{-3} = \frac{1}{10^3} = \frac{1}{1000}$
39. $7^{-3} = \frac{1}{7^3} = \frac{1}{343}$
41. $a^{-3} = \frac{1}{a^3}$
43. $\frac{1}{8^{-2}} = 8^2 = 64$
45. $\frac{1}{y^{-4}} = y^4$
47. $\frac{1}{z^{-n}} = z^n$
49. $\frac{1}{4^3} = 4^{-3}$
51. $\frac{1}{x^3} = x^{-3}$
53. $\frac{1}{a^5} = a^{-5}$
55. $2^4 \cdot 2^3 = 2^{4+3} = 2^7$
57. $8^5 \cdot 8^9 = 8^{5+9} = 8^{14}$
59. $x^4 \cdot x^3 = x^{4+3} = x^7$
61. $9^{17} \cdot 9^{21} = 9^{17+21} = 9^{38}$
63. $(3y)^4(3y)^8 = (3y)^{4+8} = (3y)^{12}$
65. $(7y)^1(7y)^{16} = (7y)^{1+16} = (7y)^{17}$
67. $3^{-5} \cdot 3^8 = 3^{-5+8} = 3^3$
69. $x^{-2} \cdot x = x^{-2+1} = x^{-1} = \frac{1}{x}$
71. $x^{14} \cdot x^3 = x^{14+3} = x^{17}$
73. $x^{-7} \cdot x^{-6} = x^{-7+(-6)} = x^{-13} = \frac{1}{x^{13}}$
75. $a^{11} \cdot a^{-3} \cdot a^{-18} = a^{11+(-3)+(-18)} = a^{-10} = \frac{1}{a^{10}}$
77. $t^8 \cdot t^{-8} = t^{8+(-8)} = t^0 = 1$
79. $\frac{7^5}{7^2} = 7^{5-2} = 7^3$
81. $\frac{8^{12}}{8^6} = 8^{12-6} = 8^6$
83. $\frac{y^9}{y^5} = y^{9-5} = y^4$
85. $\frac{16^2}{16^8} = 16^{2-8} = 16^{-6} = \frac{1}{16^6}$
87. $\frac{m^6}{m^{12}} = m^{6-12} = m^{-6} = \frac{1}{m^6}$
89. $\frac{(8x)^6}{(8x)^{10}} = (8x)^{6-10} = (8x)^{-4} = \frac{1}{(8x)^4}$

91. $\dfrac{(2y)^9}{(2y)^9} = (2y)^{9-9} = (2y)^0 = 1$

93. $\dfrac{x}{x^{-1}} = x^{1-(-1)} = x^2$

95. $\dfrac{x^7}{x^{-2}} = x^{7-(-2)} = x^9$

97. $\dfrac{z^{-6}}{z^{-2}} = z^{-6-(-2)} = z^{-4} = \dfrac{1}{z^4}$

99. $\dfrac{x^{-5}}{x^{-8}} = x^{-5-(-8)} = x^3$

101. $\dfrac{m^{-9}}{m^{-9}} = m^{-9-(-9)} = m^0 = 1$

103. $5^2 = 5 \cdot 5 = 25$

$5^{-2} = \dfrac{1}{5^2} = \dfrac{1}{25}$

$\left(\dfrac{1}{5}\right)^2 = \dfrac{1}{5} \cdot \dfrac{1}{5} = \dfrac{1}{25}$

$\left(\dfrac{1}{5}\right)^{-2} = \dfrac{1}{\left(\dfrac{1}{5}\right)^2} = \dfrac{1}{\dfrac{1}{25}} = 1 \cdot \dfrac{25}{1} = 25$

$-5^2 = -(5)(5) = -25$

$(-5)^2 = (-5)(-5) = 25$

$-\left(-\dfrac{1}{5}\right)^2 = -\left(-\dfrac{1}{5}\right)\left(-\dfrac{1}{5}\right) = -\dfrac{1}{25}$

$\left(-\dfrac{1}{5}\right)^{-2} = \dfrac{1}{\left(-\dfrac{1}{5}\right)^2} = \dfrac{1}{\dfrac{1}{25}} = 1 \cdot \dfrac{25}{1} = 25$

105. Discussion and Writing Exercise

107. $64\%t$, or $0.64t$

109.
```
              6 4
   2 4.3∧⟌ 1 5 5 5.2 ∧
          1 4 5 8
            9 7 2
            9 7 2
                0
```
The answer is 64.

111. $3x - 4 + 5x - 10x = x - 8$

$-2x - 4 = x - 8$ Collecting like terms

$-2x - 4 + 4 = x - 8 + 4$ Adding 4

$-2x = x - 4$

$-2x - x = x - 4 - x$ Subtracting x

$-3x = -4$

$\dfrac{-3x}{-3} = \dfrac{-4}{-3}$ Dividing by -3

$x = \dfrac{4}{3}$

The solution is $\dfrac{4}{3}$.

113. Familiarize. Let x = the length of the shorter piece. Then $2x$ = the length of the longer piece.

Translate.

Length of shorter piece plus length of longer piece is 12 in.

$x + 2x = 12$

Solve.

$x + 2x = 12$

$3x = 12$

$\dfrac{3x}{3} = \dfrac{12}{3}$

$x = 4$

If $x = 4$, $2x = 2 \cdot 4 = 8$.

Check. The longer piece, 8 in., is twice as long as the shorter piece, 4 in. Also, 4 in. + 8 in. = 12 in., the total length of the sandwich. The answer checks.

State. The lengths of the pieces are 4 in. and 8 in.

115. Let $y_1 = (x+1)^2$ and $y_2 = x^2 + 1$. A graph of the equations or a table of values shows that $(x+1)^2 = x^2 + 1$ is not correct.

117. Let $y_1 = (5x)^0$ and $y_2 = 5x^0$. A graph of the equations or a table of values shows that $(5x)^0 = 5x^0$ is not correct.

119. $(y^{2x})(y^{3x}) = y^{2x+3x} = y^{5x}$

121. $\dfrac{a^{6t}(a^{7t})}{a^{9t}} = \dfrac{a^{6t+7t}}{a^{9t}} = \dfrac{a^{13t}}{a^{9t}} = a^{13t-9t} = a^{4t}$

123. $\dfrac{(0.8)^5}{(0.8)^3(0.8)^2} = \dfrac{(0.8)^5}{(0.8)^{3+2}} = \dfrac{(0.8)^5}{(0.8)^5} = 1$

125. Since the bases are the same, the expression with the larger exponent is larger. Thus, $3^5 > 3^4$.

127. Since the exponents are the same, the expression with the larger base is larger. Thus, $4^3 < 5^3$.

129. Choose any number except 0. For example, let $x = 1$.

$3x^2 = 3 \cdot 1^2 = 3 \cdot 1 = 3$, but

$(3x)^2 = (3 \cdot 1)^2 = 3^2 = 9$.

Exercise Set 4.2

1. $(2^3)^2 = 2^{3 \cdot 2} = 2^6$

3. $(5^2)^{-3} = 5^{2(-3)} = 5^{-6} = \dfrac{1}{5^6}$

5. $(x^{-3})^{-4} = x^{(-3)(-4)} = x^{12}$

7. $(a^{-2})^9 = a^{-2 \cdot 9} = a^{-18} = \dfrac{1}{a^{18}}$

9. $(t^{-3})^{-6} = t^{(-3)(-6)} = t^{18}$

11. $(t^4)^{-3} = t^{4(-3)} = t^{-12} = \dfrac{1}{t^{12}}$

13. $(x^{-2})^{-4} = x^{(-2)(-4)} = x^8$

Exercise Set 4.2 63

15. $(ab)^3 = a^3b^3$ Raising each factor to the third power

17. $(ab)^{-3} = a^{-3}b^{-3} = \dfrac{1}{a^3b^3}$

19. $(mn^2)^{-3} = m^{-3}(n^2)^{-3} = m^{-3}n^{2(-3)} = m^{-3}n^{-6} = \dfrac{1}{m^3n^6}$

21. $(4x^3)^2 = 4^2(x^3)^2$ Raising each factor to the second power
$= 16x^6$

23. $(3x^{-4})^2 = 3^2(x^{-4})^2 = 3^2x^{-4\cdot 2} = 9x^{-8} = \dfrac{9}{x^8}$

25. $(x^4y^5)^{-3} = (x^4)^{-3}(y^5)^{-3} = x^{4(-3)}y^{5(-3)} = x^{-12}y^{-15} = \dfrac{1}{x^{12}y^{15}}$

27. $(x^{-6}y^{-2})^{-4} = (x^{-6})^{-4}(y^{-2})^{-4} = x^{(-6)(-4)}y^{(-2)(-4)} = x^{24}y^8$

29. $(a^{-2}b^7)^{-5} = (a^{-2})^{-5}(b^7)^{-5} = a^{10}b^{-35} = \dfrac{a^{10}}{b^{35}}$

31. $(5r^{-4}t^3)^2 = 5^2(r^{-4})^2(t^3)^2 = 25r^{-4\cdot 2}t^{3\cdot 2} = 25r^{-8}t^6 = \dfrac{25t^6}{r^8}$

33. $(a^{-5}b^7c^{-2})^3 = (a^{-5})^3(b^7)^3(c^{-2})^3 = a^{-5\cdot 3}b^{7\cdot 3}c^{-2\cdot 3} = a^{-15}b^{21}c^{-6} = \dfrac{b^{21}}{a^{15}c^6}$

35. $(3x^3y^{-8}z^{-3})^2 = 3^2(x^3)^2(y^{-8})^2(z^{-3})^2 = 9x^6y^{-16}z^{-6} = \dfrac{9x^6}{y^{16}z^6}$

37. $\left(\dfrac{y^3}{2}\right)^2 = \dfrac{(y^3)^2}{2^2} = \dfrac{y^6}{4}$

39. $\left(\dfrac{a^2}{b^3}\right)^4 = \dfrac{(a^2)^4}{(b^3)^4} = \dfrac{a^8}{b^{12}}$

41. $\left(\dfrac{y^2}{2}\right)^{-3} = \dfrac{(y^2)^{-3}}{2^{-3}} = \dfrac{y^{-6}}{2^{-3}} = \dfrac{\frac{1}{y^6}}{\frac{1}{2^3}} = \dfrac{1}{y^6} \cdot \dfrac{2^3}{1} = \dfrac{8}{y^6}$

43. $\left(\dfrac{7}{x^{-3}}\right)^2 = \dfrac{7^2}{(x^{-3})^2} = \dfrac{49}{x^{-6}} = 49x^6$

45. $\left(\dfrac{x^2y}{z}\right)^3 = \dfrac{(x^2)^3y^3}{z^3} = \dfrac{x^6y^3}{z^3}$

47. $\left(\dfrac{a^2b}{cd^3}\right)^{-2} = \dfrac{(a^2)^{-2}b^{-2}}{c^{-2}(d^3)^{-2}} = \dfrac{a^{-4}b^{-2}}{c^{-2}d^{-6}} = \dfrac{\frac{1}{a^4}\cdot\frac{1}{b^2}}{\frac{1}{c^2}\cdot\frac{1}{d^6}} = \dfrac{\frac{1}{a^4b^2}}{\frac{1}{c^2d^6}} = \dfrac{1}{a^4b^2}\cdot\dfrac{c^2d^6}{1} = \dfrac{c^2d^6}{a^4b^2}$

49. 2.8,000,000,000.
   └──────────┘ 10 places
   Large number, so the exponent is positive.
   $28,000,000,000 = 2.8 \times 10^{10}$

51. 9.07,000,000,000,000,000.
   └────────────────┘ 17 places
   Large number, so the exponent is positive.
   $907,000,000,000,000,000 = 9.07 \times 10^{17}$

53. 0.000003.04
   └─────┘ 6 places
   Small number, so the exponent is negative.
   $0.00000304 = 3.04 \times 10^{-6}$

55. 0.00000001.8
   └───────┘ 8 places
   Small number, so the exponent is negative.
   $0.000000018 = 1.8 \times 10^{-8}$

57. 1.00,000,000,000.
   └──────────┘ 11 places
   Large number, so the exponent is positive.
   $100,000,000,000 = 1.0 \times 10^{11} = 10^{11}$

59. 281 million = 281,000,000
   2.81,000,000.
   └────────┘ 8 places
   Large number, so the exponent is positive.
   281 million $= 2.81 \times 10^8$

61. $\dfrac{1}{10,000,000} = 0.0000001$
   0.0000001.
   └──────┘ 7 places
   Small number, so the exponent is negative.
   $\dfrac{1}{10,000,000} = 1 \times 10^{-7}$, or 10^{-7}

63. 8.74×10^7
   Positive exponent, so the answer is a large number.
   8.7400000.
   └─────┘ 7 places
   $8.74 \times 10^7 = 87,400,000$

65. 5.704×10^{-8}
   Negative exponent, so the answer is a small number.
   0.00000005.704
   └───────┘ 8 places
   $5.704 \times 10^{-8} = 0.00000005704$

67. $10^7 = 1 \times 10^7$

Positive exponent, so the answer is a large number.

1.0000000.

    ⌞_____⌝ 7 places

$10^7 = 10,000,000$

69. $10^{-5} = 1 \times 10^{-5}$

Negative exponent, so the answer is a small number.

0.00001.

  ⌞____⌝ 5 places

$10^{-5} = 0.00001$

71. $(3 \times 10^4)(2 \times 10^5) = (3 \cdot 2) \times (10^4 \cdot 10^5)$
$$= 6 \times 10^9$$

73. $(5.2 \times 10^5)(6.5 \times 10^{-2}) = (5.2 \cdot 6.5) \times (10^5 \cdot 10^{-2})$
$$= 33.8 \times 10^3$$

The answer at this stage is 33.8×10^3 but this is not scientific notation since 33.8 is not a number between 1 and 10. We convert 33.8 to scientific notation and simplify.

$33.8 \times 10^3 = (3.38 \times 10^1) \times 10^3 = 3.38 \times (10^1 \times 10^3) = 3.38 \times 10^4$

The answer is 3.38×10^4.

75. $(9.9 \times 10^{-6})(8.23 \times 10^{-8}) = (9.9 \cdot 8.23) \times (10^{-6} \cdot 10^{-8})$
$$= 81.477 \times 10^{-14}$$

The answer at this stage is 81.477×10^{-14}. We convert 81.477 to scientific notation and simplify.

$81.477 \times 10^{-14} = (8.1477 \times 10^1) \times 10^{-14} = 8.1477 \times (10^1 \times 10^{-14}) = 8.1477 \times 10^{-13}$.

The answer is 8.1477×10^{-13}.

77. $\dfrac{8.5 \times 10^8}{3.4 \times 10^{-5}} = \dfrac{8.5}{3.4} \times \dfrac{10^8}{10^{-5}}$
$$= 2.5 \times 10^{8-(-5)}$$
$$= 2.5 \times 10^{13}$$

79. $(3.0 \times 10^6) \div (6.0 \times 10^9) = \dfrac{3.0 \times 10^6}{6.0 \times 10^9}$
$$= \dfrac{3.0}{6.0} \times \dfrac{10^6}{10^9}$$
$$= 0.5 \times 10^{6-9}$$
$$= 0.5 \times 10^{-3}$$

The answer at this stage is 0.5×10^{-3}. We convert 0.5 to scientific notation and simplify.

$0.5 \times 10^{-3} = (5.0 \times 10^{-1}) \times 10^{-3} = 5.0 \times (10^{-1} \times 10^{-3}) = 5.0 \times 10^{-4}$

81. $\dfrac{7.5 \times 10^{-9}}{2.5 \times 10^{12}} = \dfrac{7.5}{2.5} \times \dfrac{10^{-9}}{10^{12}}$
$$= 3.0 \times 10^{-9-12}$$
$$= 3.0 \times 10^{-21}$$

83. There are 60 seconds in one minute and 60 minutes in one hour, so there are 60(60), or 3600 seconds in one hour. There are 24 hours in one day and 365 days in one year, so there are 3600(24)(365), or 31,536,000 seconds in one year.

$4,200,000 \times 31,536,000$
$= (4.2 \times 10^6) \times (3.1536 \times 10^7)$
$= (4.2 \times 3.1536) \times (10^6 \times 10^7)$
$\approx 13.25 \times 10^{13}$
$\approx (1.325 \times 10) \times 10^{13}$
$\approx 1.325 \times (10 \times 10^{13})$
$\approx 1.325 \times 10^{14}$

About 1.325×10^{14} cubic feet of water is discharged from the Amazon River in 1 yr.

85. $\dfrac{1.908 \times 10^{24}}{6 \times 10^{21}} = \dfrac{1.908}{6} \times \dfrac{10^{24}}{10^{21}}$
$$= 0.318 \times 10^3$$
$$= (3.18 \times 10^{-1}) \times 10^3$$
$$= 3.18 \times (10^{-1} \times 10^3)$$
$$= 3.18 \times 10^2$$

The mass of Jupiter is 3.18×10^2 times the mass of Earth.

87. 10 billion trillion $= 1 \times 10 \times 10^9 \times 10^{12}$
$$= 1 \times 10^{22}$$

There are 1×10^{22} stars in the known universe.

89. We divide the mass of the sun by the mass of earth.

$\dfrac{1.998 \times 10^{27}}{6 \times 10^{21}} = 0.333 \times 10^6$
$= (3.33 \times 10^{-1}) \times 10^6$
$= 3.33 \times 10^5$

The mass of the sun is 3.33×10^5 times the mass of Earth.

91. First we divide the distance from the earth to the moon by 3 days to find the number of miles per day the space vehicle travels. Note that $240,000 = 2.4 \times 10^5$.

$\dfrac{2.4 \times 10^5}{3} = 0.8 \times 10^5 = 8 \times 10^4$

The space vehicle travels 8×10^4 miles per day. Now divide the distance from the earth to Mars by 8×10^4 to find how long it will take the space vehicle to reach Mars. Note that $35,000,000 = 3.5 \times 10^7$.

$\dfrac{3.5 \times 10^7}{8 \times 10^4} = 0.4375 \times 10^3 = 4.375 \times 10^2$

It takes 4.375×10^2 days for the space vehicle to travel from the earth to Mars.

93. Discussion and Writing Exercise

95. $9x - 36 = 9 \cdot x - 9 \cdot 4 = 9(x - 4)$

97. $3s + 3t + 24 = 3 \cdot s + 3 \cdot t + 3 \cdot 8 = 3(s + t + 8)$

Exercise Set 4.3

99. $2x - 4 - 5x + 8 = x - 3$
$-3x + 4 = x - 3$ Collecting like terms
$-3x + 4 - 4 = x - 3 - 4$ Subtracting 4
$-3x = x - 7$
$-3x - x = x - 7 - x$ Subtracting x
$-4x = -7$
$\dfrac{-4x}{-4} = \dfrac{-7}{-4}$ Dividing by -4
$x = \dfrac{7}{4}$

The solution is $\dfrac{7}{4}$.

101. $8(2x + 3) - 2(x - 5) = 10$
$16x + 24 - 2x + 10 = 10$ Removing parentheses
$14x + 34 = 10$ Collecting like terms
$14x + 34 - 34 = 10 - 34$ Subtracting 34
$14x = -24$
$\dfrac{14x}{14} = \dfrac{-24}{14}$ Dividing by 14
$x = -\dfrac{12}{7}$ Simplifying

The solution is $-\dfrac{12}{7}$.

103. $y = x - 5$

The equation is equivalent to $y = x + (-5)$. The y-intercept is $(0, -5)$. We find two other points.

When $x = 2$, $y = 2 - 5 = -3$.
When $x = 4$, $y = 4 - 5 = -1$.

x	y
0	-5
2	-3
4	-1

Plot these points, draw the line they determine, and label the graph $y = x - 5$.

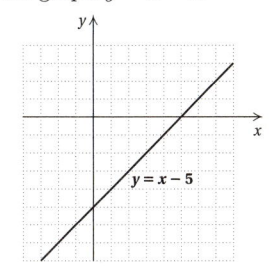

105. $\dfrac{(5.2 \times 10^6)(6.1 \times 10^{-11})}{1.28 \times 10^{-3}} = \dfrac{(5.2 \cdot 6.1)}{1.28} \times \dfrac{(10^6 \cdot 10^{-11})}{10^{-3}}$
$= 24.78125 \times 10^{-2}$
$= (2.478125 \times 10^1) \times 10^{-2}$
$= 2.478125 \times 10^{-1}$

107. $\dfrac{(5^{12})^2}{5^{25}} = \dfrac{5^{24}}{5^{25}} = 5^{24-25} = 5^{-1} = \dfrac{1}{5}$

109. $\dfrac{(3^5)^4}{3^5 \cdot 3^4} = \dfrac{3^{5 \cdot 4}}{3^{5+4}} = \dfrac{3^{20}}{3^9} = 3^{20-9} = 3^{11}$

111. $\dfrac{49^{18}}{7^{35}} = \dfrac{(7^2)^{18}}{7^{35}} = \dfrac{7^{36}}{7^{35}} = 7$

113. $\dfrac{(0.4)^5}{\left((0.4)^3\right)^2} = \dfrac{(0.4)^5}{(0.4)^6} = (0.4)^{-1} = \dfrac{1}{0.4}$, or 2.5

115. False; let $x = 2$, $y = 3$, $m = 4$, and $n = 2$:
$2^4 \cdot 3^2 = 16 \cdot 9 = 144$, but
$(2 \cdot 3)^{4 \cdot 2} = 6^8 = 1{,}679{,}616$

117. False; let $x = 5$, $y = 3$, and $m = 2$:
$(5 - 3)^2 = 2^2 = 4$, but
$5^2 - 3^2 = 25 - 9 = 16$

Exercise Set 4.3

1. $-5x + 2 = -5 \cdot 4 + 2 = -20 + 2 = -18$;
$-5x + 2 = -5(-1) + 2 = 5 + 2 = 7$

3. $2x^2 - 5x + 7 = 2 \cdot 4^2 - 5 \cdot 4 + 7 = 2 \cdot 16 - 20 + 7 = 32 - 20 + 7 = 19$;
$2x^2 - 5x + 7 = 2(-1)^2 - 5(-1) + 7 = 2 \cdot 1 + 5 + 7 = 2 + 5 + 7 = 14$

5. $x^3 - 5x^2 + x = 4^3 - 5 \cdot 4^2 + 4 = 64 - 5 \cdot 16 + 4 = 64 - 80 + 4 = -12$;
$x^3 - 5x^2 + x = (-1)^3 - 5(-1)^2 + (-1) = -1 - 5 \cdot 1 - 1 = -1 - 5 - 1 = -7$

7. $3x + 5 = 3(-2) + 5 = -6 + 5 = -1$;
$3x + 5 = 3 \cdot 0 + 5 = 0 + 5 = 5$

9. $x^2 - 2x + 1 = (-2)^2 - 2(-2) + 1 = 4 + 4 + 1 = 9$;
$x^2 - 2x + 1 = 0^2 - 2 \cdot 0 + 1 = 0 - 0 + 1 = 1$

11. $-3x^3 + 7x^2 - 3x - 2 = -3(-2)^3 + 7(-2)^2 - 3(-2) - 2 = -3(-8) + 7(4) - 3(-2) - 2 = 24 + 28 + 6 - 2 = 56$;
$-3x^3 + 7x^2 - 3x - 2 = -3 \cdot 0^3 + 7 \cdot 0^2 - 3 \cdot 0 - 2 = -3 \cdot 0 + 7 \cdot 0 - 0 - 2 = 0 + 0 - 0 - 2 = -2$

13. We evaluate the polynomial for $t = 10$:
$S = 11.12t^2 = 11.12(10)^2 = 11.12(100) = 1112$

The skydiver has fallen approximately 1112 ft.

15. a) In 2002, $t = 0$.
$E = 0.19(0) + 3.93 = 0 + 3.93 = 3.93$

The consumption in 2000 was 3.93 million gigawatt hours.

In 2001, $t = 2001 - 2000 = 1$.
$E = 0.19(1) + 3.93 = 0.19 + 3.93 = 4.12$

The consumption in 2001 was 4.12 million gigawatt hours.

In 2003, $t = 2003 - 2000 = 3$.
$E = 0.19(3) + 3.93 = 0.57 + 3.93 = 4.5$

The consumption in 2003 will be 4.5 million gigawatt hours.

In 2005, $t = 2005 - 2000 = 5$.

$E = 0.19(5) + 3.93 = 0.95 + 3.93 = 4.88$

The consumption in 2005 will be 4.88 million gigawatt hours.

In 2008, $t = 2008 - 2000 = 8$.

$E = 0.19(8) + 3.93 = 1.52 + 3.93 = 5.45$

The consumption in 2008 will be 5.45 million gigawatt hours.

In 2010, $t = 2010 - 2000 = 10$.

$E = 0.19(10) + 3.93 = 1.9 + 3.93 = 5.83$

The consumption in 2010 will be 5.83 million gigawatt hours.

b) It appears that the points $(0, 3.93)$, $(1, 4.12)$, $(3, 4.5)$, $(5, 4.88)$, $(8, 5.45)$, and $(10, 5.83)$ are on the graph, so the results check.

17. We evaluate the polynomial for $x = 75$:

$$\begin{aligned} R = 280x - 0.4x^2 &= 280(75) - 0.4(75)^2 \\ &= 280(75) - 0.4(5625) \\ &= 21{,}000 - 2250 \\ &= 18{,}750 \end{aligned}$$

The total revenue from the sale of 75 TVs is \$18,750.

We evaluate the polynomial for $x = 100$:

$$\begin{aligned} R = 280x - 0.4x^2 &= 280(100) - 0.4(100)^2 \\ &= 280(100) - 0.4(10{,}000) \\ &= 28{,}000 - 4000 \\ &= 24{,}000 \end{aligned}$$

The total revenue from the sale of 100 TVs is \$24,000.

19. Locate -3 on the x-axis. Then move vertically to the graph and horizontally to the y-axis. It appears that the y-value that is paired with -3 is -4. Thus, the value of $y = 5 - x^2$ is -4 when $x = -3$.

Locate -1 on the x-axis. Then move vertically to the graph and horizontally to the y-axis. It appears that the y-value that is paired with -1 is 4. Thus, the value of $y = 5 - x^2$ is 4 when $x = -1$.

Locate 0 on the x-axis. Then move vertically to the graph. We arrive at a point on the y-axis with the y-value 5. Thus, the value of $5 - x^2$ is 5 when $x = 0$.

Locate 1.5 on the x-axis. Then move vertically to the graph and horizontally to the y-axis. It appears that the y-value that is paired with 1.5 is 2.75. Thus, the value of $y = 5 - x^2$ is 2.75 when $x = 1.5$.

Locate 2 on the x-axis. Then move vertically to the graph and horizontally to the y-axis. It appears that the y-value that is paired with 2 is 1. Thus, the value of $y = 5 - x^2$ is 1 when $x = 2$.

21. We evaluate the polynomial for $x = 20$:

$$\begin{aligned} N &= -0.00006(20)^3 + 0.006(20)^2 - 0.1(20) + 1.9 \\ &= -0.00006(8000) + 0.006(400) - 0.1(20) + 1.9 \\ &= -0.48 + 2.4 - 2.0 + 1.9 \\ &= 1.82 \end{aligned}$$

There are about 1.82 million or 1,820,000 hearing-impaired Americans of age 20.

We evaluate the polynomial for $x = 40$:

$$\begin{aligned} N &= -0.00006(40)^3 + 0.006(40)^2 - 0.1(40) + 1.9 \\ &= -0.00006(64{,}000) + 0.006(1600) - 0.1(40) + 1.9 \\ &= -3.84 + 9.6 - 4.0 + 1.9 \\ &= 3.66 \end{aligned}$$

There are about 3.66 million, or 3,660,000, hearing-impaired Americans of age 40.

23. Locate 10 on the horizontal axis. From there move vertically to the graph and then horizontally to the M-axis. This locates an M-value of about 9. Thus, about 9 words were memorized in 10 minutes.

25. Locate 8 on the horizontal axis. From there move vertically to the graph and then horizontally to the M-axis. This locates an M-value of about 6. Thus, the value of $-0.001t^3 + 0.1t^2$ for $t = 8$ is approximately 6.

27. Locate 13 on the horizontal axis. It is halfway between 12 and 14. From there move vertically to the graph and then horizontally to the M-axis. This locates an M-value of about 15. Thus, the value of $-0.001t^3 + 0.1t^2$ when t is 13 is approximately 15.

29. $2 - 3x + x^2 = 2 + (-3x) + x^2$

The terms are 2, $-3x$, and x^2.

31. $5x^3 + 6x^2 - 3x^2$

Like terms: $6x^2$ and $-3x^2$ Same variable and exponent

33. $2x^4 + 5x - 7x - 3x^4$

Like terms: $2x^4$ and $-3x^4$ Same variable and
Like terms: $5x$ and $-7x$ exponent

35. $3x^5 - 7x + 8 + 14x^5 - 2x - 9$

Like terms: $3x^5$ and $14x^5$
Like terms: $-7x$ and $-2x$
Like terms: 8 and -9 Constant terms are like terms.

37. $-3x + 6$

The coefficient of $-3x$, the first term, is -3.

The coefficient of 6, the second term, is 6.

39. $5x^2 + 3x + 3$

The coefficient of $5x^2$, the first term, is 5.

The coefficient of $3x$, the second term, is 3.

The coefficient of 3, the third term, is 3.

41. $-5x^4 + 6x^3 - 3x^2 + 8x - 2$

The coefficient of $-5x^4$, the first term, is -5.

The coefficient of $6x^3$, the second term, is 6.

The coefficient of $-3x^2$, the third term, is -3.

The coefficient of $8x$, the fourth term, is 8.

The coefficient of -2, the fifth term, is -2.

Exercise Set 4.3

43. $2x - 5x = (2-5)x = -3x$

45. $x - 9x = 1x - 9x = (1-9)x = -8x$

47. $5x^3 + 6x^3 + 4 = (5+6)x^3 + 4 = 11x^3 + 4$

49. $5x^3 + 6x - 4x^3 - 7x = (5-4)x^3 + (6-7)x =$
$1x^3 + (-1)x = x^3 - x$

51. $6b^5 + 3b^2 - 2b^5 - 3b^2 = (6-2)b^5 + (3-3)b^2 =$
$4b^5 + 0b^2 = 4b^5$

53. $\frac{1}{4}x^5 - 5 + \frac{1}{2}x^5 - 2x - 37 =$
$\left(\frac{1}{4} + \frac{1}{2}\right)x^5 - 2x + (-5 - 37) = \frac{3}{4}x^5 - 2x - 42$

55. $6x^2 + 2x^4 - 2x^2 - x^4 - 4x^2 =$
$6x^2 + 2x^4 - 2x^2 - 1x^4 - 4x^2 =$
$(6-2-4)x^2 + (2-1)x^4 = 0x^2 + 1x^4 =$
$0 + x^4 = x^4$

57. $\frac{1}{4}x^3 - x^2 - \frac{1}{6}x^2 + \frac{3}{8}x^3 + \frac{5}{16}x^3 =$
$\frac{1}{4}x^3 - 1x^2 - \frac{1}{6}x^2 + \frac{3}{8}x^3 + \frac{5}{16}x^3 =$
$\left(\frac{1}{4} + \frac{3}{8} + \frac{5}{16}\right)x^3 + \left(-1 - \frac{1}{6}\right)x^2 =$
$\left(\frac{4}{16} + \frac{6}{16} + \frac{5}{16}\right)x^3 + \left(-\frac{6}{6} - \frac{1}{6}\right)x^2 = \frac{15}{16}x^3 - \frac{7}{6}x^2$

59. $x^5 + x + 6x^3 + 1 + 2x^2 = x^5 + 6x^3 + 2x^2 + x + 1$

61. $5y^3 + 15y^9 + y - y^2 + 7y^8 =$
$15y^9 + 7y^8 + 5y^3 - y^2 + y$

63. $3x^4 - 5x^6 - 2x^4 + 6x^6 = x^4 + x^6 = x^6 + x^4$

65. $-2x + 4x^3 - 7x + 9x^3 + 8 = -9x + 13x^3 + 8 =$
$13x^3 - 9x + 8$

67. $3x + 3x + 3x - x^2 - 4x^2 = 9x - 5x^2 = -5x^2 + 9x$

69. $-x + \frac{3}{4} + 15x^4 - x - \frac{1}{2} - 3x^4 = -2x + \frac{1}{4} + 12x^4 =$
$12x^4 - 2x + \frac{1}{4}$

71. $2x - 4 = 2x^1 - 4x^0$

The degree of $2x$ is 1.

The degree of -4 is 0.

The degree of the polynomial is 1, the largest exponent.

73. $3x^2 - 5x + 2 = 3x^2 - 5x^1 + 2x^0$

The degree of $3x^2$ is 2.

The degree of $-5x$ is 1.

The degree of 2 is 0.

The degree of the polynomial is 2, the largest exponent.

75. $-7x^3 + 6x^2 + 3x + 7 = -7x^3 + 6x^2 + 3x^1 + 7x^0$

The degree of $-7x^3$ is 3.

The degree of $6x^2$ is 2.

The degree of $3x$ is 1.

The degree of 7 is 0.

The degree of the polynomial is 3, the largest exponent.

77. $x^2 - 3x + x^6 - 9x^4 = x^2 - 3x^1 + x^6 - 9x^4$

The degree of x^2 is 2.

The degree of $-3x$ is 1.

The degree of x^6 is 6.

The degree of $-9x^4$ is 4.

The degree of the polynomial is 6, the largest exponent.

79. See the answer section in the text.

81. In the polynomial $x^3 - 27$, there are no x^2 or x terms. The x^2 term (or second-degree term) and the x term (or first-degree term) are missing.

83. In the polynomial $x^4 - x$, there are no x^3, x^2, or x^0 terms. The x^3 term (or third-degree term), the x^2 term (or second-degree term), and the x^0 term (or zero-degree term) are missing.

85. No terms are missing in the polynomial $2x^3 - 5x^2 + x - 3$.

87. $x^3 - 27 = x^3 + 0x^2 + 0x - 27$
$x^3 - 27 = x^3 \qquad\qquad - 27$

89. $x^4 - x = x^4 + 0x^3 + 0x^2 - x + 0x^0$
$x^4 - x = x^4 \qquad\qquad\quad - x$

91. There are no missing terms.

93. The polynomial $x^2 - 10x + 25$ is a *trinomial* because it has just three terms.

95. The polynomial $x^3 - 7x^2 + 2x - 4$ is *none of these* because it has more than three terms.

97. The polynomial $4x^2 - 25$ is a *binomial* because it has just two terms.

99. The polynomial $40x$ is a *monomial* because it has just one term.

101. Discussion and Writing Exercise

103. *Familiarize.* Let a = the number of apples the campers had to begin with. Then the first camper ate $\frac{1}{3}a$ apples and $a - \frac{1}{3}a$, or $\frac{2}{3}a$, apples were left. The second camper ate $\frac{1}{3}\left(\frac{2}{3}a\right)$, or $\frac{2}{9}a$, apples, and $\frac{2}{3}a - \frac{2}{9}a$, or $\frac{4}{9}a$, apples were left. The third camper ate $\frac{1}{3}\left(\frac{4}{9}a\right)$, or $\frac{4}{27}a$, apples, and $\frac{4}{9}a - \frac{4}{27}a$, or $\frac{8}{27}a$, apples were left.

Translate. We write an equation for the number of apples left after the third camper eats.

$$\underbrace{\text{Number of apples left}}_{\frac{8}{27}a} \underbrace{\text{is}}_{=} \underbrace{8.}_{8}$$

Solve. We solve the equation.
$$\frac{8}{27}a = 8$$
$$a = \frac{27}{8} \cdot 8$$
$$a = 27$$

Check. If the campers begin with 27 apples, then the first camper eats $\frac{1}{3} \cdot 27$, or 9, and $27 - 9$, or 18, are left. The second camper then eats $\frac{1}{3} \cdot 18$, or 6 apples and $18 - 6$, or 12, are left. Finally, the third camper eats $\frac{1}{3} \cdot 12$, or 4 apples and $12 - 4$, or 8, are left. The answer checks.

State. The campers had 27 apples to begin with.

105. $\frac{1}{8} - \frac{5}{6} = \frac{1}{8} + \left(-\frac{5}{6}\right)$, LCM is 24
$$= \frac{1}{8} \cdot \frac{3}{3} + \left(-\frac{5}{6}\right)\left(\frac{4}{4}\right)$$
$$= \frac{3}{24} + \left(-\frac{20}{24}\right)$$
$$= -\frac{17}{24}$$

107. $5.6 - 8.2 = 5.6 + (-8.2) = -2.6$

109. $\quad C = ab - r$
$\quad C + r = ab \quad$ Adding r
$\quad \frac{C+r}{a} = \frac{ab}{a} \quad$ Dividing by a
$\quad \frac{C+r}{a} = b \quad$ Simplifying

111. $3x - 15y + 63 = 3 \cdot x - 3 \cdot 5y + 3 \cdot 21 = 3(x - 5y + 21)$

113. $\quad (3x^2)^3 + 4x^2 \cdot 4x^4 - x^4(2x)^2 + [(2x)^2]^3 - 100x^2(x^2)^2$
$= 27x^6 + 4x^2 \cdot 4x^4 - x^4 \cdot 4x^2 + (2x)^6 - 100x^2 \cdot x^4$
$= 27x^6 + 16x^6 - 4x^6 + 64x^6 - 100x^6$
$= 3x^6$

115. $(5m^5)^2 = 5^2 m^{5 \cdot 2} = 25 m^{10}$
The degree is 10.

117. Graph $y = 5 - x^2$. Then use VALUE from the CALC menu to find the y-values that correspond to $x = -3$, $x = -1$, x=0, $x = 1.5$, and $x = 2$. As before, we find that these values are $-4, 4, 5, 2.75$, and 1, respectively.

119. Graph $y = -0.00006x^3 + 0.006x^2 - 0.1x + 1.9$. Then use VALUE from the CALC menu to find the y-values that correspond to $x = 20$ and $x = 40$. As before, we find that these values are 1.82 and 3.66, respectively. These results represent 1,820,000 and 3,660,000 hearing-impaired Americans.

Exercise Set 4.4

1. $(3x + 2) + (-4x + 3) = (3 - 4)x + (2 + 3) = -x + 5$

3. $(-6x + 2) + (x^2 + x - 3) =$
$x^2 + (-6 + 1)x + (2 - 3) = x^2 - 5x - 1$

5. $(x^2 - 9) + (x^2 + 9) = (1 + 1)x^2 + (-9 + 9) = 2x^2$

7. $(3x^2 - 5x + 10) + (2x^2 + 8x - 40) =$
$(3 + 2)x^2 + (-5 + 8)x + (10 - 40) = 5x^2 + 3x - 30$

9. $(1.2x^3 + 4.5x^2 - 3.8x) + (-3.4x^3 - 4.7x^2 + 23) =$
$(1.2 - 3.4)x^3 + (4.5 - 4.7)x^2 - 3.8x + 23 =$
$-2.2x^3 - 0.2x^2 - 3.8x + 23$

11. $(1 + 4x + 6x^2 + 7x^3) + (5 - 4x + 6x^2 - 7x^3) =$
$(1 + 5) + (4 - 4)x + (6 + 6)x^2 + (7 - 7)x^3 =$
$6 + 0x + 12x^2 + 0x^3 = 6 + 12x^2$, or $12x^2 + 6$

13. $\left(\frac{1}{4}x^4 + \frac{2}{3}x^3 + \frac{5}{8}x^2 + 7\right) + \left(-\frac{3}{4}x^4 + \frac{3}{8}x^2 - 7\right) =$
$\left(\frac{1}{4} - \frac{3}{4}\right)x^4 + \frac{2}{3}x^3 + \left(\frac{5}{8} + \frac{3}{8}\right)x^2 + (7 - 7) =$
$-\frac{2}{4}x^4 + \frac{2}{3}x^3 + \frac{8}{8}x^2 + 0 =$
$-\frac{1}{2}x^4 + \frac{2}{3}x^3 + x^2$

15. $(0.02x^5 - 0.2x^3 + x + 0.08) + (-0.01x^5 + x^4 - 0.8x - 0.02) =$
$(0.02 - 0.01)x^5 + x^4 - 0.2x^3 + (1 - 0.8)x + (0.08 - 0.02) =$
$0.01x^5 + x^4 - 0.2x^3 + 0.2x + 0.06$

17. $9x^8 - 7x^4 + 2x^2 + 5) + (8x^7 + 4x^4 - 2x) +$
$(-3x^4 + 6x^2 + 2x - 1) = 9x^8 + 8x^7 + (-7 + 4 - 3)x^4 +$
$(2 + 6)x^2 + (-2 + 2)x + (5 - 1) =$
$9x^8 + 8x^7 - 6x^4 + 8x^2 + 4$

19. Rewrite the problem so the coefficients of like terms have the same number of decimal places.

$$\begin{array}{r}
0.15x^4 + 0.10x^3 - 0.90x^2 \\
- 0.01x^3 + 0.01x^2 + x \\
1.25x^4 + 0.11x^2 + 0.01 \\
0.27x^3 + 0.99 \\
-0.35x^4 + 15.00x^2 - 0.03 \\
\hline
1.05x^4 + 0.36x^3 + 14.22x^2 + x + 0.97
\end{array}$$

21. We change the sign of the term inside the parentheses.
$-(-5x) = 5x$

23. We change the sign of every term inside the parentheses.
$-(-x^2 + 10x - 2) = x^2 - 10x + 2$

Exercise Set 4.4

25. We change the sign of every term inside the parentheses.
$-(12x^4 - 3x^3 + 3) = -12x^4 + 3x^3 - 3$

27. We change the sign of every term inside parentheses.
$-(3x - 7) = -3x + 7$

29. We change the sign of every term inside parentheses.
$-(4x^2 - 3x + 2) = -4x^2 + 3x - 2$

31. We change the sign of every term inside parentheses.
$-\left(-4x^4 + 6x^2 + \frac{3}{4}x - 8\right) = 4x^4 - 6x^2 - \frac{3}{4}x + 8$

33. $(3x + 2) - (-4x + 3) = 3x + 2 + 4x - 3$
　　　　　　　　　　　Changing the sign of every
　　　　　　　　　　　term inside parentheses
$= 7x - 1$

35. $(-6x + 2) - (x^2 + x - 3) = -6x + 2 - x^2 - x + 3$
$= -x^2 - 7x + 5$

37. $(x^2 - 9) - (x^2 + 9) = x^2 - 9 - x^2 - 9 = -18$

39. $(6x^4 + 3x^3 - 1) - (4x^2 - 3x + 3)$
$= 6x^4 + 3x^3 - 1 - 4x^2 + 3x - 3$
$= 6x^4 + 3x^3 - 4x^2 + 3x - 4$

41. $(1.2x^3 + 4.5x^2 - 3.8x) - (-3.4x^3 - 4.7x^2 + 23)$
$= 1.2x^3 + 4.5x^2 - 3.8x + 3.4x^3 + 4.7x^2 - 23$
$= 4.6x^3 + 9.2x^2 - 3.8x - 23$

43. $\frac{5}{8}x^3 - \frac{1}{4}x - \frac{1}{3} - \left(-\frac{1}{8}x^3 + \frac{1}{4}x - \frac{1}{3}\right)$
$= \frac{5}{8}x^3 - \frac{1}{4}x - \frac{1}{3} + \frac{1}{8}x^3 - \frac{1}{4}x + \frac{1}{3}$
$= \frac{6}{8}x^3 - \frac{2}{4}x$
$= \frac{3}{4}x^3 - \frac{1}{2}x$

45. $(0.08x^3 - 0.02x^2 + 0.01x) - (0.02x^3 + 0.03x^2 - 1)$
$= 0.08x^3 - 0.02x^2 + 0.01x - 0.02x^3 - 0.03x^2 + 1$
$= 0.06x^3 - 0.05x^2 + 0.01x + 1$

47. $x^2 + 5x + 6$
$\underline{x^2 + 2x}$

$x^2 + 5x + 6$
$\underline{-x^2 - 2x}$　Changing signs
$3x + 6$　Adding

49. $5x^4 + 6x^3 - 9x^2$
$\underline{-6x^4 - 6x^3 + 8x + 9}$

$5x^4 + 6x^3 - 9x^2$
$\underline{6x^4 + 6x^3 - 8x - 9}$　Changing signs
$11x^4 + 12x^3 - 9x^2 - 8x - 9$　Adding

51. $x^5 - 1$
$\underline{x^5 - x^4 + x^3 - x^2 + x - 1}$

$x^5 - 1$
$\underline{-x^5 + x^4 - x^3 + x^2 - x + 1}$　Changing signs
$x^4 - x^3 + x^2 - x$　Adding

53.

A	x		B	x
$3x$			x	

C	x		D	x
x			4	

The area of a rectangle is the product of the length and width. The sum of the areas is found as follows:

$\begin{array}{c}\text{Area}\\\text{of }A\end{array} + \begin{array}{c}\text{Area}\\\text{of }B\end{array} + \begin{array}{c}\text{Area}\\\text{of }C\end{array} + \begin{array}{c}\text{Area}\\\text{of }D\end{array}$
$= 3x \cdot x + x \cdot x + x \cdot x + 4 \cdot x$
$= 3x^2 + x^2 + x^2 + 4x$
$= 5x^2 + 4x$

A polynomial for the sum of the areas is $5x^2 + 4x$.

55. We add the lengths of the sides:
$4a + 7 + a + \frac{1}{2}a + 3 + a + 2a + 3a$
$= \left(4 + 1 + \frac{1}{2} + 1 + 2 + 3\right)a + (7 + 3)$
$= 11\frac{1}{2}a + 10,\text{ or }\frac{23}{2}a + 10$

57.

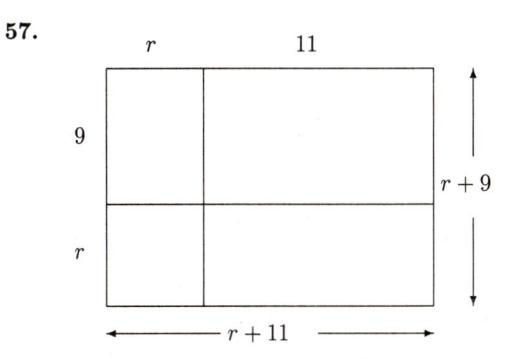

The length and width of the figure can be expressed as $r + 11$ and $r + 9$, respectively. The area of this figure (a rectangle) is the product of the length and width. An algebraic expression for the area is $(r + 11) \cdot (r + 9)$.

70 Chapter 4: Polynomials: Operations

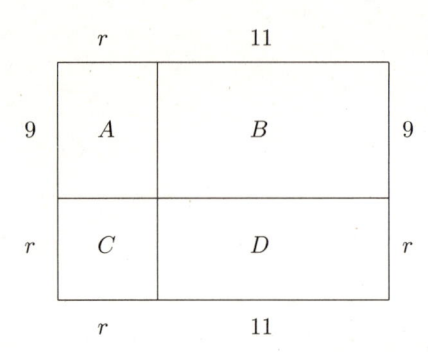

```
      r        11
   ┌─────┬──────────┐
 9 │  A  │    B     │ 9
   ├─────┼──────────┤
 r │  C  │    D     │ r
   └─────┴──────────┘
      r        11
```

The area of the figure can also be found by adding the areas of the four rectangles A, B, C, and D. The area of a rectangle is the product of the length and the width.

$$\begin{aligned}&\text{Area} \\ &\text{of } A\end{aligned} + \begin{aligned}&\text{Area} \\ &\text{of } B\end{aligned} + \begin{aligned}&\text{Area} \\ &\text{of } C\end{aligned} + \begin{aligned}&\text{Area} \\ &\text{of } D\end{aligned}$$
$$= 9 \cdot r + 11 \cdot 9 + r \cdot r + 11 \cdot r$$
$$= 9r + 99 + r^2 + 11r$$

A second algebraic expression for the area of the figure is $9r + 99 + r^2 + 11r$, or $r^2 + 20r + 99$.

59.

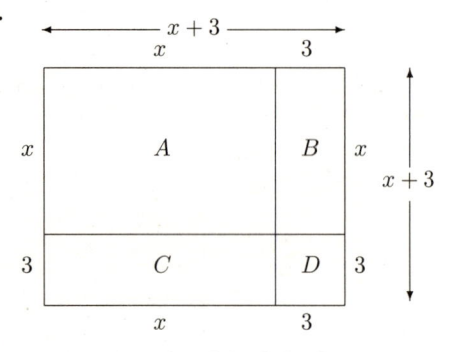

```
       ←——— x + 3 ———→
         x        3
       ┌──────┬──────┐
     x │   A  │   B  │ x
       ├──────┼──────┤   x + 3
     3 │   C  │   D  │ 3
       └──────┴──────┘
         x        3
```

The length and width of the figure can each be expressed as $x + 3$. The area can be expressed as $(x+3) \cdot (x+3)$, or $(x+3)^2$.

Another way to express the area is to find an expression for the sum of the areas of the four rectangles A, B, C, and D. The area of each rectangle is the product of its length and width.

$$\begin{aligned}&\text{Area} \\ &\text{of } A\end{aligned} + \begin{aligned}&\text{Area} \\ &\text{of } B\end{aligned} + \begin{aligned}&\text{Area} \\ &\text{of } C\end{aligned} + \begin{aligned}&\text{Area} \\ &\text{of } D\end{aligned}$$
$$= x \cdot x + 3 \cdot x + 3 \cdot x + 3 \cdot 3$$
$$= x^2 + 3x + 3x + 9$$

Then a second algebraic expression for the area of the figure is $x^2 + 3x + 3x + 9$, or $x^2 + 6x + 9$.

61.

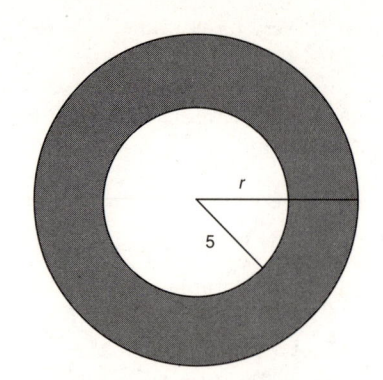

Familiarize. Recall that the area of a circle is the product of π and the square of the radius, r^2.

$$A = \pi r^2$$

Translate.

$$\begin{aligned}&\text{Area of circle} \\ &\text{with radius } r\end{aligned} - \begin{aligned}&\text{Area of circle} \\ &\text{with radius } 5\end{aligned} = \begin{aligned}&\text{Shaded} \\ &\text{area}\end{aligned}$$
$$\pi \cdot r^2 \quad - \quad \pi \cdot 5^2 \quad = \text{Shaded area}$$

Carry out. We simplify the expression.
$$\pi \cdot r^2 - \pi \cdot 5^2 = \pi r^2 - 25\pi$$

Check. We can go over our calculations. We can also assign some value to r, say 7, and carry out the computation in two ways.

Difference of areas: $\pi \cdot 7^2 - \pi \cdot 5^2 = 49\pi - 25\pi = 24\pi$

Substituting in the polynomial: $\pi \cdot 7^2 - 25\pi = 49\pi - 25\pi = 24\pi$

Since the results are the same, our solution is probably correct.

State. A polynomial for the shaded area is $\pi r^2 - 25\pi$.

63. *Familiarize.* We label the figure with additional information.

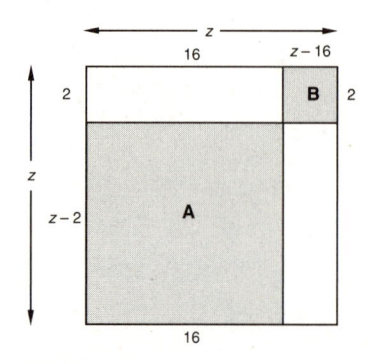

```
          ←——— z ———→
           16    z−16
        ┌───────┬───┐
      2 │       │ B │ 2
        │   ┌───┴───┤
      z │   │       │
        │   │   A   │
    z−2 │   │       │
        │   │       │
        └───┴───────┘
             16
```

Translate.

Area of shaded sections = Area of A + Area of B

Area of shaded sections = $16(z - 2) + 2(z - 16)$

Carry out. We simplify the expression.

$16(z - 2) + 2(z - 16) = 16z - 32 + 2z - 32 = 18z - 64$

Check. We can go over the calculations. We can also assign some value to z, say 30, and carry out the computation in two ways.

Exercise Set 4.4

Sum of areas:
$$16 \cdot 28 + 2 \cdot 14 = 448 + 28 = 476$$

Substituting in the polynomial:
$$18 \cdot 30 - 64 = 540 - 64 = 476$$

Since the results are the same, our solution is probably correct.

State. A polynomial for the shaded area is $18z - 64$.

65. Discussion and Writing Exercise

67.
$$8x + 3x = 66$$
$$11x = 66 \quad \text{Collecting like terms}$$
$$\frac{11x}{11} = \frac{66}{11} \quad \text{Dividing by 11}$$
$$x = 6$$

The solution is 6.

69. $\frac{3}{8}x + \frac{1}{4} - \frac{3}{4}x = \frac{11}{16} + x$, LCM is 16

$$16\left(\frac{3}{8}x + \frac{1}{4} - \frac{3}{4}x\right) = 16\left(\frac{11}{16} + x\right) \quad \text{Clearing fractions}$$
$$6x + 4 - 12x = 11 + 16x$$
$$-6x + 4 = 11 + 16x \quad \text{Collecting like terms}$$
$$-6x + 4 - 4 = 11 + 16x - 4 \quad \text{Subtracting 4}$$
$$-6x = 7 + 16x$$
$$-6x - 16x = 7 + 16x - 16x \quad \text{Subtracting } 16x$$
$$-22x = 7$$
$$\frac{-22x}{-22} = \frac{7}{-22} \quad \text{Dividing by } -22$$
$$x = -\frac{7}{22}$$

The solution is $-\frac{7}{22}$.

71.
$$1.5x - 2.7x = 22 - 5.6x$$
$$10(1.5x - 2.7x) = 10(22 - 5.6x) \quad \text{Clearing decimals}$$
$$15x - 27x = 220 - 56x$$
$$-12x = 220 - 56x \quad \text{Collecting like terms}$$
$$44x = 220 \quad \text{Adding } 56x$$
$$x = \frac{220}{44} \quad \text{Dividing by 44}$$
$$x = 5 \quad \text{Simplifying}$$

The solution is 5.

73.
$$6(y - 3) - 8 = 4(y + 2) + 5$$
$$6y - 18 - 8 = 4y + 8 + 5 \quad \text{Removing parentheses}$$
$$6y - 26 = 4y + 13 \quad \text{Collecting like terms}$$
$$6y - 26 + 26 = 4y + 13 + 26 \quad \text{Adding 26}$$
$$6y = 4y + 39$$
$$6y - 4y = 4y + 39 - 4y \quad \text{Subtracting } 4y$$
$$2y = 39$$
$$\frac{2y}{2} = \frac{39}{2} \quad \text{Dividing by 2}$$
$$y = \frac{39}{2}$$

The solution is $\frac{39}{2}$.

75.
$$3x - 7 \leq 5x + 13$$
$$-2x - 7 \leq 13 \quad \text{Subtracting } 5x$$
$$-2x \leq 20 \quad \text{Adding 7}$$
$$x \geq -10 \quad \text{Dividing by } -2 \text{ and reversing the inequality symbol}$$

The solution set is $\{x | x \geq -10\}$.

77. Familiarize. The surface area is $2lw + 2lh + 2wh$, where $l = $ length, $w = $ width, and $h = $ height of the rectangular solid. Here we have $l = 3$, $w = w$, and $h = 7$.

Translate. We substitute in the formula above.
$$2 \cdot 3 \cdot w + 2 \cdot 3 \cdot 7 + 2 \cdot w \cdot 7$$

Carry out. We simplify the expression.
$$2 \cdot 3 \cdot w + 2 \cdot 3 \cdot 7 + 2 \cdot w \cdot 7$$
$$= 6w + 42 + 14w$$
$$= 20w + 42$$

Check. We can go over the calculations. We can also assign some value to w, say 6, and carry out the computation in two ways.

Using the formula: $2 \cdot 3 \cdot 6 + 2 \cdot 3 \cdot 7 + 2 \cdot 6 \cdot 7 = 36 + 42 + 84 = 162$

Substituting in the polynomial: $20 \cdot 6 + 42 = 120 + 42 = 162$

Since the results are the same, our solution is probably correct.

State. A polynomial for the surface area is $20w + 42$.

79. Familiarize. The surface area is $2lw + 2lh + 2wh$, where $l = $ length, $w = $ width, and $h = $ height of the rectangular solid. Here we have $l = x$, $w = x$, and $h = 5$.

Translate. We substitute in the formula above.
$$2 \cdot x \cdot x + 2 \cdot x \cdot 5 + 2 \cdot x \cdot 5$$

Carry out. We simplify the expression.
$$2 \cdot x \cdot x + 2 \cdot x \cdot 5 + 2 \cdot x \cdot 5$$
$$= 2x^2 + 10x + 10x$$
$$= 2x^2 + 20x$$

Check. We can go over the calculations. We can also assign some value to x, say 3, and carry out the computation in two ways.

Using the formula: $2 \cdot 3 \cdot 3 + 2 \cdot 3 \cdot 5 + 2 \cdot 3 \cdot 5 = 18 + 30 + 30 = 78$

Substituting in the polynomial: $2 \cdot 3^2 + 20 \cdot 3 = 2 \cdot 9 + 60 = 18 + 60 = 78$

Since the results are the same, our solution is probably correct.

State. A polynomial for the surface area is $2x^2 + 20x$.

81.

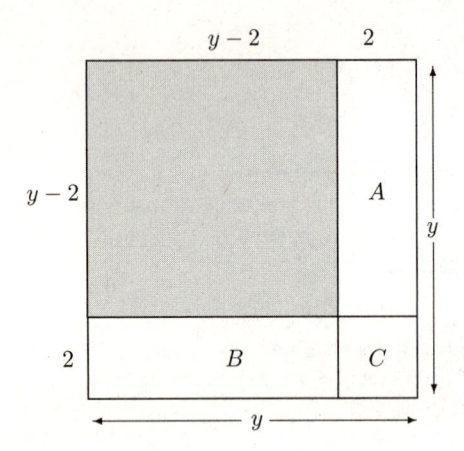

The shaded area is $(y-2)^2$. We find it as follows:

$$\begin{aligned}\text{Shaded area} &= \text{Area of square} - \text{Area of } A - \text{Area of } B - \text{Area of } C\\ (y-2)^2 &= y^2 - 2(y-2) - 2(y-2) - 2\cdot 2\\ (y-2)^2 &= y^2 - 2y + 4 - 2y + 4 - 4\\ (y-2)^2 &= y^2 - 4y + 4\end{aligned}$$

83. $(7y^2 - 5y + 6) - (3y^2 + 8y - 12) + (8y^2 - 10y + 3)$
$= 7y^2 - 5y + 6 - 3y^2 - 8y + 12 + 8y^2 - 10y + 3$
$= 12y^2 - 23y + 21$

85. $(-y^4 - 7y^3 + y^2) + (-2y^4 + 5y - 2) - (-6y^3 + y^2)$
$= -y^4 - 7y^3 + y^2 - 2y^4 + 5y - 2 + 6y^3 - y^2$
$= -3y^4 - y^3 + 5y - 2$

Exercise Set 4.5

1. $(8x^2)(5) = (8 \cdot 5)x^2 = 40x^2$

3. $(-x^2)(-x) = (-1x^2)(-1x) = (-1)(-1)(x^2 \cdot x) = x^3$

5. $(8x^5)(4x^3) = (8 \cdot 4)(x^5 \cdot x^3) = 32x^8$

7. $(0.1x^6)(0.3x^5) = (0.1)(0.3)(x^6 \cdot x^5) = 0.03x^{11}$

9. $\left(-\dfrac{1}{5}x^3\right)\left(-\dfrac{1}{3}x\right) = \left(-\dfrac{1}{5}\right)\left(-\dfrac{1}{3}\right)(x^3 \cdot x) = \dfrac{1}{15}x^4$

11. $(-4x^2)(0) = 0$ Any number multiplied by 0 is 0.

13. $(3x^2)(-4x^3)(2x^6) = (3)(-4)(2)(x^2 \cdot x^3 \cdot x^6) = -24x^{11}$

15. $2x(-x+5) = 2x(-x) + 2x(5)$
$= -2x^2 + 10x$

17. $-5x(x-1) = -5x(x) - 5x(-1)$
$= -5x^2 + 5x$

19. $x^2(x^3 + 1) = x^2(x^3) + x^2(1)$
$= x^5 + x^2$

21. $3x(2x^2 - 6x + 1) = 3x(2x^2) + 3x(-6x) + 3x(1)$
$= 6x^3 - 18x^2 + 3x$

23. $-6x^2(x^2 + x) = -6x^2(x^2) - 6x^2(x)$
$= -6x^4 - 6x^3$

25. $3y^2(6y^4 + 8y^3) = 3y^2(6y^4) + 3y^2(8y^3)$
$= 18y^6 + 24y^5$

27. $(x+6)(x+3) = (x+6)x + (x+6)3$
$= x \cdot x + 6 \cdot x + x \cdot 3 + 6 \cdot 3$
$= x^2 + 6x + 3x + 18$
$= x^2 + 9x + 18$

29. $(x+5)(x-2) = (x+5)x + (x+5)(-2)$
$= x \cdot x + 5 \cdot x + x(-2) + 5(-2)$
$= x^2 + 5x - 2x - 10$
$= x^2 + 3x - 10$

31. $(x-4)(x-3) = (x-4)x + (x-4)(-3)$
$= x \cdot x - 4 \cdot x + x(-3) - 4(-3)$
$= x^2 - 4x - 3x + 12$
$= x^2 - 7x + 12$

33. $(x+3)(x-3) = (x+3)x + (x+3)(-3)$
$= x \cdot x + 3 \cdot x + x(-3) + 3(-3)$
$= x^2 + 3x - 3x - 9$
$= x^2 - 9$

35. $(5-x)(5-2x) = (5-x)5 + (5-x)(-2x)$
$= 5 \cdot 5 - x \cdot 5 + 5(-2x) - x(-2x)$
$= 25 - 5x - 10x + 2x^2$
$= 25 - 15x + 2x^2$

37. $(2x+5)(2x+5) = (2x+5)2x + (2x+5)5$
$= 2x \cdot 2x + 5 \cdot 2x + 2x \cdot 5 + 5 \cdot 5$
$= 4x^2 + 10x + 10x + 25$
$= 4x^2 + 20x + 25$

39. $\left(x - \dfrac{5}{2}\right)\left(x + \dfrac{2}{5}\right) = \left(x - \dfrac{5}{2}\right)x + \left(x - \dfrac{5}{2}\right)\dfrac{2}{5}$
$= x \cdot x - \dfrac{5}{2} \cdot x + x \cdot \dfrac{2}{5} - \dfrac{5}{2} \cdot \dfrac{2}{5}$
$= x^2 - \dfrac{5}{2}x + \dfrac{2}{5}x - 1$
$= x^2 - \dfrac{25}{10}x + \dfrac{4}{10}x - 1$
$= x^2 - \dfrac{21}{10}x - 1$

41. $(x - 2.3)(x + 4.7) = (x - 2.3)x + (x - 2.3)4.7$
$= x \cdot x - 2.3 \cdot x + x \cdot 4.7 - 2.3(4.7)$
$= x^2 - 2.3x + 4.7x - 10.81$
$= x^2 + 2.4x - 10.81$

43. Illustrate $x(x+5)$ as the area of a rectangle with width x and length $x+5$.

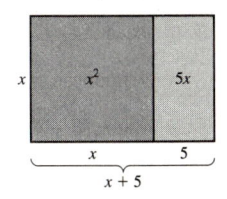

45. Illustrate $(x+1)(x+2)$ as the area of a rectangle with width $x+1$ and length $x+2$.

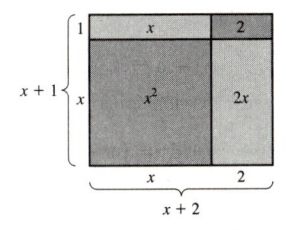

47. Illustrate $(x+5)(x+3)$ as the area of a rectangle with length $x+5$ and width $x+3$.

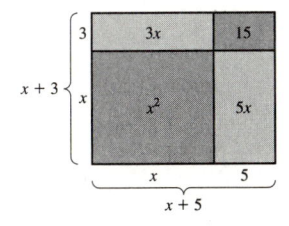

49. Illustrate $(3x+2)(3x+2)$ as the area of a square with sides of length $3x+2$.

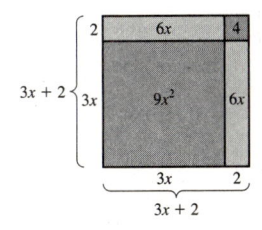

51. $(x^2 + x + 1)(x - 1)$
$= (x^2 + x + 1)x + (x^2 + x + 1)(-1)$
$= x^2 \cdot x + x \cdot x + 1 \cdot x + x^2(-1) + x(-1) + 1(-1)$
$= x^3 + x^2 + x - x^2 - x - 1$
$= x^3 - 1$

53. $(2x+1)(2x^2 + 6x + 1)$
$= 2x(2x^2 + 6x + 1) + 1(2x^2 + 6x + 1)$
$= 2x \cdot 2x^2 + 2x \cdot 6x + 2x \cdot 1 + 1 \cdot 2x^2 + 1 \cdot 6x + 1 \cdot 1$
$= 4x^3 + 12x^2 + 2x + 2x^2 + 6x + 1$
$= 4x^3 + 14x^2 + 8x + 1$

55. $(y^2 - 3)(3y^2 - 6y + 2)$
$= y^2(3y^2 - 6y + 2) - 3(3y^2 - 6y + 2)$
$= y^2 \cdot 3y^2 + y^2(-6y) + y^2 \cdot 2 - 3 \cdot 3y^2 - 3(-6y) - 3 \cdot 2$
$= 3y^4 - 6y^3 + 2y^2 - 9y^2 + 18y - 6$
$= 3y^4 - 6y^3 - 7y^2 + 18y - 6$

57. $(x^3 + x^2)(x^3 + x^2 - x)$
$= x^3(x^3 + x^2 - x) + x^2(x^3 + x^2 - x)$
$= x^3 \cdot x^3 + x^3 \cdot x^2 + x^3(-x) + x^2 \cdot x^3 + x^2 \cdot x^2 + x^2(-x)$
$= x^6 + x^5 - x^4 + x^5 + x^4 - x^3$
$= x^6 + 2x^5 - x^3$

59. $(-5x^3 - 7x^2 + 1)(2x^2 - x)$
$= (-5x^3 - 7x^2 + 1)2x^2 + (-5x^3 - 7x^2 + 1)(-x)$
$= -5x^3 \cdot 2x^2 - 7x^2 \cdot 2x^2 + 1 \cdot 2x^2 - 5x^3(-x) - 7x^2(-x) + 1(-x)$
$= -10x^5 - 14x^4 + 2x^2 + 5x^4 + 7x^3 - x$
$= -10x^5 - 9x^4 + 7x^3 + 2x^2 - x$

61. $\begin{array}{r} 1 + x + x^2 \\ -1 - x + x^2 \\ \hline x^2 + x^3 + x^4 \\ -x - x^2 - x^3 \\ -1 - x - x^2 \\ \hline -1 - 2x - x^2 + x^4 \end{array}$ Line up like terms in columns
Multiplying the top row by x^2
Multiplying by $-x$
Multiplying by -1

63. $\begin{array}{r} 2t^2 - t - 4 \\ 3t^2 + 2t - 1 \\ \hline -2t^2 + t + 4 \\ 4t^3 - 2t^2 - 8t \\ 6t^4 - 3t^3 - 12t^2 \\ \hline 6t^4 + t^3 - 16t^2 - 7t + 4 \end{array}$ Multiplying by -1
Multiplying by $2t$
Multiplying by $3t^2$

65. $\begin{array}{r} x - x^3 + x^5 \\ -1 + x^2 + x^4 \\ \hline x^5 - x^7 + x^9 \\ x^3 - x^5 + x^7 \\ -x + x^3 - x^5 \\ \hline -x + 2x^3 - x^5 + x^9 \end{array}$ Rewriting in ascending order
Multiplying by x^4
Multiplying by x^2
Multiplying by -1

67. $\begin{array}{r} x^3 + x^2 + x + 1 \\ x - 1 \\ \hline -x^3 - x^2 - x - 1 \\ x^4 + x^3 + x^2 + x \\ \hline x^4 - 1 \end{array}$

69. We will multiply horizontally while still aligning like terms.
$(x+1)(x^3 + 7x^2 + 5x + 4)$
$= x^4 + 7x^3 + 5x^2 + 4x$ Multiplying by x
$ + x^3 + 7x^2 + 5x + 4$ Multiplying by 1
$= x^4 + 8x^3 + 12x^2 + 9x + 4$

71. We will multiply horizontally while still aligning like terms.

$$\left(x - \frac{1}{2}\right)\left(2x^3 - 4x^2 + 3x - \frac{2}{5}\right)$$

$$= 2x^4 - 4x^3 + 3x^2 - \frac{2}{5}x$$

$$ - x^3 + 2x^2 - \frac{3}{2}x + \frac{1}{5}$$

$$\overline{2x^4 - 5x^3 + 5x^2 - \frac{19}{10}x + \frac{1}{5}}$$

73. Discussion and Writing Exercise

75. $-\frac{1}{4} - \frac{1}{2} = -\frac{1}{4} - \frac{1}{2} \cdot \frac{2}{2} = -\frac{1}{4} - \frac{2}{4} = -\frac{3}{4}$

77. $(10-2)(10+2) = 8 \cdot 12 = 96$

79. $15x - 18y + 12 = 3 \cdot 5x - 3 \cdot 6y + 3 \cdot 4 = 3(5x - 6y + 4)$

81. $-9x - 45y + 15 = -3 \cdot 3x - 3 \cdot 15y - 3(-5) = -3(3x + 15y - 5)$

83. $y = \frac{1}{2}x - 3$

The equation is equivalent to $y = \frac{1}{2}x + (-3)$. The y-intercept is $(0, -3)$. We find two other points, using multiples of 2 for x to avoid fractions.

When $x = -2$, $y = \frac{1}{2}(-2) - 3 = -1 - 3 = -4$.

When $x = 4$, $y = \frac{1}{2} \cdot 4 - 3 = 2 - 3 = -1$.

x	y
0	-3
-2	-4
4	-1

Plot these points, draw the line they determine, and label the graph $y = \frac{1}{2}x - 3$.

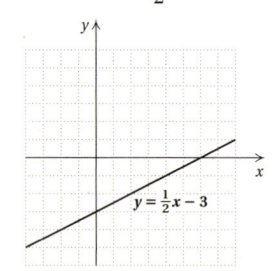

85. The shaded area is the area of the large rectangle, $6y(14y - 5)$ less the area of the unshaded rectangle, $3y(3y + 5)$. We have:

$$6y(14y - 5) - 3y(3y + 5)$$
$$= 84y^2 - 30y - 9y^2 - 15y$$
$$= 75y^2 - 45y$$

87.

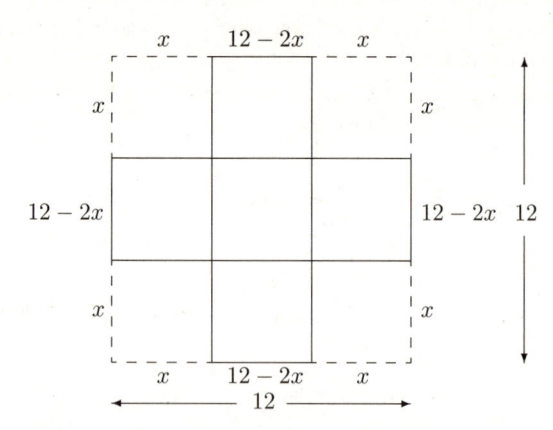

The dimensions, in inches, of the box are $12 - 2x$ by $12 - 2x$ by x. The volume is the product of the dimensions (volume = length × width × height):

Volume $= (12 - 2x)(12 - 2x)x$
$= (144 - 48x + 4x^2)x$
$= 144x - 48x^2 + 4x^3$ in^3, or
$4x^3 - 48x^2 + 144x$ in^3

The outside surface area is the sum of the area of the bottom and the areas of the four sides. The dimensions, in inches, of the bottom are $12 - 2x$ by $12 - 2x$, and the dimensions, in inches, of each side are x by $12 - 2x$.

Surface area = Area of bottom + 4 · Area of each side
$= (12 - 2x)(12 - 2x) + 4 \cdot x(12 - 2x)$
$= 144 - 24x - 24x + 4x^2 + 48x - 8x^2$
$= 144 - 48x + 4x^2 + 48x - 8x^2$
$= 144 - 4x^2$ in^2, or $-4x^2 + 144$ in^2

89. Let $n =$ the missing number.

The area of the figure is $x^2 + 3x + nx + 3n$. This is equivalent to $x^2 + 8x + 15$, so we have $3x + nx = 8x$ and $3n = 15$. Solving either equation for n, we find that the missing number is 5.

91. We have a rectangular solid with dimensions x m by x m by $x+2$ m with a rectangular solid piece with dimensions 6 m by 5 m by 7 m cut out of it.

Volume = Volume of large solid − Volume of small solid
$= (x \text{ m})(x \text{ m})(x + 2 \text{ m}) - (6 \text{ m})(5 \text{ m})(7 \text{ m})$
$= x^2(x + 2) \text{ m}^3 - 210 \text{ m}^3$
$= x^3 + 2x^2 - 210 \text{ m}^3$

Exercise Set 4.6

93. $(x-2)(x-7) - (x-7)(x-2)$

First observe that, by the commutative law of multiplication, $(x-2)(x-7)$ and $(x-7)(x-2)$ are equivalent expressions. Then when we subtract $(x-7)(x-2)$ from $(x-2)(x-7)$, the result is 0.

95. $(x-a)(x-b)\cdots(x-x)(x-y)(x-z)$
$= (x-a)(x-b)\cdots 0 \cdot (x-y)(x-z)$
$= 0$

Exercise Set 4.6

1. $(x+1)(x^2+3)$
$\ \ \text{F} \text{O} \text{I} \text{L}$
$= x \cdot x^2 + x \cdot 3 + 1 \cdot x^2 + 1 \cdot 3$
$= x^3 + 3x + x^2 + 3$

3. $(x^3+2)(x+1)$
$\ \ \text{F} \text{O} \text{I} \text{L}$
$= x^3 \cdot x + x^3 \cdot 1 + 2 \cdot x + 2 \cdot 1$
$= x^4 + x^3 + 2x + 2$

5. $(y+2)(y-3)$
$\ \ \text{F} \text{O} \text{I} \text{L}$
$= y \cdot y + y \cdot (-3) + 2 \cdot y + 2 \cdot (-3)$
$= y^2 - 3y + 2y - 6$
$= y^2 - y - 6$

7. $(3x+2)(3x+2)$
$\ \ \text{F} \text{O} \text{I} \text{L}$
$= 3x \cdot 3x + 3x \cdot 2 + 2 \cdot 3x + 2 \cdot 2$
$= 9x^2 + 6x + 6x + 4$
$= 9x^2 + 12x + 4$

9. $(5x-6)(x+2)$
$\ \ \text{F} \text{O} \text{I} \text{L}$
$= 5x \cdot x + 5x \cdot 2 + (-6) \cdot x + (-6) \cdot 2$
$= 5x^2 + 10x - 6x - 12$
$= 5x^2 + 4x - 12$

11. $(3t-1)(3t+1)$
$\ \ \text{F} \text{O} \text{I} \text{L}$
$= 3t \cdot 3t + 3t \cdot 1 + (-1) \cdot 3t + (-1) \cdot 1$
$= 9t^2 + 3t - 3t - 1$
$= 9t^2 - 1$

13. $(4x-2)(x-1)$
$\ \ \text{F} \text{O} \text{I} \text{L}$
$= 4x \cdot x + 4x \cdot (-1) + (-2) \cdot x + (-2) \cdot (-1)$
$= 4x^2 - 4x - 2x + 2$
$= 4x^2 - 6x + 2$

15. $\left(p - \dfrac{1}{4}\right)\left(p + \dfrac{1}{4}\right)$
$\ \ \text{F} \text{O} \text{I} \text{L}$
$= p \cdot p + p \cdot \dfrac{1}{4} + \left(-\dfrac{1}{4}\right) \cdot p + \left(-\dfrac{1}{4}\right) \cdot \dfrac{1}{4}$
$= p^2 + \dfrac{1}{4}p - \dfrac{1}{4}p - \dfrac{1}{16}$
$= p^2 - \dfrac{1}{16}$

17. $(x-0.1)(x+0.1)$
$\ \ \text{F} \text{O} \text{I} \text{L}$
$= x \cdot x + x \cdot (0.1) + (-0.1) \cdot x + (-0.1)(0.1)$
$= x^2 + 0.1x - 0.1x - 0.01$
$= x^2 - 0.01$

19. $(2x^2+6)(x+1)$
$\ \ \text{F} \text{O} \text{I} \text{L}$
$= 2x^3 + 2x^2 + 6x + 6$

21. $(-2x+1)(x+6)$
$\ \ \text{F} \text{O} \text{I} \text{L}$
$= -2x^2 - 12x + x + 6$
$= -2x^2 - 11x + 6$

23. $(a+7)(a+7)$
$\ \ \text{F} \text{O} \text{I} \text{L}$
$= a^2 + 7a + 7a + 49$
$= a^2 + 14a + 49$

25. $(1+2x)(1-3x)$
$\ \ \text{F} \text{O} \text{I} \text{L}$
$= 1 - 3x + 2x - 6x^2$
$= 1 - x - 6x^2$

27. $(x^2+3)(x^3-1)$
$\ \ \text{F} \text{O} \text{I} \text{L}$
$= x^5 - x^2 + 3x^3 - 3$

29. $(3x^2-2)(x^4-2)$
$\ \ \text{F} \text{O} \text{I} \text{L}$
$= 3x^6 - 6x^2 - 2x^4 + 4$

31. $(2.8x-1.5)(4.7x+9.3)$
$\ \ \text{F} \text{O} \text{I} \text{L}$
$= 2.8x(4.7x) + 2.8x(9.3) - 1.5(4.7x) - 1.5(9.3)$
$= 13.16x^2 + 26.04x - 7.05x - 13.95$
$= 13.16x^2 + 18.99x - 13.95$

33. $(3x^5+2)(2x^2+6)$
$\ \ \text{F} \text{O} \text{I} \text{L}$
$= 6x^7 + 18x^5 + 4x^2 + 12$

35. $(8x^3+1)(x^3+8)$
$\ \ \text{F} \text{O} \text{I} \text{L}$
$= 8x^6 + 64x^3 + x^3 + 8$
$= 8x^6 + 65x^3 + 8$

37. $(4x^2 + 3)(x - 3)$
   F   O   I   L
   $= 4x^3 - 12x^2 + 3x - 9$

39. $(4y^4 + y^2)(y^2 + y)$
   F   O   I   L
   $= 4y^6 + 4y^5 + y^4 + y^3$

41. $(x + 4)(x - 4)$ Product of sum and difference of two terms
   $= x^2 - 4^2$
   $= x^2 - 16$

43. $(2x + 1)(2x - 1)$ Product of sum and difference of two terms
   $= (2x)^2 - 1^2$
   $= 4x^2 - 1$

45. $(5m - 2)(5m + 2)$ Product of sum and difference of two terms
   $= (5m)^2 - 2^2$
   $= 25m^2 - 4$

47. $(2x^2 + 3)(2x^2 - 3)$ Product of sum and difference of two terms
   $= (2x^2)^2 - 3^2$
   $= 4x^4 - 9$

49. $(3x^4 - 4)(3x^4 + 4)$
   $= (3x^4)^2 - 4^2$
   $= 9x^8 - 16$

51. $(x^6 - x^2)(x^6 + x^2)$
   $= (x^6)^2 - (x^2)^2$
   $= x^{12} - x^4$

53. $(x^4 + 3x)(x^4 - 3x)$
   $= (x^4)^2 - (3x)^2$
   $= x^8 - 9x^2$

55. $(x^{12} - 3)(x^{12} + 3)$
   $= (x^{12})^2 - 3^2$
   $= x^{24} - 9$

57. $(2y^8 + 3)(2y^8 - 3)$
   $= (2y^8)^2 - 3^2$
   $= 4y^{16} - 9$

59. $\left(\dfrac{5}{8}x - 4.3\right)\left(\dfrac{5}{8}x + 4.3\right)$
   $= \left(\dfrac{5}{8}x\right)^2 - (4.3)^2$
   $= \dfrac{25}{64}x^2 - 18.49$

61. $(x + 2)^2 = x^2 + 2 \cdot x \cdot 2 + 2^2$ Square of a binomial sum
   $= x^2 + 4x + 4$

63. $(3x^2 + 1)$ Square of a binomial sum
   $= (3x^2)^2 + 2 \cdot 3x^2 \cdot 1 + 1^2$
   $= 9x^4 + 6x^2 + 1$

65. $\left(a - \dfrac{1}{2}\right)^2$ Square of a binomial sum
   $= a^2 - 2 \cdot a \cdot \dfrac{1}{2} + \left(\dfrac{1}{2}\right)^2$
   $= a^2 - a + \dfrac{1}{4}$

67. $(3 + x)^2 = 3^2 + 2 \cdot 3 \cdot x + x^2$
   $= 9 + 6x + x^2$

69. $(x^2 + 1)^2 = (x^2)^2 + 2 \cdot x^2 \cdot 1 + 1^2$
   $= x^4 + 2x^2 + 1$

71. $(2 - 3x^4)^2 = 2^2 - 2 \cdot 2 \cdot 3x^4 + (3x^4)^2$
   $= 4 - 12x^4 + 9x^8$

73. $(5 + 6t^2)^2 = 5^2 + 2 \cdot 5 \cdot 6t^2 + (6t^2)^2$
   $= 25 + 60t^2 + 36t^4$

75. $\left(x - \dfrac{5}{8}\right)^2 = x^2 - 2 \cdot x \cdot \dfrac{5}{8} + \left(\dfrac{5}{8}\right)^2$
   $= x^2 - \dfrac{5}{4}x + \dfrac{25}{64}$

77. $(3 - 2x^3)^2 = 3^2 - 2 \cdot 3 \cdot 2x^3 + (2x^3)^2$
   $= 9 - 12x^3 + 4x^6$

79. $4x(x^2 + 6x - 3)$ Product of a monomial and a trinomial
   $= 4x \cdot x^2 + 4x \cdot 6x + 4x(-3)$
   $= 4x^3 + 24x^2 - 12x$

81. $\left(2x^2 - \dfrac{1}{2}\right)\left(2x^2 - \dfrac{1}{2}\right)$ Square of a binomial difference
   $= (2x^2)^2 - 2 \cdot 2x^2 \cdot \dfrac{1}{2} + \left(\dfrac{1}{2}\right)^2$
   $= 4x^4 - 2x^2 + \dfrac{1}{4}$

83. $(-1 + 3p)(1 + 3p)$
   $= (3p - 1)(3p + 1)$ Product of the sum and difference of two terms
   $= (3p)^2 - 1^2$
   $= 9p^2 - 1$

85. $3t^2(5t^3 - t^2 + t)$ Product of a monomial and a trinomial
   $= 3t^2 \cdot 5t^3 + 3t^2(-t^2) + 3t^2 \cdot t$
   $= 15t^5 - 3t^4 + 3t^3$

Exercise Set 4.6

87. $(6x^4 + 4)^2$ Square of a binomial sum
$= (6x^4)^2 + 2 \cdot 6x^4 \cdot 4 + 4^2$
$= 36x^8 + 48x^4 + 16$

89. $(3x + 2)(4x^2 + 5)$ Product of two binomials; use FOIL
$= 3x \cdot 4x^2 + 3x \cdot 5 + 2 \cdot 4x^2 + 2 \cdot 5$
$= 12x^3 + 15x + 8x^2 + 10$

91. $(8 - 6x^4)^2$ Square of a binomial difference
$= 8^2 - 2 \cdot 8 \cdot 6x^4 + (6x^4)^2$
$= 64 - 96x^4 + 36x^8$

93.
$$\begin{array}{r} t^2+t+1 \\ t-1 \\ \hline -t^2-t-1 \\ t^3+t^2+t \\ \hline t^3 \qquad -1 \end{array}$$

95. $3^2 + 4^2 = 9 + 16 = 25$
$(3 + 4)^2 = 7^2 = 49$

97. $9^2 - 5^2 = 81 - 25 = 56$
$(9 - 5)^2 = 4^2 = 16$

99.

1	A	B
a	D	C
	a	1

We can find the shaded area in two ways.

Method 1: The figure is a square with side $a + 1$, so the area is $(a + 1)^2 = a^2 + 2a + 1$.

Method 2: We add the areas of A, B, C, and D.
$1 \cdot a + 1 \cdot 1 + 1 \cdot a + a \cdot a = a + 1 + a + a^2 = a^2 + 2a + 1$.

Either way we find that the total shaded area is $a^2 + 2a + 1$.

101.

	t	6
t	A	B
4	D	C

We can find the shaded area in two ways.

Method 1: The figure is a rectangle with dimensions $t + 6$ by $t + 4$, so the area is $(t + 6)(t + 4) = t^2 + 4t + 6t + 24 = t^2 + 10t + 24$.

Method 2: We add the areas of A, B, C, and D.
$t \cdot t + t \cdot 6 + 6 \cdot 4 + 4 \cdot t = t^2 + 6t + 24 + 4t = t^2 + 10t + 24$.

Either way, we find that the total shaded area is $t^2 + 10t + 24$.

103. Discussion and Writing Exercise

105. Familiarize. Let t = the number of watts used by the television set. Then $10t$ = the number of watts used by the lamps, and $40t$ = the number of watts used by the air conditioner.

Translate.

$$\underbrace{\text{Lamp watts}}_{10t} + \underbrace{\text{Air conditioner watts}}_{40t} + \underbrace{\text{Television watts}}_{t} = \underbrace{\text{Total watts}}_{2550}$$

Solve. We solve the equation.
$$10t + 40t + t = 2550$$
$$51t = 2550$$
$$t = 50$$

The possible solution is:

Television, t: 50 watts

Lamps, $10t$: $10 \cdot 50$, or 500 watts

Air conditioner, $40t$: $40 \cdot 50$, or 2000 watts

Check. The number of watts used by the lamps, 500, is 10 times 50, the number used by the television. The number of watts used by the air conditioner, 2000, is 40 times 50, the number used by the television. Also, $50 + 500 + 2000 = 2550$, the total wattage used.

State. The television uses 50 watts, the lamps use 500 watts, and the air conditioner uses 2000 watts.

107.
$$3(x - 2) = 5(2x + 7)$$
$$3x - 6 = 10x + 35 \quad \text{Removing parentheses}$$
$$3x - 6 + 6 = 10x + 35 + 6 \quad \text{Adding 6}$$
$$3x = 10x + 41$$
$$3x - 10x = 10x + 41 - 10x \quad \text{Subtracting } 10x$$
$$-7x = 41$$
$$\frac{-7x}{-7} = \frac{41}{-7} \quad \text{Dividing by } -7$$
$$x = -\frac{41}{7}$$

The solution is $-\frac{41}{7}$.

109. $3x - 2y = 12$
$-2y = -3x + 12$ Subtracting $3x$
$\dfrac{-2y}{-2} = \dfrac{-3x + 12}{-2}$ Dividing by -2
$y = \dfrac{3x - 12}{2}$, or
$y = \dfrac{3}{2}x - 6$

111. $5x(3x - 1)(2x + 3)$
$= 5x(6x^2 + 7x - 3)$ Using FOIL
$= 30x^3 + 35x^2 - 15x$

113. $[(a - 5)(a + 5)]^2$
$= (a^2 - 25)^2$ Finding the product of a sum and difference of same two terms
$= a^4 - 50a^2 + 625$ Squaring a binomial

115. $(3t^4 - 2)^2 1(3t^4 + 2)^2$
$= [(3t^4 - 2)(3t^4 + 2)]^2$
$= (9t^8 - 4)^2$
$= 81t^{16} - 72t^8 + 16$

117. $(x + 2)(x - 5) = (x + 1)(x - 3)$
$x^2 - 5x + 2x - 10 = x^2 - 3x + x - 3$
$x^2 - 3x - 10 = x^2 - 2x - 3$
$-3x - 10 = -2x - 3$ Adding $-x^2$
$-3x + 2x = 10 - 3$ Adding $2x$ and 10
$-x = 7$
$x = -7$

The solution is -7.

119. See the answer section in the text.

121. Enter $y_1 = (x - 1)^2$ and $y_2 = x^2 - 2x + 1$. Then compare the graphs or the y_1-and y_2-values in a table. It appears that the graphs are the same and that the y_1-and y_2-values are the same, so $(x - 1)^2 = x^2 - 2x + 1$ is correct.

123. Enter $y_1 = (x - 3)(x + 3)$ and $y_2 = x^2 - 6$. Then compare the graphs or the y_1-and y_2-values in a table. The graphs are not the same nor are the y_1-and y_2-values, so $(x - 3)(x + 3) = x^2 - 6$ is not correct.

Exercise Set 4.7

1. We replace x by 3 and y by -2.
$x^2 - y^2 + xy = 3^2 - (-2)^2 + 3(-2) = 9 - 4 - 6 = -1$

3. We replace x by 3 and y by -2.
$x^2 - 3y^2 + 2xy = 3^2 - 3(-2)^2 + 2 \cdot 3(-2) =$
$9 - 3 \cdot 4 + 2 \cdot 3(-2) = 9 - 12 - 12 = -15$

5. We replace x by 3, y by -2, and z by -5.
$8xyz = 8 \cdot 3 \cdot (-2) \cdot (-5) = 240$

7. We replace x by 3, y by -2, and z by -5.
$xyz^2 - z = 3(-2)(-5)^2 - (-5) = 3(-2)(25) - (-5) =$
$-150 + 5 = -145$

9. We replace h by 165 and A by 20.
$C = 0.041h - 0.018A - 2.69$
$= 0.041(165) - 0.018(20) - 2.69$
$= 6.765 - 0.36 - 2.69$
$= 6.405 - 2.69$
$= 3.715$

The lung capacity of a 20-year-old woman who is 165 cm tall is 3.715 liters.

11. Evaluate the polynomial for $h = 50$, $v = 40$, and $t = 2$.
$h = h_0 + vt - 4.9t^2$
$= 50 + 40 \cdot 2 - 4.9(2)^2$
$= 50 + 80 - 19.6$
$= 110.4$

The rocket will be 110.4 m above the ground 2 seconds after blast off.

13. Replace h by 4.7, r by 1.2, and π by 3.14.
$S = 2\pi rh + 2\pi r^2$
$\approx 2(3.14)(1.2)(4.7) + 2(3.14)(1.2)^2$
$\approx 2(3.14)(1.2)(4.7) + 2(3.14)(1.44)$
$\approx 35.4192 + 9.0432$
≈ 44.46

The surface area of the can is about 44.46 in^2.

15. Evaluate the polynomial for $h = 7\dfrac{1}{2}$, or $\dfrac{15}{2}$, $r = 1\dfrac{1}{4}$, or $\dfrac{5}{4}$, and $\pi \approx 3.14$.
$S = 2\pi rh + \pi r^2$
$\approx 2(3.14)\left(\dfrac{5}{4}\right)\left(\dfrac{15}{2}\right) + (3.14)\left(\dfrac{5}{4}\right)^2$
$\approx 2(3.14)\left(\dfrac{5}{4}\right)\left(\dfrac{15}{2}\right) + (3.14)\left(\dfrac{25}{16}\right)$
$\approx 58.875 + 4.90625$
≈ 63.78125

The surface area is about 63.78125 in^2.

17. $x^3y - 2xy + 3x^2 - 5$

Term	Coefficient	Degree	
x^3y	1	4	(Think: $x^3y = x^3y^1$)
$-2xy$	-2	2	(Think: $-2xy = -2x^1y^1$)
$3x^2$	3	2	
-5	-5	0	(Think: $-5 = -5x^0$)

The degree of the polynomial is the degree of the term of highest degree. The term of highest degree is x^3y. Its degree is 4. The degree of the polynomial is 4.

19. $17x^2y^3 - 3x^3yz - 7$

Term	Coefficient	Degree	
$17x^2y^3$	17	5	
$-3x^3yz$	-3	5	(Think: $-3x^3yz = -3x^3y^1z^1$)
-7	-7	0	(Think: $-7 = -7x^0$)

The terms of highest degree are $17x^2y^3$ and $-3x^3yz$. Each has degree 5. The degree of the polynomial is 5.

Exercise Set 4.7

21. $a + b - 2a - 3b = (1-2)a + (1-3)b = -a - 2b$

23. $3x^2y - 2xy^2 + x^2$

There are *no* like terms, so none of the terms can be collected.

25. $\quad 6au + 3av + 14au + 7av$
$= (6 + 14)au + (3 + 7)av$
$= 20au + 10av$

27. $\quad 2u^2v - 3uv^2 + 6u^2v - 2uv^2$
$= (2 + 6)u^2v + (-3 - 2)uv^2$
$= 8u^2v - 5uv^2$

29. $\quad (2x^2 - xy + y^2) + (-x^2 - 3xy + 2y^2)$
$= (2 - 1)x^2 + (-1 - 3)xy + (1 + 2)y^2$
$= x^2 - 4xy + 3y^2$

31. $\quad (r - 2s + 3) + (2r + s) + (s + 4)$
$= (1 + 2)r + (-2 + 1 + 1)s + (3 + 4)$
$= 3r + 0s + 7$
$= 3r + 7$

33. $\quad (b^3a^2 - 2b^2a^3 + 3ba + 4) + (b^2a^3 - 4b^3a^2 + 2ba - 1)$
$= (1 - 4)b^3a^2 + (-2 + 1)b^2a^3 + (3 + 2)ba + (4 - 1)$
$= -3b^3a^2 - b^2a^3 + 5ba + 3$, or
$-a^3b^2 - 3a^2b^3 + 5ab + 3$

35. $\quad (a^3 + b^3) - (a^2b - ab^2 + b^3 + a^3)$
$= a^3 + b^3 - a^2b + ab^2 - b^3 - a^3$
$= (1 - 1)a^3 - a^2b + ab^2 + (1 - 1)b^3$
$= -a^2b + ab^2$

37. $\quad (xy - ab - 8) - (xy - 3ab - 6)$
$= xy - ab - 8 - xy + 3ab + 6$
$= (1 - 1)xy + (-1 + 3)ab + (-8 + 6)$
$= 2ab - 2$

39. $\quad (-2a + 7b - c) - (-3b + 4c - 8d)$
$= -2a + 7b - c + 3b - 4c + 8d$
$= -2a + (7 + 3)b + (-1 - 4)c + 8d$
$= -2a + 10b - 5c + 8d$

41. $\quad (3z - u)(2z + 3u) \overset{\text{F O I L}}{=} 6z^2 + 9zu - 2uz - 3u^2$
$= 6z^2 + 7zu - 3u^2$

43. $\quad (a^2b - 2)(a^2b - 5) \overset{\text{F O I L}}{=} a^4b^2 - 5a^2b - 2a^2b + 10$
$= a^4b^2 - 7a^2b + 10$

45. $\quad (a^3 + bc)(a^3 - bc) = (a^3)^2 - (bc)^2$
$\quad\quad\quad\quad [(A + B)(A - B) = A^2 - B^2]$
$= a^6 - b^2c^2$

47.
$\quad\quad\quad\quad y^4x + y^2 + 1$
$\quad\quad\quad\quad\quad\quad\quad y^2 + 1$
$\quad\quad\quad\quad\overline{\quad y^4x + y^2 + 1}$
$\quad y^6x + y^4 \quad\quad + y^2$
$\overline{y^6x + y^4 + y^4x + 2y^2 + 1}$

49. $(3xy - 1)(4xy + 2)$
$\overset{\text{F O I L}}{=} 12x^2y^2 + 6xy - 4xy - 2$
$= 12x^2y^2 + 2xy - 2$

51. $\quad (3 - c^2d^2)(4 + c^2d^2)$
$\overset{\text{F O I L}}{=} 12 + 3c^2d^2 - 4c^2d^2 - c^4d^4$
$= 12 - c^2d^2 - c^4d^4$

53. $\quad (m^2 - n^2)(m + n)$
$\overset{\text{F O I L}}{=} m^3 + m^2n - mn^2 - n^3$

55. $\quad (xy + x^5y^5)(x^4y^4 - xy)$
$\overset{\text{F O I L}}{=} x^5y^5 - x^2y^2 + x^9y^9 - x^6y^6$
$= x^9y^9 - x^6y^6 + x^5y^5 - x^2y^2$

57. $\quad (x + h)^2$
$= x^2 + 2xh + h^2 \quad [(A + B)^2 = A^2 + 2AB + B^2]$

59. $\quad (r^3t^2 - 4)^2$
$= (r^3t^2)^2 - 2 \cdot r^3t^2 \cdot 4 + 4^2$
$\quad\quad\quad\quad [(A - B)^2 = A^2 - 2AB + B^2]$
$= r^6t^4 - 8r^3t^2 + 16$

61. $\quad (p^4 + m^2n^2)^2$
$= (p^4)^2 + 2 \cdot p^4 \cdot m^2n^2 + (m^2n^2)^2$
$\quad\quad\quad\quad [(A + B)^2 = A^2 + 2AB + B^2]$
$= p^8 + 2p^4m^2n^2 + m^4n^4$

63. $\quad \left(2a^3 - \frac{1}{2}b^3\right)^2$
$= (2a^3)^2 - 2 \cdot 2a^3 \cdot \frac{1}{2}b^3 + \left(\frac{1}{2}b^3\right)^2$
$\quad\quad\quad\quad [(A - B)^2 = A^2 - 2AB + B^2]$
$= 4a^6 - 2a^3b^3 + \frac{1}{4}b^6$

65. $3a(a - 2b)^2 = 3a(a^2 - 4ab + 4b^2)$
$= 3a^3 - 12a^2b + 12ab^2$

67. $(2a - b)(2a + b) = (2a)^2 - b^2 = 4a^2 - b^2$

69. $(c^2 - d)(c^2 + d) = (c^2)^2 - d^2$
$= c^4 - d^2$

71. $(ab + cd^2)(ab - cd^2) = (ab)^2 - (cd^2)^2$
$= a^2b^2 - c^2d^4$

73. $\quad (x + y - 3)(x + y + 3)$
$= [(x + y) - 3][(x + y) + 3]$
$= (x + y)^2 - 3^2$
$= x^2 + 2xy + y^2 - 9$

75. $\quad [x + y + z][x - (y + z)]$
$= [x + (y + z)][x - (y + z)]$
$= x^2 - (y + z)^2$
$= x^2 - (y^2 + 2yz + z^2)$
$= x^2 - y^2 - 2yz - z^2$

77. $\quad (a + b + c)(a - b - c)$
$= [a + (b + c)][a - (b + c)]$
$= a^2 - (b + c)^2$
$= a^2 - (b^2 + 2bc + c^2)$
$= a^2 - b^2 - 2bc - c^2$

79. Discussion and Writing Exercise

81. The first coordinate is positive and the second coordinate is negative, so $(2, -5)$ is in quadrant IV.

83. Both coordinates are positive, so $(16, 23)$ is in quadrant I.

85. $2x = -10$
$x = -5$

Any ordered pair $(-5, y)$ is a solution. The variable x must be -5, but y can be any number we choose. A few solutions are listed below. Plot these points and draw the line.

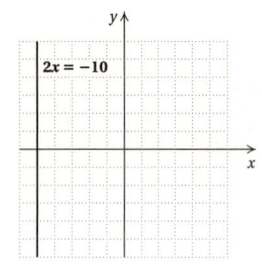

x	y
-5	-3
-5	0
-5	4

87. $8y - 16 = 0$
$8y = 16$
$y = 2$

Any ordered pair $(x, 2)$ is a solution. The variable y must be 2, but x can be any number we choose. A few solutions are listed below. Plot these points and draw the line.

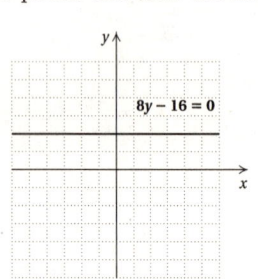

x	y
-4	2
0	2
3	2

89. It is helpful to add additional labels to the figure.

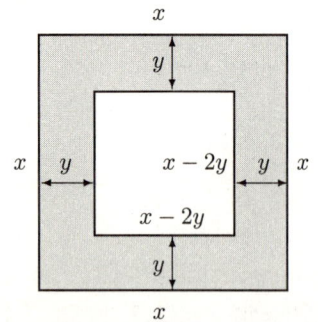

The area of the large square is $x \cdot x$, or x^2. The area of the small square is $(x - 2y)(x - 2y)$, or $(x - 2y)^2$.

$$\begin{aligned}\text{Area of shaded region} &= \text{Area of large square} - \text{Area of small square} \\ \text{Area of shaded region} &= x^2 - (x - 2y)^2 \\ &= x^2 - (x^2 - 4xy + 4y^2) \\ &= x^2 - x^2 + 4xy - 4y^2 \\ &= 4xy - 4y^2\end{aligned}$$

91. It is helpful to add additional labels to the figure.

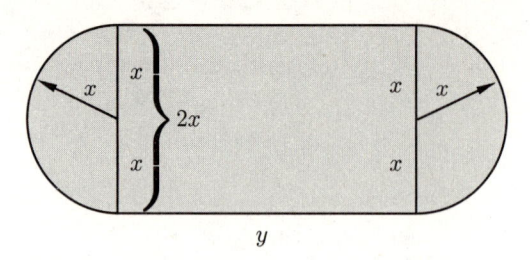

The two semicircles make a circle with radius x. The area of that circle is πx^2. The area of the rectangle is $2x \cdot y$. The sum of the two regions, $\pi x^2 + 2xy$, is the area of the shaded region.

93. The lateral surface area of the outer portion of the solid is the lateral surface area of a right circular cylinder with radius n and height h. The lateral surface area of the inner portion is the lateral surface area of a right circular cylinder with radius m and height h. Recall that the formula for the lateral surface area of a right circular cylinder with radius r and height h is $2\pi rh$.

The surface area of the top is the area of a circle with radius n less the area of a circle with radius m. The surface area of the bottom is the same as the surface area of the top.

Thus, the surface area of the solid is

$2\pi nh + 2\pi mh + 2\pi n^2 - 2\pi m^2$.

95. The height of the observatory is 40 ft and its radius is $30/2$, or 15 ft, so the surface area is $2\pi rh + \pi r^2 \approx 2(3.14)(15)(40) + (3.14)(15)^2 \approx 4474.5$ ft^2. Since 4474.5 ft$^2 / 250$ ft$^2 = 17.898$, 18 gallons of paint should be purchased.

97. Substitute \$10,400 for P, 8.5% or 0.085 for r, and 5 for t.

$$\begin{aligned}P(1+r)^t & \\ &= \$10,400(1+0.085)^5 \\ &= \$10,400(1.085)^5 \\ &\approx \$15,638.03\end{aligned}$$

Exercise Set 4.8

1. $\dfrac{24x^4}{8} = \dfrac{24}{8} \cdot x^4 = 3x^4$

Check: We multiply.
$3x^4 \cdot 8 = 24x^4$

3. $\dfrac{25x^3}{5x^2} = \dfrac{25}{5} \cdot \dfrac{x^3}{x^2} = 5x^{3-2} = 5x$

Check: We multiply.
$5x \cdot 5x^2 = 25x^3$

5. $\dfrac{-54x^{11}}{-3x^8} = \dfrac{-54}{-3} \cdot \dfrac{x^{11}}{x^8} = 18x^{11-8} = 18x^3$

Check: We multiply.
$18x^3(-3x^8) = -54x^{11}$

Exercise Set 4.8

7. $\dfrac{64a^5b^4}{16a^2b^3} = \dfrac{64}{16} \cdot \dfrac{a^5}{a^2} \cdot \dfrac{b^4}{b^3} = 4a^{5-2}b^{4-3} = 4a^3b$

   Check: We multiply.
   $(4a^3b)(16a^2b^3) = 64a^5b^4$

9. $\dfrac{24x^4 - 4x^3 + x^2 - 16}{8}$

   $= \dfrac{24x^4}{8} - \dfrac{4x^3}{8} + \dfrac{x^2}{8} - \dfrac{16}{8}$

   $= 3x^4 - \dfrac{1}{2}x^3 + \dfrac{1}{8}x^2 - 2$

   Check: We multiply.
   $$\begin{array}{r} 3x^4 - \dfrac{1}{2}x^3 + \dfrac{1}{8}x^2 - 2 \\ \times\, 8 \\ \hline 24x^4 - 4x^3 + x^2 - 16 \end{array}$$

11. $\dfrac{u - 2u^2 - u^5}{u}$

    $= \dfrac{u}{u} - \dfrac{2u^2}{u} - \dfrac{u^5}{u}$

    $= 1 - 2u - u^4$

    Check: We multiply.
    $$\begin{array}{r} 1 - 2u - u^4 \\ \times\, u \\ \hline u - 2u^2 - u^5 \end{array}$$

13. $(15t^3 + 24t^2 - 6t) \div (3t)$

    $= \dfrac{15t^3 + 24t^2 - 6t}{3t}$

    $= \dfrac{15t^3}{3t} + \dfrac{24t^2}{3t} - \dfrac{6t}{3t}$

    $= 5t^2 + 8t - 2$

    Check: We multiply.
    $$\begin{array}{r} 5t^2 + 8t - 2 \\ \times\, 3t \\ \hline 15t^3 + 24t^2 - 6t \end{array}$$

15. $(20x^6 - 20x^4 - 5x^2) \div (-5x^2)$

    $= \dfrac{20x^6 - 20x^4 - 5x^2}{-5x^2}$

    $= \dfrac{20x^6}{-5x^2} - \dfrac{20x^4}{-5x^2} - \dfrac{5x^2}{-5x^2}$

    $= -4x^4 - (-4x^2) - (-1)$

    $= -4x^4 + 4x^2 + 1$

    Check: We multiply.
    $$\begin{array}{r} -4x^4 + 4x^2 + 1 \\ \times\, -5x^2 \\ \hline 20x^6 - 20x^4 - 5x^2 \end{array}$$

17. $(24x^5 - 40x^4 + 6x^3) \div (4x^3)$

    $= \dfrac{24x^5 - 40x^4 + 6x^3}{4x^3}$

    $= \dfrac{24x^5}{4x^3} - \dfrac{40x^4}{4x^3} + \dfrac{6x^3}{4x^3}$

    $= 6x^2 - 10x + \dfrac{3}{2}$

    Check: We multiply.
    $$\begin{array}{r} 6x^2 - 10x + \dfrac{3}{2} \\ \times\, 4x^3 \\ \hline 24x^5 - 40x^4 + 6x^3 \end{array}$$

19. $\dfrac{18x^2 - 5x + 2}{2}$

    $= \dfrac{18x^2}{2} - \dfrac{5x}{2} + \dfrac{2}{2}$

    $= 9x^2 - \dfrac{5}{2}x + 1$

    Check: We multiply.
    $$\begin{array}{r} 9x^2 - \dfrac{5}{2}x + 1 \\ \times\, 2 \\ \hline 18x^2 - 5x + 2 \end{array}$$

21. $\dfrac{12x^3 + 26x^2 + 8x}{2x}$

    $= \dfrac{12x^3}{2x} + \dfrac{26x^2}{2x} + \dfrac{8x}{2x}$

    $= 6x^2 + 13x + 4$

    Check: We multiply.
    $$\begin{array}{r} 6x^2 + 13x + 4 \\ \times\, 2x \\ \hline 12x^3 + 26x^2 + 8x \end{array}$$

23. $\dfrac{9r^2s^2 + 3r^2s - 6rs^2}{3rs}$

    $= \dfrac{9r^2s^2}{3rs} + \dfrac{3r^2s}{3rs} - \dfrac{6rs^2}{3rs}$

    $= 3rs + r - 2s$

    Check: We multiply.
    $$\begin{array}{r} 3rs + r - 2s \\ \times\, 3rs \\ \hline 9r^2s^2 + 3r^2s - 6rs^2 \end{array}$$

25. $$\begin{array}{r} x + 2 \\ x+2\overline{\smash{)}x^2+4x+4} \\ \underline{x^2+2x} \\ 2x+4 \\ \underline{2x+4} \\ 0 \end{array}$$ ← $(x^2 + 4x) - (x^2 + 2x)$

    ← $(2x + 4) - (2x + 4)$

    The answer is $x + 2$.

27.
$$\begin{array}{r} x-5 \\ x-5\overline{\smash{\big)}\,x^2-10x-25} \\ \underline{x^2-5x} \\ -5x-25 \leftarrow (x^2-10x)-(x^2-5x) \\ \underline{-5x+25} \\ -50 \leftarrow (-5x-25)-(-5x+25) \end{array}$$

The answer is $x-5+\dfrac{-50}{x-5}$.

29.
$$\begin{array}{r} x-2 \\ x+6\overline{\smash{\big)}\,x^2+4x-14} \\ \underline{x^2+6x} \\ -2x-14 \leftarrow (x^2+4x)-(x^2+6x) \\ \underline{-2x-12} \\ -2 \leftarrow (-2x-14)-(-2x-12) \end{array}$$

The answer is $x-2+\dfrac{-2}{x+6}$.

31.
$$\begin{array}{r} x-3 \\ x+3\overline{\smash{\big)}\,x^2+0x-9} \leftarrow \text{Filling in the missing term} \\ \underline{x^2+3x} \\ -3x-9 \leftarrow x^2-(x^2+3x) \\ \underline{-3x-9} \\ 0 \leftarrow (-3x-9)-(-3x-9) \end{array}$$

The answer is $x-3$.

33.
$$\begin{array}{r} x^4-x^3+x^2-x+1 \\ x+1\overline{\smash{\big)}\,x^5+0x^4+0x^3+0x^2+0x+1} \leftarrow \text{Filling in missing terms} \\ \underline{x^5+x^4} \\ -x^4 \leftarrow x^5-(x^5+x^4) \\ \underline{-x^4-x^3} \\ x^3 \leftarrow -x^4-(-x^4-x^3) \\ \underline{x^3+x^2} \\ -x^2 \leftarrow x^3-(x^3+x^2) \\ \underline{-x^2-x} \\ x+1 \leftarrow -x^2-(-x^2-x) \\ \underline{x+1} \\ 0 \leftarrow (x+1)-(x+1) \end{array}$$

The answer is $x^4-x^3+x^2-x+1$.

35.
$$\begin{array}{r} 2x^2-7x+4 \\ 4x+3\overline{\smash{\big)}\,8x^3-22x^2-5x+12} \\ \underline{8x^3+6x^2} \\ -28x^2-5x \leftarrow (8x^3-22x^2)-(8x^3+6x^2) \\ \underline{-28x^2-21x} \\ 16x+12 \leftarrow (-28x^2-5x)- \\ (-28x^2-21x) \\ \underline{16x+12} \\ 0 \leftarrow (16x+12)-(16x+12) \end{array}$$

The answer is $2x^2-7x+4$.

37.
$$\begin{array}{r} x^3-6 \\ x^3-7\overline{\smash{\big)}\,x^6-13x^3+42} \\ \underline{x^6-7x^3} \\ -6x^3+42 \leftarrow (x^6-13x^3)-(x^6-7x^3) \\ \underline{-6x^3+42} \\ 0 \leftarrow (-6x^3+42)-(-6x^3+42) \end{array}$$

The answer is x^3-6.

39.
$$\begin{array}{r} x^3+2x^2+4x+8 \\ x-2\overline{\smash{\big)}\,x^4+0x^3+0x^2+0x-16} \\ \underline{x^4-2x^3} \\ 2x^3 \leftarrow x^4-(x^4-2x^3) \\ \underline{2x^3-4x^2} \\ 4x^2 \leftarrow 2x^3-(2x^3-4x^2) \\ \underline{4x^2-8x} \\ 8x-16 \leftarrow 4x^2-(4x^2-8x) \\ \underline{8x-16} \\ 0 \leftarrow (8x-16)-(8x-16) \end{array}$$

The answer is x^3+2x^2+4x+8.

41.
$$\begin{array}{r} t^2+1 \\ t-1\overline{\smash{\big)}\,t^3-t^2+t-1} \\ \underline{t^3-t^2} \leftarrow (t^3-t^2)-(t^3-t^2) \\ 0+t-1 \\ \underline{t-1} \leftarrow (t-1)-(t-1) \\ 0 \end{array}$$

The answer is t^2+1.

43. Discussion and Writing Exercise

45. $17-45 = 17+(-45) = -28$

47. $-2.3-(-9.1) = -2.3+9.1 = 6.8$

49. Familiarize. Let w = the width. Then $w+15$ = the length. We draw a picture.

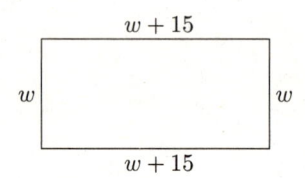

```
           w + 15
      ┌─────────────┐
    w │             │ w
      └─────────────┘
           w + 15
```

We will use the fact that the perimeter is 640 ft to find w (the width). Then we can find $w+15$ (the length) and multiply the length and the width to find the area.

Translate.

Width + Width + Length + Length = Perimeter
$w + w + (w+15) + (w+15) = 640$

Solve.
$$w+w+(w+15)+(w+15) = 640$$
$$4w+30 = 640$$
$$4w = 610$$
$$w = 152.5$$

If the width is 152.5, then the length is $152.5+15$, or 167.5. The area is $(167.5)(152.5)$, or $25{,}543.75$ ft^2.

Check. The length, 167.5 ft, is 15 ft greater than the width, 152.5 ft. The perimeter is $152.5+152.5+167.5+167.5$, or 640 ft. We should also recheck the computation we used to find the area. The answer checks.

State. The area is $25{,}543.75$ ft^2.

Exercise Set 4.8

51.
$$-6(2-x) + 10(5x-7) = 10$$
$$-12 + 6x + 50x - 70 = 10$$
$$56x - 82 = 10 \quad \text{Collecting like terms}$$
$$56x - 82 + 82 = 10 + 82 \quad \text{Adding 82}$$
$$56x = 92$$
$$\frac{56x}{56} = \frac{92}{56} \quad \text{Dividing by 56}$$
$$x = \frac{23}{14}$$

The solution is $\frac{23}{14}$.

53. $4x - 12 + 24y = 4 \cdot x - 4 \cdot 3 + 4 \cdot 6y = 4(x - 3 + 6y)$

55.
$$\begin{array}{r} x^2 + 5 \\ x^2 + 4 \overline{) x^4 + 9x^2 + 20} \\ \underline{x^4 + 4x^2} \\ 5x^2 + 20 \\ \underline{5x^2 + 20} \\ 0 \end{array}$$

The answer is $x^2 + 5$.

57.
$$\begin{array}{r} a + 3 \\ 5a^2 - 7a - 2 \overline{) 5a^3 + 8a^2 - 23a - 1} \\ \underline{5a^3 - 7a^2 - 2a} \\ 15a^2 - 21a - 1 \\ \underline{15a^2 - 21a - 6} \\ 5 \end{array}$$

The answer is $a + 3 + \dfrac{5}{5a^2 - 7a - 2}$.

59. We rewrite the dividend in descending order.
$$\begin{array}{r} 2x^2 + x - 3 \\ 3x^3 - 2x - 1 \overline{) 6x^5 + 3x^4 - 13x^3 - 4x^2 + 5x + 3} \\ \underline{6x^5 \qquad - 4x^3 - 2x^2} \\ 3x^4 - 9x^3 - 2x^2 + 5x \\ \underline{3x^4 \qquad - 2x^2 - x} \\ -9x^3 \qquad + 6x + 3 \\ \underline{-9x^3 \qquad + 6x + 3} \\ 0 \end{array}$$

The answer is $2x^2 + x - 3$.

61.
$$\begin{array}{r} a^5 + a^4b + a^3b^2 + a^2b^3 + ab^4 + b^5 \\ a - b \overline{) a^6 + 0a^5b + 0a^4b^2 + 0a^3b^3 + 0a^2b^4 + 0ab^5 - b^6} \\ \underline{a^6 - a^5b} \\ a^5b \\ \underline{a^5b - a^4b^2} \\ a^4b^2 \\ \underline{a^4b^2 - a^3b^3} \\ a^3b^3 \\ \underline{a^3b^3 - a^2b^4} \\ a^2b^4 \\ \underline{a^2b^4 - ab^5} \\ ab^5 - b^6 \\ \underline{ab^5 - b^6} \\ 0 \end{array}$$

The answer is $a^5 + a^4b + a^3b^2 + a^2b^3 + ab^4 + b^5$.

63.
$$\begin{array}{r} x + 5 \\ x - 1 \overline{) x^2 + 4x + c} \\ \underline{x^2 - x} \\ 5x + c \\ \underline{5x - 5} \\ c + 5 \end{array}$$

We set the remainder equal to 0.
$$c + 5 = 0$$
$$c = -5$$

Thus, c must be -5.

65.
$$\begin{array}{r} c^2x + (-2c + c^2) \\ x - 1 \overline{) c^2x^2 - 2cx + 1} \\ \underline{c^2x^2 - c^2x} \\ (-2c + c^2)x + 1 \\ \underline{(-2c + c^2)x - (-2c + c^2)} \\ 1 + (-2c + c^2) \end{array}$$

We set the remainder equal to 0.
$$c^2 - 2c + 1 = 0$$
$$(c - 1)^2 = 0$$
$$c = 1$$

Thus, c must be 1.

Chapter 5

Polynomials: Factoring

Exercise Set 5.1

1. Answers may vary. $8x^3 = (4x^2)(2x) = (-8)(-x^3) = (2x^2)(4x)$

3. Answers may vary. $-10a^6 = (-5a^5)(2a) = (10a^3)(-a^3) = (-2a^2)(5a^4)$

5. Answers may vary. $24x^4 = (6x)(4x^3) = (-3x^2)(-8x^2) = (2x^3)(12x)$

7. $x^2 - 6x = x \cdot x - x \cdot 6$ Factoring each term
   $ = x(x-6)$ Factoring out the common factor x

9. $2x^2 + 6x = 2x \cdot x + 2x \cdot 3$ Factoring each term
   $ = 2x(x+3)$ Factoring out the common factor $2x$

11. $x^3 + 6x^2 = x^2 \cdot x + x^2 \cdot 6$ Factoring each term
    $ = x^2(x+6)$ Factoring out x^2

13. $8x^4 - 24x^2 = 8x^2 \cdot x^2 - 8x^2 \cdot 3$
    $ = 8x^2(x^2 - 3)$ Factoring out $8x^2$

15. $2x^2 + 2x - 8 = 2 \cdot x^2 + 2 \cdot x - 2 \cdot 4$
    $ = 2(x^2 + x - 4)$ Factoring out 2

17. $17x^5y^3 + 34x^3y^2 + 51xy$
    $= 17xy \cdot x^4y^2 + 17xy \cdot 2x^2y + 17xy \cdot 3$
    $= 17xy(x^4y^2 + 2x^2y + 3)$

19. $6x^4 - 10x^3 + 3x^2 = x^2 \cdot 6x^2 - x^2 \cdot 10x + x^2 \cdot 3$
    $ = x^2(6x^2 - 10x + 3)$

21. $x^5y^5 + x^4y^3 + x^3y^3 - x^2y^2$
    $= x^2y^2 \cdot x^3y^3 + x^2y^2 \cdot x^2y + x^2y^2 \cdot xy + x^2y^2(-1)$
    $= x^2y^2(x^3y^3 + x^2y + xy - 1)$

23. $2x^7 - 2x^6 - 64x^5 + 4x^3$
    $= 2x^3 \cdot x^4 - 2x^3 \cdot x^3 - 2x^3 \cdot 32x^2 + 2x^3 \cdot 2$
    $= 2x^3(x^4 - x^3 - 32x^2 + 2)$

25. $1.6x^4 - 2.4x^3 + 3.2x^2 + 6.4x$
    $= 0.8x(2x^3) - 0.8x(3x^2) + 0.8x(4x) + 0.8x(8)$
    $= 0.8x(2x^3 - 3x^2 + 4x + 8)$

27. $\frac{5}{3}x^6 + \frac{4}{3}x^5 + \frac{1}{3}x^4 + \frac{1}{3}x^3$
    $= \frac{1}{3}x^3(5x^3) + \frac{1}{3}x^3(4x^2) + \frac{1}{3}x^3(x) + \frac{1}{3}x^3(1)$
    $= \frac{1}{3}x^3(5x^3 + 4x^2 + x + 1)$

29. Factor: $x^2(x+3) + 2(x+3)$
    The binomial $x+3$ is common to both terms:
    $x^2(x+3) + 2(x+3) = (x^2 + 2)(x+3)$

31. $5a^3(2a - 7) - (2a - 7)$
    $= 5a^3(2a - 7) - 1(2a - 7)$
    $= (5a^3 - 1)(2a - 7)$

33. $x^3 + 3x^2 + 2x + 6$
    $= (x^3 + 3x^2) + (2x + 6)$
    $= x^2(x+3) + 2(x+3)$ Factoring each binomial
    $= (x^2 + 2)(x+3)$ Factoring out the common factor $x+3$

35. $2x^3 + 6x^2 + x + 3$
    $= (2x^3 + 6x^2) + (x + 3)$
    $= 2x^2(x+3) + 1(x+3)$ Factoring each binomial
    $= (2x^2 + 1)(x+3)$

37. $8x^3 - 12x^2 + 6x - 9 = 4x^2(2x-3) + 3(2x-3)$
    $ = (4x^2 + 3)(2x - 3)$

39. $12x^3 - 16x^2 + 3x - 4$
    $= 4x^2(3x - 4) + 1(3x - 4)$ Factoring 1 out of the second binomial
    $= (4x^2 + 1)(3x - 4)$

41. $5x^3 - 5x^2 - x + 1$
    $= (5x^3 - 5x^2) + (-x + 1)$
    $= 5x^2(x-1) - 1(x-1)$ Check: $-1(x-1) = -x+1$
    $= (5x^2 - 1)(x - 1)$

43. $x^3 + 8x^2 - 3x - 24 = x^2(x+8) - 3(x+8)$
    $ = (x^2 - 3)(x + 8)$

45. $2x^3 - 8x^2 - 9x + 36 = 2x^2(x - 4) - 9(x - 4)$
    $ = (2x^2 - 9)(x - 4)$

47. Discussion and Writing Exercise

49. $-2x < 48$
    $x > -24$ Dividing by -2 and reversing the inequality symbol
    The solution set is $\{x | x > -24\}$.

51. $\dfrac{-108}{-4} = 27$ (The quotient of two negative numbers is positive.)

53. $(y+5)(y+7) = y^2 + 7y + 5y + 35$ Using FOIL
    $ = y^2 + 12y + 35$

55. $(y+7)(y-7) = y^2 - 7^2 = y^2 - 49$
$$[(A+B)(A-B) = A^2 - B^2]$$

57. $x + y = 4$

To find the x-intercept, let $y = 0$. Then solve for x.
$$x + y = 4$$
$$x + 0 = 4$$
$$x = 4$$
The x-intercept is $(4, 0)$.

To find the y-intercept, let $x = 0$. Then solve for y.
$$x + y = 4$$
$$0 + y = 4$$
$$y = 4$$
The y-intercept is $(0, 4)$.

Plot these points and draw the line.

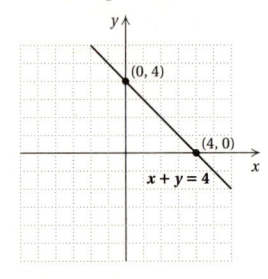

A third point should be used as a check. We substitute any value for x and solve for y. We let $x = 2$. Then
$$x + y = 4$$
$$2 + y = 4$$
$$y = 2$$
The point $(2, 2)$ is on the graph, so the graph is probably correct.

59. $5x - 3y = 15$

To find the x-intercept, let $y = 0$. Then solve for x.
$$5x - 3y = 15$$
$$5x - 3 \cdot 0 = 15$$
$$5x = 15$$
$$x = 3$$
The x-intercept is $(3, 0)$.

To find the y-intercept, let $x = 0$. Then solve for y.
$$5x - 3y = 15$$
$$5 \cdot 0 - 3y = 15$$
$$-3y = 15$$
$$y = -5$$
The y-intercept is $(0, -5)$.

Plot these points and draw the line.

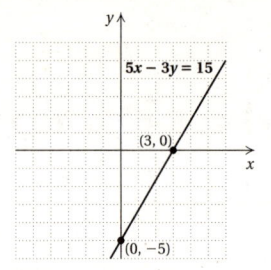

A third point should be used as a check. We substitute any value for x and solve for y. We let $x = 6$. Then
$$5x - 3y = 15$$
$$5 \cdot 6 - 3y = 15$$
$$30 - 3y = 15$$
$$-3y = -15$$
$$y = 5$$
The point $(6, 5)$ is on the graph, so the graph is probably correct.

61. $4x^5 + 6x^3 + 6x^2 + 9 = 2x^3(2x^2 + 3) + 3(2x^2 + 3)$
$$= (2x^3 + 3)(2x^2 + 3)$$

63. $x^{12} + x^7 + x^5 + 1 = x^7(x^5 + 1) + (x^5 + 1)$
$$= (x^7 + 1)(x^5 + 1)$$

65. $p^3 + p^2 - 3p + 10 = p^2(p + 1) - (3p - 10)$

This polynomial is not factorable using factoring by grouping.

Exercise Set 5.2

1. $x^2 + 8x + 15$

Since the constant term and coefficient of the middle term are both positive, we look for a factorization of 15 in which both factors are positive. Their sum must be 8.

Pairs of factors	Sums of factors
1, 15	16
3, 5	8

The numbers we want are 3 and 5.

$x^2 + 8x + 15 = (x + 3)(x + 5)$.

3. $x^2 + 7x + 12$

Since the constant term is positive and the coefficient of the middle term is positive, we look for a factorization of 12 in which both factors are positive. Their sum must be 7.

Pairs of factors	Sums of factors
1, 12	13
2, 6	8
3, 4	7

The numbers we want are 3 and 4.

$x^2 + 7x + 12 = (x + 3)(x + 4)$.

Exercise Set 5.2

5. $x^2 - 6x + 9$

Since the constant term is positive and the coefficient of the middle term is negative, we look for a factorization of 9 in which both factors are negative. Their sum must be -6.

Pairs of factors	Sums of factors
$-1, -9$	-10
$-3, -3$	-6

The numbers we want are -3 and -3.
$x^2 - 6x + 9 = (x-3)(x-3)$, or $(x-3)^2$.

7. $x^2 - 5x - 14$

Since the constant term is negative, we look for a factorization of -14 in which one factor is positive and one factor is negative. Their sum must be -5, the coefficient of the middle term.

Pairs of factors	Sums of factors
$-1, 14$	13
$1, -14$	-13
$-2, 7$	5
$2, -7$	-5

The numbers we want are 2 and -7.
$x^2 - 5x - 14 = (x+2)(x-7)$.

9. $b^2 + 5b + 4$

Since the constant term is positive and the coefficient of the middle term is positive, we look for a factorization of 4 in which both factors are positive. Their sum must be 5.

Pairs of factors	Sums of factors
1, 4	5
2, 2	4

The numbers we want are 1 and 4.
$b^2 + 5b + 4 = (b+1)(b+4)$.

11. $x^2 + \dfrac{2}{3}x + \dfrac{1}{9}$

Since the constant term is positive and the coefficient of the middle term is positive, we look for a factorization of $\dfrac{1}{9}$ in which both factors are positive. Their sum must be $\dfrac{2}{3}$.

Pairs of factors	Sums of factors
$1, \dfrac{1}{9}$	$\dfrac{10}{9}$
$\dfrac{1}{3}, \dfrac{1}{3}$	$\dfrac{2}{3}$

The numbers we want are $\dfrac{1}{3}$ and $\dfrac{1}{3}$.
$x^2 + \dfrac{2}{3}x + \dfrac{1}{9} = \left(x + \dfrac{1}{3}\right)\left(x + \dfrac{1}{3}\right)$, or $\left(x + \dfrac{1}{3}\right)^2$.

13. $d^2 - 7d + 10$

Since the constant term is positive and the coefficient of the middle term is negative, we look for a factorization of 10 in which both factors are negative. Their sum must be -7.

Pairs of factors	Sums of factors
$-1, -10$	-11
$-2, -5$	-7

The numbers we want are -2 and -5.
$d^2 - 7d + 10 = (d-2)(d-5)$.

15. $y^2 - 11y + 10$

Since the constant term is positive and the coefficient of the middle term is negative, we look for a factorization of 10 in which both factors are negative. Their sum must be -11.

Pairs of factors	Sums of factors
$-1, -10$	-11
$-2, -5$	-7

The numbers we want are -1 and -10.
$y^2 - 11y + 10 = (y-1)(y-10)$.

17. $x^2 + x + 1$

Since the constant term and the coefficient of the middle term are both positive, we look for a factorization of 1 in which both factors are positive. The sum must be 1. The only possible pair of factors is 1 and 1, but their sum is not 1. Thus, this polynomial is not factorable into binomials. It is prime.

19. $x^2 - 7x - 18$

Since the constant term is negative, we look for a factorization of -18 in which one factor is positive and one factor is negative. Their sum must be -7, the coefficient of the middle term.

Pairs of factors	Sums of factors
$-1, 18$	17
$1, -18$	-17
$-2, 9$	7
$2, -9$	-7
$-3, 6$	3
$3, -6$	-3

The numbers we want are 2 and -9.
$x^2 - 7x - 18 = (x+2)(x-9)$.

21. $x^3 - 6x^2 - 16x = x(x^2 - 6x - 16)$

After factoring out the common factor, x, we consider $x^2 - 6x - 16$. Since the constant term is negative, we look for a factorization of -16 in which one factor is positive and one factor is negative. Their sum must be -6, the coefficient of the middle term.

Pairs of factors	Sums of factors
−1, 16	15
1, −16	−15
−2, 8	6
2, −8	−6
−4, 4	0

The numbers we want are 2 and −8.
Then $x^2 - 6x - 16 = (x+2)(x-8)$, so $x^3 - 6x^2 - 16x = x(x+2)(x-8)$.

23. $y^3 - 4y^2 - 45y = y(y^2 - 4y - 45)$

After factoring out the common factor, y, we consider $y^2 - 4y - 45$. Since the constant term is negative, we look for a factorization of −45 in which one factor is positive and one factor is negative. Their sum must be −4, the coefficient of the middle term.

Pairs of factors	Sums of factors
−1, 45	44
1, −45	−44
−3, 15	12
3, −15	−12
−5, 9	4
5, −9	−4

The numbers we want are 5 and −9.
Then $y^2 - 4y - 45 = (y+5)(y-9)$, so $y^3 - 4y^2 - 45y = y(y+5)(y-9)$.

25. $-2x - 99 + x^2 = x^2 - 2x - 99$

Since the constant term is negative, we look for a factorization of −99 in which one factor is positive and one factor is negative. Their sum must be −2, the coefficient of the middle term.

Pairs of factors	Sums of factors
−1, 99	98
1, −99	−98
−3, 33	30
3, −33	−30
−9, 11	2
9, −11	−2

The numbers we want are 9 and −11.
$-2x - 99 + x^2 = (x+9)(x-11)$.

27. $c^4 + c^2 - 56$

Consider this trinomial as $(c^2)^2 + c^2 - 56$. We look for numbers p and q such that $c^4 + c^2 - 56 = (c^2 + p)(c^2 + q)$. Since the constant term is negative, we look for a factorization of −56 in which one factor is positive and one factor is negative. Their sum must be 1.

Pairs of factors	Sums of factors
−1, 56	55
1, −56	−55
−2, 28	26
2, −28	−26
−4, 14	12
4, −14	−12
−7, 8	1
7, −8	−1

The numbers we want are −7 and 8.
$c^4 + c^2 - 56 = (c^2 - 7)(c^2 + 8)$.

29. $a^4 + 2a^2 - 35$

Consider this trinomial as $(a^2)^2 + 2a^2 - 35$. We look for numbers p and q such that $a^4 + 2a^2 - 35 = (a^2 + p)(a^2 + q)$. Since the constant term is negative, we look for a factorization of −35 in which one factor is positive and one factor is negative. Their sum must be 2.

Pairs of factors	Sums of factors
−1, 35	34
1, −35	−34
−5, 7	2
5, −7	−2

The numbers we want are −5 and 7.
$a^4 + 2a^2 - 35 = (a^2 - 5)(a^2 + 7)$.

31. $x^2 + x - 42$

Since the constant term is negative, we look for a factorization of −42 in which one factor is positive and one factor is negative. Their sum must be 1, the coefficient of the middle term.

Pairs of factors	Sums of factors
−1, 42	41
1, −42	−41
−2, 21	19
2, −21	−19
−3, 14	11
3, −14	−11
−6, 7	1
6, −7	−1

The numbers we want are −6 and 7.
$x^2 + x - 42 = (x-6)(x+7)$.

33. $7 - 2p + p^2 = p^2 - 2p + 7$

Since the constant term is positive and the coefficient of the middle term is negative, we look for a factorization of 7 in which both factors are negative. The sum must be −2. The only possible pair of factors is −1 and −7, but their sum is not −2. Thus, this polynomial is not factorable into binomials. It is prime.

Exercise Set 5.2

35. $x^2 + 20x + 100$

We look for two factors, both positive, whose product is 100 and whose sum is 20.

They are 10 and 10. $10 \cdot 10 = 100$ and $10 + 10 = 20$.

$x^2 + 20x + 100 = (x+10)(x+10)$, or $(x+10)^2$.

37. $30 + 7x - x^2 = -x^2 + 7x + 30 = -1(x^2 - 7x - 30)$

Now we factor $x^2 - 7x - 30$. Since the constant term is negative, we look for a factorization of -30 in which one factor is positive and one factor is negative. Their sum must be -7, the coefficient of the middle term.

Pairs of factors	Sums of factors
-1, 30	29
1, -30	-29
-2, 15	13
2, -15	-13
-3, 10	7
3, -10	-7
-5, 6	1
5, -6	-1

The numbers we want are 3 and -10. Then $x^2 - 7x - 30 = (x+3)(x-10)$, so we have:

$-x^2 + 7x + 30$
$= -1(x+3)(x-10)$
$= (-x-3)(x-10)$ Multiplying $x+3$ by -1
$= (x+3)(-x+10)$ Multiplying $x-10$ by -1

39. $24 - a^2 - 10a = -a^2 - 10a + 24 = -1(a^2 + 10a - 24)$

Now we factor $a^2 + 10a - 24$. Since the constant term is negative, we look for a factorization of -24 in which one factor is positive and one factor is negative. Their sum must be 10, the coefficient of the middle term.

Pairs of factors	Sums of factors
-1, 24	23
1, -24	-23
-2, 12	10
2, -12	-10
-3, 8	5
3, -8	-5
-4, 6	2
4, -6	-2

The numbers we want are -2 and 12. Then $a^2 + 10a - 24 = (a-2)(a+12)$, so we have:

$-a^2 - 10a + 24$
$= -1(a-2)(a+12)$
$= (-a+2)(a+12)$ Multiplying $a-2$ by -1
$= (a-2)(-a-12)$ Multiplying $a+12$ by -1

41. $x^4 - 21x^3 - 100x^2 = x^2(x^2 - 21x - 100)$

After factoring out the common factor, x^2, we consider $x^2 - 21x - 100$. We look for two factors, one positive and one negative, whose product is -100 and whose sum is -21. They are 4 and -25. $4 \cdot (-25) = -100$ and $4 + (-25) = -21$.

Then $x^2 - 21x - 100 = (x+4)(x-25)$, so $x^4 - 21x^3 - 100x^2 = x^2(x+4)(x-25)$.

43. $x^2 - 21x - 72$

We look for two factors, one positive and one negative, whose product is -72 and whose sum is -21. They are 3 and -24.

$x^2 - 21x - 72 = (x+3)(x-24)$.

45. $x^2 - 25x + 144$

We look for two factors, both negative, whose product is 144 and whose sum is -25. They are -9 and -16.

$x^2 - 25x + 144 = (x-9)(x-16)$.

47. $a^2 + a - 132$

We look for two factors, one positive and one negative, whose product is -132 and whose sum is 1. They are -11 and 12.

$a^2 + a - 132 = (a-11)(a+12)$.

49. $120 - 23x + x^2 = x^2 - 23x + 120$

We look for two factors, both negative, whose product is 120 and whose sum is -23. They are -8 and -15.

$x^2 - 23x + 120 = (x-8)(x-15)$.

51. First write the polynomial in descending order and factor out -1.

$108 - 3x - x^2 = -x^2 - 3x + 108 = -1(x^2 + 3x - 108)$

Now we factor the polynomial $x^2 + 3x - 108$. We look for two factors, one positive and one negative, whose product is -108 and whose sum is 3. They are -9 and 12.

$x^2 + 3x - 108 = (x-9)(x+12)$

The final answer must include -1 which was factored out above.

$-x^2 - 3x + 108$
$= -1(x-9)(x+12)$
$= (-x+9)(x+12)$ Multiplying $x-9$ by -1
$= (x-9)(-x-12)$ Multiplying $x+12$ by -1

53. $y^2 - 0.2y - 0.08$

We look for two factors, one positive and one negative, whose product is -0.08 and whose sum is -0.2. They are -0.4 and 0.2.

$y^2 - 0.2y - 0.08 = (y-0.4)(y+0.2)$.

55. $p^2 + 3pq - 10q^2 = p^2 + 3pq - 10q^2$

Think of $3q$ as a "coefficient" of p. Then we look for factors of $-10q^2$ whose sum is $3q$. They are $5q$ and $-2q$.

$p^2 + 3pq - 10q^2 = (p+5q)(p-2q)$.

57. $84 - 8t - t^2 = -t^2 - 8t + 84 = -1(t^2 + 8t - 84)$

Now we factor $t^2 + 8t - 84$. We look for two factors, one positive and one negative, whose product is -84 and whose sum is 8. They are 14 and -6.

Then $t^2 + 8t - 84 = (t+14)(t-6)$, so we have:
$$-t^2 - 8t + 84$$
$$= -1(t+14)(t-6)$$
$$= (-t-14)(t-6) \quad \text{Multiplying } t+14 \text{ by } -1$$
$$= (t+14)(-t+6) \quad \text{Multiplying } t-6 \text{ by } -1$$

59. $m^2 + 5mn + 4n^2 = m^2 + 5nm + 4n^2$

We look for factors of $4n^2$ whose sum is $5n$. They are $4n$ and n.
$$m^2 + 5mn + 4n^2 = (m+4n)(m+n)$$

61. $s^2 - 2st - 15t^2 = s^2 - 2ts - 15t^2$

We look for factors of $-15t^2$ whose sum is $-2t$. They are $-5t$ and $3t$.
$$s^2 - 2st - 15t^2 = (s-5t)(s+3t)$$

63. $6a^{10} - 30a^9 - 84a^8 = 6a^8(a^2 - 5a - 14)$

After factoring out the common factor, $6a^8$, we consider $a^2 - 5a - 14$. We look for two factors, one positive and one negative, whose product is -14 and whose sum is -5. They are 2 and -7.

$a^2 - 5a - 14 = (a+2)(a-7)$, so $6a^{10} - 30a^9 - 84a^8 = 6a^8(a+2)(a-7)$.

65. Discussion and Writing Exercise

67. Discussion and Writing Exercise

69. $8x(2x^2 - 6x + 1) = 8x \cdot 2x^2 - 8x \cdot 6x + 8x \cdot 1 = 16x^3 - 48x^2 + 8x$

71. $(7w + 6)^2 = (7w)^2 + 2 \cdot 7w \cdot 6 + 6^2 = 49w^2 + 84w + 36$

73. $(4w - 11)(4w + 11) = (4w)^2 - (11)^2 = 16w^2 - 121$

75.
$$3x - 8 = 0$$
$$3x = 8 \quad \text{Adding 8 on both sides}$$
$$x = \frac{8}{3} \quad \text{Dividing by 3 on both sides}$$
The solution is $\frac{8}{3}$.

77. Familiarize. Let $n =$ the number of people arrested the year before.

Translate. We reword the problem.

$\underbrace{\text{Number arrested the year before}}_{n} \;\; \underbrace{\text{less}}_{-} \;\; \underbrace{1.2\%}_{1.2\%} \;\; \underbrace{\text{of}}_{\cdot} \;\; \underbrace{\text{that number}}_{n} \;\; \underbrace{\text{is}}_{=} \;\; \underbrace{29,200.}_{29,200}$

Carry out. We solve the equation.
$$n - 1.2\% \cdot n = 29,200$$
$$1 \cdot n - 0.012n = 29,200$$
$$0.988n = 29,200$$
$$n \approx 29,555 \quad \text{Rounding}$$

Check. 1.2% of $29,555$ is $0.012(29,555) \approx 355$ and $29,555 - 355 = 29,200$. The answer checks.

State. Approximately $29,555$ people were arrested the year before.

79. $y^2 + my + 50$

We look for pairs of factors whose product is 50. The sum of each pair is represented by m.

Pairs of factors whose product is -50	Sums of factors
$1, \; 50$	51
$-1, -50$	-51
$2, \; 25$	27
$-2, -25$	-27
$5, \; 10$	15
$-5, -10$	-15

The polynomial $y^2 + my + 50$ can be factored if m is 51, -51, 27, -27, 15, or -15.

81. $x^2 - \dfrac{1}{2}x - \dfrac{3}{16}$

We look for two factors, one positive and one negative, whose product is $-\dfrac{3}{16}$ and whose sum is $-\dfrac{1}{2}$.

They are $-\dfrac{3}{4}$ and $\dfrac{1}{4}$.

$-\dfrac{3}{4} \cdot \dfrac{1}{4} = -\dfrac{3}{16}$ and $-\dfrac{3}{4} + \dfrac{1}{4} = -\dfrac{2}{4} = -\dfrac{1}{2}$.

$$x^2 - \dfrac{1}{2}x - \dfrac{3}{16} = \left(x - \dfrac{3}{4}\right)\left(x + \dfrac{1}{4}\right)$$

83. $x^2 + \dfrac{30}{7}x - \dfrac{25}{7}$

We look for two factors, one positive and one negative, whose product is $-\dfrac{25}{7}$ and whose sum is $\dfrac{30}{7}$.

They are 5 and $-\dfrac{5}{7}$.

$5 \cdot \left(-\dfrac{5}{7}\right) = -\dfrac{25}{7}$ and $5 + \left(-\dfrac{5}{7}\right) = \dfrac{35}{7} + \left(-\dfrac{5}{7}\right) = \dfrac{30}{7}$.

$$x^2 + \dfrac{30}{7}x - \dfrac{25}{7} = (x+5)\left(x - \dfrac{5}{7}\right)$$

85. $b^{2n} + 7b^n + 10$

Consider this trinomial as $(b^n)^2 + 7b^n + 10$. We look for numbers p and q such that $b^{2n} + 7b^n + 10 = (b^n + p)(b^n + q)$. We find two factors, both positive, whose product is 10 and whose sum is 7. They are 5 and 2.
$$b^{2n} + 7b^n + 10 = (b^n + 5)(b^n + 2)$$

87. We first label the drawing with additional information.

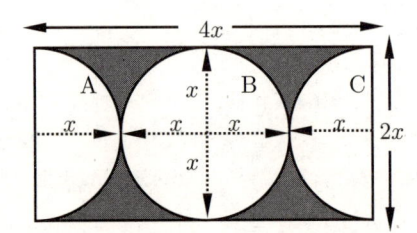

$4x$ represents the length of the rectangle and $2x$ the width. The area of the rectangle is $4x \cdot 2x$, or $8x^2$.

The area of semicircle A is $\frac{1}{2}\pi x^2$.

The area of circle B is πx^2.

The area of semicircle C is $\frac{1}{2}\pi x^2$.

$$\begin{aligned}\text{Area of} \atop \text{shaded region} &= \text{Area of} \atop \text{rectangle} - \text{Area} \atop \text{of } A - \text{Area} \atop \text{of } B - \text{Area} \atop \text{of } C \\ \text{Area of} \atop \text{shaded region} &= 8x^2 - \frac{1}{2}\pi x^2 - \pi x^2 - \frac{1}{2}\pi x^2 \\ &= 8x^2 - 2\pi x^2 \\ &= 2x^2(4-\pi)\end{aligned}$$

The shaded area can be represented by $2x^2(4-\pi)$.

89. First consider all the factorizations of 36 that contain three factors. We also find the sum of the factors in each factorization.

Factorization	Sum of Factors
$1 \cdot 1 \cdot 36$	38
$1 \cdot 2 \cdot 18$	21
$1 \cdot 3 \cdot 12$	16
$1 \cdot 4 \cdot 9$	14
$1 \cdot 6 \cdot 6$	13
$2 \cdot 2 \cdot 9$	13
$2 \cdot 3 \cdot 6$	11
$3 \cdot 3 \cdot 4$	10

We can conclude that the number on the house next door is 13, because two sums are 13. This is what causes the census taker to be puzzled. She cannot determine which trio of factors gives the children's ages. When the mother supplies the additional information that there is an oldest child, the census taker knows that the ages of the children cannot be 1, 6, and 6 because there is not an oldest child in this group. Therefore, the children's ages must be 2, 2, and 9.

Exercise Set 5.3

1. $2x^2 - 7x - 4$

(1) Look for a common factor. There is none (other than 1 or -1).

(2) Factor the first term, $2x^2$. The only possibility is $2x, x$. The desired factorization is of the form:
$$(2x+\quad)(x+\quad)$$

(3) Factor the last term, -4, which is negative. The possibilities are $-4, 1$ and $4, -1$ and $2, -2$.

(4) Look for combinations of factors from steps (2) and (3) such that the sum of their products is the middle term, $-7x$. We try some possibilities:
$$\begin{aligned}(2x-4)(x+1) &= 2x^2 - 2x - 4 \\ (2x+4)(x-1) &= 2x^2 + 2x - 4 \\ (2x+2)(x-2) &= 2x^2 - 2x - 4 \\ (2x+1)(x-4) &= 2x^2 - 7x - 4\end{aligned}$$

The factorization is $(2x+1)(x-4)$.

3. $5x^2 - x - 18$

(1) There is no common factor (other than 1 or -1).

(2) Factor the first term, $5x^2$. The only possibility is $5x, x$. The desired factorization is of the form:
$$(5x+\quad)(x+\quad)$$

(3) Factor the last term, -18. The possibilities are $-18, 1$ and $18, -1$ and $-9, 2$ and $9, -2$ and $-6, 3$ and $6, -3$.

(4) Look for combinations of factors from steps (2) and (3) such that the sum of their products is the middle term, x. We try some possibilities:
$$\begin{aligned}(5x-18)(x+1) &= 5x^2 - 13x - 18 \\ (5x+18)(x-1) &= 5x^2 + 13x - 18 \\ (5x+9)(x-2) &= 5x^2 - x - 18\end{aligned}$$

The factorization is $(5x+9)(x-2)$.

5. $6x^2 + 23x + 7$

(1) There is no common factor (other than 1 or -1).

(2) Factor the first term, $6x^2$. The possibilities are $6x, x$ and $3x, 2x$. We have these as possibilities for factorizations:
$$(6x+\quad)(x+\quad) \text{ and } (3x+\quad)(2x+\quad)$$

(3) Factor the last term, 7. The possibilities are 7, 1 and $-7, -1$.

(4) Look for combinations of factors from steps (2) and (3) such that the sum of their products is the middle term, $23x$. Since all signs are positive, we need consider only plus signs. We try some possibilities:
$$\begin{aligned}(6x+7)(x+1) &= 6x^2 + 13x + 7 \\ (3x+7)(2x+1) &= 6x^2 + 17x + 7 \\ (6x+1)(x+7) &= 6x^2 + 43x + 7 \\ (3x+1)(2x+7) &= 6x^2 + 23x + 7\end{aligned}$$

The factorization is $(3x+1)(2x+7)$.

7. $3x^2 + 4x + 1$

(1) There is no common factor (other than 1 or -1).

(2) Factor the first term, $3x^2$. The only possibility is $3x, x$. The desired factorization is of the form:
$$(3x+\quad)(x+\quad)$$

(3) Factor the last term, 1. The possibilities are 1, 1 and $-1, -1$.

(4) Look for combinations of factors from steps (2) and (3) such that the sum of their products is the middle term, $4x$. Since all signs are positive, we need consider only plus signs. There is only one such possibility:
$$(3x+1)(x+1) = 3x^2 + 4x + 1$$

The factorization is $(3x+1)(x+1)$.

9. $4x^2 + 4x - 15$

(1) There is no common factor (other than 1 or -1).

(2) Factor the first term, $4x^2$. The possibilities are $4x$, x and $2x$, $2x$. We have these as possibilities for factorizations:
$$(4x+\quad)(x+\quad) \text{ and } (2x+\quad)(2x+\quad)$$

(3) Factor the last term, -15. The possibilities are 15, -1 and -15, 1 and 5, -3 and -5, 3.

(4) We try some possibilities:
$$(4x+15)(x-1) = 4x^2 + 11x - 15$$
$$(2x+15)(2x-1) = 4x^2 + 28x - 15$$
$$(4x-15)(x+1) = 4x^2 - 11x - 15$$
$$(2x-15)(2x+1) = 4x^2 - 28x - 15$$
$$(4x+5)(x-3) = 4x^2 - 7x - 15$$
$$(2x+5)(2x-3) = 4x^2 + 4x - 15$$

The factorization is $(2x+5)(2x-3)$.

11. $2x^2 - x - 1$

(1) There is no common factor (other than 1 or -1).

(2) Factor the first term, $2x^2$. The only possibility is $2x$, x. The desired factorization is of the form:
$$(2x+\quad)(x+\quad)$$

(3) Factor the last term, -1. The only possibility is -1, 1.

(4) We try the possibilities:
$$(2x-1)(x+1) = 2x^2 + x - 1$$
$$(2x+1)(x-1) = 2x^2 - x - 1$$

The factorization is $(2x+1)(x-1)$.

13. $9x^2 + 18x - 16$

(1) There is no common factor (other than 1 or -1).

(2) Factor the first term, $9x^2$. The possibilities are $9x$, x and $3x$, $3x$. We have these as possibilities for factorizations:
$$(9x+\quad)(x+\quad) \text{ and } (3x+\quad)(3x+\quad)$$

(3) Factor the last term, -16. The possibilities are 16, -1 and -16, 1 and 8, -2 and -8, 2 and 4, -4.

(4) We try some possibilities:
$$(9x+16)(x-1) = 9x^2 + 7x - 16$$
$$(3x+16)(3x-1) = 9x^2 + 45x - 16$$
$$(9x-16)(x+1) = 9x^2 - 7x - 16$$
$$(3x-16)(3x+1) = 9x^2 - 45x - 16$$
$$(9x+8)(x-2) = 9x^2 - 10x - 16$$
$$(3x+8)(3x-2) = 9x^2 + 18x - 16$$

The factorization is $(3x+8)(3x-2)$.

15. $3x^2 - 5x - 2$

(1) There is no common factor (other than 1 or -1).

(2) Factor the first term, $3x^2$. The only possibility is $3x$, x. The desired factorization is of the form:
$$(3x+\quad)(x+\quad)$$

(3) Factor the last term, -2. The possibilities are 2, -1 and -2 and 1.

(4) We try some possibilities:
$$(3x+2)(x-1) = 3x^2 - x - 2$$
$$(3x-2)(x+1) = 3x^2 + x - 2$$
$$(3x-1)(x+2) = 3x^2 + 5x - 2$$
$$(3x+1)(x-2) = 3x^2 - 5x - 2$$

The factorization is $(3x+1)(x-2)$.

17. $12x^2 + 31x + 20$

(1) There is no common factor (other than 1 or -1).

(2) Factor the first term, $12x^2$. The possibilities are $12x$, x and $6x$, $2x$ and $4x$, $3x$. We have these as possibilities for factorizations:
$$(12x+\quad)(x+\quad) \text{ and } (6x+\quad)(2x+\quad) \text{ and}$$
$$(4x+\quad)(3x+\quad)$$

(3) Factor the last term, 20. Since all signs are positive, we need consider only positive pairs of factors. Those factor pairs are 20, 1 and 10, 2 and 5, 4.

(4) We can immediately reject all possibilities in which either factor has a common factor, such as $(12x+20)$ or $(6x+4)$, because we determined at the outset that there are no common factors. We try some of the remaining possibilities:
$$(12x+1)(x+20) = 12x^2 + 241x + 20$$
$$(12x+5)(x+4) = 12x^2 + 53x + 20$$
$$(6x+1)(2x+20) = 12x^2 + 122x + 20$$
$$(4x+5)(3x+4) = 12x^2 + 31x + 20$$

The factorization is $(4x+5)(3x+4)$.

19. $14x^2 + 19x - 3$

(1) There is no common factor (other than 1 or -1).

(2) Factor the first term, $14x^2$. The possibilities are $14x$, x and $7x$, $2x$. We have these as possibilities for factorizations:
$$(14x+\quad)(x+\quad) \text{ and } (7x+\quad)(2x+\quad)$$

(3) Factor the last term, -3. The possibilities are -1, 3 and 3, -1.

(4) We try some possibilities:
$$(14x-1)(x+3) = 14x^2 + 41x - 3$$
$$(7x-1)(2x+3) = 7x^2 + 19x - 3$$

The factorization is $(7x-1)(2x+3)$.

Exercise Set 5.3

21. $9x^2 + 18x + 8$

(1) There is no common factor (other than 1 or -1).

(2) Factor the first term, $9x^2$. The possibilities are $9x$, x and $3x$, $3x$. We have these as possibilities for factorizations:
$$(9x+\ )(x+\ ) \text{ and } (3x+\ )(3x+\ )$$

(3) Factor the last term, 8. Since all signs are positive, we need consider only positive pairs of factors. Those factor pairs are 8, 1 and 4, 2.

(4) We try some possibilities:
$$(9x+8)(x+1) = 9x^2 + 17x + 8$$
$$(3x+8)(3x+1) = 9x^2 + 27x + 8$$
$$(9x+4)(x+2) = 9x^2 + 22x + 8$$
$$(3x+4)(3x+2) = 9x^2 + 18x + 8$$

The factorization is $(3x+4)(3x+2)$.

23. $49 - 42x + 9x^2 = 9x^2 - 42x + 49$

(1) There is no common factor (other than 1 or -1).

(2) Factor the first term, $9x^2$. The possibilities are $9x$, x and $3x$, $3x$. We have these as possibilities for factorizations:
$$(9x+\ )(x+\ ) \text{ and } (3x+\ )(3x+\ )$$

(3) Factor 49. Since 49 is positive and the middle term is negative, we need consider only negative pairs of factors. Those factor pairs are $-49, -1$ and $-7, -7$.

(4) We try some possibilities:
$$(9x-49)(x-1) = 9x^2 - 58x + 49$$
$$(3x-49)(3x-1) = 9x^2 - 150x + 49$$
$$(9x-7)(x-7) = 9x^2 - 70x + 49$$
$$(3x-7)(3x-7) = 9x^2 - 42x + 49$$

The factorization is $(3x-7)(3x-7)$, or $(3x-7)^2$. This can also be expressed as follows:
$$(3x-7)^2 = (-1)^2(3x-7)^2 = [-1 \cdot (3x-7)]^2 =$$
$$(-3x+7)^2, \text{ or } (7-3x)^2$$

25. $24x^2 + 47x - 2$

(1) There is no common factor (other than 1 or -1).

(2) Factor the first term, $24x^2$. The possibilities are $24x$, x and $12x$, $2x$ and $6x$, $4x$ and $3x$, $8x$. We have these as possibilities for factorizations:
$$(24x+\ )(x+\ ) \text{ and } (12x+\ )(2x+\ ) \text{ and}$$
$$(6x+\ )(4x+\ ) \text{ and } (3x+\ )(8x+\ )$$

(3) Factor the last term, -2. The possibilities are 2, -1 and -2, 1.

(4) We can immediately reject all possibilities in which either factor has a common factor, such as $(24x+2)$ or $(12x-2)$, because we determined at the outset that there are no common factors. We try some of the remaining possibilities:
$$(24x-1)(x+2) = 24x^2 + 47x - 2$$

The factorization is $(24x-1)(x+2)$.

27. $35x^2 - 57x - 44$

(1) There is no common factor (other than 1 or -1).

(2) Factor the first term, $35x^2$. The possibilities are $35x$, x and $7x$, $5x$. We have these as possibilities for factorizations:
$$(35x+\ )(x+\ ) \text{ and } (7x+\ )(5x+\ )$$

(3) Factor the last term, -44. The possibilities are 1, -44 and -1, 44 and 2, -22 and -2, 22 and 4, -11, and -4, 11.

(4) We try some possibilities:
$$(35x+1)(x-44) = 35x^2 - 1539x - 44$$
$$(7x+1)(5x-44) = 35x^2 - 303x - 44$$
$$(35x+2)(x-22) = 35x^2 - 768x - 44$$
$$(7x+2)(5x-22) = 35x^2 - 144x - 44$$
$$(35x+4)(x-11) = 35x^2 - 381x - 44$$
$$(7x+4)(5x-11) = 35x^2 - 57x - 44$$

The factorization is $(7x+4)(5x-11)$.

29. $20 + 6x - 2x^2 = -2x^2 + 6x + 20$

We factor out the common factor, -2. Factoring out -2 rather than 2 gives us a positive leading coefficient.
$$-2(x^2 - 3x - 10)$$

Then we factor the trinomial $x^2 - 3x - 10$. We look for a pair of factors whose product is -10 and whose sum is -3. The numbers are -5 and 2. The factorization of $x^2 - 3x - 10$ is $(x-5)(x+2)$. Then $20 + 6x - 2x^2 = -2(x-5)(x+2)$. If we think of -2 and $-1 \cdot 2$ then we can write other correct factorizations:
$$20 + 6x - 2x^2$$
$$= 2(-x+5)(x+2) \quad \text{Multiplying } x-5 \text{ by } -1$$
$$= 2(x-5)(-x-2) \quad \text{Multplying } x+2 \text{ by } -1$$

Note that we can also express $2(-x+5)(x+2)$ as $2(5-x)(x+2)$ since $-x+5 = 5-x$ by the commutative law of addition.

31. $12x^2 + 28x - 24$

(1) We factor out the common factor, 4:
$$4(3x^2 + 7x - 6)$$

Then we factor the trinomial $3x^2 + 7x - 6$.

(2) Factor $3x^2$. The only possibility is $3x$, x. The desired factorization is of the form:
$$(3x+\ )(x+\ )$$

(3) Factor -6. The possibilities are 6, -1 and -6, 1 and 3, -2 and -3, 2.

(4) We can immediately reject all possibilities in which either factor has a common factor, such as $(3x+6)$ or $(3x-3)$, because we factored out the largest common factor at the outset. We try some of the remaining possibilities:
$$(3x-1)(x+6) = 3x^2 + 17x - 6$$
$$(3x-2)(x+3) = 3x^2 + 7x - 6$$

The factorization of $3x^2 + 7x - 6$ is $(3x-2)(x+3)$. We must include the common factor in order to get a factorization of the original trinomial.
$$12x^2 + 28x - 24 = 4(3x-2)(x+3)$$

33. $30x^2 - 24x - 54$

(1) We factor out the common factor, 6:
$6(5x^2 - 4x - 9)$

Then we factor the trinomial $5x^2 - 4x - 9$.

(2) Factor $5x^2$. The only possibility is $5x, x$. The desired factorization is of the form:
$$(5x+\quad)(x+\quad)$$

(3) Factor -9. The possibilities are $9, -1$ and $-9, 1$ and $3, -3$.

(4) We try some possibilities:
$$(5x+9)(x-1) = 5x^2 + 4x - 9$$
$$(5x-9)(x+1) = 5x^2 - 4x - 9$$

The factorization of $5x^2 - 4x - 9$ is $(5x-9)(x+1)$. We must include the common factor in order to get a factorization of the original trinomial.
$$30x^2 - 24x - 54 = 6(5x-9)(x+1)$$

35. $4y + 6y^2 - 10 = 6y^2 + 4y - 10$

(1) We factor out the common factor, 2:
$2(3y^2 + 2y - 5)$

Then we factor the trinomial $3y^2 + 2y - 5$.

(2) Factor $3y^2$. The only possibility is $3y, y$. The desired factorization is of the form:
$$(3y+\quad)(y+\quad)$$

(3) Factor -5. The possibilities are $5, -1$ and $-5, 1$.

(4) We try some possibilities:
$$(3y+5)(y-1) = 3y^2 + 2y - 5$$

Then $3y^2 + 2y - 5 = (3y+5)(y-1)$, so $6y^2 + 4y - 10 = 2(3y+5)(y-1)$.

37. $3x^2 - 4x + 1$

(1) There is no common factor (other than 1 or -1).

(2) Factor the first term, $3x^2$. The only possibility is $3x, x$. The desired factorization is of the form:
$$(3x+\quad)(x+\quad)$$

(3) Factor the last term, 1. Since 1 is positive and the middle term is negative, we need consider only negative factor pairs. The only such pair is $-1, -1$.

(4) There is only one possibility:
$$(3x-1)(x-1) = 3x^2 - 4x + 1$$

The factorization is $(3x-1)(x-1)$.

39. $12x^2 - 28x - 24$

(1) We factor out the common factor, 4:
$4(3x^2 - 7x - 6)$

Then we factor the trinomial $3x^2 - 7x - 6$.

(2) Factor $3x^2$. The only possibility is $3x, x$. The desired factorization is of the form:
$$(3x+\quad)(x+\quad)$$

(3) Factor -6. The possibilities are $6, -1$ and $-6, 1$ and $3, -2$ and $-3, 2$.

(4) We can immediately reject all possibilities in which either factor has a common factor, such as $(3x-6)$ or $(3x+3)$, because we factored out the largest common factor at the outset. We try some of the remaining possibilities:
$$(3x-1)(x+6) = 3x^2 + 17x - 6$$
$$(3x-2)(x+3) = 3x^2 + 7x - 6$$
$$(3x+2)(x-3) = 3x^2 - 7x - 6$$

Then $3x^2 - 7x - 6 = (3x+2)(x-3)$, so $12x^2 - 28x - 24 = 4(3x+2)(x-3)$.

41. $-1 + 2x^2 - x = 2x^2 - x - 1$

(1) There is no common factor (other than 1 or -1).

(2) Factor the first term, $2x^2$. The only possibility is $2x, x$. The desired factorization is of the form:
$$(2x+\quad)(x+\quad)$$

(3) Factor -1. The only possibility is $1, -1$.

(4) We try some possibilities:
$$(2x+1)(x-1) = 2x^2 - x - 1$$

The factorization is $(2x+1)(x-1)$.

43. $9x^2 - 18x - 16$

(1) There is no common factor (other than 1 or -1).

(2) Factor the first term, $9x^2$. The possibilities are $9x, x$ and $3x, 3x$. We have these as possibilities for factorizations:
$$(9x+\quad)(x+\quad) \text{ and } (3x+\quad)(3x+\quad)$$

(3) Factor the last term, -16. The possibilities are $16, -1$ and $-16, 1$ and $8, -2$ and $-8, 2$ and $4, -4$.

(4) We try some possibilities:
$$(9x+16)(x-1) = 9x^2 + 7x - 16$$
$$(3x+16)(3x-1) = 9x^2 + 45x - 16$$

Exercise Set 5.3

$(9x + 8)(x - 2) = 9x^2 - 10x - 16$
$(3x + 8)(3x - 2) = 9x^2 + 18x - 16$
$(3x - 8)(3x + 2) = 9x^2 - 18x - 16$

The factorization is $(3x - 8)(3x + 2)$.

45. $15x^2 - 25x - 10$

(1) Factor out the common factor, 5:
$5(3x^2 - 5x - 2)$

Then we factor the trinomial $3x^2 - 5x - 2$. This was done in Exercise 15. We know that $3x^2 - 5x - 2 = (3x + 1)(x - 2)$, so $15x^2 - 25x - 10 = 5(3x + 1)(x - 2)$.

47. $12p^3 + 31p^2 + 20p$

(1) We factor out the common factor, p:
$p(12p^2 + 31p + 20)$

Then we factor the trinomial $12p^2 + 31p + 20$. This was done in Exercise 17 although the variable is x in that exercise. We know that $12p^2 + 31p + 20 = (3p + 4)(4p + 5)$, so $12p^3 + 31p^2 + 20p = p(3p + 4)(4p + 5)$.

49. $16 + 18x - 9x^2 = -9x^2 + 18x + 16$
$= -1(9x^2 - 18x - 16)$
$= -1(3x - 8)(3x + 2)$ Using the result from Exercise 43

Other correct factorizations are:
$16 + 18x - 9x^2$
$= (-3x + 8)(3x + 2)$ Multiplying $3x - 8$ by -1
$= (3x - 8)(-3x - 2)$ Multiplying $3x + 2$ by -1

We can also express $(-3x + 8)(3x + 2)$ as $(8 - 3x)(3x + 2)$ since $-3x + 8 = 8 - 3x$ by the commutative law of addition.

51. $-15x^2 + 19x - 6 = -1(15x^2 - 19x + 6)$

Now we factor $15x^2 - 19x + 6$.

(1) There is no common factor (other than 1 or -1).

(2) Factor the first term, $15x^2$. The possibilities are $15x$, x and $5x$, $3x$. We have these as possibilities for factorizations:

$(15x+\quad)(x+\quad)$ and $(5x+\quad)(3x+\quad)$

(3) Factor the last term, 6. The possibilities are 6, 1 and -6, -1 and 3, 2 and -3, -2.

(4) We try some possibilities:

$(15x + 1)(x + 6) = 15x^2 + 91x + 6$
$(5x + 3)(3x + 2) = 15x^2 + 19x + 6$
$(5x - 3)(3x - 2) = 15x^2 - 19x + 6$

The factorization of $15x^2 - 19x + 6$ is $(5x - 3)(3x - 2)$. Then $-15x^2 + 19x - 6 = -1(5x - 3)(3x - 2)$. Other correct factorizations are:
$-15x^2 + 19x - 6$
$= (-5x + 3)(3x - 2)$ Multiplying $5x - 3$ by -1
$= (5x - 3)(-3x + 2)$ Multiplying $3x - 2$ by -1

Note that we can also express $(-5x + 3)(3x - 2)$ as $(3 - 5x)(3x - 2)$ since $-5x + 3 = 3 - 5x$ by the commutative law of addition. Similarly, we can express $(5x - 3)(-3x + 2)$ as $(5x - 3)(2 - 3x)$.

53. $14x^4 + 19x^3 - 3x^2$

(1) Factor out the common factor, x^2: $x^2(14x^2 + 19x - 3)$

Then we factor the trinomial $14x^2 + 19x - 3$. This was done in Exercise 19. We know that $14x^2 + 19x - 3 = (7x - 1)(2x + 3)$, so $14x^4 + 19x^3 - 3x^2 = x^2(7x - 1)(2x + 3)$.

55. $168x^3 - 45x^2 + 3x$

(1) Factor out the common factor, $3x$:
$3x(56x^2 - 15x + 1)$

Then we factor the trinomial $56x^2 - 15x + 1$.

(2) Factor $56x^2$. The possibilities are $56x$, x and $28x$, $2x$ and $14x$, $4x$ and $7x$, $8x$. We have these as possibilities for factorizations:

$(56x+\quad)(x+\quad)$ and $(28x+\quad)(2x+\quad)$ and
$(14x+\quad)(4x+\quad)$ and $(7x+\quad)(8x+\quad)$

(3) Factor 1. Since 1 is positive and the middle term is negative we need consider only the negative factor pair $-1, -1$.

(4) We try some possibilities:

$(56x - 1)(x - 1) = 56x^2 - 57x + 1$
$(28x - 1)(2x - 1) = 56x^2 - 30x + 1$
$(14x - 1)(4x - 1) = 56x^2 - 18x + 1$
$(7x - 1)(8x - 1) = 56x^2 - 15x + 1$

Then $56x^2 - 15x + 1 = (7x - 1)(8x - 1)$, so $168x^3 - 45x^2 + 3x = 3x(7x - 1)(8x - 1)$.

57. $15x^4 - 19x^2 + 6 = 15(x^2)^2 - 19x^2 + 6$

(1) There is no common factor (other than 1 or -1).

(2) Factor the first term, $15x^4$. The possibilities are $15x^2$, x^2 and $5x^2$, $3x^2$. We have these as possibilities for factorizations:

$(15x^2+\quad)(x^2+\quad)$ and $(5x^2+\quad)(3x^2+\quad)$

(3) Factor 6. Since 6 is positive and the middle term is negative, we need consider only negative factor pairs. Those pairs are -6, -1 and -3, -2.

(4) We can immediately reject all possibilities in which either factor has a common factor, such as $(15x^2 - 6)$ or $(3x^2 - 3)$, because we determined at the outset that there is no common factor. We try some of the remaining possibilities:

$(15x^2 - 1)(x^2 - 6) = 15x^4 - 91x^2 + 6$
$(15x^2 - 2)(x^2 - 3) = 15x^4 - 47x^2 + 6$
$(5x^2 - 6)(3x^2 - 1) = 15x^4 - 23x^2 + 6$
$(5x^2 - 3)(3x^2 - 2) = 15x^4 - 19x^2 + 6$

The factorization is $(5x^2 - 3)(3x^2 - 2)$.

59. $25t^2 + 80t + 64$

(1) There is no common factor (other than 1 or −1).

(2) Factor the first term, $25t^2$. The possibilities are $25t$, t and $5t$, $5t$. We have these as possibilities for factorizations:
$$(25t+\quad)(t+\quad) \text{ and } (5t+\quad)(5t+\quad)$$

(3) Factor the last term, 64. Since all signs are positive, we need consider only positive pairs of factors. Those factor pairs are 64, 1 and 32, 2 and 16, 4 and 8, 8.

(4) We try some possibilities:
$$(25t + 64)(t + 1) = 25t^2 + 89t + 64$$
$$(5t + 32)(5t + 2) = 25t^2 + 170t + 64$$
$$(25t + 16)(t + 4) = 25t^2 + 116t + 64$$
$$(5t + 8)(5t + 8) = 25t^2 + 80t + 64$$

The factorization is $(5t + 8)(5t + 8)$ or $(5t + 8)^2$.

61. $6x^3 + 4x^2 - 10x$

(1) Factor out the common factor, $2x$: $2x(3x^2 + 2x - 5)$

Then we factor the trinomial $3x^2 + 2x - 5$. We did this in Exercise 35 (after we factored 2 out of the original trinomial). We know that $3x^2 + 2x - 5 = (3x + 5)(x - 1)$, so $6x^3 + 4x^2 - 10x = 2x(3x + 5)(x - 1)$.

63. $25x^2 + 79x + 64$

We follow the same procedure as in Exercise 59. None of the possibilities works. Thus, $25x^2 + 79x + 64$ is not factorable. It is prime.

65. $6x^2 - 19x - 5$

(1) There is no common factor (other than 1 or −1).

(2) Factor the first term, $6x^2$. The possibilities are $6x$, x and $3x$, $2x$. We have these as possibilities for factorizations:
$$(6x+\quad)(x+\quad) \text{ and } (3x+\quad)(2x+\quad)$$

(3) Factor the last term, −5. The possibilities are −5, 1 and 5, −1.

(4) We try some possibilities:
$$(6x - 5)(x + 1) = 6x^2 + x - 5$$
$$(6x + 5)(x - 1) = 6x^2 - x - 5$$
$$(6x + 1)(x - 5) = 6x^2 - 29x - 5$$
$$(6x - 1)(x + 5) = 6x^2 + 29x - 5$$
$$(3x - 5)(2x + 1) = 6x^2 - 7x - 5$$
$$(3x + 5)(2x - 1) = 6x^2 + 7x - 5$$
$$(3x + 1)(2x - 5) = 6x^2 - 13x - 5$$
$$(3x - 1)(2x + 5) = 6x^2 + 13x - 5$$

None of the possibilities works. Thus, $6x^2 - 19x - 5$ is not factorable. It is prime.

67. $12m^2 - mn - 20n^2$

(1) There is no common factor (other than 1 or −1).

(2) Factor the first term, $12m^2$. The possibilities are $12m$, m and $6m$, $2m$ and $3m$, $4m$. We have these as possibilities for factorizations:
$$(12m+\quad)(m+\quad) \text{ and } (6m+\quad)(2m+\quad)$$
$$\text{and } (3m+\quad)(4m+\quad)$$

(3) Factor the last term, $-20n^2$. The possibilities are $20n$, $-n$ and $-20n$, n and $10n$, $-2n$ and $-10n$, $2n$ and $5n$, $-4n$ and $-5n$, $4n$.

(4) We can immediately reject all possibilities in which either factor has a common factor, such as $(12m + 20n)$ or $(4m - 2n)$, because we determined at the outset that there is no common factor. We try some of the remaining possibilities:
$$(12m - n)(m + 20n) = 12m^2 + 239mn - 20n^2$$
$$(12m + 5n)(m - 4n) = 12m^2 - 43mn - 20n^2$$
$$(3m - 20n)(4m + n) = 12m^2 - 77mn - 20n^2$$
$$(3m - 4n)(4m + 5n) = 12m^2 - mn - 20n^2$$

The factorization is $(3m - 4n)(4m + 5n)$.

69. $6a^2 - ab - 15b^2$

(1) There is no common factor (other than 1 or −1).

(2) Factor the first term, $6a^2$. The possibilities are $6a$, a and $3a$, $2a$. We have these as possibilities for factorizations:
$$(6a+\quad)(a+\quad) \text{ and } (3a+\quad)(2a+\quad)$$

(3) Factor the last term, $-15b^2$. The possibilities are $15b$, $-b$ and $-15b$, b and $5b$, $-3b$ and $-5b$, $3b$.

(4) We can immediately reject all possibilities in which either factor has a common factor, such as $(6a+15b)$ or $(3a - 3b)$, because we determined at the outset that there is no common factor. We try some of the remaining possibilities:
$$(6a - b)(a + 15b) = 6a^2 + 89ab - 15b^2$$
$$(3a - b)(2a + 15b) = 6a^2 + 43ab - 15b^2$$
$$(6a + 5b)(a - 3b) = 6a^2 - 13ab - 15b^2$$
$$(3a + 5b)(2a - 3b) = 6a^2 + ab - 15b^2$$
$$(3a - 5b)(2a + 3b) = 6a^2 - ab - 15b^2$$

The factorization is $(3a - 5b)(2a + 3b)$.

71. $9a^2 + 18ab + 8b^2$

(1) There is no common factor (other than 1 or −1).

(2) Factor the first term, $9a^2$. The possibilities are $9a$, a and $3a$, $3a$. We have these as possibilities for factorizations:
$$(9a+\quad)(a+\quad) \text{ and } (3a+\quad)(3a+\quad)$$

(3) Factor $8b^2$. Since all signs are positive, we need consider only pairs of factors with positive coefficients. Those factor pairs are $8b$, b and $4b$, $2b$.

(4) We try some possibilities:

$(9a+8b)(a+b) = 9a^2 + 17ab + 8b^2$

$(3a+8b)(3a+b) = 9a^2 + 27ab + 8b^2$

$(9a+4b)(a+2b) = 9a^2 + 22ab + 8b^2$

$(3a+4b)(3a+2b) = 9a^2 + 18ab + 8b^2$

The factorization is $(3a+4b)(3a+2b)$.

73. $35p^2 + 34pq + 8q^2$

(1) There is no common factor (other than 1 or -1).

(2) Factor the first term, $35p^2$. The possibilities are $35p, p$ and $7p, 5p$. We have these as possibilities for factorizations:

$(35p+\ )(p+\ )$ and $(7p+\ )(5p+\ )$

(3) Factor $8q^2$. Since all signs are positive, we need consider only pairs of factors with positive coefficients. Those factor pairs are $8q, q$ and $4q, 2q$.

(4) We try some possibilities:

$(35p+8q)(p+q) = 35p^2 + 43pq + 8q^2$

$(7p+8q)(5p+q) = 35p^2 + 47pq + 8q^2$

$(35p+4q)(p+2q) = 35p^2 + 74pq + 8q^2$

$(7p+4q)(5p+2q) = 35p^2 + 34pq + 8p^2$

The factorization is $(7p+4q)(5p+2q)$.

75. $18x^2 - 6xy - 24y^2$

(1) Factor out the common factor, 6:

$6(3x^2 - xy - 4y^2)$

Then we factor the trinomial $3x^2 - xy - 4y^2$.

(2) Factor $3x^2$. The only possibility is $3x, x$. The desired factorization is of the form:

$(3x+\ )(x+\ )$

(3) Factor $-4y^2$. The possibilities are $4y, -y$ and $-4y, y$ and $2y, -2y$.

(4) We try some possibilities:

$(3x+4y)(x-y) = 3x^2 + xy - 4y^2$

$(3x-4y)(x+y) = 3x^2 - xy - 4y^2$

Then $3x^2 - xy - 4y^2 = (3x-4y)(x+y)$, so $18x^2 - 6xy - 24y^2 = 6(3x-4y)(x+y)$.

77. Discussion and Writing Exercise

79. $A = pq - 7$

$A + 7 = pq$ Adding 7

$\dfrac{A+7}{p} = q$ Dividing by p

81. $3x + 2y = 6$

$2y = 6 - 3x$ Subtracting $3x$

$y = \dfrac{6-3x}{2}$ Dividing by 2

83. $5 - 4x < -11$

$-4x < -16$ Subtracting 5

$x > 4$ Dividing by -4 and reversing the inequality symbol

The solution set is $\{x | x > 4\}$.

85. Graph: $y = \dfrac{2}{5}x - 1$

Because the equation is in the form $y = mx + b$, we know the y-intercept is $(0, -1)$. We find two other points on the line, substituting multiples of 5 for x to avoid fractions.

When $x = -5$, $y = \dfrac{2}{5}(-5) - 1 = -2 - 1 = -3$.

When $x = 5$, $y = \dfrac{2}{5}(5) - 1 = 2 - 1 = 1$.

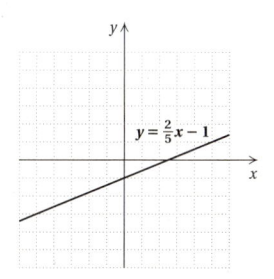

x	y
0	-1
-5	-3
5	1

87. $(3x-5)(3x+5) = (3x)^2 - 5^2 = 9x^2 - 25$

89. $20x^{2n} + 16x^n + 3 = 20(x^n)^2 + 16x^n + 3$

(1) There is no common factor (other than 1 and -1).

(2) Factor the first term, $20x^{2n}$. The possibilities are $20x^n, x^n$ and $10x^n, 2x^n$ and $5x^n, 4x^n$. We have these as possibilities for factorizations:

$(20x^n+\ )(x^n+\ )$ and $(10x^n+\ )(2x^n+\ )$ and $(5x^n+\ )(4x^n+\ )$

(3) Factor the last term, 3. Since all signs are positive, we need consider only the positive factor pair 3, 1.

(4) We try some possibilities:

$(20x^n+3)(x^n+1) = 20x^{2n} + 23x^n + 3$

$(10x^n+3)(2x^n+1) = 20x^{2n} + 16x^n + 3$

The factorization is $(10x^n + 3)(2x^n + 1)$.

91. $3x^{6a} - 2x^{3a} - 1 = 3(x^{3a})^2 - 2x^{3a} - 1$

(1) There is no common factor (other than 1 or -1).

(2) Factor the first term, $3x^{6a}$. The only possibility is $3x^{3a}, x^{3a}$. The desired factorization is of the form:

$(3x^{3a}+\ )(x^{3a}+\ )$

(3) Factor the last term, -1. The only possibility is $-1, 1$.

(4) We try the possibilities:

$(3x^{3a} - 1)(x^{3a} + 1) = 3x^{6a} + 2x^{3a} - 1$

$(3x^{3a} + 1)(x^{3a} - 1) = 3x^{6a} - 2x^{3a} - 1$

The factorization is $(3x^{3a} + 1)(x^{3a} - 1)$.

93.–101. Left to the student

Exercise Set 5.4

1. $x^2 + 2x + 7x + 14 = x(x+2) + 7(x+2)$
$= (x+7)(x+2)$

3. $x^2 - 4x - x + 4 = x(x-4) - 1(x-4)$
$= (x-1)(x-4)$

5. $6x^2 + 4x + 9x + 6 = 2x(3x+2) + 3(3x+2)$
$= (2x+3)(3x+2)$

7. $3x^2 - 4x - 12x + 16 = x(3x-4) - 4(3x-4)$
$= (x-4)(3x-4)$

9. $35x^2 - 40x + 21x - 24 = 5x(7x-8) + 3(7x-8)$
$= (5x+3)(7x-8)$

11. $4x^2 + 6x - 6x - 9 = 2x(2x+3) - 3(2x+3)$
$= (2x-3)(2x+3)$

13. $2x^4 + 6x^2 + 5x^2 + 15 = 2x^2(x^2+3) + 5(x^2+3)$
$= (2x^2+5)(x^2+3)$

15. $2x^2 + 7x - 4$

(1) First factor out a common factor, if any. There is none (other than 1 or −1).

(2) Multiply the leading coefficient, 2 and the constant, −4: $2(-4) = -8$.

(3) Look for a factorization of −8 in which the sum of the factors is the coefficient of the middle term, 7.

Pairs of factors	Sums of factors
−1, 8	7
1, −8	−7
−2, 4	2
2, −4	−2

(4) Split the middle term: $7x = -1x + 8x$

(5) Factor by grouping:
$2x^2 + 7x - 4 = 2x^2 - x + 8x - 4$
$= x(2x-1) + 4(2x-1)$
$= (x+4)(2x-1)$

17. $3x^2 - 4x - 15$

(1) First factor out a common factor, if any. There is none (other than 1 or −1).

(2) Multiply the leading coefficient, 3, and the constant, −15: $3(-15) = -45$.

(3) Look for a factorization of −45 in which the sum of the factors is the coefficient of the middle term, −4.

Pairs of factors	Sums of factors
−1, 45	44
1, −45	−44
−3, 15	12
3, −15	−12
−5, 9	4
5, −9	−4

(4) Split the middle term: $-4x = 5x - 9x$

(5) Factor by grouping:
$3x^2 - 4x - 15 = 3x^2 + 5x - 9x - 15$
$= x(3x+5) - 3(3x+5)$
$= (x-3)(3x+5)$

19. $6x^2 + 23x + 7$

(1) First factor out a common factor, if any. There is none (other than 1 or −1).

(2) Multiply the leading coefficient, 6, and the constant, 7: $6 \cdot 7 = 42$.

(3) Look for a factorization of 42 in which the sum of the factors is the coefficient of the middle term, 23. We only need to consider positive factors.

Pairs of factors	Sums of factors
1, 42	43
2, 21	23
3, 14	17
6, 7	13

(4) Split the middle term: $23x = 2x + 21x$

(5) Factor by grouping:
$6x^2 + 23x + 7 = 6x^2 + 2x + 21x + 7$
$= 2x(3x+1) + 7(3x+1)$
$= (2x+7)(3x+1)$

21. $3x^2 - 4x + 1$

(1) First factor out a common factor, if any. There is none (other than 1 or −1).

(2) Multiply the leading coefficient, 3, and the constant, 1: $3 \cdot 1 = 3$.

(3) Look for a factorization of 3 in which the sum of the factors is the coefficient of the middle term, −4. The numbers we want are −1 and −3: $-1 \cdot (-3) = 3$ and $-1 + (-3) = -4$.

(4) Split the middle term: $-4x = -1x - 3x$

(5) Factor by grouping:
$3x^2 - 4x + 1 = 3x^2 - x - 3x + 1$
$= x(3x-1) - 1(3x-1)$
$= (x-1)(3x-1)$

23. $4x^2 - 4x - 15$

(1) First factor out a common factor, if any. There is none (other than 1 or −1).

(2) Multiply the leading coefficient, 4, and the constant, −15: $4(-15) = -60$.

(3) Look for a factorization of −60 in which the sum of the factors is the coefficient of the middle term, −4.

Exercise Set 5.4

Pairs of factors	Sums of factors
−1, 60	59
1, −60	−59
−2, 30	28
2, −30	−28
−3, 20	17
3, −20	−17
−4, 15	11
4, −15	−11
−5, 12	7
5, −12	−7
−6, 10	4
6, −10	−4

(4) Split the middle term: $-4x = 6x - 10x$

(5) Factor by grouping:
$$4x^2 - 4x - 15 = 4x^2 + 6x - 10x - 15$$
$$= 2x(2x+3) - 5(2x+3)$$
$$= (2x-5)(2x+3)$$

25. $2x^2 + x - 1$

(1) First factor out a common factor, if any. There is none (other than 1 or −1).

(2) Multiply the leading coefficient, 2, and the constant, −1: $2(-1) = -2$.

(3) Look for a factorization of −2 in which the sum of the factors is the coefficient of the middle term, 1. The numbers we want are 2 and −1: $2(-1) = -2$ and $2 - 1 = 1$.

(4) Split the middle term: $x = 2x - 1x$

(5) Factor by grouping:
$$2x^2 + x - 1 = 2x^2 + 2x - x - 1$$
$$= 2x(x+1) - 1(x+1)$$
$$= (2x-1)(x+1)$$

27. $9x^2 - 18x - 16$

(1) First factor out a common factor, if any. There is none (other than 1 or −1).

(2) Multiply the leading coefficient, 9, and the constant, −16: $9(-16) = -144$.

(3) Look for a factorization of −144, so the sum of the factors is the coefficient of the middle term, −18.

Pairs of factors	Sums of factors
−1, 144	143
1, −144	−143
−2, 72	70
2, −72	−70
−3, 48	45
3, −48	−45
−4, 36	32
4, −36	−32
−6, 24	18
6, −24	−18
−8, 18	10
8, −18	−10
−9, 16	7
9, −16	−7
−12, 12	0

(4) Split the middle term: $-18x = 6x - 24x$

(5) Factor by grouping:
$$9x^2 - 18x - 16 = 9x^2 + 6x - 24x - 16$$
$$= 3x(3x+2) - 8(3x+2)$$
$$= (3x-8)(3x+2)$$

29. $3x^2 + 5x - 2$

(1) First factor out a common factor, if any. There is none (other than 1 or −1).

(2) Multiply the leading coefficient, 3, and the constant, −2: $3(-2) = -6$.

(3) Look for a factorization of −6 in which the sum of the factors is the coefficient of the middle term, 5. The numbers we want are −1 and 6: $-1(6) = -6$ and $-1 + 6 = 5$.

(4) Split the middle term: $5x = -1x + 6x$

(5) Factor by grouping:
$$3x^2 + 5x - 2 = 3x^2 - x + 6x - 2$$
$$= x(3x-1) + 2(3x-1)$$
$$= (x+2)(3x-1)$$

31. $12x^2 - 31x + 20$

(1) First factor out a common factor, if any. There is none (other than 1 or −1).

(2) Multiply the leading coefficient, 12, and the constant, 20: $12 \cdot 20 = 240$.

(3) Look for a factorization of 240 in which the sum of the factors is the coefficient of the middle term, −31. We only need to consider negative factors.

Pairs of factors	Sums of factors
$-1, -240$	-241
$-2, -120$	-122
$-3, -8$	-83
$-4, -60$	-64
$-5, -48$	-53
$-6, -40$	-46
$-8, -30$	-38
$-10, -24$	-34
$-12, -20$	-32
$-15, -16$	-31

(4) Split the middle term: $-31x = -15x - 16x$

(5) Factor by grouping:
$$12x^2 - 31x + 20 = 12x^2 - 15x - 16x + 20$$
$$= 3x(4x - 5) - 4(4x - 5)$$
$$= (3x - 4)(4x - 5)$$

33. $14x^2 - 19x - 3$

(1) First factor out a common factor, if any. There is none (other than 1 or -1).

(2) Multiply the leading coefficient, 14, and the constant, -3: $14(-3) = -42$.

(3) Look for a factorization of -42 so that the sum of the factors is the coefficient of the middle term, -19.

Pairs of factors	Sums of factors
$-1, 42$	41
$1, -42$	-41
$-2, 21$	19
$2, -21$	-19
$-3, 14$	11
$3, -14$	-11
$-6, 7$	1
$6, -7$	-1

(4) Split the middle term: $-19x = 2x - 21x$

(5) Factor by grouping:
$$14x^2 - 19x - 3 = 14x^2 + 2x - 21x - 3$$
$$= 2x(7x + 1) - 3(7x + 1)$$
$$= (2x - 3)(7x + 1)$$

35. $9x^2 + 18x + 8$

(1) First factor out a common factor, if any. There is none (other than 1 or -1).

(2) Multiply the leading coefficient, 9, and the constant, 8: $9 \cdot 8 = 72$.

(3) Look for a factorization of 72 in which the sum of the factors is the coefficient of the middle term, 18. We only need to consider positive factors.

Pairs of factors	Sums of factors
$1, 72$	73
$2, 36$	38
$3, 24$	27
$4, 18$	22
$6, 12$	18
$8, 9$	17

(4) Split the middle term: $18x = 6x + 12x$

(5) Factor by grouping:
$$9x^2 + 18x + 8 = 9x^2 + 6x + 12x + 8$$
$$= 3x(3x + 2) + 4(3x + 2)$$
$$= (3x + 4)(3x + 2)$$

37. $49 - 42x + 9x^2 = 9x^2 - 42x + 49$

(1) First factor out a common factor, if any. There is none (other than 1 or -1).

(2) Multiply the leading coefficient, 9, and the constant, 49: $9 \cdot 49 = 441$.

(3) Look for a factorization of 441 in which the sum of the factors is the coefficient of the middle term, -42. We only need to consider negative factors.

Pairs of factors	Sums of factors
$-1, -441$	-442
$-3, -147$	-150
$-7, -63$	-70
$-9, -49$	-58
$-21, -21$	-42

(4) Split the middle term: $-42x = -21x - 21x$

(5) Factor by grouping:
$$9x^2 - 42x + 49 = 9x^2 - 21x - 21x + 49$$
$$= 3x(3x - 7) - 7(3x - 7)$$
$$= (3x - 7)(3x - 7), \text{ or}$$
$$(3x - 7)^2$$

39. $24x^2 - 47x - 2$

(1) First factor out a common factor, if any. There is none (other than 1 or -1).

(2) Multiply the leading coefficient, 24, and the constant, -2: $24(-2) = -48$.

(3) Look for a factorization of -48 in which the sum of the factors is the coefficient of the middle term, -47. The numbers we want are -48 and 1: $-48 \cdot 1 = -48$ and $-48 + 1 = -47$.

(4) Split the middle term: $-47x = -48x + 1x$

(5) Factor by grouping:
$$24x^2 - 47x - 2 = 24x^2 - 48x + x - 2$$
$$= 24x(x - 2) + 1(x - 2)$$
$$= (24x + 1)(x - 2)$$

41. $5 - 9a^2 - 12a = -9a^2 - 12a + 5 = -1(9a^2 + 12a - 5)$

Now we factor $9a^2 + 12a - 5$.

(1) We have already factored out the common factor, -1, to make the leading coefficient positive.

(2) Multiply the leading coefficient, 9, and the constant, -5: $9(-5) = -45$.

(3) Look for a factorization of -45 in which the sum of the factors is the coefficient of the middle term, 12. The numbers we want are 15 and -3: $15(-3) = -45$ and $15 + (-3) = 12$.

(4) Split the middle term: $12a = 15a - 3a$

(5) Factor by grouping:
$9a^2 + 12a - 5 = 9a^2 + 15a - 3a - 5$
$= 3a(3a + 5) - (3a + 5)$
$= (3a - 1)(3a + 5)$

Then we have
$5 - 9a^2 - 12a$
$= -1(3a - 1)(3a + 5)$
$= (-3a + 1)(3a + 5)$ Multiplying $3a-1$ by -1
$= (3a - 1)(-3a - 5)$ Multiplying $3a+5$ by -1

Note that we can also express $(-3a + 1)(3a + 5)$ as $(1 - 3a)(3a + 5)$ since $-3a + 1 = 1 - 3a$ by the commutative law of addition.

43. $20 + 6x - 2x^2 = -2x^2 + 6x + 20$

(1) Factor out the common factor -2. We factor out -2 rather than 2 in order to make the leading coefficient of the trinomial factor positive.
$-2x^2 + 6x + 20 = -2(x^2 - 3x - 10)$

To factor $x^2 - 3x - 10$, we look for two factors of -10 whose sum is -3. The numbers we want are -5 and 2. Then $x^2 - 3x - 10 = (x - 5)(x + 2)$, so we have:
$20 + 6x - 2x^2$
$= -2(x - 5)(x + 2)$
$= 2(-x + 5)(x + 2)$ Multiplying $x - 5$ by -1
$= 2(x - 5)(-x - 2)$ Multiplying $x + 2$ by -1

Note that we can also express $2(-x + 5)(x + 2)$ as $2(5 - x)(x + 2)$ since $-x + 5 = 5 - x$ by the commutative law of addition.

45. $12x^2 + 28x - 24$

(1) Factor out the common factor, 4:
$12x^2 + 28x - 24 = 4(3x^2 + 7x - 6)$

(2) Now we factor the trinomial $3x^2 + 7x - 6$. Multiply the leading coefficient, 3, and the constant, -6: $3(-6) = -18$.

(3) Look for a factorization of -18 in which the sum of the factors is the coefficient of the middle term, 7. The numbers we want are 9 and -2: $9(-2) = -18$ and $9 + (-2) = 7$.

(4) Split the middle term: $7x = 9x - 2x$

(5) Factor by grouping:
$3x^2 + 7x - 6 = 3x^2 + 9x - 2x - 6$
$= 3x(x + 3) - 2(x + 3)$
$= (3x - 2)(x + 3)$

We must include the common factor to get a factorization of the original trinomial.
$12x^2 + 28x - 24 = 4(3x - 2)(x + 3)$

47. $30x^2 - 24x - 54$

(1) Factor out the common factor, 6.
$30x^2 - 24x - 54 = 6(5x^2 - 4x - 9)$

(2) Now we factor the trinomial $5x^2 - 4x - 9$. Multiply the leading coefficient, 5, and the constant, -9: $5(-9) = -45$.

(3) Look for a factorization of -45 in which the sum of the factors is the coefficient of the middle term, -4. The numbers we want are -9 and 5: $-9 \cdot 5 = -45$ and $-9 + 5 = -4$.

(4) Split the middle term: $-4x = -9x + 5x$

(5) Factor by grouping:
$5x^2 - 4x - 9 = 5x^2 - 9x + 5x - 9$
$= x(5x - 9) + (5x - 9)$
$= (x + 1)(5x - 9)$

We must include the common factor to get a factorization of the original trinomial.
$30x^2 - 24x - 54 = 6(x + 1)(5x - 9)$

49. $4y + 6y^2 - 10 = 6y^2 + 4y - 10$

(1) Factor out the common factor, 2.
$6y^2 + 4y - 10 = 2(3y^2 + 2y - 5)$

(2) Now we factor the trinomial $3y^2 + 2y - 5$. Multiply the leading coefficient, 3, and the constant, -5: $3(-5) = -15$.

(3) Look for a factorization of -15 in which the sum of the factors is the coefficient of the middle term, 2. The numbers we want are 5 and -3: $5(-3) = -15$ and $5 + (-3) = 2$.

(4) Split the middle term: $2y = 5y - 3y$

(5) Factor by grouping:
$3y^2 + 2y - 5 = 3y^2 + 5y - 3y - 5$
$= y(3y + 5) - (3y + 5)$
$= (y - 1)(3y + 5)$

We must include the common factor to get a factorization of the original trinomial.
$4y + 6y^2 - 10 = 2(y - 1)(3y + 5)$

51. $3x^2 - 4x + 1$

(1) There is no common factor (other than 1 or -1).

(2) Multiply the leading coefficient, 3, and the constant, 1: $3 \cdot 1 = 3$.

(3) Look for a factorization of 3 in which the sum of the factors is the coefficient of the middle term, -4. The numbers we want are -1 and -3: $-1(-3) = 3$ and $-1 + (-3) = -4$.

(4) Split the middle term: $-4x = -1x - 3x$

(5) Factor by grouping:
$$3x^2 - 4x + 1 = 3x^2 - x - 3x + 1$$
$$= x(3x - 1) - (3x - 1)$$
$$= (x - 1)(3x - 1)$$

53. $12x^2 - 28x - 24$

    (1) Factor out the common factor, 4:
    $12x^2 - 28x - 24 = 4(3x^2 - 7x - 6)$

    (2) Now we factor the trinomial $3x^2 - 7x - 6$. Multiply the leading coefficient, 3, and the constant, -6: $3(-6) = -18$.

    (3) Look for a factorization of -18 in which the sum of the factors is the coefficient of the middle term, -7. The numbers we want are -9 and 2: $-9 \cdot 2 = -18$ and $-9 + 2 = -7$.

    (4) Split the middle term: $-7x = -9x + 2x$

    (5) Factor by grouping:
    $$3x^2 - 7x - 6 = 3x^2 - 9x + 2x - 6$$
    $$= 3x(x - 3) + 2(x - 3)$$
    $$= (3x + 2)(x - 3)$$
    We must include the common factor to get a factorization of the original trinomial.
    $12x^2 - 28x - 24 = 4(3x + 2)(x - 3)$

55. $-1 + 2x^2 - x = 2x^2 - x - 1$

    (1) There is no common factor (other than 1 or -1).

    (2) Multiply the leading coefficient, 2, and the constant, -1: $2(-1) = -2$.

    (3) Look for a factorization of -2 in which the sum of the factors is the coefficient of the middle term, -1. The numbers we want are -2 and 1: $-2 \cdot 1 = -2$ and $-2 + 1 = -1$.

    (4) Split the middle term: $-x = -2x + 1x$

    (5) Factor by grouping:
    $$2x^2 - x - 1 = 2x^2 - 2x + x - 1$$
    $$= 2x(x - 1) + (x - 1)$$
    $$= (2x + 1)(x - 1)$$

57. $9x^2 + 18x - 16$

    (1) There is no common factor (other than 1 or -1).

    (2) Multiply the leading coefficient, 9, and the constant, -16: $9(-16) = -144$.

    (3) Look for a factorization of -144 in which the sum of the factors is the coefficient of the middle term, 18. The numbers we want are 24 and -6: $24(-6) = -144$ and $24 + (-6) = 18$.

    (4) Split the middle term: $18x = 24x - 6x$

    (5) Factor by grouping:
    $$9x^2 + 18x - 16 = 9x^2 + 24x - 6x - 16$$
    $$= 3x(3x + 8) - 2(3x + 8)$$
    $$= (3x - 2)(3x + 8)$$

59. $15x^2 - 25x - 10$

    (1) Factor out the common factor, 5:
    $15x^2 - 25x - 10 = 5(3x^2 - 5x - 2)$

    (2) Now we factor the trinomial $3x^2 - 5x - 2$. Multiply the leading coefficient, 3, and the constant, -2: $3(-2) = -6$.

    (3) Look for a factorization of -6 in which the sum of the factors is the coefficient of the middle term, -5. The numbers we want are -6 and 1: $-6 \cdot 1 = -6$ and $-6 + 1 = -5$.

    (4) Split the middle term: $-5x = -6x + 1x$

    (5) Factor by grouping:
    $$3x^2 - 5x - 2 = 3x^2 - 6x + x - 2$$
    $$= 3x(x - 2) + (x - 2)$$
    $$= (3x + 1)(x - 2)$$
    We must include the common factor to get a factorization of the original trinomial.
    $15x^2 - 25x - 10 = 5(3x + 1)(x - 2)$

61. $12p^3 + 31p^2 + 20p$

    (1) Factor out the common factor, p:
    $12p^3 + 31p^2 + 20p = p(12p^2 + 31p + 20)$

    (2) Now we factor the trinomial $12p^2 + 31p + 20$. Multiply the leading coefficient, 12, and the constant, 20: $12 \cdot 20 = 240$.

    (3) Look for a factorization of 240 in which the sum of the factors is the coefficient of the middle term, 31. The numbers we want are 15 and 16: $15 \cdot 16 = 240$ and $15 + 16 = 31$.

    (4) Split the middle term: $31p = 15p + 16p$

    (5) Factor by grouping:
    $$12p^2 + 31p + 20 = 12p^2 + 15p + 16p + 20$$
    $$= 3p(4p + 5) + 4(4p + 5)$$
    $$= (3p + 4)(4p + 5)$$
    We must include the common factor to get a factorization of the original trinomial.
    $12p^3 + 31p^2 + 20p = p(3p + 4)(4p + 5)$

63. $4 - x - 5x^2 = -5x^2 - x + 4$

    (1) Factor out -1 to make the leading coefficient positive:
    $-5x^2 - x + 4 = -1(5x^2 + x - 4)$

    (2) Now we factor the trinomial $5x^2 + x - 4$. Multiply the leading coefficient, 5, and the constant, -4: $5(-4) = -20$.

Exercise Set 5.4

(3) Look for a factorization of -20 in which the sum of the factors is the coefficient of the middle term, 1. The numbers we want are 5 and -4: $5(-4) = -20$ and $5 + (-4) = 1$.

(4) Split the middle term: $x = 5x - 4x$

(5) Factor by grouping:
$$5x^2 + x - 4 = 5x^2 + 5x - 4x - 4$$
$$= 5x(x+1) - 4(x+1)$$
$$= (5x-4)(x+1)$$

We must include the common factor to get a factorization of the original trinomial.
$$4 - x - 5x^2$$
$$= -1(5x-4)(x+1)$$
$$= (-5x+4)(x+1) \quad \text{Multiplying } 5x-4 \text{ by } -1$$
$$= (5x-4)(-x-1) \quad \text{Multiplying } x+1 \text{ by } -1$$

Note that we can also express $(-5x+4)(x+1)$ as $(4-5x)(x+1)$ since $-5x+4 = 4-5x$ by the commutative law of addition.

65. $33t - 15 - 6t^2 = -6t^2 + 33t - 15$

(1) Factor out the common factor, -3. We factor out -3 rather than 3 in order to make the leading coefficient of the trinomial factor positive.
$$-6t^2 + 33t - 15 = -3(2t^2 - 11t + 5)$$

(2) Now we factor the trinomial $2t^2 - 11t + 5$. Multiply the leading coefficient, 2, and the constant, 5: $2 \cdot 5 = 10$.

(3) Look for a factorization of 10 in which the sum of the factors is the coefficient of the middle term, -11. The numbers we want are -1 and -10: $-1(-10) = 10$ and $-1 + (-10) = -11$.

(4) Split the middle term: $-11t = -1t - 10t$

(5) Factor by grouping:
$$2t^2 - 11t + 5 = 2t^2 - t - 10t + 5$$
$$= t(2t-1) - 5(2t-1)$$
$$= (t-5)(2t-1)$$

We must include the common factor to get a factorization of the original trinomial.
$$33t - 15 - 6t^2$$
$$= -3(t-5)(2t-1)$$
$$= 3(-t+5)(2t-1) \quad \text{Multiplying } t-5 \text{ by } -1$$
$$= 3(t-5)(-2t+1) \quad \text{Multiplying } 2t-1 \text{ by } -1$$

Note that we can also express $3(-t+5)(2t-1)$ as $3(5-t)(2t-1)$ since $-t+5 = 5-t$ by the commutative law of addition. Similarly, we can express $3(t-5)(-2t+1)$ as $3(t-5)(1-2t)$.

67. $14x^4 + 19x^3 - 3x^2$

(1) Factor out the common factor, x^2:
$$14x^4 + 19x^3 - 3x^2 = x^2(14x^2 + 19x - 3)$$

(2) Now we factor the trinomial $14x^2 + 19x - 3$. Multiply the leading coefficient, 14, and the constant, -3: $14(-3) = -42$.

(3) Look for a factorization of -42 in which the sum of the factors is the coefficient of the middle term, 19. The numbers we want are 21 and -2: $21(-2) = -42$ and $21 + (-2) = 19$.

(4) Split the middle term: $19x = 21x - 2x$

(5) Factor by grouping:
$$14x^2 + 19x - 3 = 14x^2 + 21x - 2x - 3$$
$$= 7x(2x+3) - (2x+3)$$
$$= (7x-1)(2x+3)$$

We must include the common factor to get a factorization of the original trinomial.
$$14x^4 + 19x^3 - 3x^2 = x^2(7x-1)(2x+3)$$

69. $168x^3 - 45x^2 + 3x$

(1) Factor out the common factor, $3x$:
$$168x^3 - 45x^2 + 3x = 3x(56x^2 - 15x + 1)$$

(2) Now we factor the trinomial $56x^2 - 15x + 1$. Multiply the leading coefficient, 56, and the constant, 1: $56 \cdot 1 = 56$.

(3) Look for a factorization of 56 in which the sum of the factors is the coefficient of the middle term, -15. The numbers we want are -7 and -8: $-7(-8) = 56$ and $-7 + (-8) = -15$.

(4) Split the middle term: $-15x = -7x - 8x$

(5) Factor by grouping:
$$56x^2 - 15x + 1 = 56x^2 - 7x - 8x + 1$$
$$= 7x(8x-1) - (8x-1)$$
$$= (7x-1)(8x-1)$$

We must include the common factor to get a factorization of the original trinomial.
$$168x^3 - 45x^2 + 3x = 3x(7x-1)(8x-1)$$

71. $15x^4 - 19x^2 + 6$

(1) There are no common factors (other than 1 or -1).

(2) Multiply the leading coefficient, 15, and the constant, 6: $15 \cdot 6 = 90$.

(3) Look for a factorization of 90 in which the sum of the factors is the coefficient of the middle term, -19. The numbers we want are -9 and -10: $-9(-10) = 90$ and $-9 + (-10) = -19$.

(4) Split the middle term: $-19x^2 = -9x^2 - 10x^2$

(5) Factor by grouping:
$$15x^4 - 19x^2 + 6 = 15x^4 - 9x^2 - 10x^2 + 6$$
$$= 3x^2(5x^2 - 3) - 2(5x^2 - 3)$$
$$= (3x^2 - 2)(5x^2 - 3)$$

73. $25t^2 + 80t + 64$

(1) There are no common factors (other than 1 or -1).

(2) Multiply the leading coefficient, 25, and the constant, 64: $25 \cdot 64 = 1600$.

(3) Look for a factorization of 1600 in which the sum of the factors is the coefficient of the middle term, 80. The numbers we want are 40 and 40: $40 \cdot 40 = 1600$ and $40 + 40 = 80$.

(4) Split the middle term: $80t = 40t + 40t$

(5) Factor by grouping:
$$25t^2 + 80t + 64 = 25t^2 + 40t + 40t + 64$$
$$= 5t(5t + 8) + 8(5t + 8)$$
$$= (5t + 8)(5t + 8), \text{ or}$$
$$(5t + 8)^2$$

75. $6x^3 + 4x^2 - 10x$

(1) Factor out the common factor, $2x$:
$$6x^3 + 4x^2 - 10x = 2x(3x^2 + 2x - 5)$$

(2) - (5) Now we factor the trinomial $3x^2 + 2x - 5$. We did this in Exercise 49, using the variable y rather than x. We found that $3x^2 + 2x - 5 = (x-1)(3x+5)$. We must include the common factor to get a factorization of the original trinomial.
$$6x^3 + 4x^2 - 10x = 2x(x - 1)(3x + 5)$$

77. $25x^2 + 79x + 64$

(1) There are no common factors (other than 1 or -1).

(2) Multiply the leading coefficient, 25, and the constant, 64: $25 \cdot 64 = 1600$.

(3) Look for a factorization of 1600 in which the sum of the factors is the coefficient of the middle term, 79. It is not possible to find such a pair of numbers. Thus, $25x^2 + 79x + 64$ cannot be factored into a product of binomial factors. It is prime.

79. $6x^2 - 19x - 5$

(1) There are no common factors (other than 1 or -1).

(2) Multiply the leading coefficient, 6, and the constant, -5: $6(-5) = -30$.

(3) Look for a factorization of -30 in which the sum of the factors is the coefficient of the middle term, -19. There is no such pair of numbers. Thus, $6x^2 - 19x - 5$ cannot be factored into a product of binomial factors. It is prime.

81. $12m^2 - mn - 20n^2$

(1) There are no common factors (other than 1 or -1).

(2) Multiply the leading coefficient, 12, and the constant, -20: $12(-20) = -240$.

(3) Look for a factorization of -240 in which the sum of the factors is the coefficient of the middle term, -1. The numbers we want are 15 and -16: $15(-16) = -240$ and $15 + (-16) = -1$.

(4) Split the middle term: $-mn = 15mn - 16mn$

(5) Factor by grouping:
$$12m^2 - mn - 20n^2$$
$$= 12m^2 + 15mn - 16mn - 20n^2$$
$$= 3m(4m + 5n) - 4n(4m + 5n)$$
$$= (3m - 4n)(4m + 5n)$$

83. $6a^2 - ab - 15b^2$

(1) There are no common factors (other than 1 or -1).

(2) Multiply the leading coefficient, 6, and the constant, -15: $6(-15) = -90$.

(3) Look for a factorization of -90 in which the sum of the factors is the coefficient of the middle term, -1. The numbers we want are -10 and 9: $-10 \cdot 9 = -90$ and $-10 + 9 = -1$.

(4) Split the middle term: $-ab = -10ab + 9ab$

(5) Factor by grouping:
$$6a^2 - ab - 15b^2 = 6a^2 - 10ab + 9ab - 15b^2$$
$$= 2a(3a - 5b) + 3b(3a - 5b)$$
$$= (2a + 3b)(3a - 5b)$$

85. $9a^2 - 18ab + 8b^2$

(1) There are no common factors (other than 1 or -1).

(2) Multiply the leading coefficient, 9, and the constant, 8: $9 \cdot 8 = 72$.

(3) Look for a factorization of 72 in which the sum of the factors is the coefficient of the middle term, -18. The numbers we want are -6 and -12: $-6(-12) = 72$ and $-6 + (-12) = -18$.

(4) Split the middle term: $-18ab = -6ab - 12ab$

(5) Factor by grouping:
$$9a^2 - 18ab + 8b^2 = 9a^2 - 6ab - 12ab + 8b^2$$
$$= 3a(3a - 2b) - 4b(3a - 2b)$$
$$= (3a - 4b)(3a - 2b)$$

87. $35p^2 + 34pq + 8q^2$

(1) There are no common factors (other than 1 or -1).

(2) Multiply the leading coefficient, 35, and the constant, 8: $35 \cdot 8 = 280$.

(3) Look for a factorization of 280 in which the sum of the factors is the coefficient of the middle term, 34. The numbers we want are 14 and 20: $14 \cdot 20 = 280$ and $14 + 20 = 34$.

(4) Split the middle term: $34pq = 14pq + 20pq$

(5) Factor by grouping:
$$35p^2 + 34pq + 8q^2 = 35p^2 + 14pq + 20pq + 8q^2$$
$$= 7p(5p + 2q) + 4q(5p + 2q)$$
$$= (7p + 4q)(5p + 2q)$$

89. $18x^2 - 6xy - 24y^2$

(1) Factor out the common factor, 6.
$$18x^2 - 6xy - 24y^2 = 6(3x^2 - xy - 4y^2)$$

(2) Now we factor the trinomial $3x^2 - xy - 4y^2$. Multiply the leading coefficient, 3, and the constant, -4: $3(-4) = -12$.

(3) Look for a factorization of -12 in which the sum of the factors is the coefficient of the middle term, -1. The numbers we want are -4 and 3: $-4 \cdot 3 = -12$ and $-4 + 3 = -1$.

Exercise Set 5.4

(4) Split the middle term: $-xy = -4xy + 3xy$

(5) Factor by grouping:
$$3x^2 - xy - 4y^2 = 3x^2 - 4xy + 3xy - 4y^2$$
$$= x(3x - 4y) + y(3x - 4y)$$
$$= (x + y)(3x - 4y)$$

We must include the common factor to get a factorization of the original trinomial.
$$18x^2 - 6xy - 24y^2 = 6(x + y)(3x - 4y)$$

91. $60x + 18x^2 - 6x^3 = -6x^3 + 18x^2 + 60x$

(1) Factor out the common factor, $-6x$. We factor out $-6x$ rather than $6x$ in order to have a positive leading coefficient in the trinomial factor.
$$-6x^3 + 18x^2 + 60x = -6x(x^2 - 3x - 10)$$

(2) - (5) We factor $x^2 - 3x - 10$ as we did in Exercise 43, getting the $(x - 5)(x + 2)$. Then we have:
$$60x + 18x^2 - 6x^3$$
$$= -6x(x - 5)(x + 2)$$
$$= 6x(-x + 5)(x + 2)$$
Multiplying $x - 5$ by -1
$$= 6x(x - 5)(-x - 2)$$

Note that we can express $6x(-x + 5)(x + 2)$ as $6x(5 - x)(x + 2)$ since $-x + 5 = 5 - x$ by the commutative law of addition.

93. $35x^5 - 57x^4 - 44x^3$

(1) We first factor out the common factor, x^3.
$$x^3(35x^2 - 57x - 44)$$

(2) Now we factor the trinomial $35x^2 - 57x - 44$. Multiply the leading coefficient, 35, and the constant, -44: $35(-44) = -1540$.

(3) Look for a factorization of -1540 in which the sum of the factors is the coefficient of the middle term, -57.

Pairs of factors	Sums of factors
7, −220	−213
10, −154	−144
11, −140	−129
14, −110	−96
20, −77	−57

(4) Split the middle term: $-57x = 20x - 77x$

(5) Factor by grouping:
$$35x^2 - 57x - 44 = 35x^2 + 20x - 77x - 44$$
$$= 5x(7x + 4) - 11(7x + 4)$$
$$= (5x - 11)(7x + 4)$$

We must include the common factor to get a factorization of the original trinomial.
$$35x^5 - 57x^4 - 44x^3 = x^3(5x - 11)(7x + 4)$$

95. Discussion and Writing Exercise

97. $-10x > 1000$

$\dfrac{-10x}{-10} < \dfrac{1000}{-10}$ Dividing by -10 and reversing the inequality symbol

$x < -100$

The solution set is $\{x | x < -100\}$.

99. $6 - 3x \geq -18$

$-3x \geq -24$ Subtracting 6

$x \leq 8$ Dividing by -3 and reversing the inequality symbol

The solution set is $\{x | x \leq 8\}$.

101. $\dfrac{1}{2}x - 6x + 10 \leq x - 5x$

$2\left(\dfrac{1}{2}x - 6x + 10\right) \leq 2(x - 5x)$ Multiplying by 2 to clear the fraction

$x - 12x + 20 \leq 2x - 10x$

$-11x + 20 \leq -8x$ Collecting like terms

$20 \leq 3x$ Adding $11x$

$\dfrac{20}{3} \leq x$ Dividing by 3

The solution set is $\left\{x | x \geq \dfrac{20}{3}\right\}$.

103. $3x - 6x + 2(x - 4) > 2(9 - 4x)$

$3x - 6x + 2x - 8 > 18 - 8x$ Removing parentheses

$-x - 8 > 18 - 8x$ Collecting like terms

$7x > 26$ Adding $8x$ and 8

$x > \dfrac{26}{7}$ Dividing by 7

The solution set is $\left\{x | x > \dfrac{26}{7}\right\}$.

105. Familiarize. We will use the formula $C = 2\pi r$, where C is circumference and r is radius, to find the radius in kilometers. Then we will multiply that number by 0.62 to find the radius in miles.

Translate.

$\underbrace{\text{Circumference}}_{40,000} = \underbrace{2 \cdot \pi \cdot \text{radius}}_{2(3.14)r}$

$40,000 \approx 2(3.14)r$

Solve. First we solve the equation.

$40,000 \approx 2(3.14)r$

$40,000 \approx 6.28r$

$6369 \approx r$

Then we multiply to find the radius in miles:

$6369(0.62) \approx 3949$

Check. If $r = 6369$, then $2\pi r = 2(3.14)(6369) \approx 40,000$. We should also recheck the multiplication we did to find the radius in miles. Both values check.

State. The radius of the earth is about 6369 km or 3949 mi. (These values may differ slightly if a different approximation is used for π.)

107. $9x^{10} - 12x^5 + 4$

(a) First factor out a common factor, if any. There is none (other than 1 or −1).

(b) Multiply the leading coefficient, 9, and the constant, 4: $9 \cdot 4 = 36$.

(c) Look for a factorization of 36 in which the sum of the factors is the coefficient of the middle term, −12. The factors we want are −6 and −6.

(d) Split the middle term: $-12x^5 = -6x^5 - 6x^5$

(e) Factor by grouping:
$$9x^{10} - 12x^5 + 4 = 9x^{10} - 6x^5 - 6x^5 + 4$$
$$= 3x^5(3x^5 - 2) - 2(3x^5 - 2)$$
$$= (3x^5 - 2)(3x^5 - 2), \text{ or}$$
$$= (3x^5 - 2)^2$$

109. $16x^{10} + 8x^5 + 1$

(a) First factor out a common factor, if any. There is none (other than 1 or −1).

(b) Multiply the leading coefficient, 16, and the constant, 1: $16 \cdot 1 = 16$.

(c) Look for a factorization of 16 in which the sum of the factors is the coefficient of the middle term, 8. The factors we want are 4 and 4.

(d) Split the middle term: $8x^5 = 4x^5 + 4x^5$

(e) Factor by grouping:
$$16x^{10} + 8x^5 + 1 = 16x^{10} + 4x^5 + 4x^5 + 1$$
$$= 4x^5(4x^5 + 1) + 1(4x^5 + 1)$$
$$= (4x^5 + 1)(4x^5 + 1), \text{ or}$$
$$= (4x^5 + 1)^2$$

111.–119. Left to the student

Exercise Set 5.5

1. $x^2 - 14x + 49$

(a) We know that x^2 and 49 are squares.

(b) There is no minus sign before either x^2 or 49.

(c) If we multiply the square roots, x and 7, and double the product, we get $2 \cdot x \cdot 7 = 14x$. This is the opposite of the remaining term, $-14x$.

Thus, $x^2 - 14x + 49$ is a trinomial square.

3. $x^2 + 16x - 64$

Both x^2 and 64 are squares, but there is a minus sign before 64. Thus, $x^2 + 16x - 64$ is not a trinomial square.

5. $x^2 - 2x + 4$

(a) Both x^2 and 4 are squares.

(b) There is no minus sign before either x^2 or 4.

(c) If we multiply the square roots, x and 2, and double the product, we get $2 \cdot x \cdot 2 = 4x$. This is neither the remaining term nor its opposite.

Thus, $x^2 - 2x + 4$ is not a trinomial square.

7. $9x^2 - 36x + 24$

Only one term is a square. Thus, $9x^2 - 36x + 24$ is not a trinomial square.

9. $x^2 - 14x + 49 = x^2 - 2 \cdot x \cdot 7 + 7^2 = (x - 7)^2$
$ \uparrow \uparrow \uparrow \uparrow \uparrow$
$ = A^2 - 2 A B + B^2 = (A - B)^2$

11. $x^2 + 16x + 64 = x^2 + 2 \cdot x \cdot 8 + 8^2 = (x + 8)^2$
$ \uparrow \uparrow \uparrow \uparrow \uparrow$
$ = A^2 + 2 A B + B^2 = (A + B)^2$

13. $x^2 - 2x + 1 = x^2 - 2 \cdot x \cdot 1 + 1^2 = (x - 1)^2$

15. $4 + 4x + x^2 = x^2 + 4x + 4$ Changing the order
$ = x^2 + 2 \cdot x \cdot 2 + 2^2$
$ = (x + 2)^2$

17. $q^4 - 6q^2 + 9 = (q^2)^2 - 2 \cdot q^2 \cdot 3 + 3^2 = (q^2 - 3)^2$

19. $49 + 56y + 16y^2 = 16y^2 + 56y + 49$
$ = (4y)^2 + 2 \cdot 4y \cdot 7 + 7^2$
$ = (4y + 7)^2$

21. $2x^2 - 4x + 2 = 2(x^2 - 2x + 1)$
$ = 2(x^2 - 2 \cdot x \cdot 1 + 1^2)$
$ = 2(x - 1)^2$

23. $x^3 - 18x^2 + 81x = x(x^2 - 18x + 81)$
$ = x(x^2 - 2 \cdot x \cdot 9 + 9^2)$
$ = x(x - 9)^2$

25. $12q^2 - 36q + 27 = 3(4q^2 - 12q + 9)$
$ = 3[(2q)^2 - 2 \cdot 2q \cdot 3 + 3^2]$
$ = 3(2q - 3)^2$

27. $49 - 42x + 9x^2 = 7^2 - 2 \cdot 7 \cdot 3x + (3x)^2$
$ = (7 - 3x)^2$

29. $5y^4 + 10y^2 + 5 = 5(y^4 + 2y^2 + 1)$
$ = 5[(y^2)^2 + 2 \cdot y^2 \cdot 1 + 1^2]$
$ = 5(y^2 + 1)^2$

31. $1 + 4x^4 + 4x^2 = 1^2 + 2 \cdot 1 \cdot 2x^2 + (2x^2)^2$
$ = (1 + 2x^2)^2$

33. $4p^2 + 12pq + 9q^2 = (2p)^2 + 2 \cdot 2p \cdot 3q + (3q)^2$
$ = (2p + 3q)^2$

35. $a^2 - 6ab + 9b^2 = a^2 - 2 \cdot a \cdot 3b + (3b)^2$
$ = (a - 3b)^2$

37. $81a^2 - 18ab + b^2 = (9a)^2 - 2 \cdot 9a \cdot b + b^2$
$ = (9a - b)^2$

39. $36a^2 + 96ab + 64b^2 = 4(9a^2 + 24ab + 16b^2)$
$ = 4[(3a)^2 + 2 \cdot 3a \cdot 4b + (4b)^2]$
$ = 4(3a + 4b)^2$

Exercise Set 5.5

41. $x^2 - 4$

(a) The first expression is a square: x^2
The second expression is a square: $4 = 2^2$

(b) The terms have different signs.
$x^2 - 4$ is a difference of squares.

43. $x^2 + 25$

The terms do not have different signs.
$x^2 + 25$ is not a difference of squares.

45. $x^2 - 45$

The number 45 is not a square.
$x^2 - 45$ is not a difference of squares.

47. $16x^2 - 25y^2$

(a) The first expression is a square: $16x^2 = (4x)^2$
The second expression is a square: $25y^2 = (5y)^2$

(b) The terms have different signs.
$16x^2 - 25y^2$ is a difference of squares.

49. $y^2 - 4 = y^2 - 2^2 = (y+2)(y-2)$

51. $p^2 - 9 = p^2 - 3^2 = (p+3)(p-3)$

53. $-49 + t^2 = t^2 - 49 = t^2 - 7^2 = (t+7)(t-7)$

55. $a^2 - b^2 = (a+b)(a-b)$

57. $25t^2 - m^2 = (5t)^2 - m^2 = (5t+m)(5t-m)$

59. $100 - k^2 = 10^2 - k^2 = (10+k)(10-k)$

61. $16a^2 - 9 = (4a)^2 - 3^2 = (4a+3)(4a-3)$

63. $4x^2 - 25y^2 = (2x)^2 - (5y)^2 = (2x+5y)(2x-5y)$

65. $8x^2 - 98 = 2(4x^2 - 49) = 2[(2x)^2 - 7^2] =$
$2(2x+7)(2x-7)$

67. $36x - 49x^3 = x(36 - 49x^2) = x[6^2 - (7x)^2] =$
$x(6+7x)(6-7x)$

69. $49a^4 - 81 = (7a^2)^2 - 9^2 = (7a^2+9)(7a^2-9)$

71. $a^4 - 16$
$= (a^2)^2 - 4^2$
$= (a^2+4)(a^2-4)$ Factoring a difference of squares
$= (a^2+4)(a+2)(a-2)$ Factoring further: $a^2 - 4$ is a difference of squares.

73. $5x^4 - 405$
$5(x^4 - 81)$
$= 5[(x^2)^2 - 9^2]$
$= 5(x^2+9)(x^2-9)$
$= 5(x^2+9)(x+3)(x-3)$ Factoring $x^2 - 9$

75. $1 - y^8$
$= 1^2 - (y^4)^2$
$= (1+y^4)(1-y^4)$
$= (1+y^4)(1+y^2)(1-y^2)$ Factoring $1 - y^4$
$= (1+y^4)(1+y^2)(1+y)(1-y)$ Factoring $1 - y^2$

77. $x^{12} - 16$
$= (x^6)^2 - 4^2$
$= (x^6+4)(x^6-4)$
$= (x^6+4)(x^3+2)(x^3-2)$ Factoring $x^6 - 4$

79. $y^2 - \dfrac{1}{16} = y^2 - \left(\dfrac{1}{4}\right)^2$
$= \left(y+\dfrac{1}{4}\right)\left(y-\dfrac{1}{4}\right)$

81. $25 - \dfrac{1}{49}x^2 = 5^2 - \left(\dfrac{1}{7}x\right)^2$
$= \left(5+\dfrac{1}{7}x\right)\left(5-\dfrac{1}{7}x\right)$

83. $16m^4 - t^4$
$= (4m^2)^2 - (t^2)^2$
$= (4m^2+t^2)(4m^2-t^2)$
$= (4m^2+t^2)(2m+t)(2m-t)$ Factoring $4m^2 - t^2$

85. Discussion and Writing exercise

87. $-110 \div 10$ The quotient of a negative number and a positive number is negative.
$-110 \div 10 = -11$

89. $-\dfrac{2}{3} \div \dfrac{4}{5} = -\dfrac{2}{3} \cdot \dfrac{5}{4} = -\dfrac{10}{12} = -\dfrac{2 \cdot 5}{2 \cdot 6} = -\dfrac{\cancel{2} \cdot 5}{\cancel{2} \cdot 6} = -\dfrac{5}{6}$

91. $-64 \div (-32)$ The quotient of two negative numbers is a positive number.
$-64 \div (-32) = 2$

93. The shaded region is a square with sides of length $x - y - y$, or $x - 2y$. Its area is $(x-2y)(x-2y)$, or $(x-2y)^2$. Multiplying, we get the polynomial $x^2 - 4xy + 4y^2$.

95. $y^5 \cdot y^7 = y^{5+7} = y^{12}$

97. $y - 6x = 6$

To find the x-intercept, let $y = 0$. Then solve for x.
$y - 6x = 6$
$0 - 6x = 6$
$-6x = 6$
$x = -1$

The x-intercept is $(-1, 0)$.

To find the y-intercept, let $x = 0$. Then solve for y.
$y - 6x = 6$
$y - 6 \cdot 0 = 6$
$y = 6$

The y-intercept is $(0, 6)$.

Plot these points and draw the line.

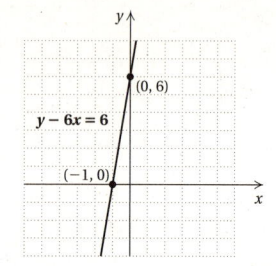

A third point should be used as a check. We substitute any value for x and solve for y. We let $x = -2$. Then
$$y - 6x = 6$$
$$y - 6(-2) = 6$$
$$y + 12 = 6$$
$$y = -6$$
The point $(-2, -6)$ is on the graph, so the graph is probably correct.

99. $49x^2 - 216$

There is no common factor. Also, $49x^2$ is a square, but 216 is not so this expression is not a difference of squares. It is not factorable. It is prime.

101. $x^2 + 22x + 121 = x^2 + 2 \cdot x \cdot 11 + 11^2$
$$= (x + 11)^2$$

103. $18x^3 + 12x^2 + 2x = 2x(9x^2 + 6x + 1)$
$$= 2x[(3x)^2 + 2 \cdot 3x \cdot 1 + 1^2]$$
$$= 2x(3x + 1)^2$$

105. $x^8 - 2^8$
$$= (x^4 + 2^4)(x^4 - 2^4)$$
$$= (x^4 + 2^4)(x^2 + 2^2)(x^2 - 2^2)$$
$$= (x^4 + 2^4)(x^2 + 2^2)(x + 2)(x - 2), \text{ or}$$
$$= (x^4 + 16)(x^2 + 4)(x + 2)(x - 2)$$

107. $3x^5 - 12x^3 = 3x^3(x^2 - 4) = 3x^3(x + 2)(x - 2)$

109. $18x^3 - \dfrac{8}{25}x = 2x\left(9x^2 - \dfrac{4}{25}\right) = 2x\left(3x + \dfrac{2}{5}\right)\left(3x - \dfrac{2}{5}\right)$

111. $0.49p - p^3 = p(0.49 - p^2) = p(0.7 + p)(0.7 - p)$

113. $0.64x^2 - 1.21 = (0.8x)^2 - (1.1)^2 = (0.8x + 1.1)(0.8x - 1.1)$

115. $(x+3)^2 - 9 = [(x+3)+3][(x+3)-3] = (x+6)x$, or $x(x+6)$

117. $x^2 - \left(\dfrac{1}{x}\right)^2 = \left(x + \dfrac{1}{x}\right)\left(x - \dfrac{1}{x}\right)$

119. $81 - b^{4k} = 9^2 - (b^{2k})^2$
$$= (9 + b^{2k})(9 - b^{2k})$$
$$= (9 + b^{2k})[3^2 - (b^k)^2]$$
$$= (9 + b^{2k})(3 + b^k)(3 - b^k)$$

121. $9b^{2n} + 12b^n + 4 = (3b^n)^2 + 2 \cdot 3b^n \cdot 2 + 2^2 = (3b^n + 2)^2$

123. $(y+3)^2 + 2(y+3) + 1$
$$= (y+3)^2 + 2 \cdot (y+3) \cdot 1 + 1^2$$
$$= [(y+3) + 1]^2$$
$$= (y+4)^2$$

125. If $cy^2 + 6y + 1$ is the square of a binomial, then $2 \cdot a \cdot 1 = 6$ where $a^2 = c$. Then $a = 3$, so $c = a^2 = 3^2 = 9$. (The polynomial is $9y^2 + 6y + 1$.)

127. Enter $y_1 = x^2 + 9$ and $y_2 = (x+3)(x+3)$ and look at a table of values. The y_1-and y_2-values are not the same, so the factorization is not correct.

129. Enter $y_1 = x^2 + 9$ and $y_2 = (x+3)^2$ and look at a table of values. The y_1-and y_2-values are not the same, so the factorization is not correct.

Exercise Set 5.6

1. $z^3 + 27 = z^3 + 3^3$
$$= (z+3)(z^2 - 3z + 9)$$
$$A^3 + B^3 = (A+B)(A^2 - AB + B^2)$$

3. $x^3 - 1 = x^3 - 1^3$
$$= (x-1)(x^2 + x + 1)$$
$$A^3 - B^3 = (A-B)(A^2 + AB + B^2)$$

5. $y^3 + 125 = y^3 + 5^3$
$$= (y+5)(y^2 - 5y + 25)$$
$$A^3 + B^3 = (A+B)(A^2 - AB + B^2)$$

7. $8a^3 + 1 = (2a)^3 + 1^3$
$$= (2a+1)(4a^2 - 2a + 1)$$
$$A^3 + B^3 = (A+B)(A^2 - AB + B^2)$$

9. $y^3 - 8 = y^3 - 2^3$
$$= (y-2)(y^2 + 2y + 4)$$
$$A^3 - B^3 = (A-B)(A^2 + AB + B^2)$$

11. $8 - 27b^3 = 2^3 - (3b)^3$
$$= (2 - 3b)(4 + 6b + 9b^2)$$

13. $64y^3 + 1 = (4y)^3 + 1^3$
$$= (4y+1)(16y^2 - 4y + 1)$$

15. $8x^3 + 27 = (2x)^3 + 3^3$
$$= (2x+3)(4x^2 - 6x + 9)$$

17. $a^3 - b^3 = (a-b)(a^2 + ab + b^2)$

19. $a^3 + \dfrac{1}{8} = a^3 + \left(\dfrac{1}{2}\right)^3$
$$= \left(a + \dfrac{1}{2}\right)\left(a^2 - \dfrac{1}{2}a + \dfrac{1}{4}\right)$$

21. $2y^3 - 128 = 2(y^3 - 64)$
$$= 2(y^3 - 4^3)$$
$$= 2(y - 4)(y^2 + 4y + 16)$$

23. $24a^3 + 3 = 3(8a^3 + 1)$
$= 3[(2a)^3 + 1^3]$
$= 3(2a+1)(4a^2 - 2a + 1)$

25. $rs^3 + 64r = r(s^3 + 64)$
$= r(s^3 + 4^3)$
$= r(s+4)(s^2 - 4s + 16)$

27. $5x^3 - 40z^3 = 5(x^3 - 8z^3)$
$= 5[x^3 - (2z)^3]$
$= 5(x - 2z)(x^2 + 2xz + 4z^2)$

29. $x^3 + 0.001 = x^3 + (0.1)^3$
$= (x + 0.1)(x^2 - 0.1x + 0.01)$

31. $64x^6 - 8t^6 = 8(8x^6 - t^6)$
$= 8[(2x^2)^3 - (t^2)^3]$
$= 8(2x^2 - t^2)(4x^4 + 2x^2t^2 + t^4)$

33. $2y^4 - 128y = 2y(y^3 - 64)$
$= 2y(y^3 - 4^3)$
$= 2y(y - 4)(y^2 + 4y + 16)$

35. $z^6 - 1$
$= (z^3)^2 - 1^2$ Writing as a difference of squares
$= (z^3 + 1)(z^3 - 1)$ Factoring a difference of squares
$= (z + 1)(z^2 - z + 1)(z - 1)(z^2 + z + 1)$
   Factoring a sum and a difference of cubes

37. $t^6 + 64y^6 = (t^2)^3 + (4y^2)^3$
$= (t^2 + 4y^2)(t^4 - 4t^2y^2 + 16y^4)$

39. $(7y^{-5})^3 = 7^3(y^{-5})^3 = 343y^{-5 \cdot 3} = 343y^{-15} = \dfrac{343}{y^{15}}$

41. $\left(\dfrac{x^3}{4}\right)^{-2} = \left(\dfrac{4}{x^3}\right)^2 = \dfrac{4^2}{(x^3)^2} = \dfrac{16}{x^{3 \cdot 2}} = \dfrac{16}{x^6}$

43. $\left(w - \dfrac{1}{3}\right)^2 = w^2 - 2 \cdot w \cdot \dfrac{1}{3} + \left(\dfrac{1}{3}\right)^2 = w^2 - \dfrac{2}{3}w + \dfrac{1}{9}$

45. $(a+b)^3 = (-2+3)^3 = 1^3 = 1$
$a^3 + b^3 = (-2)^3 + 3^3 = -8 + 27 = 19$
$(a+b)(a^2 - ab + b^2) =$
$(-2+3)[(-2)^2 - (-2)(3) + 3^2] =$
$1(4 + 6 + 9) = 1 \cdot 19 = 19$
$(a+b)(a^2 + ab + b^2) =$
$(-2+3)[(-2)^2 + (-2)(3) + 3^2] =$
$1(4 - 6 + 9) = 1 \cdot 7 = 7$
$(a+b)(a+b)(a+b) =$
$(-2+3)(-2+3)(-2+3) = 1 \cdot 1 \cdot 1 = 1$

47. $x^{6a} + y^{3b} = (x^{2a})^3 + (y^b)^3$
$= (x^{2a} + y^b)(x^{4a} - x^{2a}y^b + y^{2b})$

49. $3x^{3a} + 24y^{3b} = 3(x^{3a} + 8y^{3b})$
$= 3[(x^a)^3 + (2y^b)^3]$
$= 3(x^a + 2y^b)(x^{2a} - 2x^a y^b + 4y^{2b})$

51. $\dfrac{1}{24}x^3y^3 + \dfrac{1}{3}z^3 = \dfrac{1}{3}\left(\dfrac{1}{8}x^3y^3 + z^3\right)$
$= \dfrac{1}{3}\left[\left(\dfrac{1}{2}xy\right)^3 + z^3\right]$
$= \dfrac{1}{3}\left(\dfrac{1}{2}xy + z\right)\left(\dfrac{1}{4}x^2y^2 - \dfrac{1}{2}xyz + z^2\right)$

53. $(x+y)^3 - x^3$
$= [(x+y) - x][(x+y)^2 + x(x+y) + x^2]$
$= (x + y - x)(x^2 + 2xy + y^2 + x^2 + xy + x^2)$
$= y(3x^2 + 3xy + y^2)$

55. $(a+2)^3 - (a-2)^3$
$= [(a+2) - (a-2)][(a+2)^2 + (a+2)(a-2) + (a-2)^2]$
$= (a+2-a+2)(a^2+4a+4+a^2-4+a^2-4a+4)$
$= 4(3a^2 + 4)$

Exercise Set 5.7

1. $3x^2 - 192 = 3(x^2 - 64)$ 3 is a common factor
$= 3(x^2 - 8^2)$ Difference of squares
$= 3(x+8)(x-8)$

3. $a^2 + 25 - 10a = a^2 - 10a + 25$
$= a^2 - 2 \cdot a \cdot 5 + 5^2$ Trinomial square
$= (a - 5)^2$

5. $2x^2 - 11x + 12$

There is no common factor (other than 1). This polynomial has three terms, but it is not a trinomial square. Multiply the leading coefficient and the constant, 2 and 12: $2 \cdot 12 = 24$. Try to factor 24 so that the sum of the factors is -11. The numbers we want are -3 and -8: $-3(-8) = 24$ and $-3 + (-8) = -11$. Split the middle term and factor by grouping.

$2x^2 - 11x + 12 = 2x^2 - 3x - 8x + 12$
$= x(2x - 3) - 4(2x - 3)$
$= (x - 4)(2x - 3)$

7. $x^3 + 24x^2 + 144x$
$= x(x^2 + 24x + 144)$ x is a common factor
$= x(x^2 + 2 \cdot x \cdot 12 + 12^2)$ Trinomial square
$= x(x+12)^2$

9. $x^3 + 3x^2 - 4x - 12$
$= x^2(x+3) - 4(x+3)$ Factoring by grouping
$= (x^2 - 4)(x+3)$
$= (x+2)(x-2)(x+3)$ Factoring the difference of squares

11. $48x^2 - 3 = 3(16x^2 - 1)$ 3 is a common factor
$ = 3[(4x)^2 - 1^2]$ Difference of squares
$ = 3(4x + 1)(4x - 1)$

13. $9x^3 + 12x^2 - 45x$
$= 3x(3x^2 + 4x - 15)$ $3x$ is a common factor
$= 3x(3x - 5)(x + 3)$ Factoring the trinomial

15. $x^2 + 4$ is a *sum* of squares with no common factor. It cannot be factored. It is prime.

17. $x^4 + 7x^2 - 3x^3 - 21x = x(x^3 + 7x - 3x^2 - 21)$
$ = x[x(x^2 + 7) - 3(x^2 + 7)]$
$ = x[(x - 3)(x^2 + 7)]$
$ = x(x - 3)(x^2 + 7)$

19. $x^5 - 14x^4 + 49x^3$
$= x^3(x^2 - 14x + 49)$ x^3 is a common factor
$= x^3(x^2 - 2 \cdot x \cdot 7 + 7^2)$ Trinomial square
$= x^3(x - 7)^2$

21. $20 - 6x - 2x^2$
$= -2(-10 + 3x + x^2)$ -2 is a common factor
$= -2(x^2 + 3x - 10)$ Writing in descending order
$= -2(x + 5)(x - 2)$,   Using trial and error
or $2(-x - 5)(x - 2)$,
or $2(x + 5)(-x + 2)$

23. $x^2 - 6x + 1$

There is no common factor (other than 1 or -1). This is not a trinomial square, because $-6x \neq 2 \cdot x \cdot 1$ and $-6x \neq -2 \cdot x \cdot 1$. We try factoring using the refined trial and error procedure. We look for two factors of 1 whose sum is -6. There are none. The polynomial cannot be factored. It is prime.

25. $4x^4 - 64$
$= 4(x^4 - 16)$ 4 is a common factor
$= 4[(x^2)^2 - 4^2]$ Difference of squares
$= 4(x^2 + 4)(x^2 - 4)$ Difference of squares
$= 4(x^2 + 4)(x + 2)(x - 2)$

27. $1 - y^8$ Difference of squares
$= (1 + y^4)(1 - y^4)$ Difference of squares
$= (1 + y^4)(1 + y^2)(1 - y^2)$ Difference of squares
$= (1 + y^4)(1 + y^2)(1 + y)(1 - y)$

29. $x^5 - 4x^4 + 3x^3$
$= x^3(x^2 - 4x + 3)$ x^3 is a common factor
$= x^3(x - 3)(x - 1)$ Factoring the trinomial using trial and error

31. $\dfrac{1}{81}x^6 - \dfrac{8}{27}x^3 + \dfrac{16}{9}$
$= \dfrac{1}{9}\left(\dfrac{1}{9}x^6 - \dfrac{8}{3}x^3 + 16\right)$ $\dfrac{1}{9}$ is a common factor
$= \dfrac{1}{9}\left[\left(\dfrac{1}{3}x^3\right)^2 - 2 \cdot \dfrac{1}{3}x^3 \cdot 4 + 4^2\right]$ Trinomial square
$= \dfrac{1}{9}\left(\dfrac{1}{3}x^3 - 4\right)^2$

33. $mx^2 + my^2$
$= m(x^2 + y^2)$ m is a common factor

The factor with more than one term cannot be factored further, so we have factored completely.

35. $9x^2y^2 - 36xy = 9xy(xy - 4)$

37. $2\pi rh + 2\pi r^2 = 2\pi r(h + r)$

39. $(a + b)(x - 3) + (a + b)(x + 4)$
$= (a + b)[(x - 3) + (x + 4)]$ $(a + b)$ is a common factor
$= (a + b)(2x + 1)$

41. $(x - 1)(x + 1) - y(x + 1) = (x + 1)(x - 1 - y)$
$(x + 1)$ is a common factor

43. $n^2 + 2n + np + 2p$
$= n(n + 2) + p(n + 2)$ Factoring by grouping
$= (n + p)(n + 2)$

45. $6q^2 - 3q + 2pq - p$
$= 3q(2q - 1) + p(2q - 1)$ Factoring by grouping
$= (3q + p)(2q - 1)$

47. $4b^2 + a^2 - 4ab$
$= a^2 - 4ab + 4b^2$ Rearranging
$= a^2 - 2 \cdot a \cdot 2b + (2b)^2$ Trinomial square
$= (a - 2b)^2$

(Note that if we had rewritten the polynomial as $4b^2 - 4ab + a^2$, we might have written the result as $(2b - a)^2$. The two factorizations are equivalent.)

49. $16x^2 + 24xy + 9y^2$
$= (4x)^2 + 2 \cdot 4x \cdot 3y + (3y)^2$ Trinomial square
$= (4x + 3y)^2$

51. $49m^4 - 112m^2n + 64n^2$
$= (7m^2)^2 - 2 \cdot 7m^2 \cdot 8n + (8n)^2$ Trinomial square
$= (7m^2 - 8n)^2$

53. $y^4 + 10y^2z^2 + 25z^4$
$= (y^2)^2 + 2 \cdot y^2 \cdot 5z^2 + (5z^2)^2$ Trinomial square
$= (y^2 + 5z^2)^2$

55. $\dfrac{1}{4}a^2 + \dfrac{1}{3}ab + \dfrac{1}{9}b^2$
$= \left(\dfrac{1}{2}a\right)^2 + 2 \cdot \dfrac{1}{2}a \cdot \dfrac{1}{3}b + \left(\dfrac{1}{3}b\right)^2$
$= \left(\dfrac{1}{2}a + \dfrac{1}{3}b\right)^2$

Exercise Set 5.7

57. $a^2 - ab - 2b^2 = (a - 2b)(a + b)$ Using trial and error

59. $2mn - 360n^2 + m^2$
$= m^2 + 2mn - 360n^2$ Rewriting
$= (m + 20n)(m - 18n)$ Using trial and error

61. $m^2n^2 - 4mn - 32 = (mn - 8)(mn + 4)$ Using trial and error

63. $r^5s^2 - 10r^4s + 16r^3$
$= r^3(r^2s^2 - 10rs + 16)$ r^3 is a common factor
$= r^3(rs - 2)(rs - 8)$ Using trial and error

65. $a^5 + 4a^4b - 5a^3b^2$
$= a^3(a^2 + 4ab - 5b^2)$ a^3 is a common factor
$= a^3(a + 5b)(a - b)$ Factoring the trinomial

67. $a^2 - \dfrac{1}{25}b^2$
$= a^2 - \left(\dfrac{1}{5}b\right)^2$ Difference of squares
$= \left(a + \dfrac{1}{5}b\right)\left(a - \dfrac{1}{5}b\right)$

69. $7x^6 - 7y^6$
$= 7(x^6 - y^6)$ 7 is a common factor
$= 7[(x^3)^2 - (y^3)^2]$ Difference of squares
$= 7(x^3 + y^3)(x^3 - y^3)$
$= 7(x + y)(x^2 - xy + y^2)(x - y)(x^2 + xy + y^2)$
 Factoring the sum of cubes and the difference of cubes

71. $16 - p^4q^4$
$= 4^2 - (p^2q^2)^2$ Difference of squares
$= (4 + p^2q^2)(4 - p^2q^2)$ $4 - p^2q^2$ is a difference of squares
$= (4 + p^2q^2)(2 + pq)(2 - pq)$

73. $1 - 16x^{12}y^{12}$
$= 1^2 - (4x^6y^6)^2$ Difference of squares
$= (1 + 4x^6y^6)(1 - 4x^6y^6)$ $1 - 4x^6y^6$ is a difference of squares
$= (1 + 4x^6y^6)(1 + 2x^3y^3)(1 - 2x^3y^3)$

75. $q^3 + 8q^2 - q - 8$
$= q^2(q + 8) - (q + 8)$ Factoring by grouping
$= (q^2 - 1)(q + 8)$
$= (q + 1)(q - 1)(q + 8)$ Factoring the difference of squares

77. $112xy + 49x^2 + 64y^2$
$= 49x^2 + 112xy + 64y^2$ Rearranging
$= (7x)^2 + 2 \cdot 7x \cdot 8y + (8y)^2$ Trinomial square
$= (7x + 8y)^2$

79. Discussion and Writing Exercise

81. The highest point on the graph lies above 1999, so CD sales were highest in 1999.

83. The point on the graph corresponding to 779 million lies above 1996, so CD sales were 779 million in 1996.

85. From the graph we see that 847 million CDs were sold in 1998 and 939 million were sold in 1999. First we subtract to find the amount of the increase, in millions.

$9\ 3\ 9$
$-8\ 4\ 7$
$\overline{9\ 2}$

Now we find the percent increase.

92 is what percent of 847?
$\downarrow\ \downarrow\downarrow\downarrow\ \downarrow$
$92 = p \cdot\ 847$

We solve the equation.
$92 = p \cdot 847$
$\dfrac{92}{847} = p$
$0.109 \approx p$
$10.9\% \approx p$

Sales of CDs increased about 10.9% from 1998 to 1999.

87. $\dfrac{7}{5} \div \left(-\dfrac{11}{10}\right)$
$= \dfrac{7}{5} \cdot \left(-\dfrac{10}{11}\right)$ Multiplying by the reciprocal of the divisor
$= -\dfrac{7 \cdot 10}{5 \cdot 11}$
$= -\dfrac{7 \cdot 5 \cdot 2}{5 \cdot 11} = -\dfrac{7 \cdot 2}{11} \cdot \dfrac{5}{5}$
$= -\dfrac{14}{11}$

89. $A = aX + bX - 7$
$A + 7 = aX + bX$
$A + 7 = X(a + b)$
$\dfrac{A + 7}{a + b} = X$

91. $a^4 - 2a^2 + 1 = (a^2)^2 - 2 \cdot a^2 \cdot 1 + 1^2$
$= (a^2 - 1)^2$
$= [(a + 1)(a - 1)]^2$
$= (a + 1)^2(a - 1)^2$

93. $12.25x^2 - 7x + 1 = (3.5x)^2 - 2 \cdot (3.5x) \cdot 1 + 1^2$
$= (3.5x - 1)^2$

95. $5x^2 + 13x + 7.2$

Multiply the leading coefficient and the constant, 5 and 7.2: $5(7.2) = 36$. Try to factor 36 so that the sum of the factors is 13. The numbers we want are 9 and 4. Split the middle term and factor by grouping:

$5x^2 + 13x + 7.2 = 5x^2 + 9x + 4x + 7.2$
$= 5x(x + 1.8) + 4(x + 1.8)$
$= (5x + 4)(x + 1.8)$

97. $18 + y^3 - 9y - 2y^2$
$= y^3 - 2y^2 - 9y + 18$
$= y^2(y-2) - 9(y-2)$
$= (y^2 - 9)(y - 2)$
$= (y+3)(y-3)(y-2)$

99. $a^3 + 4a^2 + a + 4 = a^2(a+4) + 1(a+4)$
$ = (a^2 + 1)(a + 4)$

101. $x^3 - x^2 - 4x + 4 = x^2(x-1) - 4(x-1)$
$ = (x^2 - 4)(x - 1)$
$ = (x+2)(x-2)(x-1)$

103. $y^2(y-1) - 2y(y-1) + (y-1)$
$= (y-1)(y^2 - 2y + 1)$
$= (y-1)(y-1)^2$
$= (y-1)^3$

105. $(y+4)^2 + 2x(y+4) + x^2$
$= (y+4)^2 + 2 \cdot (y+4) \cdot x + x^2$ Trinomial square
$= (y+4+x)^2$

Exercise Set 5.8

1. $(x+4)(x+9) = 0$
$x + 4 = 0 \quad \text{or} \quad x + 9 = 0$ Using the principle of zero products
$x = -4 \quad \text{or} \quad x = -9$ Solving the two equations separately

Check:
For -4:
$\dfrac{(x+4)(x+9) = 0}{(-4+4)(-4+9) \;?\; 0}$
$\; 0 \cdot 5$
$ 0 \;\bigg|\; \text{TRUE}$

For -9:
$\dfrac{(x+4)(x+9) = 0}{(-9+4)(-9+9) \;?\; 0}$
$\; 13 \cdot 0$
$ 0 \;\bigg|\; \text{TRUE}$

The solutions are -4 and -9.

3. $(x+3)(x-8) = 0$
$x + 3 = 0 \quad \text{or} \quad x - 8 = 0$ Using the principle of zero products
$x = -3 \quad \text{or} \quad x = 8$

Check:
For -3:
$\dfrac{(x+3)(x-8) = 0}{(-3+3)(-3-8) \;?\; 0}$
$\; 0(-11)$
$ 0 \;\bigg|\; \text{TRUE}$

For 8:
$\dfrac{(x+3)(x-8) = 0}{(8+3)(8-8) \;?\; 0}$
$\; 11 \cdot 0$
$ 0 \;\bigg|\; \text{TRUE}$

The solutions are -3 and 8.

5. $(x+12)(x-11) = 0$
$x + 12 = 0 \quad \text{or} \quad x - 11 = 0$
$x = -12 \quad \text{or} \quad x = 11$
The solutions are -12 and 11.

7. $x(x+3) = 0$
$x = 0 \quad \text{or} \quad x + 3 = 0$
$x = 0 \quad \text{or} \quad x = -3$
The solutions are 0 and -3.

9. $0 = y(y + 18)$
$y = 0 \quad \text{or} \quad y + 18 = 0$
$y = 0 \quad \text{or} \quad y = -18$
The solutions are 0 and -18.

11. $(2x+5)(x+4) = 0$
$2x + 5 = 0 \quad \text{or} \quad x + 4 = 0$
$2x = -5 \quad \text{or} \quad x = -4$
$x = -\dfrac{5}{2} \quad \text{or} \quad x = -4$
The solutions are $-\dfrac{5}{2}$ and -4.

13. $(5x+1)(4x-12) = 0$
$5x + 1 = 0 \quad \text{or} \quad 4x - 12 = 0$
$5x = -1 \quad \text{or} \quad 4x = 12$
$x = -\dfrac{1}{5} \quad \text{or} \quad x = 3$
The solutions are $-\dfrac{1}{5}$ and 3.

15. $(7x - 28)(28x - 7) = 0$
$7x - 28 = 0 \quad \text{or} \quad 28x - 7 = 0$
$7x = 28 \quad \text{or} \quad 28x = 7$
$x = 4 \quad \text{or} \quad x = \dfrac{7}{28} = \dfrac{1}{4}$
The solutions are 4 and $\dfrac{1}{4}$.

17. $2x(3x - 2) = 0$
$2x = 0 \quad \text{or} \quad 3x - 2 = 0$
$x = 0 \quad \text{or} \quad 3x = 2$
$x = 0 \quad \text{or} \quad x = \dfrac{2}{3}$
The solutions are 0 and $\dfrac{2}{3}$.

Exercise Set 5.8

19. $\left(\dfrac{1}{5}+2x\right)\left(\dfrac{1}{9}-3x\right)=0$

$\dfrac{1}{5}+2x=0 \quad$ or $\quad \dfrac{1}{9}-3x=0$

$2x=-\dfrac{1}{5} \quad$ or $\quad -3x=-\dfrac{1}{9}$

$x=-\dfrac{1}{10} \quad$ or $\quad x=\dfrac{1}{27}$

The solutions are $-\dfrac{1}{10}$ and $\dfrac{1}{27}$.

21. $(0.3x-0.1)(0.05x+1)=0$

$0.3x-0.1=0 \quad$ or $\quad 0.05x+1=0$

$0.3x=0.1 \quad$ or $\quad 0.05x=-1$

$x=\dfrac{0.1}{0.3} \quad$ or $\quad x=-\dfrac{1}{0.05}$

$x=\dfrac{1}{3} \quad$ or $\quad x=-20$

The solutions are $\dfrac{1}{3}$ and -20.

23. $9x(3x-2)(2x-1)=0$

$9x=0 \quad$ or $\quad 3x-2=0 \quad$ or $\quad 2x-1=0$

$x=0 \quad$ or $\quad 3x=2 \quad$ or $\quad 2x=1$

$x=0 \quad$ or $\quad x=\dfrac{2}{3} \quad$ or $\quad x=\dfrac{1}{2}$

The solutions are 0, $\dfrac{2}{3}$, and $\dfrac{1}{2}$.

25. $x^2+6x+5=0$

$(x+5)(x+1)=0 \quad$ Factoring

$x+5=0 \quad$ or $\quad x+1=0 \quad$ Using the principle of zero products

$x=-5 \quad$ or $\quad x=-1$

The solutions are -5 and -1.

27. $x^2+7x-18=0$

$(x+9)(x-2)=0 \quad$ Factoring

$x+9=0 \quad$ or $\quad x-2=0 \quad$ Using the principle of zero products

$x=-9 \quad$ or $\quad x=2$

The solutions are -9 and 2.

29. $x^2-8x+15=0$

$(x-5)(x-3)=0$

$x-5=0 \quad$ or $\quad x-3=0$

$x=5 \quad$ or $\quad x=3$

The solutions are 5 and 3.

31. $x^2-8x=0$

$x(x-8)=0$

$x=0 \quad$ or $\quad x-8=0$

$x=0 \quad$ or $\quad x=8$

The solutions are 0 and 8.

33. $x^2+18x=0$

$x(x+18)=0$

$x=0 \quad$ or $\quad x+18=0$

$x=0 \quad$ or $\quad x=-18$

The solutions are 0 and -18.

35. $\qquad x^2=16$

$x^2-16=0 \quad$ Subtracting 16

$(x-4)(x+4)=0$

$x-4=0 \quad$ or $\quad x+4=0$

$x=4 \quad$ or $\quad x=-4$

The solutions are 4 and -4.

37. $\qquad 9x^2-4=0$

$(3x-2)(3x+2)=0$

$3x-2=0 \quad$ or $\quad 3x+2=0$

$3x=2 \quad$ or $\quad 3x=-2$

$x=\dfrac{2}{3} \quad$ or $\quad x=-\dfrac{2}{3}$

The solutions are $\dfrac{2}{3}$ and $-\dfrac{2}{3}$.

39. $0=6x+x^2+9$

$0=x^2+6x+9 \quad$ Writing in descending order

$0=(x+3)(x+3)$

$x+3=0 \quad$ or $\quad x+3=0$

$x=-3 \quad$ or $\quad x=-3$

There is only one solution, -3.

41. $\qquad x^2+16=8x$

$x^2-8x+16=0 \quad$ Subtracting $8x$

$(x-4)(x-4)=0$

$x-4=0 \quad$ or $\quad x-4=0$

$x=4 \quad$ or $\quad x=4$

There is only one solution, 4.

43. $\qquad 5x^2=6x$

$5x^2-6x=0$

$x(5x-6)=0$

$x=0 \quad$ or $\quad 5x-6=0$

$x=0 \quad$ or $\quad 5x=6$

$x=0 \quad$ or $\quad x=\dfrac{6}{5}$

The solutions are 0 and $\dfrac{6}{5}$.

45.
$$6x^2 - 4x = 10$$
$$6x^2 - 4x - 10 = 0$$
$$2(3x^2 - 2x - 5) = 0$$
$$2(3x - 5)(x + 1) = 0$$
$$3x - 5 = 0 \quad \text{or} \quad x + 1 = 0$$
$$3x = 5 \quad \text{or} \quad x = -1$$
$$x = \frac{5}{3} \quad \text{or} \quad x = -1$$
The solutions are $\frac{5}{3}$ and -1.

47.
$$12y^2 - 5y = 2$$
$$12y^2 - 5y - 2 = 0$$
$$(4y + 1)(3y - 2) = 0$$
$$4y + 1 = 0 \quad \text{or} \quad 3y - 2 = 0$$
$$4y = -1 \quad \text{or} \quad 3y = 2$$
$$y = -\frac{1}{4} \quad \text{or} \quad y = \frac{2}{3}$$
The solutions are $-\frac{1}{4}$ and $\frac{2}{3}$.

49.
$$t(3t + 1) = 2$$
$$3t^2 + t = 2 \quad \text{Multiplying on the left}$$
$$3t^2 + t - 2 = 0 \quad \text{Subtracting 2}$$
$$(3t - 2)(t + 1) = 0$$
$$3t - 2 = 0 \quad \text{or} \quad t + 1 = 0$$
$$3t = 2 \quad \text{or} \quad t = -1$$
$$t = \frac{2}{3} \quad \text{or} \quad t = -1$$
The solutions are $\frac{2}{3}$ and -1.

51.
$$100y^2 = 49$$
$$100y^2 - 49 = 0$$
$$(10y + 7)(10y - 7) = 0$$
$$10y + 7 = 0 \quad \text{or} \quad 10y - 7 = 0$$
$$10y = -7 \quad \text{or} \quad 10y = 7$$
$$y = -\frac{7}{10} \quad \text{or} \quad y = \frac{7}{10}$$
The solutions are $-\frac{7}{10}$ and $\frac{7}{10}$.

53.
$$x^2 - 5x = 18 + 2x$$
$$x^2 - 5x - 18 - 2x = 0 \quad \text{Subtracting 18 and } 2x$$
$$x^2 - 7x - 18 = 0$$
$$(x - 9)(x + 2) = 0$$
$$x - 9 = 0 \quad \text{or} \quad x + 2 = 0$$
$$x = 9 \quad \text{or} \quad x = -2$$
The solutions are 9 and -2.

55.
$$10x^2 - 23x + 12 = 0$$
$$(5x - 4)(2x - 3) = 0$$
$$5x - 4 = 0 \quad \text{or} \quad 2x - 3 = 0$$
$$5x = 4 \quad \text{or} \quad 2x = 3$$
$$x = \frac{4}{5} \quad \text{or} \quad x = \frac{3}{2}$$
The solutions are $\frac{4}{5}$ and $\frac{3}{2}$.

57. We let $y = 0$ and solve for x.
$$0 = x^2 + 3x - 4$$
$$0 = (x + 4)(x - 1)$$
$$x + 4 = 0 \quad \text{or} \quad x - 1 = 0$$
$$x = -4 \quad \text{or} \quad x = 1$$
The x-intercepts are $(-4, 0)$ and $(1, 0)$.

59. We let $y = 0$ and solve for x.
$$0 = 2x^2 + x - 10$$
$$0 = (2x + 5)(x - 2)$$
$$2x + 5 = 0 \quad \text{or} \quad x - 2 = 0$$
$$2x = -5 \quad \text{or} \quad x = 2$$
$$x = -\frac{5}{2} \quad \text{or} \quad x = 2$$
The x-intercepts are $\left(-\frac{5}{2}, 0\right)$ and $(2, 0)$.

61. We let $y = 0$ and solve for x.
$$0 = x^2 - 2x - 15$$
$$0 = (x - 5)(x + 3)$$
$$x - 5 = 0 \quad \text{or} \quad x + 3 = 0$$
$$x = 5 \quad \text{or} \quad x = -3$$
The x-intercepts are $(5, 0)$ and $(-3, 0)$.

63. The solutions of the equation are the first coordinates of the x-intercepts of the graph. From the graph we see that the x-intercepts are $(-1, 0)$ and $(4, 0)$, so the solutions of the equation are -1 and 4.

65. The solutions of the equation are the first coordinates of the x-intercepts of the graph. From the graph we see that the x-intercepts are $(-1, 0)$ and $(3, 0)$, so the solutions of the equation are -1 and 3.

67. Discussion and Writing Exercise

69. $(a + b)^2$

71. The two numbers have different signs, so their quotient is negative.
$$144 \div -9 = -16$$

73.
$$-\frac{5}{8} \div \frac{3}{16} = -\frac{5}{8} \cdot \frac{16}{3}$$
$$= -\frac{5 \cdot 16}{8 \cdot 3}$$
$$= -\frac{5 \cdot \cancel{8} \cdot 2}{\cancel{8} \cdot 3}$$
$$= -\frac{10}{3}$$

Exercise Set 5.9   115

75.
$$b(b+9) = 4(5+2b)$$
$$b^2 + 9b = 20 + 8b$$
$$b^2 + 9b - 8b - 20 = 0$$
$$b^2 + b - 20 = 0$$
$$(b+5)(b-4) = 0$$
$$b+5 = 0 \quad \text{or} \quad b-4 = 0$$
$$b = -5 \quad \text{or} \quad b = 4$$
The solutions are -5 and 4.

77.
$$(t-3)^2 = 36$$
$$t^2 - 6t + 9 = 36$$
$$t^2 - 6t - 27 = 0$$
$$(t-9)(t+3) = 0$$
$$t-9 = 0 \quad \text{or} \quad t+3 = 0$$
$$t = 9 \quad \text{or} \quad t = -3$$
The solutions are 9 and -3.

79.
$$x^2 - \frac{1}{64} = 0$$
$$\left(x - \frac{1}{8}\right)\left(x + \frac{1}{8}\right) = 0$$
$$x - \frac{1}{8} = 0 \quad \text{or} \quad x + \frac{1}{8} = 0$$
$$x = \frac{1}{8} \quad \text{or} \quad x = -\frac{1}{8}$$
The solutions are $\frac{1}{8}$ and $-\frac{1}{8}$.

81.
$$\frac{5}{16}x^2 = 5$$
$$\frac{5}{16}x^2 - 5 = 0$$
$$5\left(\frac{1}{16}x^2 - 1\right) = 0$$
$$5\left(\frac{1}{4}x - 1\right)\left(\frac{1}{4}x + 1\right) = 0$$
$$\frac{1}{4}x - 1 = 0 \quad \text{or} \quad \frac{1}{4}x + 1 = 0$$
$$\frac{1}{4}x = 1 \quad \text{or} \quad \frac{1}{4}x = -1$$
$$x = 4 \quad \text{or} \quad x = -4$$
The solutions are 4 and -4.

83. (a)
$$x = -3 \quad \text{or} \quad x = 4$$
$$x + 3 = 0 \quad \text{or} \quad x - 4 = 0$$
$$(x+3)(x-4) = 0 \quad \text{Principle of zero products}$$
$$x^2 - x - 12 = 0 \quad \text{Multiplying}$$

(b)
$$x = -3 \quad \text{or} \quad x = -4$$
$$x + 3 = 0 \quad \text{or} \quad x + 4 = 0$$
$$(x+3)(x+4) = 0$$
$$x^2 + 7x + 12 = 0$$

(c)
$$x = \frac{1}{2} \quad \text{or} \quad x = \frac{1}{2}$$
$$x - \frac{1}{2} = 0 \quad \text{or} \quad x - \frac{1}{2} = 0$$
$$\left(x - \frac{1}{2}\right)\left(x - \frac{1}{2}\right) = 0$$
$$x^2 - x + \frac{1}{4} = 0, \quad \text{or}$$
$$4x^2 - 4x + 1 = 0 \quad \text{Multiplying by 4}$$

(d)
$$(x-5)(x+5) = 0$$
$$x^2 - 25 = 0$$

(e)
$$(x-0)(x-0.1)\left(x - \frac{1}{4}\right) = 0$$
$$x\left(x - \frac{1}{10}\right)\left(x - \frac{1}{4}\right) = 0$$
$$x\left(x^2 - \frac{7}{20}x + \frac{1}{40}\right) = 0$$
$$x^3 - \frac{7}{20}x^2 + \frac{1}{40}x = 0, \quad \text{or}$$
$$40x^3 - 14x^2 + x = 0 \quad \text{Multiplying by 40}$$

85. $2.33, 6.77$

87. $-9.15, -4.59$

89. $0, 2.74$

Exercise Set 5.9

1. Familiarize. Let $w =$ the width of the table, in feet. Then $6w =$ the length, in feet. Recall that the area of a rectangle is Length $\cdot$ Width.

Translate.

The area of the table is 24 ft^2.
$$6w \cdot w = 24$$

Solve. We solve the equation.
$$6w \cdot w = 24$$
$$6w^2 = 24$$
$$6w^2 - 24 = 0$$
$$6(w^2 - 4) = 0$$
$$6(w+2)(w-2) = 0$$
$$w + 2 = 0 \quad \text{or} \quad w - 2 = 0$$
$$w = -2 \quad \text{or} \quad w = 2$$

Check. Since the width must be positive, -2 cannot be a solution. If the width is 2 ft, then the length is $6 \cdot 2$ ft, or 12 ft, and the area is 12 ft $\cdot$ 2 ft $= 24$ ft^2. These numbers check.

State. The table is 12 ft long and 2 ft wide.

3. Familiarize. We make a drawing. Let $w =$ the width, in cm. Then $2w + 2 =$ the length, in cm.

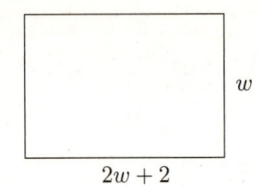

[Diagram: rectangle with width w and length $2w+2$]

Recall that the area of a rectangle is length times width.

Translate. We reword the problem.

Length times width is 144 cm².

$(2w+2) \cdot w = 144$

Solve. We solve the equation.
$$(2w+2)w = 144$$
$$2w^2 + 2w = 144$$
$$2w^2 + 2w - 144 = 0$$
$$2(w^2 + w - 72) = 0$$
$$2(w+9)(w-8) = 0$$
$$w+9 = 0 \quad or \quad w-8 = 0$$
$$w = -9 \quad or \quad w = 8$$

Check. Since the width must be positive, -9 cannot be a solution. If the width is 8 cm, then the length is $2 \cdot 8 + 2$, or 18 cm, and the area is $8 \cdot 18$, or 144 cm². Thus, 8 checks.

State. The width is 8 cm, and the length is 18 cm.

5. ***Familiarize.*** Using the labels shown on the drawing in the text, we let h = the height, in cm, and $h + 10$ = the base, in cm. Recall that the formula for the area of a triangle is $\frac{1}{2} \cdot$ (base) $\cdot$ (height).

Translate.

$\frac{1}{2}$ times base times height is 28 cm².

$\frac{1}{2} \cdot (h+10) \cdot h = 28$

Solve. We solve the equation.
$$\frac{1}{2}(h+10)h = 28$$
$$(h+10)h = 56 \quad \text{Multiplying by 2}$$
$$h^2 + 10h = 56$$
$$h^2 + 10h - 56 = 0$$
$$(h+14)(h-4) = 0$$
$$h+14 = 0 \quad or \quad h-4 = 0$$
$$h = -14 \quad or \quad h = 4$$

Check. Since the height of the triangle must be positive, -14 cannot be a solution. If the height is 4 cm, then the base is $4+10$, or 14 cm, and the area is $\frac{1}{2} \cdot 14 \cdot 4$, or 28 cm². Thus, 4 checks.

State. The height of the triangle is 4 cm, and the base is 14 cm.

7. ***Familiarize.*** Using the labels show on the drawing in the text, we let h = the height of the triangle, in meters, and $\frac{1}{2}h$ = the length of the base, in meters. Recall that the formula for the area of a triangle is $\frac{1}{2} \cdot$ (base) $\cdot$ (height).

Translate.

$\frac{1}{2}$ times base times height is 64 m².

$\frac{1}{2} \cdot \frac{1}{2}h \cdot h = 64$

Solve. We solve the equation.
$$\frac{1}{2} \cdot \frac{1}{2}h \cdot h = 64$$
$$\frac{1}{4}h^2 = 64$$
$$h^2 = 256 \quad \text{Multiplying by 4}$$
$$h^2 - 256 = 0$$
$$(h+16)(h-16) = 0$$
$$h+16 = 0 \quad or \quad h-16 = 0$$
$$h = -16 \quad or \quad h = 16$$

Check. The height of the triangle cannot be negative, so -16 cannot be a solution. If the height is 16 m, then the length of the base is $\frac{1}{2} \cdot 16$ m, or 8 m, and the area is $\frac{1}{2} \cdot 8$ m $\cdot 16$ m $= 64$ m². These numbers check.

State. The length of the base is 8 m, and the height is 16 m.

9. ***Familiarize.*** Reread Example 4 in Section 4.3.

Translate. Substitute 14 for n.
$$14^2 - 14 = N$$

Solve. We do the computation on the left.
$$14^2 - 14 = N$$
$$196 - 14 = N$$
$$182 = N$$

Check. We can redo the computation, or we can solve the equation $n^2 - n = 182$. The answer checks.

State. 182 games will be played.

11. ***Familiarize.*** Reread Example 4 in Section 4.3.

Translate. Substitute 132 for N.
$$n^2 - n = 132$$

Solve.
$$n^2 - n = 132$$
$$n^2 - n - 132 = 0$$
$$(n-12)(n+11) = 0$$
$$n-12 = 0 \quad or \quad n+11 = 0$$
$$n = 12 \quad or \quad n = -11$$

Check. The solutions of the equation are 12 and -11. Since the number of teams cannot be negative, -11 cannot be a solution. But 12 checks since $12^2 - 12 = 144 - 12 = 132$.

State. There are 12 teams in the league.

Exercise Set 5.9

13. Familiarize. We will use the formula $N = \frac{1}{2}(n^2 - n)$.

Translate. Substitute 100 for n.
$$N = \frac{1}{2}(100^2 - 100)$$

Solve. We do the computation on the right.
$$N = \frac{1}{2}(10,000 - 100)$$
$$N = \frac{1}{2}(9900)$$
$$N = 4950$$

Check. We can redo the computation, or we can solve the equation $4950 = \frac{1}{2}(n^2 - n)$. The answer checks.

State. 4950 handshakes are possible.

15. Familiarize. We will use the formula $N = \frac{1}{2}(n^2 - n)$.

Translate. Substitute 300 for N.
$$300 = \frac{1}{2}(n^2 - n)$$

Solve. We solve the equation.
$$2 \cdot 300 = 2 \cdot \frac{1}{2}(n^2 - n) \quad \text{Multiplying by 2}$$
$$600 = n^2 - n$$
$$0 = n^2 - n - 600$$
$$0 = (n + 24)(n - 25)$$
$$n + 24 = 0 \quad \text{or} \quad n - 25 = 0$$
$$n = -24 \quad \text{or} \quad n = 25$$

Check. The number of people at a meeting cannot be negative, so -24 cannot be a solution. But 25 checks since $\frac{1}{2}(25^2 - 25) = \frac{1}{2}(625 - 25) = \frac{1}{2} \cdot 600 = 300$.

State. There were 25 people at the party.

17. Familiarize. We will use the formula $N = \frac{1}{2}(n^2 - n)$, since toasts can be substituted for handshakes.

Translate. Substitute 190 for N.
$$190 = \frac{1}{2}(n^2 - n)$$

Solve.
$$190 = \frac{1}{2}(n^2 - n)$$
$$380 = n^2 - n \quad \text{Multiplying by 2}$$
$$0 = n^2 - n - 380$$
$$0 = (n - 20)(n + 19)$$
$$n - 20 = 0 \quad \text{or} \quad n + 19 = 0$$
$$n = 20 \quad \text{or} \quad n = -19$$

Check. The solutions of the equation are 20 and -19. Since the number of people cannot be negative, -19 cannot be a solution. However, 20 checks since $\frac{1}{2}(20^2 - 20) = \frac{1}{2}(400 - 20) = \frac{1}{2}(380) = 190$.

State. 20 people took part in the toast.

19. Familiarize. The page numbers on facing pages are consecutive integers. Let $x =$ the smaller integer. Then $x + 1 =$ the larger integer.

Translate. We reword the problem.

Smaller integer times larger integer is 210.
$$x \cdot (x+1) = 210$$

Solve. We solve the equation.
$$x(x+1) = 210$$
$$x^2 + x = 210$$
$$x^2 + x - 210 = 0$$
$$(x + 15)(x - 14) = 0$$
$$x + 15 = 0 \quad \text{or} \quad x - 14 = 0$$
$$x = -15 \quad \text{or} \quad x = 14$$

Check. The solutions of the equation are -15 and 14. Since a page number cannot be negative, -15 cannot be a solution of the original problem. We only need to check 14. When $x = 14$, then $x + 1 = 15$, and $14 \cdot 15 = 210$. This checks.

State. The page numbers are 14 and 15.

21. Familiarize. Let $x =$ the smaller even integer. Then $x + 2 =$ the larger even integer.

Translate. We reword the problem.

Smaller even integer times larger even integer is 168.
$$x \cdot (x+2) = 168$$

Solve.
$$x(x+2) = 168$$
$$x^2 + 2x = 168$$
$$x^2 + 2x - 168 = 0$$
$$(x + 14)(x - 12) = 0$$
$$x + 14 = 0 \quad \text{or} \quad x - 12 = 0$$
$$x = -14 \quad \text{or} \quad x = 12$$

Check. The solutions of the equation are -14 and 12. When x is -14, then $x + 2$ is -12 and $-14(-12) = 168$. The numbers -14 and -12 are consecutive even integers which are solutions of the problem. When x is 12, then $x + 2$ is 14 and $12 \cdot 14 = 168$. The numbers 12 and 14 are also consecutive even integers which are solutions of the problem.

State. We have two solutions, each of which consists of a pair of numbers: -14 and -12, and 12 and 14.

23. Familiarize. Let $x =$ the smaller odd integer. Then $x + 2 =$ the larger odd integer.

Translate. We reword the problem.

Smaller odd integer times larger odd integer is 255.
$$x \cdot (x+2) = 255$$

Solve.
$$x(x+2) = 255$$
$$x^2 + 2x = 255$$
$$x^2 + 2x - 255 = 0$$
$$(x-15)(x+17) = 0$$
$$x - 15 = 0 \quad \text{or} \quad x + 17 = 0$$
$$x = 15 \quad \text{or} \quad x = -17$$

Check. The solutions of the equation are 15 and -17. When x is 15, then $x+2$ is 17 and $15 \cdot 17 = 255$. The numbers 15 and 17 are consecutive odd integers which are solutions to the problem. When x is -17, then $x+2$ is -15 and $-17(-15) = 255$. The numbers -17 and -15 are also consecutive odd integers which are solutions to the problem.

State. We have two solutions, each of which consists of a pair of numbers: 15 and 17, and -17 and -15.

25. ***Familiarize***. We make a drawing. Let x = the length of the unknown leg. Then $x + 2$ = the length of the hypotenuse.

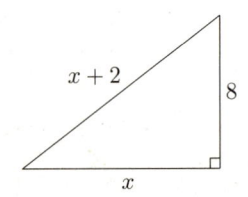

Translate. Use the Pythagorean theorem.
$$a^2 + b^2 = c^2$$
$$8^2 + x^2 = (x+2)^2$$

Solve. We solve the equation.
$$8^2 + x^2 = (x+2)^2$$
$$64 + x^2 = x^2 + 4x + 4$$
$$60 = 4x \qquad \text{Subtracting } x^2 \text{ and } 4$$
$$15 = x$$

Check. When $x = 15$, then $x + 2 = 17$ and $8^2 + 15^2 = 17^2$. Thus, 15 and 17 check.

State. The lengths of the hypotenuse and the other leg are 17 ft and 15 ft, respectively.

27. ***Familiarize***. Consider the drawing in the text. We let w = the width of Main Street, in feet.

Translate. Use the Pythagorean theorem.
$$a^2 + b^2 = c^2$$
$$24^2 + w^2 = 40^2$$

Solve. We solve the equation.
$$24^2 + w^2 = 40^2$$
$$576 + w^2 = 1600$$
$$w^2 - 1024 = 0$$
$$(w+32)(w-32) = 0$$
$$w + 32 = 0 \quad \text{or} \quad w - 32 = 0$$
$$w = -32 \quad \text{or} \quad w = 32$$

Check. The width of the street cannot be negative, so -32 cannot be a solution. If Main Street is 32 ft wide, we have $24^2 + 32^2 = 576 + 1024 = 1600$, which is 40^2. Thus, 32 ft checks.

State. Main Street is 32 ft wide.

29. ***Familiarize***. We make a drawing. Let l = the length of the cable, in ft.

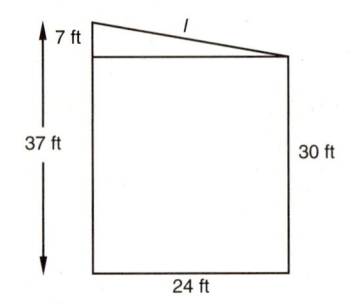

Note that we have a right triangle with hypotenuse l and legs of 24 ft and $37 - 30$, or 7 ft.

Translate. We use the Pythagorean theorem.
$$a^2 + b^2 = c^2$$
$$7^2 + 24^2 = l^2 \quad \text{Substituting}$$

Solve.
$$7^2 + 24^2 = l^2$$
$$49 + 576 = l^2$$
$$625 = l^2$$
$$0 = l^2 - 625$$
$$0 = (l+25)(l-25)$$
$$l + 25 = 0 \quad \text{or} \quad l - 25 = 0$$
$$l = -25 \quad \text{or} \quad l = 25$$

Check. The integer -25 cannot be the length of the cable, because it is negative. When $l = 25$, we have $7^2 + 24^2 = 25^2$. This checks.

State. The cable is 25 ft long.

31. ***Familiarize***. We label the drawing. Let x = the length of a side of the dining room, in ft. Then the dining room has dimensions x by x and the kitchen has dimensions x by 10. The entire rectangular space has dimension x by $x + 10$. Recall that we multiply these dimensions to find the area of the rectangle.

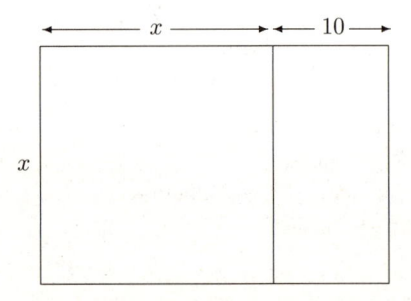

Exercise Set 5.9

Translate.

$$\underbrace{\text{The area of the rectangular space}}_{x(x+10)} \text{ is } \underbrace{264 \text{ ft}^2}_{264}.$$

Solve. We solve the equation.
$$x(x+10) = 264$$
$$x^2 + 10x = 264$$
$$x^2 + 10x - 264 = 0$$
$$(x+22)(x-12) = 0$$
$$x + 22 = 0 \quad \text{or} \quad x - 12 = 0$$
$$x = -22 \text{ or} \quad x = 12$$

Check. Since the length of a side of the dining room must be positive, -22 cannot be a solution. If x is 12 ft, then $x+10$ is 22 ft, and the area of the space is $12 \cdot 22$, or 264 ft^2. The number 12 checks.

State. The dining room is 12 ft by 12 ft, and the kitchen is 12 ft by 10 ft.

33. **Familiarize.** We will use the formula $h = 180t - 16t^2$.

    Translate. Substitute 464 for h.
    $$464 = 180t - 16t^2$$

    Solve. We solve the equation.
    $$464 = 180t - 16t^2$$
    $$16t^2 - 180t + 464 = 0$$
    $$4(4t^2 - 45t + 116) = 0$$
    $$4(4t - 29)(t - 4) = 0$$
    $$4t - 29 = 0 \quad \text{or} \quad t - 4 = 0$$
    $$4t = 29 \quad \text{or} \quad t = 4$$
    $$t = \frac{29}{4} \quad \text{or} \quad t = 4$$

    Check. The solutions of the equation are $\frac{29}{4}$, or $7\frac{1}{4}$, and 4. Since we want to find how many seconds it takes the rocket to *first* reach a height of 464 ft, we check the smaller number, 4. We substitute 4 for t in the formula.
    $$h = 180t - 16t^2$$
    $$h = 180 \cdot 4 - 16(4)^2$$
    $$h = 180 \cdot 4 - 16 \cdot 16$$
    $$h = 720 - 256$$
    $$h = 464$$
    The answer checks.

    State. The rocket will first reach a height of 464 ft after 4 seconds.

35. **Familiarize.** Let $x =$ the smaller odd positive integer. Then $x + 2 =$ the larger odd positive integer.

    Translate.

    $$\underbrace{\text{Square of the smaller odd positive integer}}_{x^2} + \underbrace{\text{Square of the larger odd positive integer}}_{(x+2)^2} \text{ is } \underbrace{74}_{74}$$

    Solve.
    $$x^2 + (x+2)^2 = 74$$
    $$x^2 + x^2 + 4x + 4 = 74$$
    $$2x^2 + 4x - 70 = 0$$
    $$2(x^2 + 2x - 35) = 0$$
    $$2(x+7)(x-5) = 0$$
    $$x + 7 = 0 \quad \text{or} \quad x - 5 = 0$$
    $$x = -7 \text{ or} \quad x = 5$$

    Check. The solutions of the equation are -7 and 5. The problem asks for odd positive integers, so -7 cannot be a solution. When x is 5, $x + 2$ is 7. The numbers 5 and 7 are consecutive odd positive integers. The sum of their squares, $25 + 49$, is 74. The numbers check.

    State. The integers are 5 and 7.

37. Discussion and Writing Exercise

39. $(3x - 5y)(3x + 5y) = (3x)^2 - (5y)^2 = 9x^2 - 25y^2$

41. $(3x + 5y)^2 = (3x)^2 + 2 \cdot 3x \cdot 5y + (5y)^2 = 9x^2 + 30xy + 25y^2$

43. $4x - 16y = 64$

    To find the x-intercept, let $y = 0$ and solve for x.
    $$4x - 16y = 64$$
    $$4x - 16 \cdot 0 = 64$$
    $$4x = 64$$
    $$x = 16$$
    The x-intercept is $(16, 0)$.

    To find the y-intercept, let $x = 0$ and solve for y.
    $$4x - 16y = 64$$
    $$4 \cdot 0 - 16y = 64$$
    $$-16y = 64$$
    $$y = -4$$
    The y-intercept is $(0, -4)$.

45. $x - 1.3y = 6.5$

    To find the x-intercept, let $y = 0$ and solve for x.
    $$x - 1.3y = 6.5$$
    $$x - 1.3(0) = 6.5$$
    $$x = 6.5$$
    The x-intercept is $(6.5, 0)$.

    To find the y-intercept, let $x = 0$ and solve for y.
    $$x - 1.3y = 6.5$$
    $$0 - 1.3y = 6.5$$
    $$-1.3y = 6.5$$
    $$y = -5$$
    The y-intercept is $(0, -5)$.

47. $y = 4 - 5x$

To find the x-intercept, let $y = 0$ and solve for x.
$$y = 4 - 5x$$
$$0 = 4 - 5x$$
$$5x = 4$$
$$x = \frac{4}{5}$$

The x-intercept is $\left(\frac{4}{5}, 0\right)$.

To find the y-intercept, let $x = 0$ and solve for y.
$$y = 4 - 5x$$
$$y = 4 - 5 \cdot 0$$
$$y = 4$$

The y-intercept is $(0, 4)$.

49. Familiarize. First we can use the Pythagorean theorem to find x, in ft. Then the height of the telephone pole is $x + 5$.

Translate. We use the Pythagorean theorem.
$$a^2 + b^2 = c^2$$
$$\left(\frac{1}{2}x + 1\right)^2 + x^2 = 34^2$$

Solve. We solve the equation.
$$\left(\frac{1}{2}x + 1\right)^2 + x^2 = 34^2$$
$$\frac{1}{4}x^2 + x + 1 + x^2 = 1156$$
$$x^2 + 4x + 4 + 4x^2 = 4624 \quad \text{Multiplying by 4}$$
$$5x^2 + 4 + 4 = 4624$$
$$5x^2 + 4x - 4620 = 0$$
$$(5x + 154)(x - 30) = 0$$
$$5x + 154 = 0 \quad \text{or} \quad x - 30 = 0$$
$$5x = -154 \quad \text{or} \quad x = 30$$
$$x = -30.8 \quad \text{or} \quad x = 30$$

Check. Since the length x must be positive, -30.8 cannot be a solution. If x is 30 ft, then $\frac{1}{2}x + 1$ is $\frac{1}{2} \cdot 30 + 1$, or 16 ft. Since $16^2 + 30^2 = 1156 = 34^2$, the number 30 checks. When x is 30 ft, then $x + 5$ is 35 ft.

State. The height of the telephone pole is 35 ft.

51. Familiarize. Using the labels shown on the drawing in the text, we let x = the width of the walk. Then the length and width of the rectangle formed by the pool and walk together are $40 + 2x$ and $20 + 2x$, respectively.

Translate.

Area is length times width.
$$1500 = (40 + 2x) \cdot (20 + 2x)$$

Solve. We solve the equation.
$$1500 = (40 + 2x)(20 + 2x)$$
$$1500 = 2(20 + x) \cdot 2(10 + x) \quad \text{Factoring 2 out of each factor on the right}$$
$$1500 = 4 \cdot (20 + x)(10 + x)$$
$$375 = (20 + x)(10 + x) \quad \text{Dividing by 4}$$
$$375 = 200 + 30x + x^2$$
$$0 = x^2 + 30x - 175$$
$$0 = (x + 35)(x - 5)$$
$$x + 35 = 0 \quad \text{or} \quad x - 5 = 0$$
$$x = -35 \quad \text{or} \quad x = 5$$

Check. The solutions of the equation are -35 and 5. Since the width of the walk cannot be negative, -35 is not a solution. When $x = 5$, $40 + 2x = 40 + 2 \cdot 5$, or 50 and $20 + 2x = 20 + 2 \cdot 5$, or 30. The total area of the pool and walk is $50 \cdot 30$, or 1500 ft^2. This checks.

State. The width of the walk is 5 ft.

53. Familiarize. We make a drawing. Let w = the width of the piece of cardboard. Then $2w$ = the length.

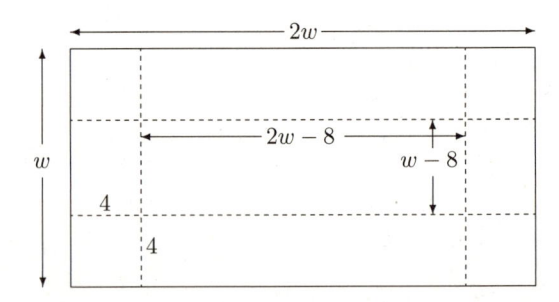

The box will have length $2w - 8$, width $w - 8$, and height 4. Recall that the formula for volume is $V =$ length $\times$ width $\times$ height.

Translate.

The volume is 616cm^3.
$$(2w - 8)(w - 8)(4) = 616$$

Solve. We solve the equation.
$$(2w - 8)(w - 8)(4) = 616$$
$$(2w^2 - 24w + 64)(4) = 616$$
$$8w^2 - 96w + 256 = 616$$
$$8w^2 - 96w - 360 = 0$$
$$8(w^2 - 12w - 45) = 0$$
$$w^2 - 12w - 45 = 0 \quad \text{Dividing by 8}$$
$$(w - 15)(w + 3) = 0$$
$$w - 15 = 0 \quad \text{or} \quad w + 3 = 0$$
$$w = 15 \quad \text{or} \quad w = -3$$

Check. The width cannot be negative, so we only need to check 15. When $w = 15$, then $2w = 30$ and the dimensions of the box are $30 - 8$ by $15 - 8$ by 4, or 22 by 7 by 4. The volume is $22 \cdot 7 \cdot 4$, or 616.

State. The cardboard is 30 cm by 15 cm.

Exercise Set 5.9

55. We add labels to the drawing in the text.

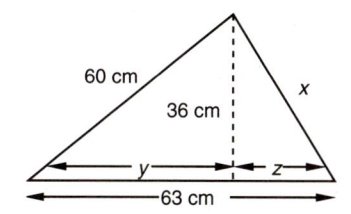

First we will use the Pythagorean theorem to find y.

$$y^2 + 36^2 = 60^2$$
$$y^2 + 1296 = 3600$$
$$y^2 - 2304 = 0$$
$$(y+48)(y-48) = 0$$

$y + 48 = 0 \quad \text{or} \quad y - 48 = 0$
$\quad y = -48 \quad \text{or} \quad \quad y = 48$

Since y cannot be negative, we use $y = 48$. Now we subtract to find z.

$$z = 63 - 48 = 15$$

Now we use the Pythagorean theorem to find x.

$$15^2 + 36^2 = x^2$$
$$225 + 1296 = x^2$$
$$1521 = x^2$$
$$0 = x^2 - 1521$$
$$0 = (x+39)(x-39)$$

$x + 39 = 0 \quad \text{or} \quad x - 39 = 0$
$\quad x = -39 \quad \text{or} \quad \quad x = 39$

Since x cannot be negative, -39 cannot be a solution. Thus, we find that x is 39 cm.

Chapter 6

Rational Expressions and Equations

Exercise Set 6.1

1. $\dfrac{-3}{2x}$

   To determine the numbers for which the rational expression is not defined, we set the denominator equal to 0 and solve:
   $$2x = 0$$
   $$x = 0$$
   The expression is not defined for the replacement number 0.

3. $\dfrac{5}{x-8}$

   To determine the numbers for which the rational expression is not defined, we set the denominator equal to 0 and solve:
   $$x - 8 = 0$$
   $$x = 8$$
   The expression is not defined for the replacement number 8.

5. $\dfrac{3}{2y+5}$

   Set the denominator equal to 0 and solve:
   $$2y + 5 = 0$$
   $$2y = -5$$
   $$y = -\dfrac{5}{2}$$
   The expression is not defined for the replacement number $-\dfrac{5}{2}$.

7. $\dfrac{x^2+11}{x^2-3x-28}$

   Set the denominator equal to 0 and solve:
   $$x^2 - 3x - 28 = 0$$
   $$(x-7)(x+4) = 0$$
   $$x - 7 = 0 \quad \text{or} \quad x + 4 = 0$$
   $$x = 7 \quad \text{or} \quad x = -4$$
   The expression is not defined for the replacement numbers 7 and -4.

9. $\dfrac{m^3-2m}{m^2-25}$

   Set the denominator equal to 0 and solve:
   $$m^2 - 25 = 0$$
   $$(m+5)(m-5) = 0$$
   $$m+5 = 0 \quad \text{or} \quad m-5 = 0$$
   $$m = -5 \quad \text{or} \quad m = 5$$
   The expression is not defined for the replacement numbers -5 and 5.

11. $\dfrac{x-4}{3}$

    Since the denominator is the constant 3, there are no replacement numbers for which the expression is not defined.

13. $\dfrac{4x}{4x} \cdot \dfrac{3x^2}{5y} = \dfrac{(4x)(3x^2)}{(4x)(5y)}$ Multiplying the numerators and the denominators

15. $\dfrac{2x}{2x} \cdot \dfrac{x-1}{x+4} = \dfrac{2x(x-1)}{2x(x+4)}$ Multiplying the numerators and the denominators

17. $\dfrac{3-x}{4-x} \cdot \dfrac{-1}{-1} = \dfrac{(3-x)(-1)}{(4-x)(-1)}$, or $\dfrac{-1(3-x)}{-1(4-x)}$

19. $\dfrac{y+6}{y+6} \cdot \dfrac{y-7}{y+2} = \dfrac{(y+6)(y-7)}{(y+6)(y+2)}$

21. $\dfrac{8x^3}{32x} = \dfrac{8 \cdot x \cdot x^2}{8 \cdot 4 \cdot x}$ Factoring numerator and denominator

    $= \dfrac{8x}{8x} \cdot \dfrac{x^2}{4}$ Factoring the rational expression

    $= 1 \cdot \dfrac{x^2}{4}$ $\left(\dfrac{8x}{8x} = 1\right)$

    $= \dfrac{x^2}{4}$ We removed a factor of 1.

23. $\dfrac{48p^7q^5}{18p^5q^4} = \dfrac{8 \cdot 6 \cdot p^5 \cdot p^2 \cdot q^4 \cdot q}{6 \cdot 3 \cdot p^5 \cdot q^4}$ Factoring numerator and denominator

    $= \dfrac{6p^5q^4}{6p^5q^4} \cdot \dfrac{8p^2q}{3}$ Factoring the rational expression

    $= 1 \cdot \dfrac{8p^2q}{3}$ $\left(\dfrac{6p^5q^4}{6p^5q^4} = 1\right)$

    $= \dfrac{8p^2q}{3}$ Removing a factor of 1

25. $\dfrac{4x-12}{4x} = \dfrac{4(x-3)}{4 \cdot x}$

    $= \dfrac{4}{4} \cdot \dfrac{x-3}{x}$

    $= 1 \cdot \dfrac{x-3}{x}$

    $= \dfrac{x-3}{x}$

27. $\dfrac{3m^2+3m}{6m^2+9m} = \dfrac{3m(m+1)}{3m(2m+3)}$

    $= \dfrac{3m}{3m} \cdot \dfrac{m+1}{2m+3}$

    $= 1 \cdot \dfrac{m+1}{2m+3}$

    $= \dfrac{m+1}{2m+3}$

29. $\dfrac{a^2-9}{a^2+5a+6} = \dfrac{(a-3)(a+3)}{(a+2)(a+3)}$
$= \dfrac{a-3}{a+2} \cdot \dfrac{a+3}{a+3}$
$= \dfrac{a-3}{a+2} \cdot 1$
$= \dfrac{a-3}{a+2}$

31. $\dfrac{a^2-10a+21}{a^2-11a+28} = \dfrac{(a-7)(a-3)}{(a-7)(a-4)}$
$= \dfrac{a-7}{a-7} \cdot \dfrac{a-3}{a-4}$
$= 1 \cdot \dfrac{a-3}{a-4}$
$= \dfrac{a-3}{a-4}$

33. $\dfrac{x^2-25}{x^2-10x+25} = \dfrac{(x-5)(x+5)}{(x-5)(x-5)}$
$= \dfrac{x-5}{x-5} \cdot \dfrac{x+5}{x-5}$
$= 1 \cdot \dfrac{x+5}{x-5}$
$= \dfrac{x+5}{x-5}$

35. $\dfrac{a^2-1}{a-1} = \dfrac{(a-1)(a+1)}{a-1}$
$= \dfrac{a-1}{a-1} \cdot \dfrac{a+1}{1}$
$= 1 \cdot \dfrac{a+1}{1}$
$= a+1$

37. $\dfrac{x^2+1}{x+1}$ cannot be simplified.
Neither the numerator nor the denominator can be factored.

39. $\dfrac{6x^2-54}{4x^2-36} = \dfrac{2 \cdot 3(x^2-9)}{2 \cdot 2(x^2-9)}$
$= \dfrac{2(x^2-9)}{2(x^2-9)} \cdot \dfrac{3}{2}$
$= 1 \cdot \dfrac{3}{2}$
$= \dfrac{3}{2}$

41. $\dfrac{6t+12}{t^2-t-6} = \dfrac{6(t+2)}{(t-3)(t+2)}$
$= \dfrac{6}{t-3} \cdot \dfrac{t+2}{t+2}$
$= \dfrac{6}{t-3} \cdot 1$
$= \dfrac{6}{t-3}$

43. $\dfrac{2t^2+6t+4}{4t^2-12t-16} = \dfrac{2(t^2+3t+2)}{4(t^2-3t-4)}$
$= \dfrac{2(t+2)(t+1)}{2 \cdot 2(t-4)(t+1)}$
$= \dfrac{2(t+1)}{2(t+1)} \cdot \dfrac{t+2}{2(t-4)}$
$= 1 \cdot \dfrac{t+2}{2(t-4)}$
$= \dfrac{t+2}{2(t-4)}$

45. $\dfrac{t^2-4}{(t+2)^2} = \dfrac{(t-2)(t+2)}{(t+2)(t+2)}$
$= \dfrac{t-2}{t+2} \cdot \dfrac{t+2}{t+2}$
$= \dfrac{t-2}{t+2} \cdot 1$
$= \dfrac{t-2}{t+2}$

47. $\dfrac{6-x}{x-6} = \dfrac{-(-6+x)}{x-6}$
$= \dfrac{-1(x-6)}{x-6}$
$= -1 \cdot \dfrac{x-6}{x-6}$
$= -1 \cdot 1$
$= -1$

49. $\dfrac{a-b}{b-a} = \dfrac{-(-a+b)}{b-a}$
$= \dfrac{-1(b-a)}{b-a}$
$= -1 \cdot \dfrac{b-a}{b-a}$
$= -1 \cdot 1$
$= -1$

51. $\dfrac{6t-12}{2-t} = \dfrac{-6(-t+2)}{2-t}$
$= \dfrac{-6(2-t)}{2-t}$
$= \dfrac{-6(2-t)}{2-t}$
$= -6$

53. $\dfrac{x^2-1}{1-x} = \dfrac{(x+1)(x-1)}{-1(-1+x)}$
$= \dfrac{(x+1)(x-1)}{-1(x-1)}$
$= \dfrac{(x+1)(\cancel{x-1})}{-1(\cancel{x-1})}$
$= -(x+1)$
$= -x-1$

Exercise Set 6.1 125

55. $\dfrac{4x^3}{3x} \cdot \dfrac{14}{x} = \dfrac{4x^3 \cdot 14}{3x \cdot x}$ Multiplying the numerators and the denominators

$= \dfrac{4 \cdot x \cdot x \cdot x \cdot 14}{3 \cdot x \cdot x}$ Factoring the numerator and the denominator

$= \dfrac{4 \cdot \cancel{x} \cdot \cancel{x} \cdot x \cdot 14}{3 \cdot \cancel{x} \cdot \cancel{x}}$ Removing a factor of 1

$= \dfrac{56x}{3}$ Simplifying

57. $\dfrac{3c}{d^2} \cdot \dfrac{4d}{6c^3} = \dfrac{3c \cdot 4d}{d^2 \cdot 6c^3}$ Multiplying the numerators and the denominators

$= \dfrac{3 \cdot c \cdot 2 \cdot 2 \cdot d}{d \cdot d \cdot 3 \cdot 2 \cdot c \cdot c \cdot c}$ Factoring the numerator and the denominator

$= \dfrac{\cancel{3} \cdot \cancel{c} \cdot \cancel{2} \cdot 2 \cdot \cancel{d}}{\cancel{d} \cdot d \cdot \cancel{3} \cdot \cancel{2} \cdot \cancel{c} \cdot c \cdot c}$

$= \dfrac{2}{dc^2}$

59. $\dfrac{x^2 - 3x - 10}{(x-2)^2} \cdot \dfrac{x-2}{x-5} = \dfrac{(x^2 - 3x - 10)(x-2)}{(x-2)^2(x-5)}$

$= \dfrac{(x-5)(x+2)(x-2)}{(x-2)(x-2)(x-5)}$

$= \dfrac{\cancel{(x-5)}(x+2)\cancel{(x-2)}}{\cancel{(x-2)}(x-2)\cancel{(x-5)}}$

$= \dfrac{x+2}{x-2}$

61. $\dfrac{a^2 - 9}{a^2} \cdot \dfrac{a^2 - 3a}{a^2 + a - 12} = \dfrac{(a-3)(a+3)(a)(a-3)}{a \cdot a(a+4)(a-3)}$

$= \dfrac{(a-3)(a+3)\cancel{(a)}(a-3)}{\cancel{a} \cdot a(a+4)\cancel{(a-3)}}$

$= \dfrac{(a-3)(a+3)}{a(a+4)}$

63. $\dfrac{4a^2}{3a^2 - 12a + 12} \cdot \dfrac{3a-6}{2a} = \dfrac{4a^2(3a-6)}{(3a^2 - 12a + 12)2a}$

$= \dfrac{2 \cdot 2 \cdot a \cdot a \cdot 3 \cdot (a-2)}{3 \cdot (a-2) \cdot (a-2) \cdot 2 \cdot a}$

$= \dfrac{\cancel{2} \cdot 2 \cdot \cancel{a} \cdot a \cdot \cancel{3} \cdot \cancel{(a-2)}}{\cancel{3} \cdot \cancel{(a-2)} \cdot (a-2) \cdot \cancel{2} \cdot \cancel{a}}$

$= \dfrac{2a}{a-2}$

65. $\dfrac{t^4 - 16}{t^4 - 1} \cdot \dfrac{t^2 + 1}{t^2 + 4}$

$= \dfrac{(t^4 - 16)(t^2 + 1)}{(t^4 - 1)(t^2 + 4)}$

$= \dfrac{(t^2 + 4)(t+2)(t-2)(t^2 + 1)}{(t^2 + 1)(t+1)(t-1)(t^2 + 4)}$

$= \dfrac{\cancel{(t^2 + 4)}(t+2)(t-2)\cancel{(t^2 + 1)}}{\cancel{(t^2 + 1)}(t+1)(t-1)\cancel{(t^2 + 4)}}$

$= \dfrac{(t+2)(t-2)}{(t+1)(t-1)}$

67. $\dfrac{(x+4)^3}{(x+2)^3} \cdot \dfrac{x^2 + 4x + 4}{x^2 + 8x + 16}$

$= \dfrac{(x+4)^3(x^2 + 4x + 4)}{(x+2)^3(x^2 + 8x + 16)}$

$= \dfrac{(x+4)(x+4)(x+4)(x+2)(x+2)}{(x+2)(x+2)(x+2)(x+4)(x+4)}$

$= \dfrac{\cancel{(x+4)}\cancel{(x+4)}(x+4)\cancel{(x+2)}\cancel{(x+2)}}{\cancel{(x+2)}\cancel{(x+2)}(x+2)\cancel{(x+4)}\cancel{(x+4)}}$

$= \dfrac{x+4}{x+2}$

69. $\dfrac{5a^2 - 180}{10a^2 - 10} \cdot \dfrac{20a + 20}{2a - 12} = \dfrac{(5a^2 - 180)(20a + 20)}{(10a^2 - 10)(2a - 12)}$

$= \dfrac{5(a+6)(a-6)(2)(10)(a+1)}{10(a+1)(a-1)(2)(a-6)}$

$= \dfrac{5(a+6)\cancel{(a-6)}\cancel{(2)}\cancel{(10)}\cancel{(a+1)}}{\cancel{10}\cancel{(a+1)}(a-1)\cancel{(2)}\cancel{(a-6)}}$

$= \dfrac{5(a+6)}{a-1}$

71. Discussion and Writing Exercise

73. Familiarize. Let $x =$ the smaller even integer. Then $x + 2 =$ the larger even integer.

Translate. We reword the problem.

$\underbrace{\text{Smaller even integer}}_{x} \cdot \underbrace{\text{times}}_{\cdot} \cdot \underbrace{\text{larger even integer}}_{(x+2)} \underbrace{\text{is}}_{=} \underbrace{360.}_{360}$

Solve.
$x(x+2) = 360$
$x^2 + 2x = 360$
$x^2 + 2x - 360 = 0$
$(x + 20)(x - 18) = 0$
$x + 20 = 0 \quad \text{or} \quad x - 18 = 0$
$x = -20 \quad \text{or} \quad x = 18$

Check. The solutions of the equation are -20 and 18. When $x = -20$, then $x + 2 = -18$ and $-20(-18) = 360$. The numbers -20 and -18 are consecutive even integers which are solutions to the problem. When $x = 18$, then $x + 2 = 20$ and $18 \cdot 20 = 360$. The numbers 18 and 20 are also consecutive even integers which are solutions to the problem.

State. We have two solutions, each of which consists of a pair of numbers: -20 and -18, and 18 and 20.

75. $x^2 - x - 56$

We look for a pair of numbers whose product is -56 and whose sum is -1. The numbers are -8 and 7.

$x^2 - x - 56 = (x - 8)(x + 7)$

77. $x^5 - 2x^4 - 35x^3 = x^3(x^2 - 2x - 35) = x^3(x - 7)(x + 5)$

79. $16 - t^4 = 4^2 - (t^2)^2$ Difference of squares
$= (4 + t^2)(4 - t^2)$
$= (4 + t^2)(2^2 - t^2)$ Difference of squares
$= (4 + t^2)(2 + t)(2 - t)$

81. $x^2 - 9x + 14$

We look for a pair of numbers whose product is 14 and whose sum is -9. The numbers are -2 and -7.
$$x^2 - 9x + 14 = (x-2)(x-7)$$

83. $16x^2 - 40xy + 25y^2$
$= (4x)^2 - 2 \cdot 4x \cdot 5y + (5y)^2$ Trinomial square
$= (4x - 5y)^2$

85. $\dfrac{x^4 - 16y^2}{(x^2 + 4y^2)(x - 2y)}$
$= \dfrac{(x^2 + 4y^2)(x + 2y)(x - 2y)}{(x^2 + 4y^2)(x - 2y)}$
$= \dfrac{\cancel{(x^2 + 4y^2)}(x + 2y)\cancel{(x - 2y)}}{\cancel{(x^2 + 4y^2)}\cancel{(x - 2y)}(1)}$
$= x + 2y$

87. $\dfrac{t^4 - 1}{t^4 - 81} \cdot \dfrac{t^2 - 9}{t^2 + 1} \cdot \dfrac{(t-9)^2}{(t+1)^2}$
$= \dfrac{(t^2 + 1)(t + 1)(t - 1)(t + 3)(t - 3)(t - 9)(t - 9)}{(t^2 + 9)(t + 3)(t - 3)(t^2 + 1)(t + 1)(t + 1)}$
$= \dfrac{\cancel{(t^2+1)}\cancel{(t+1)}(t-1)\cancel{(t+3)}\cancel{(t-3)}(t-9)(t-9)}{(t^2+9)\cancel{(t+3)}\cancel{(t-3)}\cancel{(t^2+1)}\cancel{(t+1)}(t+1)}$
$= \dfrac{(t-1)(t-9)(t-9)}{(t^2+9)(t+1)}$, or $\dfrac{(t-1)(t-9)^2}{(t^2+9)(t+1)}$

89. $\dfrac{x^2 - y^2}{(x-y)^2} \cdot \dfrac{x^2 - 2xy + y^2}{x^2 - 4xy - 5y^2}$
$= \dfrac{(x+y)(x-y)(x-y)(x-y)}{(x-y)(x-y)(x-5y)(x+y)}$
$= \dfrac{\cancel{(x+y)}\cancel{(x-y)}\cancel{(x-y)}(x-y)}{\cancel{(x-y)}\cancel{(x-y)}(x-5y)\cancel{(x+y)}}$
$= \dfrac{x-y}{x-5y}$

91. $\dfrac{5(2x+5) - 25}{10} = \dfrac{10x + 25 - 25}{10}$
$= \dfrac{10x}{10}$
$= x$

You get the same number you selected.

To do a number trick, ask someone to select a number and then perform these operations. The person will probably be surprised that the result is the original number.

Exercise Set 6.2

1. The reciprocal of $\dfrac{4}{x}$ is $\dfrac{x}{4}$ because $\dfrac{4}{x} \cdot \dfrac{x}{4} = 1$.

3. The reciprocal of $x^2 - y^2$ is $\dfrac{1}{x^2 - y^2}$ because
$\dfrac{x^2 - y^2}{1} \cdot \dfrac{1}{x^2 - y^2} = 1$.

5. The reciprocal of $\dfrac{1}{a+b}$ is $a+b$ because $\dfrac{1}{a+b} \cdot (a+b) = 1$.

7. The reciprocal of $\dfrac{x^2 + 2x - 5}{x^2 - 4x + 7}$ is $\dfrac{x^2 - 4x + 7}{x^2 + 2x - 5}$ because
$\dfrac{x^2 + 2x - 5}{x^2 - 4x + 7} \cdot \dfrac{x^2 - 4x + 7}{x^2 + 2x - 5} = 1$.

9. $\dfrac{2}{5} \div \dfrac{4}{3} = \dfrac{2}{5} \cdot \dfrac{3}{4}$ Multiplying by the reciprocal of the divisor
$= \dfrac{2 \cdot 3}{5 \cdot 4}$
$= \dfrac{2 \cdot 3}{5 \cdot 2 \cdot 2}$ Factoring the denominator
$= \dfrac{\cancel{2} \cdot 3}{5 \cdot \cancel{2} \cdot 2}$ Removing a factor of 1
$= \dfrac{3}{10}$ Simplifying

11. $\dfrac{2}{x} \div \dfrac{8}{x} = \dfrac{2}{x} \cdot \dfrac{x}{8}$ Multiplying by the reciprocal of the divisor
$= \dfrac{2 \cdot x}{x \cdot 8}$
$= \dfrac{2 \cdot x \cdot 1}{x \cdot 2 \cdot 4}$ Factoring the numerator and the denominator
$= \dfrac{\cancel{2} \cdot \cancel{x} \cdot 1}{\cancel{x} \cdot \cancel{2} \cdot 4}$ Removing a factor of 1
$= \dfrac{1}{4}$ Simplifying

13. $\dfrac{a}{b^2} \div \dfrac{a^2}{b^3} = \dfrac{a}{b^2} \cdot \dfrac{b^3}{a^2}$ Multiplying by the reciprocal of the divisor
$= \dfrac{a \cdot b^3}{b^2 \cdot a^2}$
$= \dfrac{a \cdot b^2 \cdot b}{b^2 \cdot a \cdot a}$
$= \dfrac{\cancel{a} \cdot \cancel{b^2} \cdot b}{\cancel{b^2} \cdot \cancel{a} \cdot a}$
$= \dfrac{b}{a}$

15. $\dfrac{a+2}{a-3} \div \dfrac{a-1}{a+3} = \dfrac{a+2}{a-3} \cdot \dfrac{a+3}{a-1}$
$= \dfrac{(a+2)(a+3)}{(a-3)(a-1)}$

17. $\dfrac{x^2 - 1}{x} \div \dfrac{x+1}{x-1} = \dfrac{x^2 - 1}{x} \cdot \dfrac{x-1}{x+1}$
$= \dfrac{(x^2 - 1)(x - 1)}{x(x+1)}$
$= \dfrac{(x-1)(x+1)(x-1)}{x(x+1)}$
$= \dfrac{(x-1)\cancel{(x+1)}(x-1)}{x\cancel{(x+1)}}$
$= \dfrac{(x-1)^2}{x}$

Exercise Set 6.2

19. $\dfrac{x+1}{6} \div \dfrac{x+1}{3} = \dfrac{x+1}{6} \cdot \dfrac{3}{x+1}$

$= \dfrac{(x+1) \cdot 3}{6(x+1)}$

$= \dfrac{3(x+1)}{2 \cdot 3(x+1)}$

$= \dfrac{1 \cdot \cancel{3}(\cancel{x+1})}{2 \cdot \cancel{3}(\cancel{x+1})}$

$= \dfrac{1}{2}$

21. $\dfrac{5x-5}{16} \div \dfrac{x-1}{6} = \dfrac{5x-5}{16} \cdot \dfrac{6}{x-1}$

$= \dfrac{(5x-5) \cdot 6}{16(x-1)}$

$= \dfrac{5(x-1) \cdot 2 \cdot 3}{2 \cdot 8(x-1)}$

$= \dfrac{5(\cancel{x-1}) \cdot \cancel{2} \cdot 3}{\cancel{2} \cdot 8(\cancel{x-1})}$

$= \dfrac{15}{8}$

23. $\dfrac{-6+3x}{5} \div \dfrac{4x-8}{25} = \dfrac{-6+3x}{5} \cdot \dfrac{25}{4x-8}$

$= \dfrac{(-6+3x) \cdot 25}{5(4x-8)}$

$= \dfrac{3(x-2) \cdot 5 \cdot 5}{5 \cdot 4(x-2)}$

$= \dfrac{3(\cancel{x-2}) \cdot \cancel{5} \cdot 5}{\cancel{5} \cdot 4(\cancel{x-2})}$

$= \dfrac{15}{4}$

25. $\dfrac{a+2}{a-1} \div \dfrac{3a+6}{a-5} = \dfrac{a+2}{a-1} \cdot \dfrac{a-5}{3a+6}$

$= \dfrac{(a+2)(a-5)}{(a-1)(3a+6)}$

$= \dfrac{(a+2)(a-5)}{(a-1) \cdot 3 \cdot (a+2)}$

$= \dfrac{(\cancel{a+2})(a-5)}{(a-1) \cdot 3 \cdot (\cancel{a+2})}$

$= \dfrac{a-5}{3(a-1)}$

27. $\dfrac{x^2-4}{x} \div \dfrac{x-2}{x+2} = \dfrac{x^2-4}{x} \cdot \dfrac{x+2}{x-2}$

$= \dfrac{(x^2-4)(x+2)}{x(x-2)}$

$= \dfrac{(x-2)(x+2)(x+2)}{x(x-2)}$

$= \dfrac{(\cancel{x-2})(x+2)(x+2)}{x(\cancel{x-2})}$

$= \dfrac{(x+2)^2}{x}$

29. $\dfrac{x^2-9}{4x+12} \div \dfrac{x-3}{6} = \dfrac{x^2-9}{4x+12} \cdot \dfrac{6}{x-3}$

$= \dfrac{(x^2-9) \cdot 6}{(4x+12)(x-3)}$

$= \dfrac{(x-3)(x+3) \cdot 3 \cdot 2}{2 \cdot 2(x+3)(x-3)}$

$= \dfrac{(\cancel{x-3})(\cancel{x+3}) \cdot 3 \cdot \cancel{2}}{\cancel{2} \cdot 2(\cancel{x+3})(\cancel{x-3})}$

$= \dfrac{3}{2}$

31. $\dfrac{c^2+3c}{c^2+2c-3} \div \dfrac{c}{c+1} = \dfrac{c^2+3c}{c^2+2c-3} \cdot \dfrac{c+1}{c}$

$= \dfrac{(c^2+3c)(c+1)}{(c^2+2c-3)c}$

$= \dfrac{c(c+3)(c+1)}{(c+3)(c-1)c}$

$= \dfrac{\cancel{c}(\cancel{c+3})(c+1)}{(\cancel{c+3})(c-1)\cancel{c}}$

$= \dfrac{c+1}{c-1}$

33. $\dfrac{2y^2-7y+3}{2y^2+3y-2} \div \dfrac{6y^2-5y+1}{3y^2+5y-2}$

$= \dfrac{2y^2-7y+3}{2y^2+3y-2} \cdot \dfrac{3y^2+5y-2}{6y^2-5y+1}$

$= \dfrac{(2y^2-7y+3)(3y^2+5y-2)}{(2y^2+3y-2)(6y^2-5y+1)}$

$= \dfrac{(2y-1)(y-3)(3y-1)(y+2)}{(2y-1)(y+2)(3y-1)(2y-1)}$

$= \dfrac{(\cancel{2y-1})(y-3)(\cancel{3y-1})(\cancel{y+2})}{(\cancel{2y-1})(\cancel{y+2})(\cancel{3y-1})(2y-1)}$

$= \dfrac{y-3}{2y-1}$

35. $\dfrac{x^2-1}{4x+4} \div \dfrac{2x^2-4x+2}{8x+8} = \dfrac{x^2-1}{4x+4} \cdot \dfrac{8x+8}{2x^2-4x+2}$

$= \dfrac{(x^2-1)(8x+8)}{(4x+4)(2x^2-4x+2)}$

$= \dfrac{(x+1)(x-1)(2)(4)(x+1)}{4(x+1)(2)(x-1)(x-1)}$

$= \dfrac{(\cancel{x+1})(\cancel{x-1})(\cancel{2})(\cancel{4})(x+1)}{\cancel{4}(\cancel{x+1})(\cancel{2})(x-1)(\cancel{x-1})}$

$= \dfrac{x+1}{x-1}$

37. Discussion and Writing Exercise

39. *Familiarize.* Let $s =$ Bonnie's score on the last test.

Translate. The average of the four scores must be at least 90. This means it must be greater than or equal to 90. We translate.

$$\dfrac{96+98+89+s}{4} \geq 90$$

Solve. We solve the inequality. First we multiply by 4 to clear the fraction.
$$4\left(\frac{96+98+89+s}{4}\right) \geq 4 \cdot 90$$
$$96+98+89+s \geq 360$$
$$283+s \geq 360$$
$$s \geq 77 \quad \text{Subtracting 283}$$

Check. We can do a partial check by substituting a value for s less than 77 and a value for s greater than 77.

For $s=76$: $\dfrac{96+98+89+76}{4} = 89.75 < 90$

For $s=78$: $\dfrac{96+98+89+78}{4} = 90.25 \leq 90$

Since the average is less than 90 for a value of s less than 77 and greater than or equal to 90 for a value greater than or equal to 77, the answer is probably correct.

State. The scores on the last test that will earn Bonnie an A are $\{s | s \geq 77\}$.

41. $(8x^3 - 3x^2 + 7) - (8x^2 + 3x - 5) =$
$8x^3 - 3x^2 + 7 - 8x^2 - 3x + 5 =$
$8x^3 - 11x^2 - 3x + 12$

43. $(2x^{-3}y^4)^2 = 2^2(x^{-3})^2(y^4)^2$
$\phantom{(2x^{-3}y^4)^2} = 2^2 x^{-6} y^8 \quad \text{Multiplying exponents}$
$\phantom{(2x^{-3}y^4)^2} = 4x^{-6} y^8 \quad (2^2 = 4)$
$\phantom{(2x^{-3}y^4)^2} = \dfrac{4y^8}{x^6} \quad \left(x^{-6} = \dfrac{1}{x^6}\right)$

45. $\left(\dfrac{2x^3}{y^5}\right)^2 = \dfrac{2^2(x^3)^2}{(y^5)^2}$
$\phantom{\left(\dfrac{2x^3}{y^5}\right)^2} = \dfrac{2^2 x^6}{y^{10}} \quad \text{Multiplying exponents}$
$\phantom{\left(\dfrac{2x^3}{y^5}\right)^2} = \dfrac{4x^6}{y^{10}} \quad (2^2 = 4)$

47. $\dfrac{3a^2 - 5ab - 12b^2}{3ab + 4b^2} \div (3b^2 - ab)$
$= \dfrac{3a^2 - 5ab - 12b^2}{3ab + 4b^2} \cdot \dfrac{1}{3b^2 - ab}$
$= \dfrac{(3a + 4b)(a - 3b)}{b(3a + 4b) \cdot b(3b - a)}$
$= \dfrac{(3a + 4b)(-1)(3b - a)}{b(3a + 4b) \cdot b(3b - a)}$
$= \dfrac{\cancel{(3a + 4b)}(-1)\cancel{(3b - a)}}{b\cancel{(3a + 4b)} \cdot b\cancel{(3b - a)}}$
$= -\dfrac{1}{b^2}$

49. The volume V of a rectangular solid is given by the formula $V = l \cdot w \cdot h$, where $l =$ the length, $w =$ the width, and $h =$ the height. We substitute in the formula and solve for h.
$$V = l \cdot w \cdot h$$
$$x - 3 = \dfrac{x-3}{x-7} \cdot \dfrac{x+y}{x-7} \cdot h$$
$$\dfrac{x-7}{x-3} \cdot \dfrac{x-7}{x+y} \cdot (x-3) = \dfrac{\cancel{x-7}}{\cancel{x-3}} \cdot \dfrac{\cancel{x-7}}{\cancel{x+y}} \cdot \dfrac{\cancel{x-3}}{\cancel{x-7}} \cdot \dfrac{\cancel{x+y}}{\cancel{x-7}} \cdot h$$
$$\dfrac{(x-7)^2}{x+y} = h$$

The height is $\dfrac{(x-7)^2}{x+y}$.

Exercise Set 6.3

1. $12 = 2 \cdot 2 \cdot 3$
$27 = 3 \cdot 3 \cdot 3$
LCM $= 2 \cdot 2 \cdot 3 \cdot 3 \cdot 3$, or 108

3. $8 = 2 \cdot 2 \cdot 2$
$9 = 3 \cdot 3$
LCM $= 2 \cdot 2 \cdot 2 \cdot 3 \cdot 3$, or 72

5. $6 = 2 \cdot 3$
$9 = 3 \cdot 3$
$21 = 3 \cdot 7$
LCM $= 2 \cdot 3 \cdot 3 \cdot 7$, or 126

7. $24 = 2 \cdot 2 \cdot 2 \cdot 3$
$36 = 2 \cdot 2 \cdot 3 \cdot 3$
$40 = 2 \cdot 2 \cdot 2 \cdot 5$
LCM $= 2 \cdot 2 \cdot 2 \cdot 3 \cdot 3 \cdot 5$, or 360

9. $10 = 2 \cdot 5$
$100 = 2 \cdot 2 \cdot 5 \cdot 5$
$500 = 2 \cdot 2 \cdot 5 \cdot 5 \cdot 5$
LCM $= 2 \cdot 2 \cdot 5 \cdot 5 \cdot 5$, or 500

(We might have observed at the outset that both 10 and 100 are factors of 500, so the LCM is 500.)

11. $24 = 2 \cdot 2 \cdot 2 \cdot 3$
$18 = 2 \cdot 3 \cdot 3$
LCD $= 2 \cdot 2 \cdot 2 \cdot 3 \cdot 3$, or 72
$\dfrac{7}{24} + \dfrac{11}{18} = \dfrac{7}{2 \cdot 2 \cdot 2 \cdot 3} \cdot \dfrac{3}{3} + \dfrac{11}{2 \cdot 3 \cdot 3} \cdot \dfrac{2 \cdot 2}{2 \cdot 2}$
$\phantom{\dfrac{7}{24} + \dfrac{11}{18}} = \dfrac{21}{2 \cdot 2 \cdot 2 \cdot 3 \cdot 3} + \dfrac{44}{2 \cdot 2 \cdot 2 \cdot 3 \cdot 3}$
$\phantom{\dfrac{7}{24} + \dfrac{11}{18}} = \dfrac{65}{72}$

Exercise Set 6.3

13. $\dfrac{1}{6} + \dfrac{3}{40}$

$= \dfrac{1}{2 \cdot 3} + \dfrac{3}{2 \cdot 2 \cdot 2 \cdot 5}$

$\qquad$ LCD is $2 \cdot 2 \cdot 2 \cdot 3 \cdot 5$, or 120

$= \dfrac{1}{2 \cdot 3} \cdot \dfrac{2 \cdot 2 \cdot 5}{2 \cdot 2 \cdot 5} + \dfrac{3}{2 \cdot 2 \cdot 2 \cdot 5} \cdot \dfrac{3}{3}$

$= \dfrac{20 + 9}{2 \cdot 2 \cdot 2 \cdot 3 \cdot 5}$

$= \dfrac{29}{120}$

15. $\dfrac{1}{20} + \dfrac{1}{30} + \dfrac{2}{45}$

$= \dfrac{1}{2 \cdot 2 \cdot 5} + \dfrac{1}{2 \cdot 3 \cdot 5} + \dfrac{2}{3 \cdot 3 \cdot 5}$

$\qquad$ LCD is $2 \cdot 2 \cdot 3 \cdot 3 \cdot 5$, or 180

$= \dfrac{1}{2 \cdot 2 \cdot 5} \cdot \dfrac{3 \cdot 3}{3 \cdot 3} + \dfrac{1}{2 \cdot 3 \cdot 5} \cdot \dfrac{2 \cdot 3}{2 \cdot 3} + \dfrac{2}{3 \cdot 3 \cdot 5} \cdot \dfrac{2 \cdot 2}{2 \cdot 2}$

$= \dfrac{9 + 6 + 8}{2 \cdot 2 \cdot 3 \cdot 3 \cdot 5}$

$= \dfrac{23}{180}$

17. $6x^2 = 2 \cdot 3 \cdot x \cdot x$

$12x^3 = 2 \cdot 2 \cdot 3 \cdot x \cdot x \cdot x$

LCM $= 2 \cdot 2 \cdot 3 \cdot x \cdot x \cdot x$, or $12x^3$

19. $2x^2 = 2 \cdot x \cdot x$

$6xy = 2 \cdot 3 \cdot x \cdot y$

$18y^2 = 2 \cdot 3 \cdot 3 \cdot y \cdot y$

LCM $= 2 \cdot 3 \cdot 3 \cdot x \cdot x \cdot y \cdot y$, or $18x^2y^2$

21. $2(y - 3) = 2 \cdot (y - 3)$

$6(y - 3) = 2 \cdot 3 \cdot (y - 3)$

LCM $= 2 \cdot 3 \cdot (y - 3)$, or $6(y - 3)$

23. $t, t + 2, t - 2$

The expressions are not factorable, so the LCM is their product:

LCM $= t(t + 2)(t - 2)$

25. $x^2 - 4 = (x + 2)(x - 2)$

$x^2 + 5x + 6 = (x + 3)(x + 2)$

LCM $= (x + 2)(x - 2)(x + 3)$

27. $t^3 + 4t^2 + 4t = t(t^2 + 4t + 4) = t(t + 2)(t + 2)$

$t^2 - 4t = t(t - 4)$

LCM $= t(t + 2)(t + 2)(t - 4) = t(t + 2)^2(t - 4)$

29. $a + 1 = a + 1$

$(a - 1)^2 = (a - 1)(a - 1)$

$a^2 - 1 = (a + 1)(a - 1)$

LCM $= (a + 1)(a - 1)(a - 1) = (a + 1)(a - 1)^2$

31. $m^2 - 5m + 6 = (m - 3)(m - 2)$

$m^2 - 4m + 4 = (m - 2)(m - 2)$

LCM $= (m - 3)(m - 2)(m - 2) = (m - 3)(m - 2)^2$

33. $2 + 3x = 2 + 3x$

$4 - 9x^2 = (2 + 3x)(2 - 3x)$

$2 - 3x = 2 - 3x$

LCM $= (2 + 3x)(2 - 3x)$

35. $10v^2 + 30v = 10v(v + 3) = 2 \cdot 5 \cdot v(v + 3)$

$5v^2 + 35v + 60 = 5(v^2 + 7v + 12)$

$\qquad\qquad\qquad\quad = 5(v + 4)(v + 3)$

LCM $= 2 \cdot 5 \cdot v(v + 3)(v + 4) = 10v(v + 3)(v + 4)$

37. $9x^3 - 9x^2 - 18x = 9x(x^2 - x - 2)$

$\qquad\qquad\qquad\quad = 3 \cdot 3 \cdot x(x - 2)(x + 1)$

$6x^5 - 24x^4 + 24x^3 = 6x^3(x^2 - 4x + 4)$

$\qquad\qquad\qquad\quad = 2 \cdot 3 \cdot x \cdot x \cdot x(x - 2)(x - 2)$

LCM $= 2 \cdot 3 \cdot 3 \cdot x \cdot x \cdot x(x - 2)(x - 2)(x + 1) =$
$18x^3(x - 2)^2(x + 1)$

39. $x^5 + 4x^4 + 4x^3 = x^3(x^2 + 4x + 4)$

$\qquad\qquad\qquad = x \cdot x \cdot x(x + 2)(x + 2)$

$3x^2 - 12 = 3(x^2 - 4) = 3(x + 2)(x - 2)$

$2x + 4 = 2(x + 2)$

LCM $= 2 \cdot 3 \cdot x \cdot x \cdot x(x + 2)(x + 2)(x - 2)$

$\qquad = 6x^3(x + 2)^2(x - 2)$

41. Discussion and Writing Exercise

43. $x^2 - 6x + 9 = x^2 - 2 \cdot x \cdot 3 + 3^2 \qquad$ Trinomial square

$\qquad\qquad\quad = (x - 3)^2$

45. $x^2 - 9 = x^2 - 3^2 \qquad$ Difference of squares

$\qquad\quad = (x + 3)(x - 3)$

47. $x^2 + 6x + 9 = x^2 + 2 \cdot x \cdot 3 + 3^2 \qquad$ Trinomial square

$\qquad\qquad\quad = (x + 3)^2$

49. Locate 1970 on the horizontal axis, go up to the graph, and then go over to the corresponding point on the vertical axis. We read that about 54% of those married in 1970 will divorce.

51. Locate 1990 on the horizontal axis, go up to the graph, and then go over to the corresponding point on the vertical axis. We read that about 74% of those married in 1990 will divorce.

53. Locate 50 on the vertical axis, go across to the graph, and then go down to the corresponding point on the horizontal axis. We read that the divorce percentage was about 50% in 1965.

55. The time it takes Pedro and Maria to meet again at the starting place is the LCM of the times it takes them to complete one round of the course.

$6 = 2 \cdot 3$

$8 = 2 \cdot 2 \cdot 2$

LCM $= 2 \cdot 2 \cdot 2 \cdot 3$, or 24

It takes 24 min.

Exercise Set 6.4

1. $\dfrac{5}{8} + \dfrac{3}{8} = \dfrac{5+3}{8} = \dfrac{8}{8} = 1$

3. $\dfrac{1}{3+x} + \dfrac{5}{3+x} = \dfrac{1+5}{3+x} = \dfrac{6}{3+x}$

5. $\dfrac{x^2 + 7x}{x^2 - 5x} + \dfrac{x^2 - 4x}{x^2 - 5x} = \dfrac{(x^2 + 7x) + (x^2 - 4x)}{x^2 - 5x}$

$= \dfrac{2x^2 + 3x}{x^2 - 5x}$

$= \dfrac{x(2x+3)}{x(x-5)}$

$= \dfrac{\cancel{x}(2x+3)}{\cancel{x}(x-5)}$

$= \dfrac{2x+3}{x-5}$

7. $\dfrac{2}{x} + \dfrac{5}{x^2} = \dfrac{2}{x} + \dfrac{5}{x \cdot x}$ LCD $= x \cdot x$, or x^2

$= \dfrac{2}{x} \cdot \dfrac{x}{x} + \dfrac{5}{x \cdot x}$

$= \dfrac{2x + 5}{x^2}$

9. $\begin{matrix} 6r = 2 \cdot 3 \cdot r \\ 8r = 2 \cdot 2 \cdot 2 \cdot r \end{matrix}$ LCD $= 2 \cdot 2 \cdot 2 \cdot 3 \cdot r$, or $24r$

$\dfrac{5}{6r} + \dfrac{7}{8r} = \dfrac{5}{6r} \cdot \dfrac{4}{4} + \dfrac{7}{8r} \cdot \dfrac{3}{3}$

$= \dfrac{20 + 21}{24r}$

$= \dfrac{41}{24r}$

11. $\begin{matrix} xy^2 = x \cdot y \cdot y \\ x^2 y = x \cdot x \cdot y \end{matrix}$ LCD $= x \cdot x \cdot y \cdot y$, or $x^2 y^2$

$\dfrac{4}{xy^2} + \dfrac{6}{x^2 y} = \dfrac{4}{xy^2} \cdot \dfrac{x}{x} + \dfrac{6}{x^2 y} \cdot \dfrac{y}{y}$

$= \dfrac{4x + 6y}{x^2 y^2}$

13. $\begin{matrix} 9t^3 = 3 \cdot 3 \cdot t \cdot t \cdot t \\ 6t^2 = 2 \cdot 3 \cdot t \cdot t \end{matrix}$ LCD $= 2 \cdot 3 \cdot 3 \cdot t \cdot t \cdot t$, or $18t^3$

$\dfrac{2}{9t^3} + \dfrac{1}{6t^2} = \dfrac{2}{9t^3} \cdot \dfrac{2}{2} + \dfrac{1}{6t^2} \cdot \dfrac{3t}{3t}$

$= \dfrac{4 + 3t}{18t^3}$

15. LCD $= x^2 y^2$ (See Exercise 11.)

$\dfrac{x+y}{xy^2} + \dfrac{3x+y}{x^2 y} = \dfrac{x+y}{xy^2} \cdot \dfrac{x}{x} + \dfrac{3x+y}{x^2 y} \cdot \dfrac{y}{y}$

$= \dfrac{x(x+y) + y(3x+y)}{x^2 y^2}$

$= \dfrac{x^2 + xy + 3xy + y^2}{x^2 y^2}$

$= \dfrac{x^2 + 4xy + y^2}{x^2 y^2}$

17. The denominators do not factor, so the LCD is their product, $(x-2)(x+2)$.

$\dfrac{3}{x-2} + \dfrac{3}{x+2} = \dfrac{3}{x-2} \cdot \dfrac{x+2}{x+2} + \dfrac{3}{x+2} \cdot \dfrac{x-2}{x-2}$

$= \dfrac{3(x+2) + 3(x-2)}{(x-2)(x+2)}$

$= \dfrac{3x + 6 + 3x - 6}{(x-2)(x+2)}$

$= \dfrac{6x}{(x-2)(x+2)}$

19. $\begin{matrix} 3x = 3 \cdot x \\ x+1 = x+1 \end{matrix}$ LCD $= 3x(x+1)$

$\dfrac{3}{x+1} + \dfrac{2}{3x} = \dfrac{3}{x+1} \cdot \dfrac{3x}{3x} + \dfrac{2}{3x} \cdot \dfrac{x+1}{x+1}$

$= \dfrac{9x + 2(x+1)}{3x(x+1)}$

$= \dfrac{9x + 2x + 2}{3x(x+1)}$

$= \dfrac{11x + 2}{3x(x+1)}$

21. $\begin{matrix} x^2 - 16 = (x+4)(x-4) \\ x - 4 = x - 4 \end{matrix}$ LCD $= (x+4)(x-4)$

$\dfrac{2x}{x^2 - 16} + \dfrac{x}{x-4} = \dfrac{2x}{(x+4)(x-4)} + \dfrac{x}{x-4} \cdot \dfrac{x+4}{x+4}$

$= \dfrac{2x + x(x+4)}{(x+4)(x-4)}$

$= \dfrac{2x + x^2 + 4x}{(x+4)(x-4)}$

$= \dfrac{x^2 + 6x}{(x+4)(x-4)}$

23. $\dfrac{5}{z+4} + \dfrac{3}{3z+12} = \dfrac{5}{z+4} + \dfrac{3}{3(z+4)}$ LCD $= 3(z+4)$

$= \dfrac{5}{z+4} \cdot \dfrac{3}{3} + \dfrac{3}{3(z+4)}$

$= \dfrac{15 + 3}{3(z+4)} = \dfrac{18}{3(z+4)}$

$= \dfrac{3 \cdot 6}{3(z+4)} = \dfrac{\cancel{3} \cdot 6}{\cancel{3}(z+4)}$

$= \dfrac{6}{z+4}$

Exercise Set 6.4

25. $\dfrac{3}{x-1} + \dfrac{2}{(x-1)^2}$ LCD $= (x-1)^2$

$= \dfrac{3}{x-1} \cdot \dfrac{x-1}{x-1} + \dfrac{2}{(x-1)^2}$

$= \dfrac{3(x-1) + 2}{(x-1)^2}$

$= \dfrac{3x - 3 + 2}{(x-1)^2}$

$= \dfrac{3x - 1}{(x-1)^2}$

27. $\dfrac{4a}{5a-10} + \dfrac{3a}{10a-20} = \dfrac{4a}{5(a-2)} + \dfrac{3a}{2 \cdot 5(a-2)}$

                            LCD $= 2 \cdot 5(a-2)$

$= \dfrac{4a}{5(a-2)} \cdot \dfrac{2}{2} + \dfrac{3a}{2 \cdot 5(a-2)}$

$= \dfrac{8a + 3a}{10(a-2)}$

$= \dfrac{11a}{10(a-2)}$

29. $\dfrac{x+4}{x} + \dfrac{x}{x+4}$ LCD $= x(x+4)$

$= \dfrac{x+4}{x} \cdot \dfrac{x+4}{x+4} + \dfrac{x}{x+4} \cdot \dfrac{x}{x}$

$= \dfrac{(x+4)^2 + x^2}{x(x+4)}$

$= \dfrac{x^2 + 8x + 16 + x^2}{x(x+4)}$

$= \dfrac{2x^2 + 8x + 16}{x(x+4)}$

31. $\dfrac{4}{a^2 - a - 2} + \dfrac{3}{a^2 + 4a + 3}$

$= \dfrac{4}{(a-2)(a+1)} + \dfrac{3}{(a+3)(a+1)}$

                 LCD $= (a-2)(a+1)(a+3)$

$= \dfrac{4}{(a-2)(a+1)} \cdot \dfrac{a+3}{a+3} + \dfrac{3}{(a+3)(a+1)} \cdot \dfrac{a-2}{a-2}$

$= \dfrac{4(a+3) + 3(a-2)}{(a-2)(a+1)(a+3)}$

$= \dfrac{4a + 12 + 3a - 6}{(a-2)(a+1)(a+3)}$

$= \dfrac{7a + 6}{(a-2)(a+1)(a+3)}$

33. $\dfrac{x+3}{x-5} + \dfrac{x-5}{x+3}$ LCD $= (x-5)(x+3)$

$= \dfrac{x+3}{x-5} \cdot \dfrac{x+3}{x+3} + \dfrac{x-5}{x+3} \cdot \dfrac{x-5}{x-5}$

$= \dfrac{(x+3)^2 + (x-5)^2}{(x-5)(x+3)}$

$= \dfrac{x^2 + 6x + 9 + x^2 - 10x + 25}{(x-5)(x+3)}$

$= \dfrac{2x^2 - 4x + 34}{(x-5)(x+3)}$

35. $\dfrac{a}{a^2 - 1} + \dfrac{2a}{a^2 - a}$

$= \dfrac{a}{(a+1)(a-1)} + \dfrac{2a}{a(a-1)}$

                 LCD $= a(a+1)(a-1)$

$= \dfrac{a}{(a+1)(a-1)} \cdot \dfrac{a}{a} + \dfrac{2a}{a(a-1)} \cdot \dfrac{a+1}{a+1}$

$= \dfrac{a^2 + 2a(a+1)}{a(a+1)(a-1)} = \dfrac{a^2 + 2a^2 + 2a}{a(a+1)(a-1)}$

$= \dfrac{3a^2 + 2a}{a(a+1)(a-1)} = \dfrac{a(3a+2)}{a(a+1)(a-1)}$

$= \dfrac{\cancel{a}(3a+2)}{\cancel{a}(a+1)(a-1)} = \dfrac{3a+2}{(a+1)(a-1)}$

37. $\dfrac{7}{8} + \dfrac{5}{-8} = \dfrac{7}{8} + \dfrac{5}{-8} \cdot \dfrac{-1}{-1}$

$= \dfrac{7}{8} + \dfrac{-5}{8}$

$= \dfrac{7 + (-5)}{8}$

$= \dfrac{2}{8} = \dfrac{\cancel{2} \cdot 1}{4 \cdot \cancel{2}}$

$= \dfrac{1}{4}$

39. $\dfrac{3}{t} + \dfrac{4}{-t} = \dfrac{3}{t} + \dfrac{4}{-t} \cdot \dfrac{-1}{-1}$

$= \dfrac{3}{t} + \dfrac{-4}{t}$

$= \dfrac{3 + (-4)}{t}$

$= \dfrac{-1}{t}$

$= -\dfrac{1}{t}$

41. $\dfrac{2x+7}{x-6} + \dfrac{3x}{6-x} = \dfrac{2x+7}{x-6} + \dfrac{3x}{6-x} \cdot \dfrac{-1}{-1}$

$= \dfrac{2x+7}{x-6} + \dfrac{-3x}{x-6}$

$= \dfrac{(2x+7) + (-3x)}{x-6}$

$= \dfrac{-x + 7}{x - 6}$

43.
$$\frac{y^2}{y-3} + \frac{9}{3-y} = \frac{y^2}{y-3} + \frac{9}{3-y} \cdot \frac{-1}{-1}$$
$$= \frac{y^2}{y-3} + \frac{-9}{y-3}$$
$$= \frac{y^2 + (-9)}{y-3}$$
$$= \frac{y^2 - 9}{y-3}$$
$$= \frac{(y+3)(y-3)}{y-3}$$
$$= \frac{(y+3)\cancel{(y-3)}}{1\cancel{(y-3)}}$$
$$= y + 3$$

45.
$$\frac{b-7}{b^2-16} + \frac{7-b}{16-b^2} = \frac{b-7}{b^2-16} + \frac{7-b}{16-b^2} \cdot \frac{-1}{-1}$$
$$= \frac{b-7}{b^2-16} + \frac{b-7}{b^2-16}$$
$$= \frac{(b-7)+(b-7)}{b^2-16}$$
$$= \frac{2b-14}{b^2-16}$$

47.
$$\frac{a^2}{a-b} + \frac{b^2}{b-a} = \frac{a^2}{a-b} + \frac{b^2}{b-a} \cdot \frac{-1}{-1}$$
$$= \frac{a^2}{a-b} + \frac{-b^2}{a-b}$$
$$= \frac{a^2 + (-b^2)}{a-b}$$
$$= \frac{a^2 - b^2}{a-b}$$
$$= \frac{(a+b)(a-b)}{a-b}$$
$$= \frac{(a+b)\cancel{(a-b)}}{1\cancel{(a-b)}}$$
$$= a + b$$

49.
$$\frac{x+3}{x-5} + \frac{2x-1}{5-x} + \frac{2(3x-1)}{x-5}$$
$$= \frac{x+3}{x-5} + \frac{2x-1}{5-x} \cdot \frac{-1}{-1} + \frac{2(3x-1)}{x-5}$$
$$= \frac{x+3}{x-5} + \frac{1-2x}{x-5} + \frac{2(3x-1)}{x-5}$$
$$= \frac{(x+3)+(1-2x)+(6x-2)}{x-5}$$
$$= \frac{5x+2}{x-5}$$

51.
$$\frac{2(4x+1)}{5x-7} + \frac{3(x-2)}{7-5x} + \frac{-10x-1}{5x-7}$$
$$= \frac{2(4x+1)}{5x-7} + \frac{3(x-2)}{7-5x} \cdot \frac{-1}{-1} + \frac{-10x-1}{5x-7}$$
$$= \frac{2(4x+1)}{5x-7} + \frac{-3(x-2)}{5x-7} + \frac{-10x-1}{5x-7}$$
$$= \frac{(8x+2)+(-3x+6)+(-10x-1)}{5x-7}$$
$$= \frac{-5x+7}{5x-7}$$
$$= \frac{-1(5x-7)}{5x-7}$$
$$= \frac{-1\cancel{(5x-7)}}{\cancel{5x-7}}$$
$$= -1$$

53.
$$\frac{x+1}{(x+3)(x-3)} + \frac{4(x-3)}{(x-3)(x+3)} + \frac{(x-1)(x-3)}{(3-x)(x+3)}$$
$$= \frac{x+1}{(x+3)(x-3)} + \frac{4(x-3)}{(x-3)(x+3)} + \frac{(x-1)(x-3)}{(3-x)(x+3)} \cdot \frac{-1}{-1}$$
$$= \frac{x+1}{(x+3)(x-3)} + \frac{4(x-3)}{(x-3)(x+3)} + \frac{-1(x^2-4x+3)}{(x-3)(x+3)}$$
$$= \frac{(x+1)+(4x-12)+(-x^2+4x-3)}{(x+3)(x-3)}$$
$$= \frac{-x^2+9x-14}{(x+3)(x-3)}$$

55.
$$\frac{6}{x-y} + \frac{4x}{y^2-x^2}$$
$$= \frac{6}{x-y} + \frac{4x}{(y-x)(y+x)}$$
$$= \frac{6}{x-y} + \frac{4x}{(y-x)(y+x)} \cdot \frac{-1}{-1}$$
$$= \frac{6}{x-y} + \frac{-4x}{(x-y)(x+y)}$$
$$[-1(y-x) = x-y;\ y+x = x+y]$$
$$\text{LCD} = (x-y)(x+y)$$
$$= \frac{6}{x-y} \cdot \frac{x+y}{x+y} + \frac{-4x}{(x-y)(x+y)}$$
$$= \frac{6(x+y)-4x}{(x-y)(x+y)}$$
$$= \frac{6x+6y-4x}{(x-y)(x+y)}$$
$$= \frac{2x+6y}{(x-y)(x+y)}$$

Exercise Set 6.4

57. $\dfrac{4-a}{25-a^2} + \dfrac{a+1}{a-5}$

$= \dfrac{4-a}{25-a^2} \cdot \dfrac{-1}{-1} + \dfrac{a+1}{a-5}$

$= \dfrac{a-4}{a^2-25} + \dfrac{a+1}{a-5}$

$= \dfrac{a-4}{(a+5)(a-5)} + \dfrac{a+1}{a-5}$

$\quad\quad\quad\quad$ LCD $= (a+5)(a-5)$

$= \dfrac{a-4}{(a+5)(a-5)} + \dfrac{a+1}{a-5} \cdot \dfrac{a+5}{a+5}$

$= \dfrac{a-4}{(a+5)(a-5)} + \dfrac{(a+1)(a+5)}{(a+5)(a-5)}$

$= \dfrac{(a-4)+(a+1)(a+5)}{(a+5)(a-5)}$

$= \dfrac{a-4+a^2+6a+5}{(a+5)(a-5)}$

$= \dfrac{a^2+7a+1}{(a+5)(a-5)}$

59. $\dfrac{2}{t^2+t-6} + \dfrac{3}{t^2-9}$

$= \dfrac{2}{(t+3)(t-2)} + \dfrac{3}{(t+3)(t-3)}$

$\quad\quad\quad\quad$ LCD $= (t+3)(t-2)(t-3)$

$= \dfrac{2}{(t+3)(t-2)} \cdot \dfrac{t-3}{t-3} + \dfrac{3}{(t+3)(t-3)} \cdot \dfrac{t-2}{t-2}$

$= \dfrac{2(t-3)+3(t-2)}{(t+3)(t-2)(t-3)}$

$= \dfrac{2t-6+3t-6}{(t+3)(t-2)(t-3)}$

$= \dfrac{5t-12}{(t+3)(t-2)(t-3)}$

61. Discussion and Writing Exercise

63. $(x^2+x)-(x+1) = x^2+x-x-1 = x^2-1$

65. $(2x^4y^3)^{-3} = \dfrac{1}{(2x^4y^3)^3} = \dfrac{1}{2^3(x^4)^3(y^3)^3} = \dfrac{1}{8x^{12}y^9}$

67. $\left(\dfrac{x^{-4}}{y^7}\right)^3 = \dfrac{(x^{-4})^3}{(y^7)^3} = \dfrac{x^{-12}}{y^{21}} = \dfrac{1}{x^{12}y^{21}}$

69. $y = \dfrac{1}{2}x - 5 = \dfrac{1}{2}x + (-5)$

The y-intercept is $(0, -5)$. We find two other pairs.

When $x = 2$, $y = \dfrac{1}{2} \cdot 2 - 5 = 1 - 5 = -4$.

When $x = 4$, $y = \dfrac{1}{2} \cdot 4 - 5 = 2 - 5 = -3$.

x	y
0	-5
2	-4
4	-3

Plot these points, draw the line they determine, and label the graph $y = \dfrac{1}{2}x - 5$.

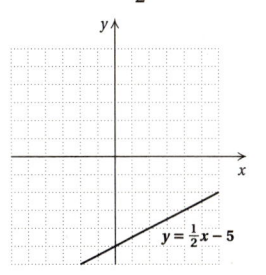

71. $y = 3$

Any ordered pair $(x, 3)$ is a solution. The variable y must be 3, but x can be any number we choose. A few solutions are listed below. Plot these points and draw the line.

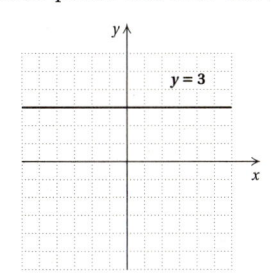

x	y
-4	3
0	3
3	3

73. $3x - 7 = 5x + 9$

$-2x - 7 = 9$ $\quad$ Subtracting $5x$

$-2x = 16$ $\quad$ Adding 7

$x = -8$ $\quad$ Dividing by -2

The solution is -8.

75. $x^2 - 8x + 15 = 0$

$(x-3)(x-5) = 0$

$x - 3 = 0$ or $x - 5 = 0$ $\quad$ Principle of zero products

$x = 3$ or $\quad x = 5$

The solutions are 3 and 5.

77. To find the perimeter we add the lengths of the sides:

$\dfrac{y+4}{3} + \dfrac{y+4}{3} + \dfrac{y-2}{5} + \dfrac{y-2}{5}$ $\quad$ LCD $= 3 \cdot 5$

$= \dfrac{y+4}{3} \cdot \dfrac{5}{5} + \dfrac{y+4}{3} \cdot \dfrac{5}{5} + \dfrac{y-2}{5} \cdot \dfrac{3}{3} + \dfrac{y-2}{5} \cdot \dfrac{3}{3}$

$= \dfrac{5y+20+5y+20+3y-6+3y-6}{3 \cdot 5}$

$= \dfrac{16y+28}{15}$

To find the area we multiply the length and the width:

$\left(\dfrac{y+4}{3}\right)\left(\dfrac{y-2}{5}\right) = \dfrac{(y+4)(y-2)}{3 \cdot 5} = \dfrac{y^2+2y-8}{15}$

79. $\dfrac{5}{z+2} + \dfrac{4z}{z^2-4} + 2$

$= \dfrac{5}{z+2} + \dfrac{4z}{(z+2)(z-2)} + \dfrac{2}{1}$ LCD $= (z+2)(z-2)$

$= \dfrac{5}{z+2} \cdot \dfrac{z-2}{z-2} + \dfrac{4z}{(z+2)(z-2)} + \dfrac{2}{1} \cdot \dfrac{(z+2)(z-2)}{(z+2)(z-2)}$

$= \dfrac{5z - 10 + 4z + 2(z^2 - 4)}{(z+2)(z-2)}$

$= \dfrac{5z - 10 + 4z + 2z^2 - 8}{(z+2)(z-2)} = \dfrac{2z^2 + 9z - 18}{(z+2)(z-2)}$

$= \dfrac{(2z-3)(z+6)}{(z+2)(z-2)}$

81. $\dfrac{3z^2}{z^4 - 4} + \dfrac{5z^2 - 3}{2z^4 + z^2 - 6}$

$= \dfrac{3z^2}{(z^2+2)(z^2-2)} + \dfrac{5z^2 - 3}{(2z^2-3)(z^2+2)}$

$\qquad$ LCD $= (z^2+2)(z^2-2)(2z^2-3)$

$= \dfrac{3z^2}{(z^2+2)(z^2-2)} \cdot \dfrac{2z^2 - 3}{2z^2 - 3} +$

$\qquad \dfrac{5z^2 - 3}{(2z^2-3)(z^2+2)} \cdot \dfrac{z^2 - 2}{z^2 - 2}$

$= \dfrac{6z^4 - 9z^2 + 5z^4 - 13z^2 + 6}{(z^2+2)(z^2-2)(2z^2-3)}$

$= \dfrac{11z^4 - 22z^2 + 6}{(z^2+2)(z^2-2)(2z^2-3)}$

83.–85. Left to the student

Exercise Set 6.5

1. $\dfrac{7}{x} - \dfrac{3}{x} = \dfrac{7-3}{x} = \dfrac{4}{x}$

3. $\dfrac{y}{y-4} - \dfrac{4}{y-4} = \dfrac{y-4}{y-4} = 1$

5. $\dfrac{2x-3}{x^2+3x-4} - \dfrac{x-7}{x^2+3x-4}$

$= \dfrac{2x - 3 - (x - 7)}{x^2 + 3x - 4}$

$= \dfrac{2x - 3 - x + 7}{x^2 + 3x - 4}$

$= \dfrac{x + 4}{x^2 + 3x - 4}$

$= \dfrac{x + 4}{(x+4)(x-1)}$

$= \dfrac{\cancel{(x+4)} \cdot 1}{\cancel{(x+4)}(x-1)}$

$= \dfrac{1}{x-1}$

7. $\dfrac{a-2}{10} - \dfrac{a+1}{5} = \dfrac{a-2}{10} - \dfrac{a+1}{5} \cdot \dfrac{2}{2}$ LCD $= 10$

$= \dfrac{a-2}{10} - \dfrac{2(a+1)}{10}$

$= \dfrac{(a-2) - 2(a+1)}{10}$

$= \dfrac{a - 2 - 2a - 2}{10}$

$= \dfrac{-a - 4}{10}$

9. $\dfrac{4z-9}{3z} - \dfrac{3z-8}{4z} = \dfrac{4z-9}{3z} \cdot \dfrac{4}{4} - \dfrac{3z-8}{4z} \cdot \dfrac{3}{3}$

$\qquad$ LCD $= 3 \cdot 4 \cdot z$, or $12z$

$= \dfrac{16z - 36}{12z} - \dfrac{9z - 24}{12z}$

$= \dfrac{16z - 36 - (9z - 24)}{12z}$

$= \dfrac{16z - 36 - 9z + 24}{12z}$

$= \dfrac{7z - 12}{12z}$

11. $\dfrac{4x + 2t}{3xt^2} - \dfrac{5x - 3t}{x^2 t}$ LCD $= 3x^2 t^2$

$= \dfrac{4x + 2t}{3xt^2} \cdot \dfrac{x}{x} - \dfrac{5x - 3t}{x^2 t} \cdot \dfrac{3t}{3t}$

$= \dfrac{4x^2 + 2tx}{3x^2 t^2} - \dfrac{15xt - 9t^2}{3x^2 t^2}$

$= \dfrac{4x^2 + 2tx - (15xt - 9t^2)}{3x^2 t^2}$

$= \dfrac{4x^2 + 2tx - 15xt + 9t^2}{3x^2 t^2}$

$= \dfrac{4x^2 - 13xt + 9t^2}{3x^2 t^2}$

13. $\dfrac{5}{x+5} - \dfrac{3}{x-5}$ LCD $= (x+5)(x-5)$

$= \dfrac{5}{x+5} \cdot \dfrac{x-5}{x-5} - \dfrac{3}{x-5} \cdot \dfrac{x+5}{x+5}$

$= \dfrac{5x - 25}{(x+5)(x-5)} - \dfrac{3x + 15}{(x+5)(x-5)}$

$= \dfrac{5x - 25 - (3x + 15)}{(x+5)(x-5)}$

$= \dfrac{5x - 25 - 3x - 15}{(x+5)(x-5)}$

$= \dfrac{2x - 40}{(x+5)(x-5)}$

Exercise Set 6.5

15. $\dfrac{3}{2t^2 - 2t} - \dfrac{5}{2t - 2}$

$= \dfrac{3}{2t(t-1)} - \dfrac{5}{2(t-1)}$ LCD $= 2t(t-1)$

$= \dfrac{3}{2t(t-1)} - \dfrac{5}{2(t-1)} \cdot \dfrac{t}{t}$

$= \dfrac{3}{2t(t-1)} - \dfrac{5t}{2t(t-1)}$

$= \dfrac{3 - 5t}{2t(t-1)}$

17. $\dfrac{2s}{t^2 - s^2} - \dfrac{s}{t-s}$ LCD $= (t-s)(t+s)$

$= \dfrac{2s}{(t-s)(t+s)} - \dfrac{s}{t-s} \cdot \dfrac{t+s}{t+s}$

$= \dfrac{2s}{(t-s)(t+s)} - \dfrac{st + s^2}{(t-s)(t+s)}$

$= \dfrac{2s - (st + s^2)}{(t-s)(t+s)}$

$= \dfrac{2s - st - s^2}{(t-s)(t+s)}$

19. $\dfrac{y-5}{y} - \dfrac{3y-1}{4y} = \dfrac{y-5}{y} \cdot \dfrac{4}{4} - \dfrac{3y-1}{4y}$ LCD $= 4y$

$= \dfrac{4y - 20}{4y} - \dfrac{3y - 1}{4y}$

$= \dfrac{4y - 20 - (3y - 1)}{4y}$

$= \dfrac{4y - 20 - 3y + 1}{4y}$

$= \dfrac{y - 19}{4y}$

21. $\dfrac{a}{x+a} - \dfrac{a}{x-a}$ LCD $= (x+a)(x-a)$

$= \dfrac{a}{x+a} \cdot \dfrac{x-a}{x-a} - \dfrac{a}{x-a} \cdot \dfrac{x+a}{x+a}$

$= \dfrac{ax - a^2}{(x+a)(x-a)} - \dfrac{ax + a^2}{(x+a)(x-a)}$

$= \dfrac{ax - a^2 - (ax + a^2)}{(x+a)(x-a)}$

$= \dfrac{ax - a^2 - ax - a^2}{(x+a)(x-a)}$

$= \dfrac{-2a^2}{(x+a)(x-a)}$

23. $\dfrac{11}{6} - \dfrac{5}{-6} = \dfrac{11}{6} - \dfrac{5}{-6} \cdot \dfrac{-1}{-1}$

$= \dfrac{11}{6} - \dfrac{-5}{6}$

$= \dfrac{11 - (-5)}{6}$

$= \dfrac{11 + 5}{6}$

$= \dfrac{16}{6}$

$= \dfrac{8}{3}$

25. $\dfrac{5}{a} - \dfrac{8}{-a} = \dfrac{5}{a} - \dfrac{8}{-a} \cdot \dfrac{-1}{-1}$

$= \dfrac{5}{a} - \dfrac{-8}{a}$

$= \dfrac{5 - (-8)}{a}$

$= \dfrac{5 + 8}{a}$

$= \dfrac{13}{a}$

27. $\dfrac{4}{y-1} - \dfrac{4}{1-y} = \dfrac{4}{y-1} - \dfrac{4}{1-y} \cdot \dfrac{-1}{-1}$

$= \dfrac{4}{y-1} - \dfrac{4(-1)}{(1-y)(-1)}$

$= \dfrac{4}{y-1} - \dfrac{-4}{y-1}$

$= \dfrac{4 - (-4)}{y-1}$

$= \dfrac{4 + 4}{y-1}$

$= \dfrac{8}{y-1}$

29. $\dfrac{3-x}{x-7} - \dfrac{2x-5}{7-x} = \dfrac{3-x}{x-7} - \dfrac{2x-5}{7-x} \cdot \dfrac{-1}{-1}$

$= \dfrac{3-x}{x-7} - \dfrac{(2x-5)(-1)}{(7-x)(-1)}$

$= \dfrac{3-x}{x-7} - \dfrac{5-2x}{x-7}$

$= \dfrac{(3-x) - (5-2x)}{x-7}$

$= \dfrac{3 - x - 5 + 2x}{x-7}$

$= \dfrac{x-2}{x-7}$

31. $\dfrac{a-2}{a^2-25} - \dfrac{6-a}{25-a^2} = \dfrac{a-2}{a^2-25} - \dfrac{6-a}{25-a^2} \cdot \dfrac{-1}{-1}$

$= \dfrac{a-2}{a^2-25} - \dfrac{(6-a)(-1)}{(25-a^2)(-1)}$

$= \dfrac{a-2}{a^2-25} - \dfrac{a-6}{a^2-25}$

$= \dfrac{(a-2)-(a-6)}{a^2-25}$

$= \dfrac{a-2-a+6}{a^2-25}$

$= \dfrac{4}{a^2-25}$

33. $\dfrac{4-x}{x-9} - \dfrac{3x-8}{9-x} = \dfrac{4-x}{x-9} - \dfrac{3x-8}{9-x} \cdot \dfrac{-1}{-1}$

$= \dfrac{4-x}{x-9} - \dfrac{8-3x}{x-9}$

$= \dfrac{(4-x)-(8-3x)}{x-9}$

$= \dfrac{4-x-8+3x}{x-9}$

$= \dfrac{2x-4}{x-9}$

35. $\dfrac{5x}{x^2-9} - \dfrac{4}{3-x}$

$= \dfrac{5x}{(x+3)(x-3)} - \dfrac{4}{3-x}$ $\quad x-3$ and $3-x$ are opposites

$= \dfrac{5x}{(x+3)(x-3)} - \dfrac{4}{3-x} \cdot \dfrac{-1}{-1}$

$= \dfrac{5x}{(x+3)(x-3)} - \dfrac{-4}{x-3}$ $\quad$ LCD $= (x+3)(x-3)$

$= \dfrac{5x}{(x+3)(x-3)} - \dfrac{-4}{x-3} \cdot \dfrac{x+3}{x+3}$

$= \dfrac{5x}{(x+3)(x-3)} - \dfrac{-4x-12}{(x+3)(x-3)}$

$= \dfrac{5x-(-4x-12)}{(x+3)(x-3)}$

$= \dfrac{5x+4x+12}{(x+3)(x-3)}$

$= \dfrac{9x+12}{(x+3)(x-3)}$

37. $\dfrac{t^2}{2t^2-2t} - \dfrac{1}{2t-2}$

$= \dfrac{t^2}{2t(t-1)} - \dfrac{1}{2(t-1)}$ $\quad$ LCD $= 2t(t-1)$

$= \dfrac{t^2}{2t(t-1)} - \dfrac{1}{2(t-1)} \cdot \dfrac{t}{t}$

$= \dfrac{t^2}{2t(t-1)} - \dfrac{t}{2t(t-1)}$

$= \dfrac{t^2-t}{2t(t-1)}$

$= \dfrac{t(t-1)}{2t(t-1)}$

$= \dfrac{\cancel{t}(\cancel{t-1})(1)}{\cancel{2t}(\cancel{t-1})}$

$= \dfrac{1}{2}$

39. $\dfrac{x}{x^2+5x+6} - \dfrac{2}{x^2+3x+2}$

$= \dfrac{x}{(x+3)(x+2)} - \dfrac{2}{(x+2)(x+1)}$

$\quad$ LCD $= (x+3)(x+2)(x+1)$

$= \dfrac{x}{(x+3)(x+2)} \cdot \dfrac{x+1}{x+1} - \dfrac{2}{(x+2)(x+1)} \cdot \dfrac{x+3}{x+3}$

$= \dfrac{x^2+x}{(x+3)(x+2)(x+1)} - \dfrac{2x+6}{(x+3)(x+2)(x+1)}$

$= \dfrac{x^2+x-(2x+6)}{(x+3)(x+2)(x+1)}$

$= \dfrac{x^2+x-2x-6}{(x+3)(x+2)(x+1)}$

$= \dfrac{x^2-x-6}{(x+3)(x+2)(x+1)}$

$= \dfrac{(x-3)(x+2)}{(x+3)(x+2)(x+1)}$

$= \dfrac{(x-3)\cancel{(x+2)}}{(x+3)\cancel{(x+2)}(x+1)}$

$= \dfrac{x-3}{(x+3)(x+1)}$

41. $\dfrac{3(2x+5)}{x-1} - \dfrac{3(2x-3)}{1-x} + \dfrac{6x+1}{x-1}$

$= \dfrac{3(2x+5)}{x-1} - \dfrac{3(2x-3)}{1-x} \cdot \dfrac{-1}{-1} + \dfrac{6x-1}{x-1}$

$= \dfrac{3(2x+5)}{x-1} - \dfrac{-3(2x-3)}{x-1} + \dfrac{6x-1}{x-1}$

$= \dfrac{(6x+15)-(-6x+9)+(6x-1)}{x-1}$

$= \dfrac{6x+15+6x-9+6x-1}{x-1}$

$= \dfrac{18x+5}{x-1}$

Exercise Set 6.5

43. $\dfrac{x-y}{x^2-y^2} + \dfrac{x+y}{x^2-y^2} - \dfrac{2x}{x^2-y^2}$

$= \dfrac{x-y+x+y-2x}{x^2-y^2}$

$= \dfrac{0}{x^2-y^2}$

$= 0$

45. $\dfrac{2(x-1)}{2x-3} - \dfrac{3(x+2)}{2x-3} - \dfrac{x-1}{3-2x}$

$= \dfrac{2(x-1)}{2x-3} - \dfrac{3(x+2)}{2x-3} - \dfrac{x-1}{3-2x} \cdot \dfrac{-1}{-1}$

$= \dfrac{2(x-1)}{2x-3} - \dfrac{3(x+2)}{2x-3} - \dfrac{1-x}{2x-3}$

$= \dfrac{(2x-2) - (3x+6) - (1-x)}{2x-3}$

$= \dfrac{2x-2-3x-6-1+x}{2x-3}$

$= \dfrac{-9}{2x-3}$

47. $\dfrac{10}{2y-1} - \dfrac{6}{1-2y} + \dfrac{y}{2y-1} + \dfrac{y-4}{1-2y}$

$= \dfrac{10}{2y-1} - \dfrac{6}{1-2y} \cdot \dfrac{-1}{-1} + \dfrac{y}{2y-1} + \dfrac{y-4}{1-2y} \cdot \dfrac{-1}{-1}$

$= \dfrac{10}{2y-1} - \dfrac{-6}{2y-1} + \dfrac{y}{2y-1} + \dfrac{4-y}{2y-1}$

$= \dfrac{10 - (-6) + y + 4 - y}{2y-1}$

$= \dfrac{10 + 6 + y + 4 - y}{2y-1}$

$= \dfrac{20}{2y-1}$

49. $\dfrac{a+6}{4-a^2} - \dfrac{a+3}{a+2} + \dfrac{a-3}{2-a}$

$= \dfrac{a+6}{(2+a)(2-a)} - \dfrac{a+3}{2+a} + \dfrac{a-3}{2-a}$

$a + 2 = 2 + a;\ \text{LCD} = (2+a)(2-a)$

$= \dfrac{a+6}{(2+a)(2-a)} - \dfrac{a+3}{2+a} \cdot \dfrac{2-a}{2-a} + \dfrac{a-3}{2-a} \cdot \dfrac{2+a}{2+a}$

$= \dfrac{(a+6) - (a+3)(2-a) + (a-3)(2+a)}{(2+a)(2-a)}$

$= \dfrac{a+6 - (-a^2-a+6) + (a^2-a-6)}{(2+a)(2-a)}$

$= \dfrac{a+6+a^2+a-6+a^2-a-6}{(2+a)(2-a)}$

$= \dfrac{2a^2+a-6}{(2+a)(2-a)}$

$= \dfrac{(2a-3)(a+2)}{(2+a)(2-a)}$

$= \dfrac{(2a-3)\cancel{(2+a)}}{\cancel{(2+a)}(2-a)}$

$= \dfrac{2a-3}{2-a}$

51. $\dfrac{2z}{1-2z} + \dfrac{3z}{2z+1} - \dfrac{3}{4z^2-1}$

$= \dfrac{2z}{1-2z} \cdot \dfrac{-1}{-1} + \dfrac{3z}{2z+1} - \dfrac{3}{4z^2-1}$

$= \dfrac{-2z}{2z-1} + \dfrac{3z}{2z+1} - \dfrac{3}{(2z-1)(2z+1)}$

$\text{LCD} = (2z-1)(2z+1)$

$= \dfrac{-2z}{2z-1} \cdot \dfrac{2z+1}{2z+1} + \dfrac{3z}{2z+1} \cdot \dfrac{2z-1}{2z-1} - \dfrac{3}{(2z-1)(2z+1)}$

$= \dfrac{(-4z^2-2z) + (6z^2-3z) - 3}{(2z-1)(2z+1)}$

$= \dfrac{2z^2-5z-3}{(2z-1)(2z+1)}$

$= \dfrac{(z-3)(2z+1)}{(2z-1)(2z+1)}$

$= \dfrac{(z-3)\cancel{(2z+1)}}{(2z-1)\cancel{(2z+1)}}$

$= \dfrac{z-3}{2z-1}$

53. $\dfrac{1}{x+y} - \dfrac{1}{x-y} + \dfrac{2x}{x^2-y^2}$

$= \dfrac{1}{x+y} - \dfrac{1}{x-y} + \dfrac{2x}{(x+y)(x-y)}$

$\qquad$ LCD $= (x+y)(x-y)$

$= \dfrac{1}{x+y} \cdot \dfrac{x-y}{x-y} - \dfrac{1}{x-y} \cdot \dfrac{x+y}{x+y} + \dfrac{x+y}{x+y} +$

$\qquad \dfrac{2x}{(x+y)(x-y)}$

$= \dfrac{x-y-(x+y)+2x}{(x+y)(x-y)}$

$= \dfrac{x-y-x-y+2x}{(x+y)(x-y)}$

$= \dfrac{2x-2y}{(x+y)(x-y)}$

$= \dfrac{2(x-y)}{(x+y)(x-y)}$

$= \dfrac{2\cancel{(x-y)}}{(x+y)\cancel{(x-y)}}$

$= \dfrac{2}{x+y}$

55. Discussion and Writing Exercise

57. $\dfrac{x^8}{x^3} = x^{8-3} = x^5$

59. $(a^2 b^{-5})^{-4} = a^{2(-4)} b^{-5(-4)} = a^{-8} b^{20} = \dfrac{b^{20}}{a^8}$

61. $\dfrac{66x^2}{11x^5} = \dfrac{6 \cdot \cancel{11} \cdot \cancel{x^2}}{\cancel{11} \cdot \cancel{x^2} \cdot x^3} = \dfrac{6}{x^3}$

63. The shaded area has dimensions $x-6$ by $x-3$. Then the area is $(x-6)(x-3)$, or $x^2 - 9x + 18$.

65. $\dfrac{2x+11}{x-3} \cdot \dfrac{3}{x+4} + \dfrac{2x+1}{4+x} \cdot \dfrac{3}{3-x}$

$= \dfrac{6x+33}{(x-3)(x+4)} + \dfrac{6x+3}{(4+x)(3-x)}$

$= \dfrac{6x+33}{(x-3)(x+4)} + \dfrac{6x+3}{(4+x)(3-x)} \cdot \dfrac{-1}{-1}$

$= \dfrac{6x+33}{(x-3)(x+4)} + \dfrac{-6x-3}{(x+4)(x-3)}$

$= \dfrac{6x+33-6x-3}{(x-3)(x+4)}$

$= \dfrac{30}{(x-3)(x+4)}$

67. $\dfrac{x}{x^4-y^4} - \left(\dfrac{1}{x+y}\right)^2$

$= \dfrac{x}{(x^2+y^2)(x+y)(x-y)} - \dfrac{1}{(x+y)^2}$

$\qquad$ LCD $= (x^2+y^2)(x+y)^2(x-y)$

$= \dfrac{x}{(x^2+y^2)(x+y)(x-y)} \cdot \dfrac{x+y}{x+y} -$

$\qquad \dfrac{1}{(x+y)^2} \cdot \dfrac{(x^2+y^2)(x-y)}{(x^2+y^2)(x-y)}$

$= \dfrac{x(x+y) - (x^2+y^2)(x-y)}{(x^2+y^2)(x+y)^2(x-y)}$

$= \dfrac{x^2+xy-(x^3-x^2y+xy^2-y^3)}{(x^2+y^2)(x+y)^2(x-y)}$

$= \dfrac{x^2+xy-x^3+x^2y-xy^2+y^3}{(x^2+y^2)(x+y)^2(x-y)}$

69. Let $l =$ the length of the missing side.

$\dfrac{a^2-5a-9}{a-6} + \dfrac{a^2-6}{a-6} + l = 2a+5$

$\dfrac{2a^2-5a-15}{a-6} + l = 2a+5$

$l = 2a+5 - \dfrac{2a^2-5a-15}{a-6}$

$l = (2a+5) \cdot \dfrac{a-6}{a-6} - \dfrac{2a^2-5a-15}{a-6}$

$l = \dfrac{2a^2-7a-30}{a-6} - \dfrac{2a^2-5a-15}{a-6}$

$l = \dfrac{2a^2-7a-30-(2a^2-5a-15)}{a-6}$

$l = \dfrac{2a^2-7a-30-2a^2+5a+15}{a-6}$

$l = \dfrac{-2a-15}{a-6}$

The length of the missing side is $\dfrac{-2a-15}{a-6}$.

Now find the area.

$A = \dfrac{1}{2} \cdot b \cdot h$

$A = \dfrac{1}{2} \left(\dfrac{-2a-15}{a-6}\right)\left(\dfrac{a^2-6}{a-6}\right)$

$A = \dfrac{(-2a-15)(a^2-6)}{2(a-6)^2}$, or

$A = \dfrac{-2a^3-15a^2+12a+90}{2a^2-24a+72}$

71.–73. Left to the student

Exercise Set 6.6

1. $\dfrac{1+\dfrac{9}{16}}{1-\dfrac{3}{4}}$ LCM of the denominators is 16.

$= \dfrac{1+\dfrac{9}{16}}{1-\dfrac{3}{4}} \cdot \dfrac{16}{16}$ Multiplying by 1 using $\dfrac{16}{16}$

$= \dfrac{\left(1+\dfrac{9}{16}\right)16}{\left(1-\dfrac{3}{4}\right)16}$ Multiplying numerator and denominator by 16

$= \dfrac{1(16) + \dfrac{9}{16}(16)}{1(16) - \dfrac{3}{4}(16)}$

$= \dfrac{16+9}{16-12}$

$= \dfrac{25}{4}$

3. $\dfrac{1-\dfrac{3}{5}}{1+\dfrac{1}{5}}$

$= \dfrac{1 \cdot \dfrac{5}{5} - \dfrac{3}{5}}{1 \cdot \dfrac{5}{5} + \dfrac{1}{5}}$ Getting a common denominator in numerator and in denominator

$= \dfrac{\dfrac{5}{5} - \dfrac{3}{5}}{\dfrac{5}{5} + \dfrac{1}{5}}$

$= \dfrac{\dfrac{2}{5}}{\dfrac{6}{5}}$ Subtracting in numerator; adding in denominator

$= \dfrac{2}{5} \cdot \dfrac{5}{6}$ Multiplying by the reciprocal of the divisor

$= \dfrac{2 \cdot 5}{5 \cdot 2 \cdot 3}$

$= \dfrac{\cancel{2} \cdot \cancel{5} \cdot 1}{\cancel{5} \cdot \cancel{2} \cdot 3}$

$= \dfrac{1}{3}$

5. $\dfrac{\dfrac{1}{2}+\dfrac{3}{4}}{\dfrac{5}{8}-\dfrac{5}{6}} = \dfrac{\dfrac{1}{2} \cdot \dfrac{2}{2} + \dfrac{3}{4}}{\dfrac{5}{8} \cdot \dfrac{3}{3} - \dfrac{5}{6} \cdot \dfrac{4}{4}}$ Getting a common denominator in numerator and denominator

$= \dfrac{\dfrac{2}{4}+\dfrac{3}{4}}{\dfrac{15}{24}-\dfrac{20}{24}}$

$= \dfrac{\dfrac{5}{4}}{\dfrac{-5}{24}}$ Adding in numerator; subtracting in denominator

$= \dfrac{5}{4} \cdot \dfrac{24}{-5}$ Multiplying by the reciprocal of the divisor

$= \dfrac{5 \cdot 4 \cdot 6}{4 \cdot (-1) \cdot 5}$

$= \dfrac{\cancel{5} \cdot \cancel{4} \cdot 6}{\cancel{4} \cdot (-1) \cdot \cancel{5}}$

$= -6$

7. $\dfrac{\dfrac{1}{x}+3}{\dfrac{1}{x}-5}$ LCM of the denominators is x.

$= \dfrac{\dfrac{1}{x}+3}{\dfrac{1}{x}-5} \cdot \dfrac{x}{x}$ Multiplying by 1 using $\dfrac{x}{x}$

$= \dfrac{\left(\dfrac{1}{x}+3\right)x}{\left(\dfrac{1}{x}-5\right)x}$

$= \dfrac{\dfrac{1}{x} \cdot x + 3 \cdot x}{\dfrac{1}{x} \cdot x - 5 \cdot x}$

$= \dfrac{1+3x}{1-5x}$

9. $\dfrac{4 - \dfrac{1}{x^2}}{2 - \dfrac{1}{x}}$ LCM of the denominators is x^2.

$= \dfrac{4 - \dfrac{1}{x^2}}{2 - \dfrac{1}{x}} \cdot \dfrac{x^2}{x^2}$

$= \dfrac{\left(4 - \dfrac{1}{x^2}\right)x^2}{\left(2 - \dfrac{1}{x}\right)x^2}$

$= \dfrac{4 \cdot x^2 - \dfrac{1}{x^2} \cdot x^2}{2 \cdot x^2 - \dfrac{1}{x} \cdot x^2}$

$= \dfrac{4x^2 - 1}{2x^2 - x}$

$= \dfrac{(2x+1)(2x-1)}{x(2x-1)}$ Factoring numerator and denominator

$= \dfrac{(2x+1)\cancel{(2x-1)}}{x\cancel{(2x-1)}}$

$= \dfrac{2x+1}{x}$

11. $\dfrac{8 + \dfrac{8}{d}}{1 + \dfrac{1}{d}} = \dfrac{8 \cdot \dfrac{d}{d} + \dfrac{8}{d}}{1 \cdot \dfrac{d}{d} + \dfrac{1}{d}}$

$= \dfrac{\dfrac{8d+8}{d}}{\dfrac{d+1}{d}}$

$= \dfrac{8d+8}{d} \cdot \dfrac{d}{d+1}$

$= \dfrac{8(d+1)(d)}{d(d+1)}$

$= \dfrac{8\cancel{(d+1)}\cancel{(d)}}{\cancel{d}\cancel{(d+1)}(1)}$

$= 8$

13. $\dfrac{\dfrac{x}{8} - \dfrac{8}{x}}{\dfrac{1}{8} + \dfrac{1}{x}}$ LCM of the denominators is $8x$.

$= \dfrac{\dfrac{x}{8} - \dfrac{8}{x}}{\dfrac{1}{8} + \dfrac{1}{x}} \cdot \dfrac{8x}{8x}$

$= \dfrac{\left(\dfrac{x}{8} - \dfrac{8}{x}\right)8x}{\left(\dfrac{1}{8} + \dfrac{1}{x}\right)8x}$

$= \dfrac{\dfrac{x}{8}(8x) - \dfrac{8}{x}(8x)}{\dfrac{1}{8}(8x) + \dfrac{1}{x}(8x)}$

$= \dfrac{x^2 - 64}{x + 8}$

$= \dfrac{(x+8)(x-8)}{x+8}$

$= \dfrac{\cancel{(x+8)}(x-8)}{1\cancel{(x+8)}}$

$= x - 8$

15. $\dfrac{1 + \dfrac{1}{y}}{1 - \dfrac{1}{y^2}} = \dfrac{1 \cdot \dfrac{y}{y} + \dfrac{1}{y}}{1 \cdot \dfrac{y^2}{y^2} - \dfrac{1}{y^2}}$

$= \dfrac{\dfrac{y+1}{y}}{\dfrac{y^2-1}{y^2}}$

$= \dfrac{y+1}{y} \cdot \dfrac{y^2}{y^2-1}$

$= \dfrac{(y+1)y \cdot y}{y(y+1)(y-1)}$

$= \dfrac{\cancel{(y+1)}\cancel{y} \cdot y}{\cancel{y}\cancel{(y+1)}(y-1)}$

$= \dfrac{y}{y-1}$

Exercise Set 6.6

17. $\dfrac{\dfrac{1}{5} - \dfrac{1}{a}}{\dfrac{5-a}{5}}$ LCM of the denominators is $5a$.

$= \dfrac{\dfrac{1}{5} - \dfrac{1}{a}}{\dfrac{5-a}{5}} \cdot \dfrac{5a}{5a}$

$= \dfrac{\left(\dfrac{1}{5} - \dfrac{1}{a}\right)5a}{\left(\dfrac{5-a}{5}\right)5a}$

$= \dfrac{\dfrac{1}{5}(5a) - \dfrac{1}{a}(5a)}{a(5-a)}$

$= \dfrac{a-5}{5a-a^2}$

$= \dfrac{a-5}{-a(-5+a)}$

$= \dfrac{1(a-5)}{-a(a-5)}$

$= -\dfrac{1}{a}$

19. $\dfrac{\dfrac{1}{a} + \dfrac{1}{b}}{\dfrac{1}{a^2} - \dfrac{1}{b^2}}$ LCM of the denominators is a^2b^2.

$= \dfrac{\dfrac{1}{a} + \dfrac{1}{b}}{\dfrac{1}{a^2} - \dfrac{1}{b^2}} \cdot \dfrac{a^2b^2}{a^2b^2}$

$= \dfrac{\left(\dfrac{1}{a} + \dfrac{1}{b}\right) \cdot a^2b^2}{\left(\dfrac{1}{a^2} - \dfrac{1}{b^2}\right) \cdot a^2b^2}$

$= \dfrac{\dfrac{1}{a} \cdot a^2b^2 + \dfrac{1}{b} \cdot a^2b^2}{\dfrac{1}{a^2} \cdot a^2b^2 - \dfrac{1}{b^2} \cdot a^2b^2}$

$= \dfrac{ab^2 + a^2b}{b^2 - a^2}$

$= \dfrac{ab(b+a)}{(b+a)(b-a)}$

$= \dfrac{ab(b+a)}{(b+a)(b-a)}$

$= \dfrac{ab}{b-a}$

21. $\dfrac{\dfrac{p}{q} + \dfrac{q}{p}}{\dfrac{1}{p} + \dfrac{1}{q}}$ LCM of the denominators is pq.

$= \dfrac{\left(\dfrac{p}{q} + \dfrac{q}{p}\right) \cdot pq}{\left(\dfrac{1}{p} + \dfrac{1}{q}\right) \cdot pq}$

$= \dfrac{\dfrac{p}{q} \cdot pq + \dfrac{q}{p} \cdot pq}{\dfrac{1}{p} \cdot pq + \dfrac{1}{q} \cdot pq}$

$= \dfrac{p^2 + q^2}{q + p}$

23. $\dfrac{\dfrac{2}{a} + \dfrac{4}{a^2}}{\dfrac{5}{a^3} - \dfrac{3}{a}}$ LCD is a^3

$= \dfrac{\dfrac{2}{a} + \dfrac{4}{a^2}}{\dfrac{5}{a^3} - \dfrac{3}{a}} \cdot \dfrac{a^3}{a^3}$

$= \dfrac{\dfrac{2}{a} \cdot a^3 + \dfrac{4}{a^2} \cdot a^3}{\dfrac{5}{a^3} \cdot a^3 - \dfrac{3}{a} \cdot a^3}$

$= \dfrac{2a^2 + 4a}{5 - 3a^2}$

(Although the numerator can be factored, doing so will not enable us to simplify further.)

25. $\dfrac{\dfrac{2}{7a^4} - \dfrac{1}{14a}}{\dfrac{3}{5a^2} + \dfrac{2}{15a}} = \dfrac{\dfrac{2}{7a^4} \cdot \dfrac{2}{2} - \dfrac{1}{14a} \cdot \dfrac{a^3}{a^3}}{\dfrac{3}{5a^2} \cdot \dfrac{3}{3} + \dfrac{2}{15a} \cdot \dfrac{a}{a}}$

$= \dfrac{\dfrac{4-a^3}{14a^4}}{\dfrac{9+2a}{15a^2}}$

$= \dfrac{4-a^3}{14a^4} \cdot \dfrac{15a^2}{9+2a}$

$= \dfrac{15 \cdot a^2(4-a^3)}{14a^2 \cdot a^2(9+2a)}$

$= \dfrac{15(4-a^3)}{14a^2(9+2a)}$, or $\dfrac{60-15a^3}{126a^2 + 28a^3}$

27. $\dfrac{\frac{a}{b}+\frac{c}{d}}{\frac{b}{a}+\frac{d}{c}} = \dfrac{\frac{a}{b}\cdot\frac{d}{d}+\frac{c}{d}\cdot\frac{b}{b}}{\frac{b}{a}\cdot\frac{c}{c}+\frac{d}{c}\cdot\frac{a}{a}}$

$= \dfrac{\frac{ad+bc}{bd}}{\frac{bc+ad}{ac}}$

$= \dfrac{ad+bc}{bd}\cdot\dfrac{ac}{bc+ad}$

$= \dfrac{ac(ad+bc)}{bd(bc+ad)}$

$= \dfrac{ac}{bd}\cdot\dfrac{ad+bc}{bc+ad}$

$= \dfrac{ac}{bd}\cdot 1$

$= \dfrac{ac}{bd}$

29. $\dfrac{\frac{x}{5y^3}+\frac{3}{10y}}{\frac{3}{10y}+\frac{x}{5y^3}}$

Observe that, by the commutative law of addition, the numerator and denominator are equivalent, so the result is 1. We could also simplify this expression as follows:

$\dfrac{\frac{x}{5y^3}+\frac{3}{10y}}{\frac{3}{10y}+\frac{x}{5y^3}} = \dfrac{\frac{x}{5y^3}+\frac{3}{10y}}{\frac{3}{10y}+\frac{x}{5y^3}}\cdot\dfrac{10y^3}{10y^3}$

$= \dfrac{\frac{x}{5y^3}\cdot 10y^3 + \frac{3}{10y}\cdot 10y^3}{\frac{3}{10y}\cdot 10y^3 + \frac{x}{5y^3}\cdot 10y^3}$

$= \dfrac{2x+3y^2}{3y^2+2x}$

$= 1$

31. $\dfrac{\frac{3}{x+1}+\frac{1}{x}}{\frac{2}{x+1}+\frac{3}{x}} = \dfrac{\frac{3}{x+1}+\frac{1}{x}}{\frac{2}{x+1}+\frac{3}{x}}\cdot\dfrac{x(x+1)}{x(x+1)}$

$= \dfrac{\frac{3}{x+1}\cdot x(x+1) + \frac{1}{x}\cdot x(x+1)}{\frac{2}{x+1}\cdot x(x+1) + \frac{3}{x}\cdot x(x+1)}$

$= \dfrac{3x+x+1}{2x+3(x+1)}$

$= \dfrac{4x+1}{2x+3x+3}$

$= \dfrac{4x+1}{5x+3}$

33. Discussion and Writing Exercise

35. $(2x^3 - 4x^2 + x - 7) + (4x^4 + x^3 + 4x^2 + x)$
$= 4x^4 + 3x^3 + 2x - 7$

37. $p^2 - 10p + 25 = p^2 - 2\cdot p \cdot 5 + 5^2$ Trinomial square
$= (p-5)^2$

39. $50p^2 - 100 = 50(p^2 - 2)$ Factoring out the common factor

Since $p^2 - 2$ cannot be factored, we have factored completely.

41. **Familiarize**. Let w = the width of the rectangle. Then $w + 3$ = the length. Recall that the formula for the area of a rectangle is $A = lw$ and the formula for the perimeter of a rectangle is $P = 2l + 2w$.

Translate. We substitute in the formula for area.
$10 = lw$
$10 = (w+3)w$

Solve.
$10 = (w+3)w$
$10 = w^2 + 3w$
$0 = w^2 + 3w - 10$
$0 = (w+5)(w-2)$
$w + 5 = 0$ or $w - 2 = 0$
$w = -5$ or $w = 2$

Check. Since the width cannot be negative, we only check 2. If $w = 2$, then $w + 3 = 2 + 3$, or 5. Since $2 \cdot 5 = 10$, the given area, the answer checks. Now we find the perimeter:
$P = 2l + 2w$
$P = 2\cdot 5 + 2\cdot 2$
$P = 10 + 4$
$P = 14$

We can check this by repeating the calculation.

State. The perimeter is 14 yd.

43. $\dfrac{1}{\frac{2}{x-1}-\frac{1}{3x-2}}$

$= \dfrac{1}{\frac{2}{x-1}-\frac{1}{3x-2}}\cdot\dfrac{(x-1)(3x-2)}{(x-1)(3x-2)}$

$= \dfrac{(x-1)(3x-2)}{\left(\frac{2}{x-1}-\frac{1}{3x-2}\right)(x-1)(3x-2)}$

$= \dfrac{(x-1)(3x-2)}{\frac{2}{x-1}(x-1)(3x-2) - \frac{1}{3x-2}(x-1)(3x-2)}$

$= \dfrac{(x-1)(3x-2)}{2(3x-2)-(x-1)}$

$= \dfrac{(x-1)(3x-2)}{6x-4-x+1}$

$= \dfrac{(x-1)(3x-2)}{5x-3}$

Exercise Set 6.7   143

45. $1 + \dfrac{1}{1 + \dfrac{1}{1 + \dfrac{1}{1 + \dfrac{1}{x}}}} = 1 + \dfrac{1}{1 + \dfrac{1}{1 + \dfrac{1}{\dfrac{x+1}{x}}}}$

$= 1 + \dfrac{1}{1 + \dfrac{1}{1 + \dfrac{x}{x+1}}}$

$= 1 + \dfrac{1}{1 + \dfrac{1}{\dfrac{x+1+x}{x+1}}}$

$= 1 + \dfrac{1}{1 + \dfrac{1}{\dfrac{2x+1}{x+1}}}$

$= 1 + \dfrac{1}{1 + \dfrac{x+1}{2x+1}}$

$= 1 + \dfrac{1}{\dfrac{2x+1+x+1}{2x+1}}$

$= 1 + \dfrac{1}{\dfrac{3x+2}{2x+1}}$

$= 1 + \dfrac{2x+1}{3x+2}$

$= \dfrac{3x+2+2x+1}{3x+2}$

$= \dfrac{5x+3}{3x+2}$

Exercise Set 6.7

1. $\dfrac{4}{5} - \dfrac{2}{3} = \dfrac{x}{9}$, LCM = 45

$45\left(\dfrac{4}{5} - \dfrac{2}{3}\right) = 45 \cdot \dfrac{x}{9}$

$45 \cdot \dfrac{4}{5} - 45 \cdot \dfrac{2}{3} = 45 \cdot \dfrac{x}{9}$

$36 - 30 = 5x$

$6 = 5x$

$\dfrac{6}{5} = x$

Check:
$$\dfrac{\dfrac{4}{5} - \dfrac{2}{3} = \dfrac{x}{9}}{\begin{array}{c|c} \dfrac{4}{5} - \dfrac{2}{3} \;?\; \dfrac{\frac{6}{5}}{9} \\ \dfrac{12}{15} - \dfrac{10}{15} & \dfrac{6}{5} \cdot \dfrac{1}{9} \\ \dfrac{2}{15} & \dfrac{2}{15} \quad \text{TRUE} \end{array}}$$

This checks, so the solution is $\dfrac{6}{5}$.

3. $\dfrac{3}{5} + \dfrac{1}{8} = \dfrac{1}{x}$, LCM = $40x$

$40x\left(\dfrac{3}{5} + \dfrac{1}{8}\right) = 40x \cdot \dfrac{1}{x}$

$40x \cdot \dfrac{3}{5} + 40x \cdot \dfrac{1}{8} = 40x \cdot \dfrac{1}{x}$

$24x + 5x = 40$

$29x = 40$

$x = \dfrac{40}{29}$

Check:
$$\dfrac{\dfrac{3}{5} + \dfrac{1}{8} = \dfrac{1}{x}}{\begin{array}{c|c} \dfrac{3}{5} + \dfrac{1}{8} \;?\; \dfrac{1}{\frac{40}{29}} \\ \dfrac{24}{40} + \dfrac{5}{40} & 1 \cdot \dfrac{29}{40} \\ \dfrac{29}{40} & \dfrac{29}{40} \quad \text{TRUE} \end{array}}$$

This checks, so the solution is $\dfrac{40}{29}$.

5. $\dfrac{3}{8} + \dfrac{4}{5} = \dfrac{x}{20}$, LCM = 40

$40\left(\dfrac{3}{8} + \dfrac{4}{5}\right) = 40 \cdot \dfrac{x}{20}$

$40 \cdot \dfrac{3}{8} + 40 \cdot \dfrac{4}{5} = 40 \cdot \dfrac{x}{20}$

$15 + 32 = 2x$

$47 = 2x$

$\dfrac{47}{2} = x$

Check:
$$\dfrac{\dfrac{3}{8} + \dfrac{4}{5} = \dfrac{x}{20}}{\begin{array}{c|c} \dfrac{3}{8} + \dfrac{4}{5} \;?\; \dfrac{\frac{47}{2}}{20} \\ \dfrac{15}{40} + \dfrac{32}{40} & \dfrac{47}{2} \cdot \dfrac{1}{20} \\ \dfrac{47}{40} & \dfrac{47}{40} \quad \text{TRUE} \end{array}}$$

This checks, so the solution is $\dfrac{47}{2}$.

7. $\dfrac{1}{x} = \dfrac{2}{3} - \dfrac{5}{6}$, LCM $= 6x$

$6x \cdot \dfrac{1}{x} = 6x\left(\dfrac{2}{3} - \dfrac{5}{6}\right)$

$6x \cdot \dfrac{1}{x} = 6x \cdot \dfrac{2}{3} - 6x \cdot \dfrac{5}{6}$

$6 = 4x - 5x$

$6 = -x$

$-6 = x$

Check:
$$\dfrac{1}{x} = \dfrac{2}{3} - \dfrac{5}{6}$$
$\dfrac{1}{-6}$? $\dfrac{2}{3} - \dfrac{5}{6}$
$-\dfrac{1}{6}$ | $\dfrac{4}{6} - \dfrac{5}{6}$
     | $-\dfrac{1}{6}$ TRUE

This checks, so the solution is -6.

9. $\dfrac{1}{6} + \dfrac{1}{8} = \dfrac{1}{t}$, LCM $= 24t$

$24t\left(\dfrac{1}{6} + \dfrac{1}{8}\right) = 24t \cdot \dfrac{1}{t}$

$24t \cdot \dfrac{1}{6} + 24t \cdot \dfrac{1}{8} = 24t \cdot \dfrac{1}{t}$

$4t + 3t = 24$

$7t = 24$

$t = \dfrac{24}{7}$

Check:
$$\dfrac{1}{6} + \dfrac{1}{8} = \dfrac{1}{t}$$
$\dfrac{1}{6} + \dfrac{1}{8}$? $\dfrac{1}{24/7}$
$\dfrac{4}{24} + \dfrac{3}{24}$ | $1 \cdot \dfrac{7}{24}$
$\dfrac{7}{24}$ | $\dfrac{7}{24}$ TRUE

This checks, so the solution is $\dfrac{24}{7}$.

11. $x + \dfrac{4}{x} = -5$, LCM $= x$

$x\left(x + \dfrac{4}{x}\right) = x(-5)$

$x \cdot x + x \cdot \dfrac{4}{x} = x(-5)$

$x^2 + 4 = -5x$

$x^2 + 5x + 4 = 0$

$(x+4)(x+1) = 0$

$x + 4 = 0 \quad \text{or} \quad x + 1 = 0$

$x = -4 \quad \text{or} \quad x = -1$

Check:
$$x + \dfrac{4}{x} = -5 \qquad x + \dfrac{4}{x} = -5$$
$-4 + \dfrac{4}{-4}$? -5 | $-1 + \dfrac{4}{-1}$? -5
$-4 - 1$ | | $-1 - 4$ |
-5 | TRUE | -5 | TRUE

Both of these check, so the two solutions are -4 and -1.

13. $\dfrac{x}{4} - \dfrac{4}{x} = 0$, LCM $= 4x$

$4x\left(\dfrac{x}{4} - \dfrac{4}{x}\right) = 4x \cdot 0$

$4x \cdot \dfrac{x}{4} - 4x \cdot \dfrac{4}{x} = 4x \cdot 0$

$x^2 - 16 = 0$

$(x+4)(x-4) = 0$

$x + 4 = 0 \quad \text{or} \quad x - 4 = 0$

$x = -4 \quad \text{or} \quad x = 4$

Check:
$$\dfrac{x}{4} - \dfrac{4}{x} = 0 \qquad \dfrac{x}{4} - \dfrac{4}{x} = 0$$
$\dfrac{-4}{4} - \dfrac{4}{-4}$? 0 | $\dfrac{4}{4} - \dfrac{4}{4}$? 0
$-1 - (-1)$ | | $1 - 1$ |
$-1 + 1$ | | 0 | TRUE
0 | TRUE | |

Both of these check, so the two solutions are -4 and 4.

15. $\dfrac{5}{x} = \dfrac{6}{x} - \dfrac{1}{3}$, LCM $= 3x$

$3x \cdot \dfrac{5}{x} = 3x\left(\dfrac{6}{x} - \dfrac{1}{3}\right)$

$3x \cdot \dfrac{5}{x} = 3x \cdot \dfrac{6}{x} - 3x \cdot \dfrac{1}{3}$

$15 = 18 - x$

$-3 = -x$

$3 = x$

Check:
$$\dfrac{5}{x} = \dfrac{6}{x} - \dfrac{1}{3}$$
$\dfrac{5}{3}$? $\dfrac{6}{3} - \dfrac{1}{3}$
    | $\dfrac{5}{3}$ TRUE

This checks, so the solution is 3.

Exercise Set 6.7

17. $\dfrac{5}{3x} + \dfrac{3}{x} = 1$, LCM $= 3x$

$$3x\left(\dfrac{5}{3x} + \dfrac{3}{x}\right) = 3x \cdot 1$$

$$3x \cdot \dfrac{5}{3x} + 3x \cdot \dfrac{3}{x} = 3x \cdot 1$$

$$5 + 9 = 3x$$

$$14 = 3x$$

$$\dfrac{14}{3} = x$$

Check:
$$\dfrac{\dfrac{5}{3x} + \dfrac{3}{x} = 1}{\dfrac{5}{3 \cdot (14/3)} + \dfrac{3}{(14/3)} \;?\; 1}$$

$$\dfrac{5}{14} + \dfrac{9}{14}$$

$$\dfrac{14}{14}$$

$$1 \;\bigg|\; \text{TRUE}$$

This checks, so the solution is $\dfrac{14}{3}$.

19. $\dfrac{t-2}{t+3} = \dfrac{3}{8}$, LCM $= 8(t+3)$

$$8(t+3)\left(\dfrac{t-2}{t+3}\right) = 8(t+3)\left(\dfrac{3}{8}\right)$$

$$8(t-2) = 3(t+3)$$

$$8t - 16 = 3t + 9$$

$$5t = 25$$

$$t = 5$$

Check:
$$\dfrac{t-2}{t+3} = \dfrac{3}{8}$$

$$\dfrac{5-2}{5+3} \;?\; \dfrac{3}{8}$$

$$\dfrac{3}{8} \;\bigg|\; \text{TRUE}$$

This checks, so the solution is 5.

21. $\dfrac{2}{x+1} = \dfrac{1}{x-2}$, LCM $= (x+1)(x-2)$

$$(x+1)(x-2) \cdot \dfrac{2}{x+1} = (x+1)(x-2) \cdot \dfrac{1}{x-2}$$

$$2(x-2) = x+1$$

$$2x - 4 = x + 1$$

$$x = 5$$

This checks, so the solution is 5.

23. $\dfrac{x}{6} - \dfrac{x}{10} = \dfrac{1}{6}$, LCM $= 30$

$$30\left(\dfrac{x}{6} - \dfrac{x}{10}\right) = 30 \cdot \dfrac{1}{6}$$

$$30 \cdot \dfrac{x}{6} - 30 \cdot \dfrac{x}{10} = 30 \cdot \dfrac{1}{6}$$

$$5x - 3x = 5$$

$$2x = 5$$

$$x = \dfrac{5}{2}$$

This checks, so the solution is $\dfrac{5}{2}$.

25. $\dfrac{t+2}{5} - \dfrac{t-2}{4} = 1$, LCM $= 20$

$$20\left(\dfrac{t+2}{5} - \dfrac{t-2}{4}\right) = 20 \cdot 1$$

$$20\left(\dfrac{t+2}{5}\right) - 20\left(\dfrac{t-2}{4}\right) = 20 \cdot 1$$

$$4(t+2) - 5(t-2) = 20$$

$$4t + 8 - 5t + 10 = 20$$

$$-t + 18 = 20$$

$$-t = 2$$

$$t = -2$$

This checks, so the solution is -2.

27. $\dfrac{5}{x-1} = \dfrac{3}{x+2}$,

$$\text{LCD} = (x-1)(x+2)$$

$$(x-1)(x+2) \cdot \dfrac{5}{x-1} = (x-1)(x+2) \cdot \dfrac{3}{x+2}$$

$$5(x+2) = 3(x-1)$$

$$5x + 10 = 3x - 3$$

$$2x = -13$$

$$x = -\dfrac{13}{2}$$

This checks, so the solution is $-\dfrac{13}{2}$.

29. $\dfrac{a-3}{3a+2} = \dfrac{1}{5}$, LCM $= 5(3a+2)$

$$5(3a+2) \cdot \dfrac{a-3}{3a+2} = 5(3a+2) \cdot \dfrac{1}{5}$$

$$5(a-3) = 3a+2$$

$$5a - 15 = 3a + 2$$

$$2a = 17$$

$$a = \dfrac{17}{2}$$

This checks, so the solution is $\dfrac{17}{2}$.

31. $$\frac{x-1}{x-5} = \frac{4}{x-5}, \text{ LCM} = x-5$$
$$(x-5) \cdot \frac{x-1}{x-5} = (x-5) \cdot \frac{4}{x-5}$$
$$x - 1 = 4$$
$$x = 5$$

The number 5 is not a solution because it makes a denominator zero. Thus, there is no solution.

33. $$\frac{2}{x+3} = \frac{5}{x}, \text{ LCM} = x(x+3)$$
$$x(x+3) \cdot \frac{2}{x+3} = x(x+3) \cdot \frac{5}{x}$$
$$2x = 5(x+3)$$
$$2x = 5x + 15$$
$$-15 = 3x$$
$$-5 = x$$

This checks, so the solution is -5.

35. $$\frac{x-2}{x-3} = \frac{x-1}{x+1}, \text{ LCM} = (x-3)(x+1)$$
$$(x-3)(x+1) \cdot \frac{x-2}{x-3} = (x-3)(x+1) \cdot \frac{x-1}{x+1}$$
$$(x+1)(x-2) = (x-3)(x-1)$$
$$x^2 - x - 2 = x^2 - 4x + 3$$
$$-x - 2 = -4x + 3$$
$$3x = 5$$
$$x = \frac{5}{3}$$

This checks, so the solution is $\frac{5}{3}$.

37. $$\frac{1}{x+3} + \frac{1}{x-3} = \frac{1}{x^2-9},$$
$$\text{LCM} = (x+3)(x-3)$$
$$(x+3)(x-3)\left(\frac{1}{x+3} + \frac{1}{x-3}\right) = (x+3)(x-3) \cdot \frac{1}{(x+3)(x-3)}$$
$$(x-3) + (x+3) = 1$$
$$2x = 1$$
$$x = \frac{1}{2}$$

This checks, so the solution is $\frac{1}{2}$.

39. $$\frac{x}{x+4} - \frac{4}{x-4} = \frac{x^2+16}{x^2-16},$$
$$\text{LCM} = (x+4)(x-4)$$
$$(x+4)(x-4)\left(\frac{x}{x+4} - \frac{x}{x-4}\right) = (x+4)(x-4) \cdot \frac{x^2+16}{(x+4)(x-4)}$$
$$x(x-4) - 4(x+4) = x^2 + 16$$
$$x^2 - 4x - 4x - 16 = x^2 + 16$$
$$x^2 - 8x - 16 = x^2 + 16$$
$$-8x - 16 = 16$$
$$-8x = 32$$
$$x = -4$$

The number -4 is not a solution because it makes a denominator zero. Thus, there is no solution.

41. $$\frac{4-a}{8-a} = \frac{4}{a-8} \quad \begin{array}{l} 8-a \text{ and } a-8 \\ \text{are opposites} \end{array}$$
$$\frac{4-a}{8-a} \cdot \frac{-1}{-1} = \frac{4}{a-8}$$
$$\frac{a-4}{a-8} = \frac{4}{a-8}, \text{ LCM} = a-8$$
$$(a-8)\left(\frac{a-4}{a-8}\right) = (a-8)\left(\frac{4}{a-8}\right)$$
$$a - 4 = 4$$
$$a = 8$$

The number 8 is not a solution because it makes a denominator zero. Thus, there is no solution.

43. $$2 - \frac{a-2}{a+3} = \frac{a^2-4}{a+3}, \text{ LCM} = a+3$$
$$(a+3)\left(2 - \frac{a-2}{a+3}\right) = (a+3) \cdot \frac{a^2-4}{a+3}$$
$$2(a+3) - (a-2) = a^2 - 4$$
$$2a + 6 - a + 2 = a^2 - 4$$
$$0 = a^2 - a - 12$$
$$0 = (a-4)(a+3)$$
$$a - 4 = 0 \text{ or } a + 3 = 0$$
$$a = 4 \text{ or } \quad a = -3$$

Only 4 checks, so the solution is 4.

45. Discussion and Writing Exercise

47. $(a^2 b^5)^{-3} = \dfrac{1}{(a^2 b^5)^3} = \dfrac{1}{(a^2)^3 (b^5)^3} = \dfrac{1}{a^6 b^{15}}$

49. $\left(\dfrac{2x}{t^2}\right)^4 = \dfrac{(2x)^4}{(t^2)^4} = \dfrac{2^4 x^4}{t^8} = \dfrac{16x^4}{t^8}$

51. $4x^{-5} \cdot 8x^{11} = 4 \cdot 8 x^{-5+11} = 32x^6$

Exercise Set 6.7

53. $5x + 10y = 20$

To find the x-intercept, let $y = 0$. Then solve for x.
$$5x + 10y = 20$$
$$5x + 10 \cdot 0 = 20$$
$$5x = 20$$
$$x = 4$$

The x-intercept is $(4, 0)$.

To find the y-intercept, let $x = 0$. Then solve for y.
$$5x + 10y = 20$$
$$5 \cdot 0 + 10y = 20$$
$$10y = 20$$
$$y = 2$$

The y-intercept is $(0, 2)$.

Plot these points and draw the line.

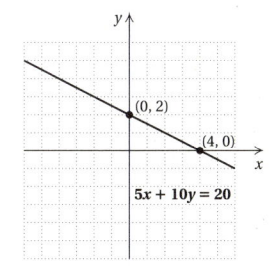

A third point should be used as a check. We substitute any value for x and solve for y. We let $x = -4$. Then
$$5x + 10y = 20$$
$$5(-4) + 10y = 20$$
$$-20 + 10y = 20$$
$$10y = 40$$
$$y = 4$$

The point $(-4, 4)$ is on the graph, so the graph is probably correct.

55. $10y - 4x = -20$

To find the x-intercept, let $y = 0$. Then solve for x.
$$10y - 4x = -20$$
$$10 \cdot 0 - 4x = -20$$
$$-4x = -20$$
$$x = 5$$

The x-intercept is $(5, 0)$.

To find the y-intercept, let $x = 0$. Then solve for y.
$$10y - 4x = -20$$
$$10y - 4 \cdot 0 = -20$$
$$10y = -20$$
$$y = -2$$

The y-intercept is $(0, -2)$.

Plot these points and draw the line.

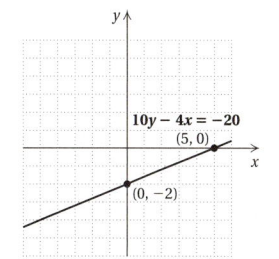

A third point should be used as a check. We substitute any value for x and solve for y. We let $x = -5$. Then
$$10y - 4x = -20$$
$$10y - 4(-5) = -20$$
$$10y + 20 = -20$$
$$10y = -40$$
$$y = -4.$$

The point $(-5, -4)$ is on the graph, so the graph is probably correct.

57.
$$\frac{4}{y-2} - \frac{2y-3}{y^2-4} = \frac{5}{y+2},$$
$$\text{LCM} = (y+2)(y-2)$$
$$(y+2)(y-2)\left(\frac{4}{y-2} - \frac{2y-3}{(y+2)(y-2)}\right) =$$
$$(y+2)(y-2) \cdot \frac{5}{y+2}$$
$$4(y+2) - (2y-3) = 5(y-2)$$
$$4y + 8 - 2y + 3 = 5y - 10$$
$$2y + 11 = 5y - 10$$
$$21 = 3y$$
$$7 = y$$

This checks, so the solution is 7.

59.
$$\frac{x+1}{x+2} = \frac{x+3}{x+4},$$
$$\text{LCM} = (x+2)(x+4)$$
$$(x+2)(x+4)\left(\frac{x+1}{x+2}\right) = (x+2)(x+4)\left(\frac{x+3}{x+4}\right)$$
$$(x+4)(x+1) = (x+2)(x+3)$$
$$x^2 + 5x + 4 = x^2 + 5x + 6$$
$$4 = 6 \quad \text{Subtracting } x^2 \text{ and } 5x$$

We get a false equation, so the original equation has no solution.

61.
$$4a - 3 = \frac{a+13}{a+1}, \quad \text{LCM} = a+1$$
$$(a+1)(4a-3) = (a+1) \cdot \frac{a+13}{a+1}$$
$$4a^2 + a - 3 = a + 13$$
$$4a^2 - 16 = 0$$
$$4(a+2)(a-2) = 0$$
$$a + 2 = 0 \quad \text{or} \quad a - 2 = 0$$
$$a = -2 \quad \text{or} \quad a = 2$$

Both of these check, so the two solutions are -2 and 2.

63.
$$\frac{y^2-4}{y+3} = 2 - \frac{y-2}{y+3}, \text{ LCM} = y+3$$
$$(y+3) \cdot \frac{y^2-4}{y+3} = (y+3)\left(2 - \frac{y-2}{y+3}\right)$$
$$y^2 - 4 = 2(y+3) - (y-2)$$
$$y^2 - 4 = 2y + 6 - y + 2$$
$$y^2 - 4 = y + 8$$
$$y^2 - y - 12 = 0$$
$$(y-4)(y+3) = 0$$
$$y - 4 = 0 \text{ or } y + 3 = 0$$
$$y = 4 \text{ or } y = -3$$

The number 4 is a solution, but −3 is not because it makes a denominator zero.

65. Left to the student

Exercise Set 6.8

1. Familiarize. The job takes Mandy 4 hours working alone and Omar 5 hours working alone. Then in 1 hour Mandy does $\frac{1}{4}$ of the job and Omar does $\frac{1}{5}$ of the job. Working together, they can do $\frac{1}{4} + \frac{1}{5}$, or $\frac{9}{20}$ of the job in 1 hour. In two hours, Mandy does $2\left(\frac{1}{4}\right)$ of the job and Omar does $2\left(\frac{1}{5}\right)$ of the job. Working together they can do $2\left(\frac{1}{4}\right) + 2\left(\frac{1}{5}\right)$, or $\frac{9}{10}$ of the job in 2 hours. In 3 hours they can do $3\left(\frac{1}{4}\right) + 3\left(\frac{1}{5}\right)$, or $1\frac{7}{20}$ of the job which is more of the job then needs to be done. The answer is somewhere between 2 hr and 3 hr.

Translate. If they work together t hours, then Mandy does $t\left(\frac{1}{4}\right)$ of the job and Omar does $t\left(\frac{1}{5}\right)$ of the job. We want some number t such that
$$t\left(\frac{1}{4}\right) + t\left(\frac{1}{5}\right) = 1, \text{ or } \frac{t}{4} + \frac{t}{5} = 1.$$

Solve. We solve the equation.
$$\frac{t}{4} + \frac{t}{5} = 1, \text{ LCM} = 20$$
$$20\left(\frac{t}{4} + \frac{t}{5}\right) = 20 \cdot 1$$
$$20 \cdot \frac{t}{4} + 20 \cdot \frac{t}{5} = 20$$
$$5t + 4t = 20$$
$$9t = 20$$
$$t = \frac{20}{9}, \text{ or } 2\frac{2}{9}$$

Check. The check can be done by repeating the computations. We also have a partial check in that we expected from our familiarization step that the answer would be between 2 hr and 3 hr.

State. Working together, it takes them $2\frac{2}{9}$ hr to complete the job.

3. Familiarize. The job takes Vern 45 min working alone and Nina 60 min working alone. Then in 1 minute Vern does $\frac{1}{45}$ of the job and Nina does $\frac{1}{60}$ of the job. Working together, they can do $\frac{1}{45} + \frac{1}{60}$, or $\frac{7}{180}$ of the job in 1 minute. In 20 minutes, Vern does $\frac{20}{45}$ of the job and Nina does $\frac{20}{60}$ of the job. Working together, they can do $\frac{20}{45} + \frac{20}{60}$, or $\frac{7}{9}$ of the job. In 30 minutes, they can do $\frac{30}{45} + \frac{30}{60}$, or $\frac{7}{6}$ of the job which is more of the job than needs to be done. The answer is somewhere between 20 minutes and 30 minutes.

Translate. If they work together t minutes, then Vern does $t\left(\frac{1}{45}\right)$ of the job and Nina does $t\left(\frac{1}{60}\right)$ of the job. We want some number t such that
$$t\left(\frac{1}{45}\right) + t\left(\frac{1}{60}\right) = 1, \text{ or } \frac{t}{45} + \frac{t}{60} = 1.$$

Solve. We solve the equation.
$$\frac{t}{45} + \frac{t}{60} = 1, \text{ LCM} = 180$$
$$180\left(\frac{t}{45} + \frac{t}{60}\right) = 180 \cdot 1$$
$$180 \cdot \frac{t}{45} + 180 \cdot \frac{t}{60} = 180$$
$$4t + 3t = 180$$
$$7t = 180$$
$$t = \frac{180}{7}, \text{ or } 25\frac{5}{7}$$

Check. The check can be done by repeating the computations. We also have a partial check in that we expected from our familiarization step that the answer would be between 20 minutes and 30 minutes.

State. It would take them $25\frac{5}{7}$ minutes to complete the job working together.

5. Familiarize. The job takes Kenny Dewitt 9 hours working alone and Betty Wohat 7 hours working alone. Then in 1 hour Kenny does $\frac{1}{9}$ of the job and Betty does $\frac{1}{7}$ of the job. Working together they can do $\frac{1}{9} + \frac{1}{7}$, or $\frac{16}{63}$ of the job in 1 hour. In two hours, Kenny does $2\left(\frac{1}{9}\right)$ of the job and Betty does $2\left(\frac{1}{7}\right)$ of the job. Working together they can do $2\left(\frac{1}{9}\right) + 2\left(\frac{1}{7}\right)$, or $\frac{32}{63}$ of the job in two hours. In five hours they can do $5\left(\frac{1}{9}\right) + 5\left(\frac{1}{7}\right)$, or $\frac{80}{63}$, or $1\frac{17}{63}$ of the job which is more of the job than needs to be done. The answer is somewhere between 2 hr and 5 hr.

Exercise Set 6.8

Translate. If they work together t hours, Kenny does $t\left(\dfrac{1}{9}\right)$ of the job and Betty does $t\left(\dfrac{1}{7}\right)$ of the job. We want some number t such that

$$t\left(\dfrac{1}{9}\right) + t\left(\dfrac{1}{7}\right) = 1, \text{ or } \dfrac{t}{9} + \dfrac{t}{7} = 1.$$

Solve. We solve the equation.

$$\dfrac{t}{9} + \dfrac{t}{7} = 1, \text{ LCM} = 63$$

$$63\left(\dfrac{t}{9} + \dfrac{t}{7}\right) = 63 \cdot 1$$

$$63 \cdot \dfrac{t}{9} + 63 \cdot \dfrac{t}{7} = 63$$

$$7t + 9t = 63$$

$$16t = 63$$

$$t = \dfrac{63}{16}, \text{ or } 3\dfrac{15}{16}$$

Check. The check can be done by repeating the computations. We also have a partial check in that we expected from our familiarization step that the answer would be between 2 hr and 5 hr.

State. Working together, it takes them $3\dfrac{15}{16}$ hr to complete the job.

7. *Familiarize*. Let t = the number of minutes it takes Nicole and Glen to weed the garden, working together.

Translate. We use the work principle.

$$t\left(\dfrac{1}{50}\right) + t\left(\dfrac{1}{40}\right) = 1, \text{ or } \dfrac{t}{50} + \dfrac{t}{40} = 1$$

Solve. We solve the equation.

$$\dfrac{t}{50} + \dfrac{t}{40} = 1, \text{ LCM} = 200$$

$$200\left(\dfrac{t}{50} + \dfrac{t}{40}\right) = 200 \cdot 1$$

$$200 \cdot \dfrac{t}{50} + 200 \cdot \dfrac{t}{40} = 200$$

$$4t + 5t = 200$$

$$9t = 200$$

$$t = \dfrac{200}{9}, \text{ or } 22\dfrac{2}{9}$$

Check. In $\dfrac{200}{9}$ min, the portion of the job done is $\dfrac{1}{50} \cdot \dfrac{200}{9} + \dfrac{1}{40} \cdot \dfrac{200}{9} = \dfrac{4}{9} + \dfrac{5}{9} = 1$. The answer checks.

State. It would take $22\dfrac{2}{9}$ min to weed the garden if Nicole and Glen worked together.

9. *Familiarize*. Let t = the number of minutes it would take the two machines to copy the dissertation, working together.

Translate. We use the work principle.

$$t\left(\dfrac{1}{12}\right) + t\left(\dfrac{1}{20}\right) = 1, \text{ or } \dfrac{t}{12} + \dfrac{t}{20} = 1$$

Solve. We solve the equation.

$$\dfrac{t}{12} + \dfrac{t}{20} = 1, \text{ LCM} = 60$$

$$60\left(\dfrac{t}{12} + \dfrac{t}{20}\right) = 60 \cdot 1$$

$$60 \cdot \dfrac{t}{12} + 60 \cdot \dfrac{t}{20} = 60$$

$$5t + 3t = 60$$

$$8t = 60$$

$$t = \dfrac{15}{2}, \text{ or } 7.5$$

Check. In $\dfrac{15}{2}$ min, the portion of the job done is $\dfrac{1}{12} \cdot \dfrac{15}{2} + \dfrac{1}{20} \cdot \dfrac{15}{2} = \dfrac{5}{8} + \dfrac{3}{8} = 1$. The answer checks.

State. It would take the two machines 7.5 min to copy the dissertation, working together.

11. *Familiarize*. We complete the table shown in the text.

	Distance	Speed	Time	
Car	150	r	t	$\rightarrow 150 = r(t)$
Truck	350	$r + 40$	t	$\rightarrow 350 = (r+40)t$

$d = r \cdot t$

Translate. We apply the formula $d = rt$ along the rows of the table to obtain two equations:

$$150 = rt,$$
$$350 = (r+40)t$$

Then we solve each equation for t and set the results equal:

Solving $150 = rt$ for t: $t = \dfrac{150}{r}$

Solving $350 = (r+40)t$ for t: $t = \dfrac{350}{r+40}$

Thus, we have

$$\dfrac{150}{r} = \dfrac{350}{r+40}.$$

Solve. We multiply by the LCM, $r(r+40)$.

$$r(r+40) \cdot \dfrac{150}{r} = r(r+40) \cdot \dfrac{350}{r+40}$$

$$150(r+40) = 350r$$

$$150r + 6000 = 350r$$

$$6000 = 200r$$

$$30 = r$$

Check. If r is 30 km/h, then $r + 40$ is 70 km/h. The time for the car is 150/30, or 5 hr. The time for the truck is 350/70, or 5 hr. The times are the same. The values check.

State. The speed of Sarah's car is 30 km/h, and the speed of Rick's truck is 70 km/h.

13. Familiarize. We complete the table shown in the text.

	Distance	Speed	Time
Freight	330	$r-14$	t
Passenger	400	r	t

Translate. From the rows of the table we have two equations:
$$330 = (r-14)t,$$
$$400 = rt$$

We solve each equation for t and set the results equal:

Solving $330 = (r-14)t$ for t: $t = \dfrac{330}{r-14}$

Solving $400 = rt$ for t: $t = \dfrac{400}{r}$

Thus, we have
$$\dfrac{330}{r-14} = \dfrac{400}{r}.$$

Solve. We multiply by the LCM, $r(r-14)$.
$$r(r-14)\cdot\dfrac{330}{r-14} = r(r-14)\cdot\dfrac{400}{r}$$
$$330r = 400(r-14)$$
$$330r = 400r - 5600$$
$$-70r = -5600$$
$$r = 80$$

Then substitute 80 for r in either equation to find t:
$$t = \dfrac{400}{r}$$
$$t = \dfrac{400}{80} \quad \text{Substituting 80 for } r$$
$$t = 5$$

Check. If $r = 80$, then $r - 14 = 66$. In 5 hr the freight train travels $66 \cdot 5$, or 330 mi, and the passenger train travels $80 \cdot 5$, or 400 mi. The values check.

State. The speed of the passenger train is 80 mph. The speed of the freight train is 66 mph.

15. Familiarize. We let r represent the speed going. Then $2r$ is the speed returning. We let t represent the time going. Then $t - 3$ represents the time returning. We organize the information in a table.

	Distance	Speed	Time
Going	120	r	t
Returning	120	$2r$	$t-3$

Translate. The rows of the table give us two equations:
$$120 = rt,$$
$$120 = 2r(t-3)$$

We can solve each equation for r and set the results equal:

Solving $120 = rt$ for r: $r = \dfrac{120}{t}$

Solving $120 = 2r(t-3)$ for r: $r = \dfrac{120}{2(t-3)}$, or
$$r = \dfrac{60}{t-3}$$

Then $\dfrac{120}{t} = \dfrac{60}{t-3}$.

Solve. We multiply on both sides by the LCM, $t(t-3)$.
$$t(t-3)\cdot\dfrac{120}{t} = t(t-3)\cdot\dfrac{60}{t-3}$$
$$120(t-3) = 60t$$
$$120t - 360 = 60t$$
$$-360 = -60t$$
$$6 = t$$

Then substitute 6 for t in either equation to find r, the speed going:
$$r = \dfrac{120}{t}$$
$$r = \dfrac{120}{6} \quad \text{Substituting 6 for } t$$
$$r = 20$$

Check. If $r = 20$ and $t = 6$, then $2r = 2 \cdot 20$, or 40 mph and $t - 3 = 6 - 3$, or 3 hr. The distance going is $6 \cdot 20$, or 120 mi. The distance returning is $40 \cdot 3$, or 120 mi. The numbers check.

State. The speed going is 20 mph.

17. Familiarize. Let r = Kelly's speed, in km/h, and t = the time the bicyclists travel, in hours. Organize the information in a table.

	Distance	Speed	Time
Hank	42	$r-5$	t
Kelly	57	r	t

Translate. We can replace the t's in the table above using the formula $r = d/t$.

	Distance	Speed	Time
Hank	42	$r-5$	$\dfrac{42}{r-5}$
Kelly	57	r	$\dfrac{57}{r}$

Since the times are the same for both bicyclists, we have the equation
$$\dfrac{42}{r-5} = \dfrac{57}{r}.$$

Solve. We first multiply by the LCD, $r(r-5)$.
$$r(r-5)\cdot\dfrac{42}{r-5} = r(r-5)\cdot\dfrac{57}{r}$$
$$42r = 57(r-5)$$
$$42r = 57r - 285$$
$$-15r = -285$$
$$r = 19$$

Exercise Set 6.8

If $r = 19$, then $r - 5 = 14$.

Check. If Hank's speed is 14 km/h and Kelly's speed is 19 km/h, then Hank bicycles 5 km/h slower than Kelly. Hank's time is 42/14, or 3 hr. Kelly's time is 57/19, or 3 hr. Since the times are the same, the answer checks.

State. Hank travels at 14 km/h, and Kelly travels at 19 km/h.

19. **Familiarize.** Let r = Ralph's speed, in km/h. Then Bonnie's speed is $r + 3$. Also set t = the time, in hours, that Ralph and Bonnie walk. We organize the information in a table.

	Distance	Speed	Time
Ralph	7.5	r	t
Bonnie	12	$r + 3$	t

Translate. We can replace the t's in the table shown above using the formula $r = d/t$.

	Distance	Speed	Time
Ralph	7.5	r	$\dfrac{7.5}{r}$
Bonnie	12	$r + 3$	$\dfrac{12}{r+3}$

Since the times are the same for both walkers, we have the equation
$$\frac{7.5}{r} = \frac{12}{r+3}.$$

Solve. We first multiply by the LCD, $r(r+3)$.
$$r(r+3) \cdot \frac{7.5}{r} = r(r+3) \cdot \frac{12}{r+3}$$
$$7.5(r+3) = 12r$$
$$7.5r + 22.5 = 12r$$
$$22.5 = 4.5r$$
$$5 = r$$

If $r = 5$, then $r + 3 = 8$.

Check. If Ralph's speed is 5 km/h and Bonnie's speed is 8 km/h, then Bonnie walks 3 km/h faster than Ralph. Ralph's time is 7.5/5, or 1.5 hr. Bonnie's time is 12/8, or 1.5 hr. Since the times are the same, the answer checks.

State. Ralph's speed is 5 km/h, and Bonnie's speed is 8 km/h.

21. **Familiarize.** Let t = the time it takes Caledonia to drive to town and organize the given information in a table.

	Distance	Speed	Time
Caledonia	15	r	t
Manley	20	r	$t+1$

Translate. We can replace the r's in the table above using the formula $r = d/t$.

	Distance	Speed	Time
Caledonia	15	$\dfrac{15}{t}$	t
Manley	20	$\dfrac{20}{t+1}$	$t+1$

Since the speeds are the same for both riders, we have the equation
$$\frac{15}{t} = \frac{20}{t+1}.$$

Solve. We multiply by the LCD, $t(t+1)$.
$$t(t+1) \cdot \frac{15}{t} = t(t+1) \cdot \frac{20}{t+1}$$
$$15(t+1) = 20t$$
$$15t + 15 = 20t$$
$$15 = 5t$$
$$3 = t$$

If $t = 3$, then $t + 1 = 3 + 1$, or 4.

Check. If Caledonia's time is 3 hr and Manley's time is 4 hr, then Manley's time is 1 hr more than Caledonia's. Caledonia's speed is 15/3, or 5 mph. Manley's speed is 20/4, or 5 mph. Since the speeds are the same, the answer checks.

State. It takes Caledonia 3 hr to drive to town.

23. $\dfrac{10 \text{ divorces}}{18 \text{ marriages}} = \dfrac{10}{18}$ divorce/marriage = $\dfrac{5}{9}$ divorce/marriage

25. $\dfrac{4.6 \text{ km}}{2 \text{ hr}} = 2.3$ km/h

27. **Familiarize.** A 120-lb person should eat at least 44 g of protein each day, and we wish to find the minimum protein required for a 180-lb person. We can set up ratios. We let p = the minimum number of grams of protein a 180-lb person should eat each day.

Translate. If we assume the rates of protein intake are the same, the ratios are the same and we have an equation.

$$\begin{array}{c} \text{Protein} \to \\ \text{Weight} \to \end{array} \frac{44}{120} = \frac{p}{180} \begin{array}{c} \leftarrow \text{Protein} \\ \leftarrow \text{Weight} \end{array}$$

Solve. We solve the proportion.
$$360 \cdot \frac{44}{120} = 360 \cdot \frac{p}{180} \quad \text{Multiplying by the LCM, 360}$$
$$3 \cdot 44 = 2 \cdot p$$
$$132 = 2p$$
$$66 = p$$

Check. $\dfrac{44}{120} = \dfrac{4 \cdot 11}{4 \cdot 30} = \dfrac{\cancel{4} \cdot 11}{\cancel{4} \cdot 30} = \dfrac{11}{30}$ and

$\dfrac{66}{180} = \dfrac{6 \cdot 11}{6 \cdot 30} = \dfrac{\cancel{6} \cdot 11}{\cancel{6} \cdot 30} = \dfrac{11}{30}$. The ratios are the same.

State. A 180-lb person should eat a minimum of 66 g of protein each day.

29. Familiarize. 10 cc of human blood contains 1.2 grams of hemoglobin, and we wish to find how many grams of hemoglobin are contained in 16 cc of the same blood. We can set up ratios. Let H = the amount of hemoglobin in 16 cc of the same blood.

Translate. Assuming the two ratios are the same, we can translate to a proportion.

$$\begin{array}{c}\text{Grams} \to \\ \text{cm}^3 \to\end{array} \frac{H}{16} = \frac{1.2}{10} \begin{array}{c}\leftarrow \text{Grams} \\ \leftarrow \text{cm}^3\end{array}$$

Solve. We solve the proportion.

We multiply by 16 to get H alone.

$$16 \cdot \frac{H}{16} = 16 \cdot \frac{1.2}{10}$$
$$H = \frac{19.2}{10}$$
$$H = 1.92$$

Check.
$$\frac{1.92}{16} = 0.12 \quad \frac{1.2}{10} = 0.12$$

The ratios are the same.

State. 16 cc of the same blood would contain 1.92 grams of hemoglobin.

31. Familiarize. Let h = the amount of honey, in pounds, that 35,000 trips to flowers would produce.

Translate. We translate to a proportion.

$$\begin{array}{c}\text{Honey} \to \\ \text{Trips} \to\end{array} \frac{1}{20,000} = \frac{h}{35,000} \begin{array}{c}\leftarrow \text{Honey} \\ \leftarrow \text{Trips}\end{array}$$

Solve. We solve the proportion.
$$35,000 \cdot \frac{1}{20,000} = 35,000 \cdot \frac{h}{35,000}$$
$$1.75 = h$$

Check. $\frac{1}{20,000} = 0.00005$ and $\frac{1.75}{35,000} = 0.00005$.

The ratios are the same.

State. 35,000 trips to gather nectar will produce 1.75 lb of honey.

33. Familiarize. The ratio of the weight of copper to the weight of zinc in a U.S. penny is $\frac{1}{39}$, and we wish to find how much copper is needed if 50 kg of zinc is being turned into pennies. We can set up a second ratio to go with the one we already have. Let C = the amount of copper needed, in kg, if 50 kg of zinc is being turned into pennies.

Translate. We translate to a proportion.
$$\frac{1}{39} = \frac{C}{50}$$

Solve. We solve the proportion.
$$50 \cdot \frac{1}{39} = 50 \cdot \frac{C}{50}$$
$$\frac{50}{39} = C, \text{ or}$$
$$1\frac{11}{39} = C$$

Check. $\frac{50/39}{50} = \frac{1}{39}$, so the ratios are the same.

State. $1\frac{11}{39}$ kg of copper is needed if 50 kg of zinc is turned into pennies.

35. (a) $\frac{96}{266} \approx 0.361$

Suzuki's batting average was 0.361.

(b) Let h = the number of hits Suzuki would get in the 162-game season. We translate to a proportion and solve it.
$$\frac{96}{58} = \frac{h}{162}$$
$$162 \cdot \frac{96}{58} = 162 \cdot \frac{h}{162}$$
$$268 \approx h$$

Suzuki would get 268 hits in the 162-game season.

(c) Let h = the number of hits Suzuki would get if he batted 560 times. We translate to a proportion and solve it.
$$\frac{96}{266} = \frac{h}{560}$$
$$560 \cdot \frac{96}{266} = 560 \cdot \frac{h}{560}$$
$$202 \approx h$$

Suzuki would get 202 hits if he batted 560 times.

37. Let h = the head circumference, in inches. We translate to a proportion and solve it.

$$\frac{6\frac{3}{4}}{21\frac{1}{5}} = \frac{7}{h}$$

$$6\frac{3}{4} \cdot h = 21\frac{1}{5} \cdot 7$$

$$\frac{27}{4} \cdot h = \frac{106}{5} \cdot 7$$

$$h = \frac{4}{27} \cdot \frac{106}{5} \cdot 7$$

$$h \approx 22$$

The head circumference is 22 in.

Now let c = the head circumference, in centimeters. We translate to a proportion and solve it.

$$\frac{6\frac{3}{4}}{53.8} = \frac{7}{c}$$

$$\frac{6.75}{53.8} = \frac{7}{c} \quad \left(6\frac{3}{4} = 6.75\right)$$

$$6.75 \cdot c = 53.8 \cdot 7$$

$$c = \frac{53.8 \cdot 7}{6.75}$$

$$c \approx 55.8$$

The head circumference is 55.8 cm.

Exercise Set 6.8

39. Let $h =$ the hat size. We translate to a proportion and solve it.

$$\frac{6\frac{3}{4}}{21\frac{1}{5}} = \frac{h}{22\frac{4}{5}}$$

$$6\frac{3}{4} \cdot 22\frac{4}{5} = 21\frac{1}{5} \cdot h$$

$$\frac{27}{4} \cdot \frac{114}{5} = \frac{106}{5} \cdot h$$

$$\frac{5}{106} \cdot \frac{27}{4} \cdot \frac{114}{5} = h$$

$$7.26 \approx h$$

$$7\frac{1}{4} \approx h$$

The hat size is $7\frac{1}{4}$.

Now let $c =$ the head circumference, in centimeters. We translate to a proportion and solve it. We use the hat size found above in the translation.

$$\frac{6\frac{3}{4}}{53.8} = \frac{7\frac{1}{4}}{c}$$

$$\frac{6.75}{53.8} = \frac{7.25}{c}$$

$$6.75 \cdot c = 53.8 \cdot 7.25$$

$$c = \frac{53.8 \cdot 7.25}{6.75}$$

$$c \approx 57.8$$

The head circumference is 57.8 cm. (Answers may vary slightly.)

41. Let $h =$ the hat size. We translate to a proportion and solve it.

$$\frac{6\frac{3}{4}}{53.8} = \frac{h}{59.8}$$

$$\frac{6.75}{53.8} = \frac{h}{59.8}$$

$$59.8 \cdot \frac{6.75}{53.8} = h$$

$$7.5 \approx h, \text{ or}$$

$$7\frac{1}{2} \approx h$$

The hat size is $7\frac{1}{2}$.

Now let $c =$ the head circumference, in inches. We translate to a proportion and solve it. We use the hat size found above in the translation.

$$\frac{6\frac{3}{4}}{21\frac{1}{5}} = \frac{7\frac{1}{2}}{c}$$

$$6\frac{3}{4} \cdot c = 21\frac{1}{5} \cdot 7\frac{1}{2}$$

$$\frac{27}{4} \cdot c = \frac{106}{5} \cdot \frac{15}{2}$$

$$c = \frac{4}{27} \cdot \frac{106}{5} \cdot \frac{15}{2}$$

$$c \approx 23.6, \text{ or}$$

$$c \approx 23\frac{3}{5}$$

The head circumference is $23\frac{3}{5}$ in.

43. Familiarize. The ratio of blue whales tagged to the total blue whale population, P, is $\frac{500}{P}$. Of the 400 blue whales checked later, 20 were tagged. The ratio of blue whales tagged to blue whales checked is $\frac{20}{400}$.

Translate. Assuming the two ratios are the same, we can translate to a proportion.

Whales tagged originally $\longrightarrow$ $\frac{500}{P} = \frac{20}{400}$ $\longleftarrow$ Tagged whales caught later
Whale population $\longleftarrow$ Whales caught later

Solve. We solve the equation.

$$400P \cdot \frac{500}{P} = 400P \cdot \frac{20}{400} \quad \text{Multiplying by the LCM, } 400P$$

$$400 \cdot 500 = P \cdot 20$$

$$200,000 = 20P$$

$$10,000 = P$$

Check.

$$\frac{500}{10,000} = \frac{1}{20} \text{ and } \frac{20}{400} = \frac{1}{20}.$$

The ratios are the same.

State. The blue whale population is about 10,000.

45. Familiarize. The ratio of the weight of an object on Mars to the weight of an object on earth is 0.4 to 1.

a) We wish to find how much a 12-ton rocket would weigh on Mars.

b) We wish to find how much a 120-lb astronaut would weigh on Mars.

We can set up ratios. We let $r =$ the weight of a 12-ton rocket and $a =$ the weight of a 120-lb astronaut on Mars.

Translate. Assuming the ratios are the same, we can translate to proportions.

a) Weight on Mars $\rightarrow$ $\frac{0.4}{1} = \frac{r}{12}$ $\leftarrow$ Weight on Mars
Weight on earth $\rightarrow$ $\leftarrow$ Weight on earth

b) $\begin{array}{l}\text{Weight} \\ \text{on Mars} \to \\ \text{Weight} \to \\ \text{on earth}\end{array} \dfrac{0.4}{1} = \dfrac{a}{120} \begin{array}{l}\leftarrow \text{on Mars} \\ \leftarrow \text{Weight} \\ \text{on earth}\end{array}$

Solve. We solve each proportion.

a) $\dfrac{0.4}{1} = \dfrac{r}{12}$ b) $\dfrac{0.4}{1} = \dfrac{1}{120}$

$12(0.4) = r \qquad\qquad 120(0.4) = a$

$4.8 = r \qquad\qquad\quad 48 = a$

Check. $\dfrac{0.4}{1} = 0.4$, $\dfrac{4.8}{12} = 0.4$, and $\dfrac{48}{120} = 0.4$.
The ratios are the same.

State. a) A 12-ton rocket would weigh 4.8 tons on Mars.

b) A 120-lb astronaut would weigh 48 lb on Mars.

47. **Familiarize**. A sample of 144 firecrackers contained 9 duds, and we wish to find how many duds could be expected in a sample of 3200 firecrackers. We can set up ratios, letting $d =$ the number of duds expected in a sample of 3200 firecrackers.

Translate. Assuming the rates of occurrence of duds are the same, we can translate to a proportion.

$\begin{array}{l}\text{Duds} \to \\ \text{Sample size} \to\end{array} \dfrac{9}{144} = \dfrac{d}{3200} \begin{array}{l}\leftarrow \text{Duds} \\ \leftarrow \text{Sample size}\end{array}$

Solve. We solve the equation. We multiply by 3200 to get d alone.

$3200 \cdot \dfrac{9}{144} = 3200 \cdot \dfrac{d}{3200}$

$\dfrac{28{,}800}{144} = d$

$200 = d$

Check.

$\dfrac{9}{144} = 0.0625$ and $\dfrac{200}{3200} = 0.0625$

The ratios are the same.

State. You would expect 200 duds in a sample of 3200 firecrackers.

49. We write a proportion and then solve it.

$\dfrac{b}{6} = \dfrac{7}{4}$

$b = \dfrac{7}{4} \cdot 6 \qquad$ Multiplying by 6

$b = \dfrac{42}{4}$

$b = \dfrac{21}{2}$, or 10.5

$\left(\text{Note that the proportions } \dfrac{6}{b} = \dfrac{4}{7},\ \dfrac{b}{7} = \dfrac{6}{4},\ \text{or}\ \dfrac{7}{b} = \dfrac{4}{6}\ \text{could also be used.}\right)$

51. We write a proportion and then solve it.

$\dfrac{4}{f} = \dfrac{6}{4}$

$4f \cdot \dfrac{4}{f} = 4f \cdot \dfrac{6}{4}$

$16 = 6f$

$\dfrac{8}{3} = f \qquad$ Simplifying

$\left(\text{One of the following proportions could also be used:} \right.$
$\left.\dfrac{f}{4} = \dfrac{4}{6},\ \dfrac{4}{f} = \dfrac{9}{6},\ \dfrac{f}{4} = \dfrac{6}{9},\ \dfrac{4}{9} = \dfrac{f}{6},\ \dfrac{9}{4} = \dfrac{6}{f}\right)$

53. We write a proportion and then solve it.

$\dfrac{h}{7} = \dfrac{10}{6}$

$h = \dfrac{10}{6} \cdot 7 \qquad$ Multiplying by 7

$h = \dfrac{70}{6}$

$h = \dfrac{35}{3} \qquad$ Simplifying

$\left(\text{Note that the proportions } \dfrac{7}{h} = \dfrac{6}{10},\ \dfrac{h}{10} = \dfrac{7}{6},\ \text{or}\ \dfrac{10}{h} = \dfrac{6}{7}\right.$
could also be used.$\left.\right)$

55. We write a proportion and then solve it.

$\dfrac{4}{10} = \dfrac{6}{l}$

$10l \cdot \dfrac{4}{10} = 10l \cdot \dfrac{6}{l}$

$4l = 60$

$l = 15$ ft

$\left(\text{One of the following proportions could also be used:}\right.$
$\left.\dfrac{4}{6} = \dfrac{10}{l},\ \dfrac{10}{4} = \dfrac{l}{6},\ \text{or}\ \dfrac{6}{4} = \dfrac{l}{10}\right)$

57. Discussion and Writing Exercise

59. $x^5 \cdot x^6 = x^{5+6} = x^{11}$

61. $x^{-5} \cdot x^{-6} = x^{-5+(-6)} = x^{-11} = \dfrac{1}{x^{11}}$

63. Graph: $y = 2x - 6$.

We select some x-values and compute y-values.

If $x = 1$, then $y = 2 \cdot 1 - 6 = -4$.

If $x = 3$, then $y = 2 \cdot 3 - 6 = 0$.

If $x = 5$, then $y = 2 \cdot 5 - 6 = 4$.

x	y	(x,y)
1	-4	$(1,-4)$
3	0	$(3,0)$
5	4	$(5,4)$

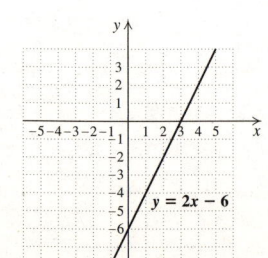

65. Graph: $3x + 2y = 12$.

We can replace either variable with a number and then calculate the other coordinate. We will find the intercepts and one other point.

If $y = 0$, we have:
$$3x + 2 \cdot 0 = 12$$
$$3x = 12$$
$$x = 4$$

The x-intercept is $(4, 0)$.

If $x = 0$, we have:
$$3 \cdot 0 + 2y = 12$$
$$2y = 12$$
$$y = 6$$

The y-intercept is $(0, 6)$.

If $y = -3$, we have:
$$3x + 2(-3) = 12$$
$$3x - 6 = 12$$
$$3x = 18$$
$$x = 6$$

The point $(6, -3)$ is on the graph.

We plot these points and draw a line through them.

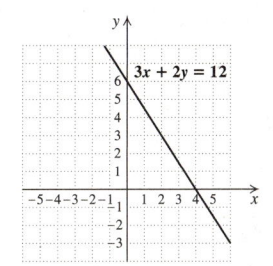

67. Graph: $y = -\frac{3}{4}x + 2$

We select some x-values and compute y-values. We use multiples of 4 to avoid fractions.

If $x = -4$, then $y = -\frac{3}{4}(-4) + 2 = 5$.

If $x = 0$, then $y = -\frac{3}{4} \cdot 0 + 2 = 2$.

If $x = 4$, then $y = -\frac{3}{4} \cdot 4 + 2 = -1$.

x	y	(x, y)
-4	5	$(-4, 5)$
0	2	$(0, 2)$
4	-1	$(4, -1)$

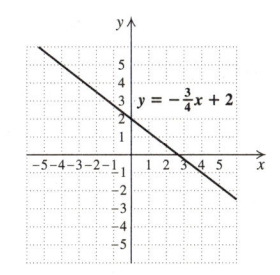

69. *Familiarize*. Let $t =$ the time it would take for Ann to complete the report working alone. Then $t + 6 =$ the time it would take Betty to complete the report working alone.

In 1 hour they would complete $\frac{1}{t} + \frac{1}{t+6}$ of the report and in 4 hours they would complete $4\left(\frac{1}{t} + \frac{1}{t+6}\right)$, or $\frac{4}{t} + \frac{4}{t+6}$ of the report.

Translate. In 4 hours one entire job is done, so we have
$$\frac{4}{t} + \frac{4}{t+6} = 1.$$

Solve. We solve the equation.
$$\frac{4}{t} + \frac{4}{t+6} = 1, \text{ LCM} = t(t+6)$$
$$t(t+6)\left(\frac{4}{t} + \frac{4}{t+6}\right) = t(t+6) \cdot 1$$
$$t(t+6) \cdot \frac{4}{t} + t(t+6) \cdot \frac{4}{t+6} = t^2 + 6t$$
$$4(t+6) + 4t = t^2 + 6t$$
$$4t + 24 + 4t = t^2 + 6t$$
$$0 = t^2 - 2t - 24$$
$$0 = (t-6)(t+4)$$
$$t - 6 = 0 \text{ or } t + 4 = 0$$
$$t = 6 \text{ or } t = -4$$

Check. The time cannot be negative, so we check only 6. If it takes Ann 6 hr to complete the report, then it would take Betty $6 + 6$, or 12 hr, to complete the report. In 4 hr Ann does $4 \cdot \frac{1}{6}$, or $\frac{2}{3}$, of the report, Betty does $4 \cdot \frac{1}{12}$, or $\frac{1}{3}$, of the report, and together they do $\frac{2}{3} + \frac{1}{3}$, or 1 entire job. The answer checks.

State. It would take Ann 6 hr and Betty 12 hr to complete the report working alone.

71. *Familiarize*. Let $t =$ the number of minutes after 5:00 at which the hands of the clock will first be together. While the minute hand moves through t minutes, the hour hand moves through $t/12$ minutes. At 5:00 the hour hand is on the 25-minute mark. We wish to find when a move of the minute hand through t minutes is equal to $25 + t/12$ minutes.

Translate. We use the last sentence of the familiarization step to write an equation.
$$t = 25 + \frac{t}{12}$$

Solve. We solve the equation.
$$t = 25 + \frac{t}{12}$$
$$12 \cdot t = 12\left(25 + \frac{t}{12}\right)$$
$$12t = 300 + t \qquad \text{Multiplying by 12}$$
$$11t = 300$$
$$t = \frac{300}{11} \text{ or } 27\frac{3}{11}$$

Check. At $27\frac{3}{11}$ minutes after 5:00, the minute hand is at the $27\frac{3}{11}$-minutes mark and the hour hand is at the

25 + $\dfrac{27\frac{3}{11}}{12}$-minute mark. Simplifying 25 + $\dfrac{27\frac{3}{11}}{12}$, we get

$$25 + \dfrac{\frac{300}{11}}{12} = 25 + \dfrac{300}{11} \cdot \dfrac{1}{12} = 25 + \dfrac{25}{11} = 25 + 2\dfrac{3}{11} = 27\dfrac{3}{11}.$$

Thus, the hands are together.

State. The hands are first together $27\dfrac{3}{11}$ minutes after 5:00.

73. $\dfrac{t}{a} + \dfrac{t}{b} = 1$, LCM = ab

$$ab\left(\dfrac{t}{a} + \dfrac{t}{b}\right) = ab \cdot 1$$

$$ab \cdot \dfrac{t}{a} + ab \cdot \dfrac{t}{b} = ab$$

$$bt + at = ab$$

$$t(b+a) = ab$$

$$t = \dfrac{ab}{b+a}$$

Exercise Set 6.9

1. $y = kx$

   $40 = k \cdot 8$ Substituting

   $5 = k$ Solving for k

   The variation constant is 5.

   The equation of variation is $y = 5x$.

3. $y = kx$

   $4 = k \cdot 30$ Substituting

   $\dfrac{4}{30} = k$, or   Solving for k

   $\dfrac{2}{15} = k$ Simplifying

   The variation constant is $\dfrac{2}{15}$.

   The equation of variation is $y = \dfrac{2}{15}x$.

5. $y = kx$

   $0.9 = k \cdot 0.4$ Substituting

   $\dfrac{0.9}{0.4} = k$, or

   $\dfrac{9}{4} = k$

   The variation constant is $\dfrac{9}{4}$.

   The equation of variation is $y = \dfrac{9}{4}x$.

7. Let p = the number of people using the cans.

   $N = kp$ N varies directly as p.

   $60,000 = k \cdot 250$ Substituting

   $\dfrac{60,000}{250} = k$ Solving for k

   $240 = k$ Variation constant

   $N = 240p$ Equation of variation

   $N = 240(1,008,000)$ Substituting

   $N = 241,920,000$

   In Dallas 241,920,000 cans are used each year.

9. $d = kw$ d varies directly as w.

   $40 = k \cdot 3$ Substituting

   $\dfrac{40}{3} = k$ Variation constant

   $d = \dfrac{40}{3}w$ Equation of variation

   $d = \dfrac{40}{3} \cdot 5$ Substituting

   $d = \dfrac{200}{3}$, or $66\dfrac{2}{3}$

   The spring is stretched $66\dfrac{2}{3}$ cm by a 5-kg barbell.

11. Let F = the number of grams of fat and w = the weight.

    $F = kw$ F varies directly as w.

    $60 = k \cdot 120$ Substituting

    $\dfrac{60}{120} = k$, or   Solving for k

    $\dfrac{1}{2} = k$ Variation constant

    $F = \dfrac{1}{2}w$ Equation of variation

    $F = \dfrac{1}{2} \cdot 180$ Substituting

    $F = 90$

    The maximum daily fat intake for a person weighing 180 lb is 90 g.

13. Let m = the mass of the body.

    $W = km$ W varies directly as m.

    $64 = k \cdot 96$ Substituting

    $\dfrac{64}{96} = k$ Solving for k

    $\dfrac{2}{3} = k$ Variation constant

    $W = \dfrac{2}{3}m$ Equation of variation

    $W = \dfrac{2}{3} \cdot 60$ Substituting

    $W = 40$

    There are 40 kg of water in a 60-kg person.

15. $y = \dfrac{k}{x}$

    $14 = \dfrac{k}{7}$ Substituting

    $7 \cdot 14 = k$ Solving for k

    $98 = k$

Exercise Set 6.9

The variation constant is 98.

The equation of variation is $y = \dfrac{98}{x}$.

17. $y = \dfrac{k}{x}$

   $3 = \dfrac{k}{12}$ Substituting

   $12 \cdot 3 = k$ Solving for k

   $36 = k$

The variation constant is 36.

The equation of variation is $y = \dfrac{36}{x}$.

19. $y = \dfrac{k}{x}$

   $0.1 = \dfrac{k}{0.5}$ Substituting

   $0.5(0.1) = k$ Solving for k

   $0.05 = k$

The variation constant is 0.05.

The equation of variation is $y = \dfrac{0.05}{x}$.

21. $T = \dfrac{k}{P}$ T varies inversely as P.

   $5 = \dfrac{k}{7}$ Substituting

   $35 = k$ Variation constant

   $T = \dfrac{35}{P}$ Equation of variation

   $T = \dfrac{35}{10}$ Substituting

   $T = 3.5$

It will take 10 bricklayers 3.5 hr to complete the job.

23. $I = \dfrac{k}{R}$ I varies inversely as R.

   $\dfrac{1}{2} = \dfrac{k}{240}$ Substituting

   $240 \cdot \dfrac{1}{2} = k$

   $120 = k$ Variation constant

   $I = \dfrac{120}{R}$ Equation of variation

   $I = \dfrac{120}{540}$ Substituting

   $I = \dfrac{2}{9}$

When the resistance is 540 ohms, the current is $\dfrac{2}{9}$ ampere.

25. $P = \dfrac{k}{W}$ P varies inversely as W.

   $330 = \dfrac{k}{3.2}$ Substituting

   $1056 = k$ Variation constant

   $P = \dfrac{1056}{W}$ Equation of variation

   $550 = \dfrac{1056}{W}$ Substituting

   $550W = 1056$ Multiplying by W

   $W = \dfrac{1056}{550}$ Dividing by 550

   $W = 1.92$ Simplifying

A tone with a pitch of 550 vibrations per second has a wavelength of 1.92 ft.

27. $V = \dfrac{k}{P}$ V varies inversely as P.

   $200 = \dfrac{k}{32}$ Substituting

   $6400 = k$ Variation constant

   $V = \dfrac{6400}{P}$ Equation of variation

   $V = \dfrac{6400}{40}$ Substituting

   $V = 160$

The volume is 160 cm^3 under a pressure of 40 kg/cm^2.

29. $y = kx^2$

   $0.15 = k(0.1)^2$ Substituting

   $0.15 = 0.01k$

   $\dfrac{0.15}{0.01} = k$

   $15 = k$

The equation of variation is $y = 15x^2$.

31. $y = \dfrac{k}{x^2}$

   $0.15 = \dfrac{k}{(0.1)^2}$ Substituting

   $0.15 = \dfrac{k}{0.01}$

   $0.15(0.01) = k$

   $0.0015 = k$

The equation of variation is $y = \dfrac{0.0015}{x^2}$.

33. $y = kxz$

   $56 = k \cdot 7 \cdot 8$ Substituting

   $56 = 56k$

   $1 = k$

The equation of variation is $y = xz$.

35.
$$y = kxz^2$$
$$105 = k \cdot 14 \cdot 5^2 \quad \text{Substituting}$$
$$105 = 350k$$
$$\frac{105}{350} = k$$
$$\frac{3}{10} = k$$
The equation of variation is $y = \frac{3}{10}xz^2$.

37.
$$y = k\frac{xz}{wp}$$
$$\frac{3}{28} = k\frac{3 \cdot 10}{7 \cdot 8} \quad \text{Substituting}$$
$$\frac{3}{28} = k \cdot \frac{30}{56}$$
$$\frac{3}{28} \cdot \frac{56}{30} = k$$
$$\frac{1}{5} = k$$
The equation of variation is $y = \frac{xz}{5wp}$.

39.
$$d = kr^2$$
$$200 = k \cdot 60^2 \quad \text{Substituting}$$
$$200 = 3600k$$
$$\frac{200}{3600} = k$$
$$\frac{1}{18} = k$$
The equation of variation is $d = \frac{1}{18}r^2$.
Substitute 72 for d and find r.
$$72 = \frac{1}{18}r^2$$
$$1296 = r^2$$
$$36 = r$$
A car can travel 36 mph and still stop in 72 ft.

41.
$$I = \frac{k}{d^2}$$
$$90 = \frac{k}{5^2} \quad \text{Substituting}$$
$$90 = \frac{k}{25}$$
$$2250 = k$$
The equation of variation is $I = \frac{2250}{d^2}$.
Substitute 40 for I and find d.
$$40 = \frac{2250}{d^2}$$
$$40d^2 = 2250$$
$$d^2 = 56.25$$
$$d = 7.5$$
The distance from 5 m to 7.5 m is $7.5 - 5$, or 2.5 m, so it is 2.5 m further to a point where the intensity is 40 W/m².

43. $E = \frac{kR}{I}$
We first find k.
$$3.18 = \frac{k \cdot 71}{201} \quad \text{Substituting}$$
$$3.18\left(\frac{201}{71}\right) = k \quad \text{Multiplying by } \frac{201}{71}$$
$$9 \approx k$$
The equation of variation is $E = \frac{9R}{I}$.
Substitute 3.18 for E and 300 for I and solve R.
$$3.18 = \frac{9R}{300}$$
$$3.18\left(\frac{300}{9}\right) = R \quad \text{Multiplying by } \frac{300}{9}$$
$$106 = R$$
Shawn Estes would have given up 106 earned runs if he had pitched 300 innings.

45. $Q = kd^2$
We first find k.
$$225 = k \cdot 5^2$$
$$225 = 25k$$
$$9 = k$$
The equation of variation is $Q = 9d^2$.
Substitute 9 for d and compute Q.
$$Q = 9 \cdot 9^2$$
$$Q = 9 \cdot 81$$
$$Q = 729$$
729 gallons of water are emptied by a pipe that is 9 in. in diameter.

47. Discussion and Writing Exercise

49. $x^2 - x - 56$
We look for a pair of numbers whose product is -56 and whose sum is -1. The numbers are -8 and 7.
$$x^2 - x - 56 = (x-8)(x+7)$$

51. $x^5 - 2x^4 - 35x^3 = x^3(x^2 - 2x - 35) = x^3(x-7)(x+5)$

53.
$$16 - t^4 = 4^2 - (t^2)^2 \quad \text{Difference of squares}$$
$$= (4+t^2)(4-t^2)$$
$$= (4+t^2)(2^2 - t^2) \quad \text{Difference of squares}$$
$$= (4+t^2)(2+t)(2-t)$$

55. $x^2 - 9x + 14$
We look for a pair of numbers whose product is 14 and whose sum is -9. The numbers are -2 and -7.
$$x^2 - 9x + 14 = (x-2)(x-7)$$

57.
$$16x^2 - 40xy + 25y^2$$
$$= (4x)^2 - 2 \cdot 4x \cdot 5y + (5y)^2 \quad \text{Trinomial square}$$
$$= (4x - 5y)^2$$

59. $3x^3 - 3y^3 = 3(x^3 - y^3) = 3(x-y)(x^2 + xy + y^2)$

Exercise Set 6.9

61. We are told $A = kd^2$, and we know $A = \pi r^2$ so we have:
$$kd^2 = \pi r^2$$

$$kd^2 = \pi \left(\frac{d}{2}\right)^2 \qquad r = \frac{d}{2}$$

$$kd^2 = \frac{\pi d^2}{4}$$

$$k = \frac{\pi}{4} \qquad \text{Variation constant}$$

63. $Q = \dfrac{kp^2}{q^3}$

Q varies directly as the square of p and inversely as the cube of q.

65. Let V represent the volume and p represent the price of a jar of peanut butter. Recall that the formula for the volume of a right circular cylinder with radius r and height h is $V = \pi r^2 h$. The diameter of the smaller jar is 3 in., so its radius is 3 in./2, or $\dfrac{3}{2}$ in. The diameter of the larger jar is 6 in., so its radius is 6 in./2, or 3 in.

$$V = kp$$
$$\pi r^2 h = kp$$
$$\pi \left(\frac{3}{2}\right)^2 (4) = k(1.2) \qquad \text{Substituting}$$
$$7.5\pi = k \qquad \text{Variation constant}$$
$$V = 7.5\pi p \qquad \text{Equation of variation}$$
$$\pi r^2 h = 7.5\pi p$$
$$\pi(3)^2(6) = 7.5\pi p \qquad \text{Substituting}$$
$$54\pi = 7.5\pi p$$
$$7.2 = p$$

The larger jar should cost $7.20.

Chapter 7

Graphs, Functions, and Applications

Exercise Set 7.1

1. Yes; each member of the domain is matched to only one member of the range.

3. Yes; each member of the domain is matched to only one member of the range.

5. Yes; each member of the domain is matched to only one member of the range.

7. No; a member of the domain is matched to more than one member of the range. In fact, each member of the domain is matched to 3 members of the range.

9. This correspondence is a function, because each person in the family has only one height.

11. The correspondence is not a function, since it is reasonable to assume that at least one avenue is intersected by more than one road.

13. This correspondence is a function, because each number in the domain, when squared and then increased by 4, corresponds to only one number in the range.

15. $f(x) = x + 5$
    a) $f(4) = 4 + 5 = 9$
    b) $f(7) = 7 + 5 = 12$
    c) $f(-3) = -3 + 5 = 2$
    d) $f(0) = 0 + 5 = 5$
    e) $f(2.4) = 2.4 + 5 = 7.4$
    f) $f\left(\dfrac{2}{3}\right) = \dfrac{2}{3} + 5 = 5\dfrac{2}{3}$

17. $h(p) = 3p$
    a) $h(-7) = 3(-7) = -21$
    b) $h(5) = 3 \cdot 5 = 15$
    c) $h(14) = 3 \cdot 14 = 42$
    d) $h(0) = 3 \cdot 0 = 0$
    e) $h\left(\dfrac{2}{3}\right) = 3 \cdot \dfrac{2}{3} = \dfrac{6}{3} = 2$
    f) $h(a+1) = 3(a+1) = 3a + 3$

19. $g(s) = 3s + 4$
    a) $g(1) = 3 \cdot 1 + 4 = 3 + 4 = 7$
    b) $g(-7) = 3(-7) + 4 = -21 + 4 = -17$
    c) $g(6.7) = 3(6.7) + 4 = 20.1 + 4 = 24.1$
    d) $g(0) = 3 \cdot 0 + 4 = 0 + 4 = 4$
    e) $g(-10) = 3(-10) + 4 = -30 + 4 = -26$
    f) $g\left(\dfrac{2}{3}\right) = 3 \cdot \dfrac{2}{3} + 4 = 2 + 4 = 6$

21. $f(x) = 2x^2 - 3x$
    a) $f(0) = 2 \cdot 0^2 - 3 \cdot 0 = 0 - 0 = 0$
    b) $f(-1) = 2(-1)^2 - 3(-1) = 2 + 3 = 5$
    c) $f(2) = 2 \cdot 2^2 - 3 \cdot 2 = 8 - 6 = 2$
    d) $f(10) = 2 \cdot 10^2 - 3 \cdot 10 = 200 - 30 = 170$
    e) $f(-5) = 2(-5)^2 - 3(-5) = 50 + 15 = 65$
    f) $f(4a) = 2(4a)^2 - 3(4a) = 32a^2 - 12a$

23. $f(x) = |x| + 1$
    a) $f(0) = |0| + 1 = 0 + 1 = 1$
    b) $f(-2) = |-2| + 1 = 2 + 1 = 3$
    c) $f(2) = |2| + 1 = 2 + 1 = 3$
    d) $f(-3) = |-3| + 1 = 3 + 1 = 4$
    e) $f(-10) = |-10| + 1 = 10 + 1 = 11$
    f) $f(a-1) = |a-1| + 1$

25. $f(x) = x^3$
    a) $f(0) = 0^3 = 0$
    b) $f(-1) = (-1)^3 = -1$
    c) $f(2) = 2^3 = 8$
    d) $f(10) = 10^3 = 1000$
    e) $f(-5) = (-5)^3 = -125$
    f) $f(-10) = (-10)^3 = -1000$

27. $F(x) = 2.75x + 71.48$
    a) $F(32) = 2.75(32) + 71.48$
       $= 88 + 71.48$
       $= 159.48$ cm
    b) $F(35) = 2.75(35) + 71.48$
       $= 96.25 + 71.48$
       $= 167.73$ cm

29. $P(d) = 1 + \dfrac{d}{33}$
    $P(20) = 1 + \dfrac{20}{33} = 1\dfrac{20}{33}$ atm
    $P(30) = 1 + \dfrac{30}{33} = 1\dfrac{10}{11}$ atm
    $P(100) = 1 + \dfrac{100}{33} = 1 + 3\dfrac{1}{33} = 4\dfrac{1}{33}$ atm

31. $W(d) = 0.112d$
    $W(16) = 0.112(16) = 1.792$ cm
    $W(25) = 0.112(25) = 2.8$ cm
    $W(100) = 0.112(100) = 11.2$ cm

33. Graph $f(x) = 3x - 1$

Make a list of function values in a table.

$f(-1) = 3(-1) - 1 = -3 - 1 = -4$
$f(0) = 3 \cdot 0 - 1 = 0 - 1 = -1$
$f(1) = 3 \cdot 1 - 1 = 3 - 1 = 2$
$f(2) = 3 \cdot 2 - 1 = 6 - 1 = 5$

x	$f(x)$
-1	-4
0	-1
1	2
2	5

Plot these points and connect them.

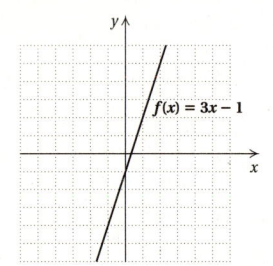

35. Graph $g(x) = -2x + 3$

Make a list of function values in a table.

$g(-1) = -2(-1) + 3 = 2 + 3 = 5$
$g(0) = -2 \cdot 0 + 3 = 0 + 3 = 3$
$g(3) = -2 \cdot 3 + 3 = -6 + 3 = -3$

x	$g(x)$
-1	5
0	3
3	-3

Plot these points and connect them.

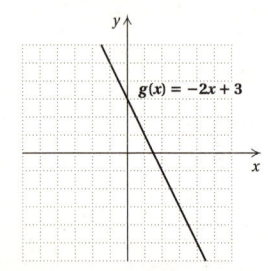

37. Graph $f(x) = \frac{1}{2}x + 1$.

Make a list of function values in a table.

$f(-2) = \frac{1}{2}(-2) + 1 = -1 + 1 = 0$
$f(0) = \frac{1}{2} \cdot 0 + 1 = 0 + 1 = 1$
$f(4) = \frac{1}{2} \cdot 4 + 1 = 2 + 1 = 3$

x	$f(x)$
-2	0
0	1
4	3

Plot these points and connect them.

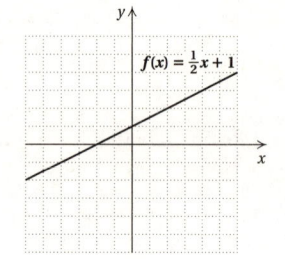

39. Graph $f(x) = 2 - |x|$.

Make a list of function values in a table.

$f(-3) = 2 - |-3| = 2 - 3 = -1$
$f(-2) = 2 - |-2| = 2 - 2 = 0$
$f(-1) = 2 - |-1| = 2 - 1 = 1$
$f(0) = 2 - |0| = 2 - 0 = 2$
$f(1) = 2 - |1| = 2 - 1 = 1$
$f(2) = 2 - |2| = 2 - 2 = 0$
$f(3) = 2 - |3| = 2 - 3 = -1$

x	$f(x)$
-3	-1
-2	0
-1	1
0	2
1	1
2	0
3	-1

Plot these points and connect them.

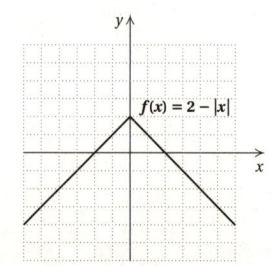

41. Graph $f(x) = x^2$.

Make a list of function values in a table.

$f(-3) = (-3)^2 = 9$
$f(-2) = (-2)^2 = 4$
$f(-1) = (-1)^2 = 1$

Exercise Set 7.1

$f(0) = 0^2 = 0$
$f(1) = 1^2 = 1$
$f(2) = 2^2 = 4$
$f(3) = 3^2 = 9$

x	$f(x)$
-3	9
-2	4
-1	1
0	0
1	1
2	4
3	9

Plot these points and connect them.

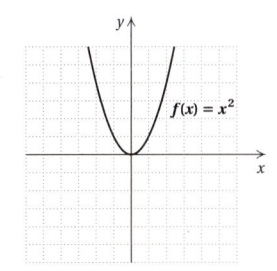

43. Graph $f(x) = x^2 - x - 2$.

Make a list of function values in a table.

$f(-3) = (-3)^2 - (-3) - 2 = 9 + 3 - 2 = 10$
$f(-2) = (-2)^2 - (-2) - 2 = 4 + 2 - 2 = 4$
$f(-1) = (-1)^2 - (-1) - 2 = 1 + 1 - 2 = 0$
$f(0) = 0^2 - 0 - 2 = -2$
$f(1) = 1^2 - 1 - 2 = 1 - 1 - 2 = -2$
$f(2) = 2^2 - 2 - 2 = 4 - 2 - 2 = 0$
$f(3) = 3^2 - 3 - 2 = 9 - 3 - 2 = 4$

x	$f(x)$
-3	10
-2	4
-1	0
0	-2
1	-2
2	0
3	4

Plot these points and connect them.

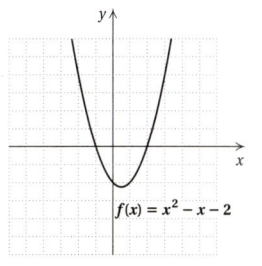

45. We can use the vertical line test:

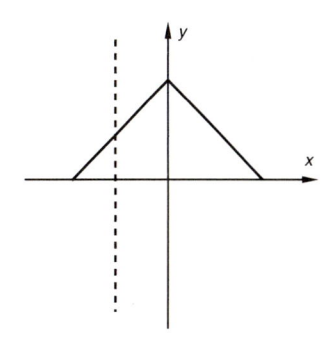

Visualize moving this vertical line across the graph. No vertical line will intersect the graph more than once. Thus, the graph is a graph of a function.

47. We can use the vertical line test:

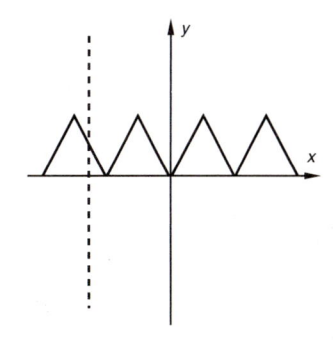

Visualize moving this vertical line across the graph. No vertical line will intersect the graph more than once. Thus, the graph is a graph of a function.

49. We can use the vertical line test.

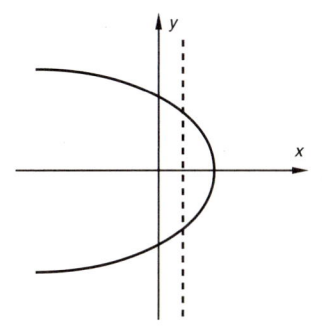

It is possible for a vertical line to intersect the graph more than once. Thus this is not the graph of a function.

51. We can use the vertical line test.

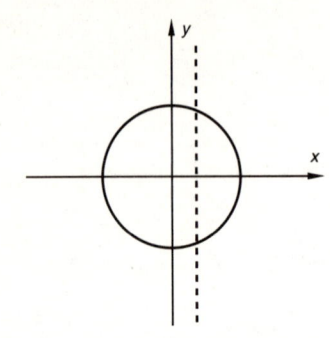

It is possible for a vertical line to intersect the graph more than once. Thus this is not a graph of a function.

53. Locate the point that is directly above 225. Then estimate its second coordinate by moving horizontally from the point to the vertical axis. The rate is about 75 per 10,000 men.

55. Locate 2 on the horizontal axis and move directly up to the graph. Then move across to the vertical axis and read the revenue. The movie revenue for week 2 was about $25 million.

57. Discussion and Writing Exercise

59. Discussion and Writing Exercise

61. *Familiarize.* If x represents the first integer, then $x + 2$ represents the second integer, and $x + 4$ represents the third.

Translate. We write an equation.

First integer plus two times the second plus three times the third is 124.

$$x + 2(x+2) + 3(x+4) = 124$$

Carry out. Solve the equation.
$$x + 2(x+2) + 3(x+4) = 124$$
$$x + 2x + 4 + 3x + 12 = 124$$
$$6x + 16 = 124$$
$$6x = 108$$
$$x = 18$$

If $x = 18$, then $x + 2 = 18 + 2$, or 20, and $x + 4 = 18 + 4$, or 22.

Check. 18, 20, and 22 are consecutive even integers. Also, $18 + 2 \cdot 20 + 3 \cdot 22 = 18 + 40 + 66 = 124$. The numbers check.

State. The integers are 18, 20, and 22.

63.
$$S = 2lh + 2lw + 2wh$$
$$S - 2wh = 2lh + 2lw$$
$$S - 2wh = l(2h + 2w)$$
$$\frac{S - 2wh}{2h + 2w} = l$$

65. 9.3×10^{-9}

Negative exponent, so the answer is a small number.

0.000000009.3

9 places

$9.3 \times 10^{-9} = 0.0000000093$

67. 1.075,000,000.

9 places

Large number, so the exponent is positive.

$1,075,000,000 = 1.075 \times 10^9$

69. $w^3 + \dfrac{1}{27} = w^3 + \left(\dfrac{1}{3}\right)^3 = \left(w + \dfrac{1}{3}\right)\left(w^2 - \dfrac{1}{3}w + \dfrac{1}{9}\right)$.

71. $xy^3 + 64x = x(y^3 + 64)$
$= x(y^3 + 4^3)$
$= x(y + 4)(y^2 - 4y + 16)$

73. To find $f(g(-4))$, we first find $g(-4)$:
$g(-4) = 2(-4) + 5 = -8 + 5 = -3$.
Then $f(g(-4)) = f(-3) = 3(-3)^2 - 1 = 3 \cdot 9 - 1 = 27 - 1 = 26$.

To find $g(f(-4))$, we first find $f(-4)$:
$f(-4) = 3(-4)^2 - 1 = 3 \cdot 16 - 1 = 48 - 1 = 47$.
Then $g(f(-4)) = g(47) = 2 \cdot 47 + 5 = 94 + 5 = 99$.

75. We know that $(-1, -7)$ and $(3, 8)$ are both solutions of $g(x) = mx + b$. Substituting, we have
$-7 = m(-1) + b$, or $-7 = -m + b$,
and $8 = m(3) + b$, or $8 = 3m + b$.

Solve the first equation for b and substitute that expression into the second equation.

$-7 = -m + b$	First equation
$m - 7 = b$	Solving for b
$8 = 3m + b$	Second equation
$8 = 3m + (m - 7)$	Substituting
$8 = 3m + m - 7$	
$8 = 4m - 7$	
$15 = 4m$	
$\dfrac{15}{4} = m$	

We know that $m - 7 = b$, so $\dfrac{15}{4} - 7 = b$, or $-\dfrac{13}{4} = b$.

We have $m = \dfrac{15}{4}$ and $b = -\dfrac{13}{4}$, so $g(x) = \dfrac{15}{4}x - \dfrac{13}{4}$.

Exercise Set 7.2

1. a) Locate 1 on the horizontal axis and then find the point on the graph for which 1 is the first coordinate. From that point, look to the vertical axis to find the corresponding y-coordinate, 3. Thus, $f(1) = 3$.

   b) The domain is the set of all x-values in the graph. It is $\{-4, -3, -2, -1, 0, 1, 2\}$.

   c) To determine which member(s) of the domain are paired with 2, locate 2 on the vertical axis. From there look left and right to the graph to find any points for which 2 is the second coordinate. Two such points exist, $(-2, 2)$ and $(0, 2)$. Thus, the x-values for which $f(x) = 2$ are -2 and 0.

   d) The range is the set of all y-values in the graph. It is $\{1, 2, 3, 4\}$.

3. a) Locate 1 on the horizontal axis and then find the point on the graph for which 1 is the first coordinate. From that point, look to the vertical axis to find the corresponding y-coordinate, 2. Thus, $f(1) = 2$.

   b) The domain is the set of all x-values in the graph. It is $\{-6, -4, -2, 0, 1, 3, 4\}$.

   c) To determine which member(s) of the domain are paired with 2, locate 2 on the vertical axis. From there look left and right to the graph to find any points for which 2 is the second coordinate. Two such points exist, $(1, 2)$ and $(3, 2)$. Thus, the x-values for which $f(x) = 2$ are 1 and 3.

   d) The range is the set of all y-values in the graph. It is $\{-5, -2, 0, 2, 5\}$.

5. a) Locate 1 on the horizontal axis and then find the point on the graph for which 1 is the first coordinate. From that point, look to the vertical axis to find the corresponding y-coordinate, about 2.5 or $\frac{5}{2}$. Thus, $f(1) \approx \frac{5}{2}$.

   b) The set of all x-values in the graph extends from -3 to 5, so the domain is $\{x| -3 \leq x \leq 5\}$, or $[-3, 5]$.

   c) To determine which member(s) of the domain are paired with 2, locate 2 on the vertical axis. From there look left and right to the graph to find any points for which 2 is the second coordinate. One such point exists. Its first coordinate appears to be about $2\frac{1}{4}$ or $\frac{9}{4}$. Thus, the x-value for which $f(x) = 2$ is about $\frac{9}{4}$.

   d) The set of all y-values in the graph extends from 1 to 4, so the range is $\{y| 1 \leq y \leq 4\}$, or $[1, 4]$.

7. a) Locate 1 on the horizontal axis and the find the point on the graph for which 1 is the first coordinate. From that point, look to the vertical axis to find the corresponding y-coordinate. It appears to be about $2\frac{1}{4}$, or $\frac{9}{4}$. Thus, $f(1) \approx \frac{9}{4}$.

   b) The set of all x-values in the graph extends from -4 to 3, so the domain is $\{x| -4 \leq x \leq 3\}$, or $[-4, 3]$.

   c) To determine which member(s) of the domain are paired with 2, locate 2 on the vertical axis. From there look left and right to the graph to find any points for which 2 is the second coordinate. One such point exists. Its first coordinate is about 0, so the x-value for which $f(x) = 2$ is about 0.

   d) The set of all y-values in the graph extends from -5 to 4, so the range is $\{y| -5 \leq y \leq 4\}$, or $[-5, 4]$.

9. a) Locate 1 on the horizontal axis and then find the point on the graph for which 1 is the first coordinate. From that point, look to the vertical axis to find the corresponding y-coordinate, 2. Thus, $f(1) = 2$.

   b) The set of all x-values in the graph extends from -5 to 4, so the domain is $\{x| -5 \leq x \leq 4\}$, or $[-5, 4]$.

   c) To determine which member(s) of the domain are paired with 2, locate 2 on the vertical axis. From there look left and right to the graph to find any points for which 2 is the second coordinate. All points in the set $\{x| 1 \leq x \leq 4\}$, or $[1, 4]$ satisfy this condition. These are the x-values for which $f(x) = 2$.

   d) The set of all y-values in the graph extends from -3 to 2, so the range is $\{y| -3 \leq y \leq 2\}$, or $[-3, 2]$.

11. a) Locate 1 on the horizontal axis and then find the point on the graph for which 1 is the first coordinate. From that point, look to the vertical axis to find the corresponding y-coordinate, -1. Thus, $f(1) = -1$.

    b) The set of all x-values in the graph extends from -6 to 5, so the domain is $\{x| -6 \leq x \leq 5\}$, or $[-6, 5]$.

    c) To determine which member(s) of the domain are paired with 2, locate 2 on the vertical axis. From there look left and right to the graph to find any points for which 2 is the second coordinate. Three such points exist, $(-4, 2)$, $(0, 2)$ and $(3, 2)$. Thus, the x-values for which $f(x) = 2$ are -4, 0, and 3.

    d) The set of all y-values in the graph extends from -2 to 2, so the range is $\{y| -2 \leq y \leq 2\}$, or $[-2, 2]$.

13. $f(x) = \dfrac{2}{x+3}$

    Since $\dfrac{2}{x+3}$ cannot be calculated when the denominator is 0, we find the x-value that causes $x + 3$ to be 0:

    $$x + 3 = 0$$
    $$x = -3 \quad \text{Subtracting 3 on both sides}$$

    Thus, -3 is not in the domain of f, while all other real numbers are. The domain of f is

    $\{x| x \text{ is a real number } and \ x \neq -3\}$, or

    $(-\infty, -3) \cup (-3, \infty)$.

15. $f(x) = 2x + 1$

    Since we can calculate $2x + 1$ for any real number x, the domain is the set of all real numbers.

17. $f(x) = x^2 + 3$

Since we can calculate $x^2 + 3$ for any real number x, the domain is the set of all real numbers.

19. $f(x) = \dfrac{8}{5x - 14}$

Since $\dfrac{8}{5x - 14}$ cannot be calculated when the denominator is 0, we find the x-value that causes $5x - 14$ to be 0:
$$5x - 14 = 0$$
$$5x = 14$$
$$x = \dfrac{14}{5}$$

Thus, $\dfrac{14}{5}$ is not in the domain of f, while all other real numbers are. The domain of f is
$$\left\{x \,\Big|\, x \text{ is a real number } and\ x \neq \dfrac{14}{5}\right\}, \text{ or }$$
$$\left(\infty, \dfrac{14}{5}\right) \cup \left(\dfrac{14}{5}, \infty\right).$$

21. $f(x) = |x| - 4$

Since we can calculate $|x| - 4$ for any real number x, the domain is the set of all real numbers.

23. $f(x) = \dfrac{4}{|2x - 3|}$

Since $\dfrac{4}{|2x - 3|}$ cannot be calculated when the denominator is 0, we find the x-values that causes $|2x - 3|$ to be 0:
$$|2x - 3| = 0$$
$$2x - 3 = 0$$
$$2x = 3$$
$$x = \dfrac{3}{2}$$

Thus, $\dfrac{3}{2}$ is not in the domain of f, while all other real numbers are. The domain of f is
$$\left\{x \,\Big|\, x \text{ is a real number } and\ x \neq \dfrac{3}{2}\right\}, \text{ or }$$
$$\left(-\infty, \dfrac{3}{2}\right) \cup \left(\dfrac{3}{2}, \infty\right).$$

25. $g(x) = \dfrac{1}{x - 1}$

Since $\dfrac{1}{x - 1}$ cannot be calculated when the denominator is 0, we find the x-value that causes $x - 1$ to be 0:
$$x - 1 = 0$$
$$x = 1$$

Thus, 1 is not in the domain of g, while all other real numbers are. The domain of g is
$$\{x | x \text{ is a real number } and\ x \neq 1\}, \text{ or } (-\infty, 1) \cup (1, \infty).$$

27. $g(x) = x^2 - 2x + 1$

Since we can calculate $x^2 - 2x + 1$ for any real number x, the domain is the set of all real numbers.

29. $g(x) = x^3 - 1$

Since we can calculate $x^3 - 1$ for any real number x, the domain is the set of all real numbers.

31. $g(x) = \dfrac{7}{20 - 8x}$

Since $\dfrac{7}{20 - 8x}$ cannot be calculated when the denominator is 0, we find the x-values that cause $20 - 8x$ to be 0:
$$20 - 8x = 0$$
$$-8x = -20$$
$$x = \dfrac{5}{2}$$

Thus, $\dfrac{5}{2}$ is not in the domain of g, while all other real numbers are. The domain of g is
$$\left\{x \,\Big|\, x \text{ is a real number } and\ x \neq \dfrac{5}{2}\right\}, \text{ or }$$
$$\left(-\infty, \dfrac{5}{2}\right) \cup \left(\dfrac{5}{2}, \infty\right).$$

33. $g(x) = |x + 7|$

Since we can calculate $|x + 7|$ for any real number x, the domain is the set of all real numbers.

35. $g(x) = \dfrac{-2}{|4x + 5|}$

Since $\dfrac{-2}{|4x + 5|}$ cannot be calculated when the denominator is 0, we find the x-value that causes $|4x + 5|$ to be 0:
$$|4x + 5| = 0$$
$$4x + 5 = 0$$
$$4x = -5$$
$$x = -\dfrac{5}{4}$$

Thus, $-\dfrac{5}{4}$ is not in the domain of g, while all other real numbers are. The domain of g is
$$\left\{x \,\Big|\, x \text{ is a real number } and\ x \neq -\dfrac{5}{4}\right\}, \text{ or }$$
$$\left(-\infty, -\dfrac{5}{4}\right) \cup \left(-\dfrac{5}{4}, \infty\right).$$

37. The input -1 has the output -8, so $f(-1) = -8$; the input 0 has the output 0, so $f(0) = 0$; the input 1 has the output -2, so $f(1) = -2$.

39. Discussion and Writing Exercise

41.
$$\dfrac{a^2 - 1}{a + 1} = \dfrac{(a + 1)(a - 1)}{a + 1}$$
$$= \dfrac{(a\!\!\!\!/+1)(a - 1)}{(a\!\!\!\!/+1)(1)}$$
$$= \dfrac{a - 1}{1}$$
$$= a - 1$$

Exercise Set 7.3

43. $\dfrac{5x-15}{x^2-x-6} = \dfrac{5(x-3)}{(x+2)(x-3)}$
$= \dfrac{5\cancel{(x-3)}}{(x+2)\cancel{(x-3)}}$
$= \dfrac{5}{x+2}$

45.
$$\begin{array}{r} w+1\\ w+3\overline{\smash{)}w^2+4w+5}\\ \underline{w^2+3w}\\ w+5\\ \underline{w+3}\\ 2 \end{array}$$
$(w^2+4w)-(w^2+3w)=w$
$(w+5)-(w+3)=2$

The answer is $w+1$, R 2, or $w+1+\dfrac{2}{w+3}$.

47. $(7x-3)(2x+9) = 7x \cdot 2x + 7x \cdot 9 - 3 \cdot 2x - 3 \cdot 9$
$= 14x^2 + 63x - 6x - 27$
$= 14x^2 + 57x - 27$

49. $(9y+10)^2 = (9y)^2 + 2 \cdot 9y \cdot 10 + 10^2$
$= 81y^2 + 180y + 100$

51. We graph each function and determine the range.

The range of $f(x) = \dfrac{2}{x+3}$ is $(-\infty, 0) \cup (0, \infty)$; the range of $f(x) = x^2 - 2x + 3$ is $[2, \infty)$; the range of $f(x) = |x| - 4$ is $[-4, \infty)$; the range of $f(x) = |x-4|$ is $[0, \infty)$.

Exercise Set 7.3

1. $y = 4x + 5$
$\uparrow \uparrow$
$y = mx + b$

The slope is 4, and the y-intercept is $(0, 5)$.

3. $f(x) = -2x - 6$
$\uparrow \uparrow$
$f(x) = mx + b$

The slope is -2, and the y-intercept is $(0, -6)$.

5. $y = -\dfrac{3}{8}x - \dfrac{1}{5}$
$\uparrow \uparrow$
$y = mx + b$

The slope is $-\dfrac{3}{8}$, and the y-intercept is $\left(0, -\dfrac{1}{5}\right)$.

7. $g(x) = 0.5x - 9$
$\uparrow \uparrow$
$g(x) = mx + b$

The slope is 0.5, and the y-intercept is $(0, -9)$.

9. First we find the slope-intercept form of the equation by solving for y. This allows us to determine the slope and y-intercept easily.
$2x - 3y = 8$
$-3y = -2x + 8$
$\dfrac{-3y}{-3} = \dfrac{-2x+8}{-3}$
$y = \dfrac{2}{3}x - \dfrac{8}{3}$

The slope is $\dfrac{2}{3}$, and the y-intercept is $\left(0, -\dfrac{8}{3}\right)$.

11. First we find the slope-intercept form of the equation by solving for y. This allows us to determine the slope and y-intercept easily.
$9x = 3y + 6$
$9x - 6 = 3y$
$\dfrac{9x-6}{3} = \dfrac{3y}{y}$
$3x - 2 = y$, or
$y = 3x - 2$

The slope is 3, and the y-intercept is $(0, -2)$.

13. First we find the slope-intercept form of the equation by solving for y. This allows us to determine the slope and y-intercept easily.
$3 - \dfrac{1}{4}y = 2x$
$-\dfrac{1}{4}y = 2x - 3$
$-4\left(-\dfrac{1}{4}y\right) = -4(2x - 3)$
$y = -8x + 12$

The slope is -8, and the y-intercept is $(0, 12)$.

15. First we find the slope-intercept form of the equation by solving for y. This allows us to determine the slope and y-intercept easily.
$17y + 4x + 3 = 7 + 4x$
$17y + 3 = 7$
$17y = 4$
$y = \dfrac{4}{17}$, or
$y = 0 \cdot x + \dfrac{4}{17}$

The slope is 0, and the y-intercept is $\left(0, \dfrac{4}{17}\right)$.

17. We can use any two points on the line, such as $(0, 3)$ and $(4, 1)$.
$\text{Slope} = \dfrac{\text{change in } y}{\text{change in } x}$
$= \dfrac{1-3}{4-0} = \dfrac{-2}{4} = -\dfrac{1}{2}$

19. We can use any two points on the line, such as $(-3, 1)$ and $(3, 3)$.
$\text{Slope} = \dfrac{\text{change in } y}{\text{change in } x}$
$= \dfrac{3-1}{3-(-3)} = \dfrac{2}{6} = \dfrac{1}{3}$

21. $\text{Slope} = \dfrac{\text{change in } y}{\text{change in } x} = \dfrac{5-9}{4-6} = \dfrac{-4}{-2} = 2$

23. $\text{Slope} = \dfrac{\text{change in } y}{\text{change in } x} = \dfrac{-8-(-4)}{3-9} = \dfrac{-4}{-6} = \dfrac{2}{3}$

25. Slope = $\dfrac{\text{change in } y}{\text{change in } x} = \dfrac{8.7 - 12.4}{-5.2 - (-16.3)} = \dfrac{-3.7}{11.1} =$

$-\dfrac{37}{111} = -\dfrac{1}{3}$

27. Slope = $\dfrac{0.4}{5} = 0.08 = 8\%$; this can also be expressed as $\dfrac{2}{25}$.

29. Slope = $\dfrac{43.33}{1238} = 0.035 = 3.5\%$

31. The rate of change can be found using the coordinates of any two points on the line. We use $(2000, 9.7)$ and $(2005, 35)$.

$\text{Rate} = \dfrac{\text{change in volume of e-mail}}{\text{corresponding change in time}}$

$= \dfrac{35 - 9.7}{2005 - 2000}$

$= \dfrac{25.3}{5}$

$= 5.06$ billion messages daily per year

33. We can use the coordinates of any two points on the line. We'll use $(0, 30)$ and $(3, 3)$.

Slope = $\dfrac{\text{change in } y}{\text{change in } x} = \dfrac{3 - 30}{3 - 0} = \dfrac{-27}{3} = -9$

The rate of change is $-\$900$ per year. That is, the value is decreasing at a rate of $\$900$ per year.

35. We can use the coordinates of any two points on the line. We'll use $(15, 470)$ and $(55, 510)$:

Slope = $\dfrac{\text{change in } y}{\text{change in } x} = \dfrac{510 - 470}{55 - 15} = \dfrac{40}{40} = 1$

The average SAT math score is increasing at a rate of 1 point per thousand dollars of family income.

37. Discussion and Writing Exercise

39. $\quad 3^2 - 24 \cdot 56 + 144 \div 12$

$= 9 - 24 \cdot 56 + 144 \div 12$

$= 9 - 1344 + 144 \div 12$

$= 9 - 1344 + 12$

$= -1335 + 12$

$= -1323$

41. $\quad 10\{2x + 3[5x - 2(-3x + y^1 - 2)]\}$

$= 10\{2x + 3[5x - 2(-3x + y - 2)]\}$

$= 10\{2x + 3[5x + 6x - 2y + 4]\}$

$= 10\{2x + 3[11x - 2y + 4]\}$

$= 10\{2x + 33x - 6y + 12\}$

$= 10\{35x - 6y + 12\}$

$= 350x - 60y + 120$

43. *Familiarize.* Let t represent the length of a side of the triangle. Then $t - 5$ represents the length of a side of the square.

Translate.

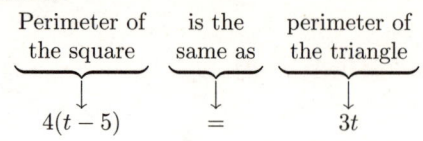

Perimeter of the square is the same as perimeter of the triangle

$4(t - 5) \quad = \quad 3t$

Solve.

$4(t - 5) = 3t$

$4t - 20 = 3t$

$t - 20 = 0$

$t = 20$

Check. If 20 is the length of a side of the triangle, then the length of a side of the square is $20 - 5$, or 15. The perimeter of the square is $4 \cdot 15$, or 60, and the perimeter of the triangle is $3 \cdot 20$, or 60. The numbers check.

State. The square and triangle have sides of length 15 yd and 20 yd, respectively.

45. $\quad c^6 - d^6 = (c^3)^2 - (d^3)^2$

$= (c^3 + d^3)(c^3 - d^3)$

$= (c + d)(c^2 - cd + d^2)(c - d)(c^2 + cd + d^2)$

47.

$$\begin{array}{r} a - 10 \\ a - 1 \overline{\smash{)}a^2 - 11a + 6} \\ \underline{a^2 - a} \\ -10a + 6 \\ \underline{-10a + 10} \\ -4 \end{array}$$

$(a^2 - 11a) - (a^2 - a) = -10a$

$(-10a + 6) - (-10a + 10) = -4$

The answer is $a - 10$, R -4, or $a - 10 + \dfrac{-4}{a - 1}$.

Exercise Set 7.4

1. $x - 2 = y$

To find the x-intercept we let $y = 0$ and solve for x. We have $x - 2 = 0$, or $x = 2$. The x-intercept is $(2, 0)$.

To find the y-intercept we let $x = 0$ and solve for y.

$x - 2 = y$

$0 - 2 = y$

$-2 = y$

The y-intercept is $(0, -2)$. We plot these points and draw the line.

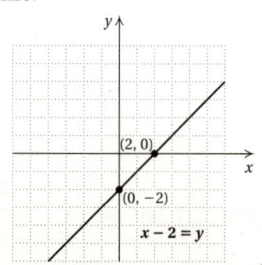

We use a third point as a check. We choose $x = 5$ and solve for y.

$5 - 2 = y$

$3 = y$

We plot $(5, 3)$ and note that it is on the line.

Exercise Set 7.4

3. $x + 3y = 6$

To find the x-intercept we let $y = 0$ and solve for x.
$$x + 3y = 6$$
$$x + 3 \cdot 0 = 6$$
$$x = 6$$

The x-intercept is $(6, 0)$.

To find the y-intercept we let $x = 0$ and solve for y.
$$x + 3y = 6$$
$$0 + 3y = 6$$
$$3y = 6$$
$$y = 2$$

The y-intercept is $(0, 2)$.

We plot these points and draw the line.

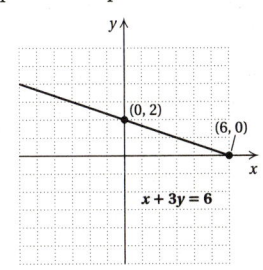

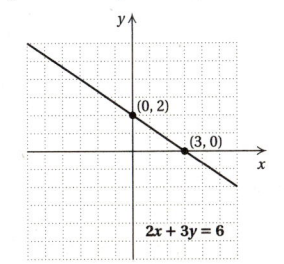

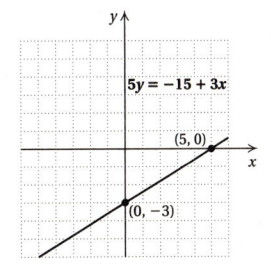

We use a third point as a check. We choose $x = 3$ and solve for y.
$$3 + 3y = 6$$
$$3y = 3$$
$$y = 1$$

We plot $(3, 1)$ and note that it is on the line.

5. $2x + 3y = 6$

To find the x-intercept we let $y = 0$ and solve for x.
$$2x + 3y = 6$$
$$2x + 3 \cdot 0 = 6$$
$$2x = 6$$
$$x = 3$$

The x-intercept is $(3, 0)$.

To find the y-intercept we let $x = 0$ and solve for y.
$$2x + 3y = 6$$
$$2 \cdot 0 + 3y = 6$$
$$3y = 6$$
$$y = 2$$

The y-intercept is $(0, 2)$.

We plot these points and draw the line.

We use a third point as a check. We choose $x = -3$ and solve for y.
$$2(-3) + 3y = 6$$
$$-6 + 3y = 6$$
$$3y = 12$$
$$y = 4$$

We plot $(-3, 4)$ and note that it is on the line.

7. $f(x) = -2 - 2x$

We can think of this equation as $y = -2 - 2x$.

To find the x-intercept we let $f(x) = 0$ and solve for x. We have $0 = -2 - 2x$, or $2x = -2$, or $x = -1$. The x-intercept is $(-1, 0)$.

To find the y-intercept we let $x = 0$ and solve for $f(x)$, or y.
$$y = -2 - 2x$$
$$y = -2 - 2 \cdot 0$$
$$y = -2$$

The y-intercept is $(0, -2)$.

We plot these points and draw the line.

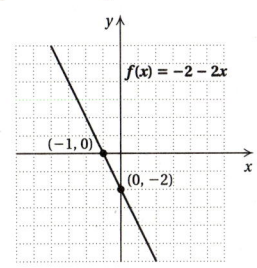

We use a third point as a check. We choose $x = -3$ and calculate y.
$$y = -2 - 2(-3) = -2 + 6 = 4$$

We plot $(-3, 4)$ and note that it is on the line.

9. $5y = -15 + 3x$

To find the x-intercept we let $y = 0$ and solve for x. We have $0 = -15 + 3x$, or $15 = 3x$, or $5 = x$. The x-intercept is $(5, 0)$.

To find the y-intercept we let $x = 0$ and solve for y.
$$5y = -15 + 3x$$
$$5y = -15 + 3 \cdot 0$$
$$5y = -15$$
$$y = -3$$

The y-intercept is $(0, -3)$.

We plot these points and draw the line.

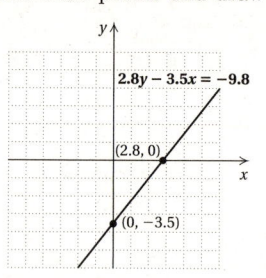

We use a third point as a check. We choose $x = -5$ and solve for y.
$$5y = -15 + 3(-5)$$
$$5y = -15 - 15$$
$$5y = -30$$
$$y = -6$$
We plot $(-5, -6)$ and note that it is on the line.

11. $2x - 3y = 6$

To find the x-intercept we let $y = 0$ and solve for x.
$$2x - 3y = 6$$
$$2x - 3 \cdot 0 = 6$$
$$2x = 6$$
$$x = 3$$
The x-intercept is $(3, 0)$.

To find the y-intercept we let $x = 0$ and solve for y.
$$2x - 3y = 6$$
$$2 \cdot 0 - 3y = 6$$
$$-3y = 6$$
$$y = -2$$
The y-intercept is $(0, -2)$.

We plot these points and draw the line.

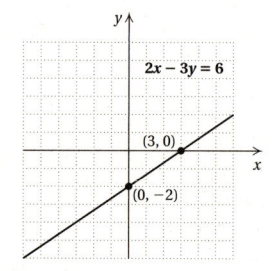

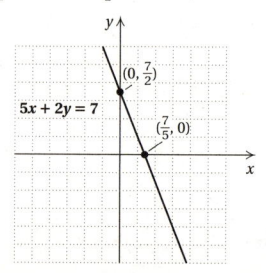

We use a third point as a check. We choose $x = -3$ and solve for y.
$$2(-3) - 3y = 6$$
$$-6 - 3y = 6$$
$$-3y = 12$$
$$y = -4$$
We plot $(-3, -4)$ and note that it is on the line.

13. $2.8y - 3.5x = -9.8$

To find the x-intercept we let $y = 0$ and solve for x.
$$2.8y - 3.5x = -9.8$$
$$2.8(0) - 3.5x = -9.8$$
$$-3.5x = -9.8$$
$$x = 2.8$$
The x-intercept is $(2.8, 0)$.

To find the y-intercept we let $x = 0$ and solve for y.
$$2.8y - 3.5x = -9.8$$
$$2.8y - 3.5(0) = -9.8$$
$$2.8y = -9.8$$
$$y = -3.5$$

The y-intercept is $(0, -3.5)$.

We plot these points and draw the line.

We use a third point as a check. We choose $x = 5$ and solve for y.
$$2.8y - 3.5(5) = -9.8$$
$$2.8y - 17.5 = -9.8$$
$$2.8y = 7.7$$
$$y = 2.75$$
We plot $(5, 2.75)$ and note that it is on the line.

15. $5x + 2y = 7$

To find the x-intercept we let $y = 0$ and solve for x.
$$5x + 2y = 7$$
$$5x + 2 \cdot 0 = 7$$
$$5x = 7$$
$$x = \frac{7}{5}$$
The x-intercept is $\left(\frac{7}{5}, 0\right)$.

To find the y-intercept we let $x = 0$ and solve for y.
$$5x + 2y = 7$$
$$5 \cdot 0 + 2y = 7$$
$$2y = 7$$
$$y = \frac{7}{2}$$
The y-intercept is $\left(0, \frac{7}{2}\right)$.

We plot these points and draw the line.

We use a third point as a check. We choose $x = 3$ and solve for y.
$$5 \cdot 3 + 2y = 7$$
$$15 + 2y = 7$$
$$2y = -8$$
$$y = -4$$
We plot $(3, -4)$ and note that it is on the line.

Exercise Set 7.4 171

17. $y = \frac{5}{2}x + 1$

First we plot the y-intercept $(0, 1)$. Then we consider the slope $\frac{5}{2}$. Starting at the y-intercept and using the slope, we find another point by moving 5 units up and 2 units to the right. We get to a new point $(2, 6)$.

We can also think of the slope as $\frac{-5}{-2}$. We again start at the y-intercept $(0, 1)$. We move 5 units down and 2 units to the left. We get to another new point $(-2, -4)$. We plot the points and draw the line.

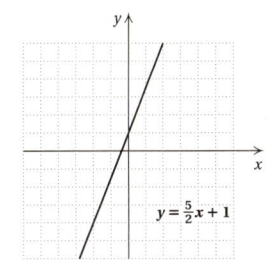

19. $f(x) = -\frac{5}{2}x - 4$

First we plot the y-intercept $(0, -4)$. We can think of the slope as $\frac{-5}{2}$. Starting at the y-intercept and using the slope, we find another point by moving 5 units down and 2 units to the right. We get to a new point $(2, -9)$.

We can also think of the slope as $\frac{5}{-2}$. We again start at the y-intercept $(0, -4)$. We move 5 units up and 2 units to the left. We get to another new point $(-2, 1)$. We plot the points and draw the line.

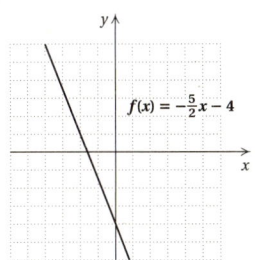

21. $x + 2y = 4$

First we write the equation in slope-intercept form by solving for y.

$$x + 2y = 4$$
$$2y = -x + 4$$
$$\frac{2y}{2} = \frac{-x+4}{2}$$
$$y = -\frac{1}{2}x + 2$$

Now we plot the y-intercept $(0, 2)$. We can think of the slope as $\frac{-1}{2}$. Starting at the y-intercept and using the slope, we find another point by moving 1 unit down and 2 units to the right. We get to a new point $(2, 1)$.

We can also think of the slope as $\frac{1}{-2}$. We again start at the y-intercept $(0, 2)$. We move 1 unit up and 2 units to the left. We get to another new point $(-2, 3)$. We plot the points and draw the line.

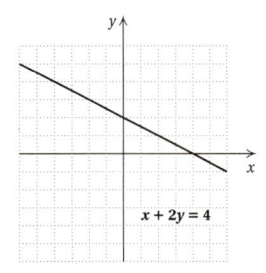

23. $4x - 3y = 12$

First we write the equation in slope-intercept form by solving for y.

$$4x - 3y = 12$$
$$-3y = -4x + 12$$
$$\frac{-3y}{-3} = \frac{-4x+12}{-3}$$
$$y = \frac{4}{3}x - 4$$

Now we plot the y-intercept $(0, -4)$ and consider the slope $\frac{4}{3}$. Starting at the y-intercept and using the slope, we find another point by moving 4 units up and 3 units to the right. We get to a new point $(3, 0)$. In a similar manner we can move from the point $(3, 0)$ to find another point $(6, 4)$. We plot these points and draw the line.

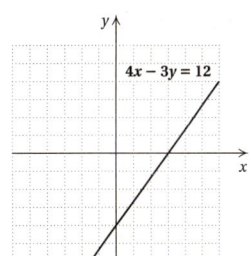

25. $f(x) = \frac{1}{3}x - 4$

First we plot the y-intercept $(0, -4)$. Then we consider the slope $\frac{1}{3}$. Starting at the y-intercept and using the slope, we find another point by moving 1 unit up and 3 units to the right. We get to a new point $(3, -3)$.

We can also think of the slope as $\frac{-1}{-3}$. We again start at the y-intercept $(0, -4)$. We move 1 unit down and 3 units to the left. We get to another new point $(-3, -5)$. We plot these points and draw the line.

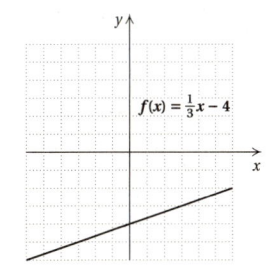

27. $5x + 4 \cdot f(x) = 4$

First we solve for $f(x)$.
$$5x + 4 \cdot f(x) = 4$$
$$4 \cdot f(x) = -5x + 4$$
$$\frac{4 \cdot f(x)}{4} = \frac{-5x + 4}{4}$$
$$f(x) = -\frac{5}{4}x + 1$$

Now we plot the y-intercept $(0, 1)$. We can think of the slope as $\frac{-5}{4}$. Starting at the y-intercept and using the slope, we find another point by moving 5 units down and 4 units to the right. We get to a new point $(4, -4)$.

We can also think of the slope as $\frac{5}{-4}$. We again start at the y-intercept $(0, 1)$. We move 5 units up and 4 units to the left. We get to another new point $(-4, 6)$. We plot these points and draw the line.

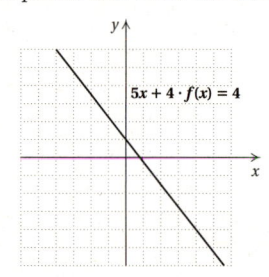

[Graph of $5x + 4 \cdot f(x) = 4$]

29. $x = 1$

Since y is missing, any number for y will do. Thus all ordered pairs $(1, y)$ are solutions. The graph is parallel to the y-axis.

x	y
1	-2
1	0	← x-intercept
1	3

↑   ↑   Choose any
x must   number for y.
be 1.

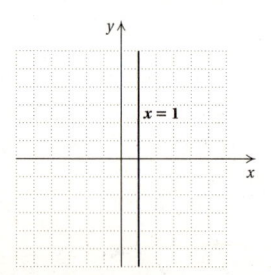

[Graph of $x = 1$]

This is a vertical line, so the slope is not defined.

31. $y = -1$

Since x is missing, any number for x will do. Thus all ordered pairs $(x, -1)$ are solutions. The graph is parallel to the x-axis.

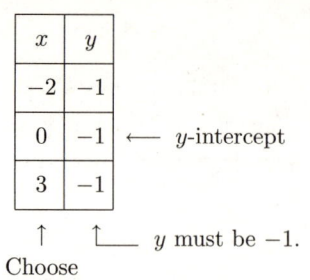

x	y
-2	-1
0	-1	← y-intercept
3	-1

↑   ↑ y must be -1.
Choose
any number
for x.

[Graph of $y = -1$]

This is a horizontal line, so the slope is 0.

33. $f(x) = -6$

Since x is missing all ordered pairs $(x, 6)$ are solutions. The graph is parallel to the x-axis.

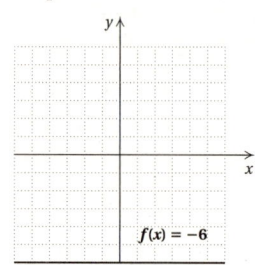

[Graph of $f(x) = -6$]

This is a horizontal line, so the slope is 0.

35. $y = 0$

Since x is missing, all ordered pairs $(x, 0)$ are solutions. The graph is the x-axis.

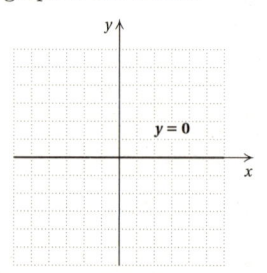

[Graph of $y = 0$]

This is a horizontal line, so the slope is 0.

37. $2 \cdot f(x) + 5 = 0$
$$2 \cdot f(x) = -5$$
$$f(x) = -\frac{5}{2}$$

Since x is missing, all ordered pairs $\left(x, -\frac{5}{2}\right)$ are solutions. The graph is parallel to the x-axis.

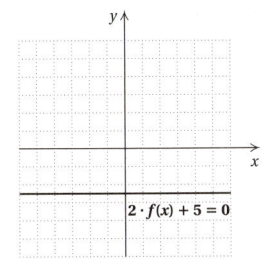

Exercise Set 7.4

[Graph: $2 \cdot f(x) + 5 = 0$, horizontal line]

This is a horizontal line, so the slope is 0.

39. $7 - 3x = 4 + 2x$
$7 - 5x = 4$
$-5x = -3$
$x = \dfrac{3}{5}$

Since y is missing, all ordered pairs $\left(\dfrac{3}{5}, y\right)$ are solutions. The graph is parallel to the y-axis.

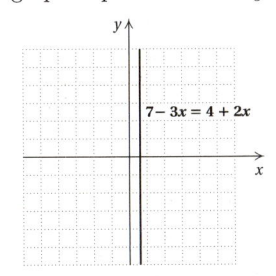

[Graph: $7 - 3x = 4 + 2x$, vertical line]

This is a vertical line, so the slope is not defined.

41. We first solve for y and determine the slope of each line.
$x + 6 = y$
$y = x + 6$ Reversing the order

The slope of $y = x + 6$ is 1.

$y - x = -2$
$y = x - 2$

The slope of $y = x - 2$ is 1.

The slopes are the same, and the y-intercepts are different. The lines are parallel.

43. We first solve for y and determine the slope of each line.
$y + 3 = 5x$
$y = 5x - 3$

The slope of $y = 5x - 3$ is 5.

$3x - y = -2$
$3x + 2 = y$
$y = 3x + 2$ Reversing the order

The slope of $y = 3x + 2$ is 3.

The slopes are not the same; the lines are not parallel.

45. We determine the slope of each line.

The slope of $y = 3x + 9$ is 3.

$2y = 6x - 2$
$y = 3x - 1$

The slope of $y = 3x - 1$ is 3.

The slopes are the same, and the y-intercepts are different. The lines are parallel.

47. We solve each equation for x.
$12x = 3$ $-7x = 10$
$x = \dfrac{1}{4}$ $x = -\dfrac{10}{7}$

We have two vertical lines, so they are parallel.

49. We determine the slope of each line.

The slope of $y = 4x - 5$ is 4.

$4y = 8 - x$
$4y = -x + 8$
$y = -\dfrac{1}{4}x + 2$

The slope of $4y = 8 - x$ is $-\dfrac{1}{4}$.

The product of their slopes is $4\left(-\dfrac{1}{4}\right)$, or -1; the lines are perpendicular.

51. We determine the slope of each line.
$x + 2y = 5$
$2y = -x + 5$
$y = -\dfrac{1}{2}x + \dfrac{5}{2}$

The slope of $x + 2y = 5$ is $-\dfrac{1}{2}$.

$2x + 4y = 8$
$4y = -2x + 8$
$y = -\dfrac{1}{2}x + 2$

The slope of $2x + 4y = 8$ is $-\dfrac{1}{2}$.

The product of their slopes is $\left(-\dfrac{1}{2}\right)\left(-\dfrac{1}{2}\right)$, or $\dfrac{1}{4}$; the lines are not perpendicular. For the lines to be perpendicular, the product must be -1.

53. We determine the slope of each line.
$2x - 3y = 7$
$-3y = -2x + 7$
$y = \dfrac{2}{3}x - \dfrac{7}{3}$

The slope of $2x - 3y = 7$ is $\dfrac{2}{3}$.

$2y - 3x = 10$
$2y = 3x + 10$
$y = \dfrac{3}{2}x + 5$

The slope of $2y - 3x = 10$ is $\dfrac{3}{2}$.

The product of their slopes is $\dfrac{2}{3} \cdot \dfrac{3}{2} = 1$; the lines are not perpendicular. For the lines to be perpendicular, the product must be -1.

55. Solving the first equation for x and the second for y, we have $x = \dfrac{3}{2}$ and $y = -2$. The graph of $x = \dfrac{3}{2}$ is a vertical line, and the graph of $y = -2$ is a horizontal line. Since one line is vertical and the other is horizontal, the lines are perpendicular.

57. Discussion and Writing Exercise

59. 5.3,000,000,000.
   ↑_____|
   10 places

Large number, so the exponent is positive.
$53,000,000,000 = 5.3 \times 10^{10}$

61. 0.01. 8
   |_↑
   2 places

Small number, so the exponent is negative.
$0.018 = 1.8 \times 10^{-2}$

63. Negative exponent, so the number is small.
   0.00002. 13
   ↑_____|
   5 places
   $2.13 \times 10^{-5} = 0.0000213$

65. Positive exponent, so the number is large.
   2.0000.
   |___↑
   4 places
   $2 \times 10^4 = 20,000$

67. $9x - 15y = 3 \cdot 3x - 3 \cdot 5y = 3(3x - 5y)$

69. $21p - 7pq + 14p = 7p \cdot 3 - 7p \cdot q + 7p \cdot 2$
$= 7p(3 - q + 2)$

71. The equation will be of the form $y = b$. Since the line passes through $(-2, 3)$, b must be 3. Thus, we have $y = 3$.

73. Find the slope of each line.
$$5y = ax + 5$$
$$y = \dfrac{a}{5}x + 1$$
The slope of $5y = ax + 5$ is $\dfrac{a}{5}$.
$$\dfrac{1}{4}y = \dfrac{1}{10}x - 1$$
$$4 \cdot \dfrac{1}{4}y = 4\left(\dfrac{1}{10}x - 1\right)$$
$$y = \dfrac{2}{5}x - 4$$
The slope of $\dfrac{1}{4}y = \dfrac{1}{10}x - 1$ is $\dfrac{2}{5}$.

In order for the graphs to be parallel, their slopes must be the same. (Note that the y-intercepts are different.)
$$\dfrac{a}{5} = \dfrac{2}{5}$$
$$a = 2 \quad \text{Multiplying by 5}$$

75. The y-intercept is $\left(0, \dfrac{2}{5}\right)$, so the equation is of the form $y = mx + \dfrac{2}{5}$. We substitute -3 for x and 0 for y in this equation to find m.
$$y = mx + \dfrac{2}{5}$$
$$0 = m(-3) + \dfrac{2}{5} \quad \text{Substituting}$$
$$0 = -3m + \dfrac{2}{5}$$
$$3m = \dfrac{2}{5} \quad \text{Adding } 3m$$
$$m = \dfrac{2}{15} \quad \text{Multiplying by } \dfrac{1}{3}$$
The equation is $y = \dfrac{2}{15}x + \dfrac{2}{5}$.

(We could also have found the slope as follows:
$$m = \dfrac{\dfrac{2}{5} - 0}{0 - (-3)} = \dfrac{\dfrac{2}{5}}{3} = \dfrac{2}{15})$$

77. All points on the x-axis are pairs of the form $(x, 0)$. Thus any number for x will do and y must be 0. The equation is $y = 0$. This equation is a function because its graph passes the vertical-line test.

79. We substitute 4 for x and 0 for y.
$$y = mx + 3$$
$$0 = m(4) + 3$$
$$-3 = 4m$$
$$-\dfrac{3}{4} = m$$

81. a) Graph II indicates that 200 mL of fluid was dripped in the first 3 hr, a rate of 200/3 mL/hr. It also indicates that 400 mL of fluid was dripped in the next 3 hr, a rate of 400/3 mL/hr, and that this rate continues until the end of the time period shown. Since the rate of 400/3 mL/hr is double the rate of 200/3 mL/hr, this graph is appropriate for the given situation.

b) Graph IV indicates that 300 mL of fluid was dripped in the first 2 hr, a rate of 300/2, or 150 mL/hr. In the next 2 hr, 200 mL was dripped. This is a rate of 200/2, or 100 mL/hr. Then 100 mL was dripped in the next 3 hr, a rate of 100/3, or $33\dfrac{1}{3}$ mL/hr. Finally, in the remaining 2 hr, 0 mL of fluid was dripped, a rate of 0/2, or 0 mL/hr. Since the rate at which the fluid was given decreased as time progressed and eventually became 0, this graph is appropriate for the given situation.

c) Graph I is the only graph that shows a constant rate for 5 hours, in this case from 3 PM to 8 PM. Thus, it is appropriate for the given situation.

d) Graph III indicates that 100 mL of fluid was dripped in the first 4 hr, a rate of 100/4, or 25 mL/hr. In the next 3 hr, 200 mL was dripped. This is a rate of 200/3, or $66\frac{2}{3}$ mL/hr. Then 100 mL was dripped in the next hour, a rate of 100 mL/hr. In the last hour 200 mL was dripped, a rate of 200 mL/hr. Since the rate at which the fluid was given gradually increased, this graph is appropriate for the given situation.

Exercise Set 7.5

1. We use the slope-intercept equation and substitute -8 for m and 4 for b.
$$y = mx + b$$
$$y = -8x + 4$$

3. We use the slope-intercept equation and substitute 2.3 for m and -1 for b.
$$y = mx + b$$
$$y = 2.3x - 1$$

5. We use the slope-intercept equation and substitute $-\frac{7}{3}$ for m and -5 for b.
$$y = mx + b$$
$$y = -\frac{7}{3}x - 5$$

7. We use the slope-intercept equation and substitute $\frac{2}{3}$ for m and $\frac{5}{8}$ for b.
$$y = mx + b$$
$$y = \frac{2}{3}x + \frac{5}{8}$$

9. Using the point-slope equation:

    Substitute 4 for x_1, 3 for y_1, and 5 for m.
$$y - y_1 = m(x - x_1)$$
$$y - 3 = 5(x - 4)$$
$$y - 3 = 5x - 20$$
$$y = 5x - 17$$

    Using the slope-intercept equation:

    Substitute 4 for x, 3 for y, and 5 for m in $y = mx + b$ and solve for b.
$$y = mx + b$$
$$3 = 5 \cdot 4 + b$$
$$3 = 20 + b$$
$$-17 = b$$

    Then we use the equation $y = mx + b$ and substitute 5 for m and -17 for b.
$$y = 5x - 17$$

11. Using the point-slope equation:

    Substitute 9 for x_1, 6 for y_1, and -3 for m.
$$y - y_1 = m(x - x_1)$$
$$y - 6 = -3(x - 9)$$
$$y - 6 = -3x + 27$$
$$y = -3x + 33$$

    Using the slope-intercept equation:

    Substitute 9 for x, 6 for y, and -3 for m in $y = mx + b$ and solve for b.
$$y = mx + b$$
$$6 = -3 \cdot 9 + b$$
$$6 = -27 + b$$
$$33 = b$$

    Then we use the equation $y = mx + b$ and substitute -3 for m and 33 for b.
$$y = -3x + 33$$

13. Using the point-slope equation:

    Substitute -1 for x_1, -7 for y_1, and 1 for m.
$$y - y_1 = m(x - x_1)$$
$$y - (-7) = 1(x - (-1))$$
$$y + 7 = 1(x + 1)$$
$$y + 7 = x + 1$$
$$y = x - 6$$

    Using the slope-intercept equation:

    Substitute -1 for x, -7 for y, and 1 for m in $y = mx + b$ and solve for b.
$$y = mx + b$$
$$-7 = 1(-1) + b$$
$$-7 = -1 + b$$
$$-6 = b$$

    Then we use the equation $y = mx + b$ and substitute 1 for m and -6 for b.
$$y = 1x - 6, \text{ or } y = x - 6$$

15. Using the point-slope equation:

    Substitute 8 for x_1, 0 for y_1, and -2 for m.
$$y - y_1 = m(x - x_1)$$
$$y - 0 = -2(x - 8)$$
$$y = -2x + 16$$

    Using the slope-intercept equation:

    Substitute 8 for x, 0 for y, and -2 for m in $y = mx + b$ and solve for b.
$$y = mx + b$$
$$0 = -2 \cdot 8 + b$$
$$0 = -16 + b$$
$$16 = b$$

    Then we use the equation $y = mx + b$ and substitute -2 for m and 16 for b.
$$y = -2x + 16$$

17. Using the point-slope equation:

Substitute 0 for x_1, -7 for y_1, and 0 for m.
$$y - y_1 = m(x - x_1)$$
$$y - (-7) = 0(x - 0)$$
$$y + 7 = 0$$
$$y = -7$$

Using the slope-intercept equation:

Substitute 0 for x, -7 for y, and 0 for m in $y = mx + b$ and solve for b.
$$y = mx + b$$
$$-7 = 0 \cdot 0 + b$$
$$-7 = b$$

Then we use the equation $y = mx + b$ and substitute 0 for m and -7 for b.
$$y = 0x - 7, \text{ or } y = -7$$

19. Using the point-slope equation:

Substitute 1 for x_1, -2 for y_1, and $\frac{2}{3}$ for m.
$$y - y_1 = m(x - x_1)$$
$$y - (-2) = \frac{2}{3}(x - 1)$$
$$y + 2 = \frac{2}{3}x - \frac{2}{3}$$
$$y = \frac{2}{3}x - \frac{8}{3}$$

Using the slope-intercept equation:

Substitute 1 for x, -2 for y and $\frac{2}{3}$ for m in $y = mx + b$ and solve for b.
$$y = mx + b$$
$$-2 = \frac{2}{3} \cdot 1 + b$$
$$-2 = \frac{2}{3} + b$$
$$-\frac{8}{3} = b$$

Then we use the equation $y = mx + b$ and substitute $\frac{2}{3}$ for m and $-\frac{8}{3}$ for b.
$$y = \frac{2}{3}x - \frac{8}{3}$$

21. First find the slope of the line:
$$m = \frac{6 - 4}{5 - 1} = \frac{2}{4} = \frac{1}{2}$$

Using the point-slope equation:

We choose to use the point $(1, 4)$ and substitute 1 for x_1, 4 for y_1, and $\frac{1}{2}$ for m.
$$y - y_1 = m(x - x_1)$$
$$y - 4 = \frac{1}{2}(x - 1)$$
$$y - 4 = \frac{1}{2}x - \frac{1}{2}$$
$$y = \frac{1}{2}x + \frac{7}{2}$$

Using the slope-intercept equation:

We choose $(1, 4)$ and substitute 1 for x, 4 for y, and $\frac{1}{2}$ for m in $y = mx + b$. Then we solve for b.
$$y = mx + b$$
$$4 = \frac{1}{2} \cdot 1 + b$$
$$4 = \frac{1}{2} + b$$
$$\frac{7}{2} = b$$

Finally, we use the equation $y = mx + b$ and substitute $\frac{1}{2}$ for m and $\frac{7}{2}$ for b.
$$y = \frac{1}{2}x + \frac{7}{2}$$

23. First find the slope of the line:
$$m = \frac{-3 - 2}{-3 - 2} = \frac{-5}{-5} = 1$$

Using the point-slope equation:

We choose to use the point $(2, 2)$ and substitute 2 for x_1, 2 for y_1, and 1 for m.
$$y - y_1 = m(x - x_1)$$
$$y - 2 = 1(x - 2)$$
$$y - 2 = x - 2$$
$$y = x$$

Using the slope-intercept equation:

We choose $(2, 2)$ and substitute 2 for x, 2 for y, and 1 for m in $y = mx + b$. Then we solve for b.
$$y = mx + b$$
$$2 = 1 \cdot 2 + b$$
$$2 = 2 + b$$
$$0 = b$$

Finally, we use the equation $y = mx + b$ and substitute 1 for m and 0 for b.
$$y = 1x + 0, \text{ or } y = x$$

25. First find the slope of the line:
$$m = \frac{0 - 7}{-4 - 0} = \frac{-7}{-4} = \frac{7}{4}$$

Using the point-slope equation:

We choose $(0, 7)$ and substitute 0 for x_1, 7 for y_1, and $\frac{7}{4}$ for m.
$$y - y_1 = m(x - x_1)$$
$$y - 7 = \frac{7}{4}(x - 0)$$
$$y - 7 = \frac{7}{4}x$$
$$y = \frac{7}{4}x + 7$$

Using the slope-intercept equation:

We choose $(0, 7)$ and substitute 0 for x, 7 for y, and $\frac{7}{4}$ for m in $y = mx + b$. Then we solve for b.

$$y = mx + b$$
$$7 = \frac{7}{4} \cdot 0 + b$$
$$7 = b$$

Finally, we use the equation $y = mx + b$ and substitute $\frac{7}{4}$ for m and 7 for b.
$$y = \frac{7}{4}x + 7$$

27. First find the slope of the line:
$$m = \frac{-6-(-3)}{-4-(-2)} = \frac{-6+3}{-4+2} = \frac{-3}{-2} = \frac{3}{2}$$

Using the point-slope equation:

We choose $(-2, -3)$ and substitute -2 for x_1, -3 for y_1, and $\frac{3}{2}$ for m.
$$y - y_1 = m(x - x_1)$$
$$y - (-3) = \frac{3}{2}(x - (-2))$$
$$y + 3 = \frac{3}{2}(x + 2)$$
$$y + 3 = \frac{3}{2}x + 3$$
$$y = \frac{3}{2}x$$

Using the slope-intercept equation:

We choose $(-2, -3)$ and substitute -2 for x, -3 for y, and $\frac{3}{2}$ for m in $y = mx + b$. Then we solve for b.
$$y = mx + b$$
$$-3 = \frac{3}{2}(-2) + b$$
$$-3 = -3 + b$$
$$0 = b$$

Finally, we use the equation $y = mx + b$ and substitute $\frac{3}{2}$ for m and 0 for b.
$$y = \frac{3}{2}x + 0, \text{ or } y = \frac{3}{2}x$$

29. First find the slope of the line:
$$m = \frac{1-0}{6-0} = \frac{1}{6}$$

Using the point-slope equation:

We choose $(0, 0)$ and substitute 0 for x_1, 0 for y_1, and $\frac{1}{6}$ for m.
$$y - y_1 = m(x - x_1)$$
$$y - 0 = \frac{1}{6}(x - 0)$$
$$y = \frac{1}{6}x$$

Using the slope-intercept equation:

We choose $(0, 0)$ and substitute 0 for x, 0 for y, and $\frac{1}{6}$ for m in $y = mx + b$. Then we solve for b.

$$y = mx + b$$
$$0 = \frac{1}{6} \cdot 0 + b$$
$$0 = b$$

Finally, we use the equation $y = mx + b$ and substitute $\frac{1}{6}$ for m and 0 for b.
$$y = \frac{1}{6}x + 0, \text{ or } y = \frac{1}{6}x$$

31. First find the slope of the line:
$$m = \frac{-\frac{1}{2} - 6}{\frac{1}{4} - \frac{3}{4}} = \frac{-\frac{13}{2}}{-\frac{1}{2}} = 13$$

Using the point-slope equation:

We choose $\left(\frac{3}{4}, 6\right)$ and substitute $\frac{3}{4}$ for x_1, 6 for y_1, and 13 for m.
$$y - y_1 = m(x - x_1)$$
$$y - 6 = 13\left(x - \frac{3}{4}\right)$$
$$y - 6 = 13x - \frac{39}{4}$$
$$y = 13x - \frac{15}{4}$$

Using the slope-intercept equation:

We choose $\left(\frac{3}{4}, 6\right)$ and substitute $\frac{3}{4}$ for x, 6 for y, and 13 for m in $y = mx + b$. Then we solve for b.
$$y = mx + b$$
$$6 = 13 \cdot \frac{3}{4} + b$$
$$6 = \frac{39}{4} + b$$
$$-\frac{15}{4} = b$$

Finally, we use the equation $y = mx + b$ and substitute 13 for m and $-\frac{15}{4}$ for b.
$$y = 13x - \frac{15}{4}$$

33. First solve the equation for y and determine the slope of the given line.
$$x + 2y = 6 \qquad \text{Given line}$$
$$2y = -x + 6$$
$$y = -\frac{1}{2}x + 3$$

The slope of the given line is $-\frac{1}{2}$. The line through $(3, 7)$ must have slope $-\frac{1}{2}$.

Using the point-slope equation:

Substitute 3 for x_1, 7 for y_1, and $-\frac{1}{2}$ for m.

$$y - y_1 = m(x - x_1)$$
$$y - 7 = -\frac{1}{2}(x - 3)$$
$$y - 7 = -\frac{1}{2}x + \frac{3}{2}$$
$$y = -\frac{1}{2}x + \frac{17}{2}$$

Using the slope-intercept equation:

Substitute 3 for x, 7 for y, and $-\frac{1}{2}$ for m and solve for b.
$$y = mx + b$$
$$7 = -\frac{1}{2} \cdot 3 + b$$
$$7 = -\frac{3}{2} + b$$
$$\frac{17}{2} = b$$

Then we use the equation $y = mx + b$ and substitute $-\frac{1}{2}$ for m and $\frac{17}{2}$ for b.
$$y = -\frac{1}{2}x + \frac{17}{2}$$

35. First solve the equation for y and determine the slope of the given line.
$$5x - 7y = 8 \quad \text{Given line}$$
$$5x - 8 = 7y$$
$$\frac{5}{7}x - \frac{8}{7} = y$$
$$y = \frac{5}{7}x - \frac{8}{7}$$

The slope of the given line is $\frac{5}{7}$. The line through $(2, -1)$ must have slope $\frac{5}{7}$.

Using the point-slope equation:

Substitute 2 for x_1, -1 for y_1, and $\frac{5}{7}$ for m.
$$y - y_1 = m(x - x_1)$$
$$y - (-1) = \frac{5}{7}(x - 2)$$
$$y + 1 = \frac{5}{7}x - \frac{10}{7}$$
$$y = \frac{5}{7}x - \frac{17}{7}$$

Using the slope-intercept equation:

Substitute 2 for x, -1 for y, and $\frac{5}{7}$ for m and solve for b.
$$y = mx + b$$
$$-1 = \frac{5}{7} \cdot 2 + b$$
$$-1 = \frac{10}{7} + b$$
$$-\frac{17}{7} = b$$

Then we use the equation $y = mx + b$ and substitute $\frac{5}{7}$ for m and $-\frac{17}{7}$ for b.
$$y = \frac{5}{7}x - \frac{17}{7}$$

37. First solve the equation for y and determine the slope of the given line.
$$3x - 9y = 2 \quad \text{Given line}$$
$$3x - 2 = 9y$$
$$\frac{1}{3}x - \frac{2}{9} = y$$

The slope of the given line is $\frac{1}{3}$. The line through $(-6, 2)$ must have slope $\frac{1}{3}$.

Using the point-slope equation:

Substitute -6 for x_1, 2 for y_1, and $\frac{1}{3}$ for m.
$$y - y_1 = m(x - x_1)$$
$$y - 2 = \frac{1}{3}(x - (-6))$$
$$y - 2 = \frac{1}{3}(x + 6)$$
$$y - 2 = \frac{1}{3}x + 2$$
$$y = \frac{1}{3}x + 4$$

Using the slope-intercept equation:

Substitute -6 for x, 2 for y, and $\frac{1}{3}$ for m and solve for b.
$$y = mx + b$$
$$2 = \frac{1}{3}(-6) + b$$
$$2 = -2 + b$$
$$4 = b$$

Then we use the equation $y = mx + b$ and substitute $\frac{1}{3}$ for m and 4 for b.
$$y = \frac{1}{3}x + 4$$

39. First solve the equation for y and determine the slope of the given line.
$$2x + y = -3 \quad \text{Given line}$$
$$y = -2x - 3$$

The slope of the given line is -2. The slope of the perpendicular line is the opposite of the reciprocal of -2. Thus, the line through $(2, 5)$ must have slope $\frac{1}{2}$.

Using the point-slope equation:

Substitute 2 for x_1, 5 for y_1, and $\frac{1}{2}$ for m.

Exercise Set 7.5

$$y - y_1 = m(x - x_1)$$
$$y - 5 = \frac{1}{2}(x - 2)$$
$$y - 5 = \frac{1}{2}x - 1$$
$$y = \frac{1}{2}x + 4$$

Using the slope-intercept equation:

Substitute 2 for x, 5 for y, and $\frac{1}{2}$ for m and solve for b.
$$y = mx + b$$
$$5 = \frac{1}{2} \cdot 2 + b$$
$$5 = 1 + b$$
$$4 = b$$

Then we use the equation $y = mx + b$ and substitute $\frac{1}{2}$ for m and 4 for b.
$$y = \frac{1}{2}x + 4$$

41. First solve the equation for y and determine the slope of the given line.
$$3x + 4y = 5 \qquad \text{Given line}$$
$$4y = -3x + 5$$
$$y = -\frac{3}{4}x + \frac{5}{4}$$

The slope of the given line is $-\frac{3}{4}$. The slope of the perpendicular line is the opposite of the reciprocal of $-\frac{3}{4}$. Thus, the line through $(3, -2)$ must have slope $\frac{4}{3}$.

Using the point-slope equation:

Substitute 3 for x_1, -2 for y_1, and $\frac{4}{3}$ for m.
$$y - y_1 = m(x - x_1)$$
$$y - (-2) = \frac{4}{3}(x - 3)$$
$$y + 2 = \frac{4}{3}x - 4$$
$$y = \frac{4}{3}x - 6$$

Using the slope-intercept equation:

Substitute 3 for x, -2 for y, and $\frac{4}{3}$ for m.
$$y = mx + b$$
$$-2 = \frac{4}{3} \cdot 3 + b$$
$$-2 = 4 + b$$
$$-6 = b$$

Then we use the equation $y = mx + b$ and substitute $\frac{4}{3}$ for m and -6 for b.
$$y = \frac{4}{3}x - 6$$

43. First solve the equation for y and determine the slope of the given line.
$$2x + 5y = 7 \qquad \text{Given line}$$
$$5y = -2x + 7$$
$$y = -\frac{2}{5}x + \frac{7}{5}$$

The slope of the given line is $-\frac{2}{5}$. The slope of the perpendicular line is the opposite of the reciprocal of $-\frac{2}{5}$. Thus, the line through $(0, 9)$ must have slope $\frac{5}{2}$.

Using the point-slope equation:

Substitute 0 for x_1, 9 for y_1, and $\frac{5}{2}$ for m.
$$y - y_1 = m(x - x_1)$$
$$y - 9 = \frac{5}{2}(x - 0)$$
$$y - 9 = \frac{5}{2}x$$
$$y = \frac{5}{2}x + 9$$

Using the slope-intercept equation:

Substitute 0 for x, 9 for y, and $\frac{5}{2}$ for m.
$$y = mx + b$$
$$9 = \frac{5}{2} \cdot 0 + b$$
$$9 = b$$

Then we use the equation $y = mx + b$ and substitute $\frac{5}{2}$ for m and 9 for b.
$$y = \frac{5}{2}x + 9$$

45. a) The problem describes a situation in which an hourly fee is charged after an initial assessment of $85. After 1 hour, the total cost is $85 + $40 · 1. After 2 hours, the total cost is $85 + $40 · 2. Then after t hours, the total cost is $C(t) = 85 + 40t$, or $C(t) = 40t + 85$, where $t > 0$.

b) For $C(t) = 40t + 85$, the y-intercept is $(0, 85)$ and the slope, or rate of change, is $40 per hour. We plot $(0, 85)$ and from there we count up $40 and to the right 1 hour. This takes us to $(1, 125)$. Then we draw a line through the points, calculating a third value as a check:
$$C(5) = 40 \cdot 5 + 85 = 285$$

c) To find the cost for $6\frac{1}{2}$ hours of moving service, we determine $C(6.5)$:

$C(6.5) = 40(6.5) + 85 = 345$

Thus, it would cost \$345 for $6\frac{1}{2}$ hours of moving service.

47. a) The problem describes a situation in which the value of the fax machine decreases at the rate of \$25 per month from an initial value of \$750. After 1 month, the value is $\$750 - \$25 \cdot 1$. After 2 months, the value is $\$750 - \$25 \cdot 2$. Then after t months, the value is $V(t) = 750 - 25t$, where $t \geq 0$.

b) For $V(t) = 750 - 25t$, or $V(t) = -25t + 750$, the y-intercept is $(0, 750)$ and the slope, or rate of change, is $-\$25$ per month. We think of the slope as $\dfrac{-25}{1}$. Plot $(0, 750)$ and from there count down \$25 and to the right 1 month. This takes us to $(1, 725)$. We draw the line through the points, calculating a third value as a check:

$V(6) = 750 - 25 \cdot 6 = 600$

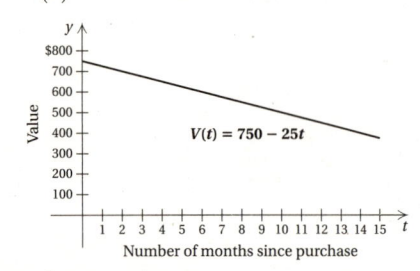

c) To find the value of the machine after 13 months, we determine $V(13)$:

$V(13) = 750 - 25 \cdot 13 = 425$

Thus, after 13 months the value of the machine is \$425.

49. a) In 1991, $x = 1991 - 1990 = 1$, so one data point is $(1, 271,000)$. In 2001, $x = 2001 - 1990 = 11$, so the other data point is $(11, 1,430,000)$.

First we find the slope of the line:

$m = \dfrac{1,430,000 - 271,000}{11 - 1} = \dfrac{1,159,000}{10} = 115,900.$

Using the point-slope equation:

We choose $(1, 271,000)$ and substitute 1 for x_1, 271,000 for y_1, and 115,900 for m.

$y - y_1 = m(x - x_1)$

$y - 271,000 = 115,900(x - 1)$

$y - 271,000 = 115,900x - 115,900$

$\qquad y = 115,900x + 155,100$, or

$\qquad S = 115,900x + 155,100$

Using the slope-intercept equation:

We choose $(1, 271,000)$ and substitute 1 for x, 271,000 for y, and 115,900 for m in $y = mx + b$.

$y = mx + b$

$271,000 = 115,900 \cdot 1 + b$

$271,000 = 115,900 + b$

$155,100 = b$

Finally, we use the equation $y = mx + b$ and substitute 115,900 for m and 155,100 for b.

$y = 115,900x + 155,100$, or

$S = 115,900x + 155,100$

b) In 2005, $x = 2005 - 1990 = 15$. Substitute 15 for x and compute S.

$S = 115,900(15) + 155,100$

$S = 1,738,500 + 155,100$

$S = 1,893,600$

We estimate the average salary in 2005 to be \$1,893,600.

In 2010, $x = 2010 - 1990 = 20$. Substitute 20 for x and compute S.

$S = 115,900(20) + 155,100$

$S = 2,318,000 + 155,100$

$S = 2,473,100$

We estimate the average salary in 2010 to be \$2,473,100.

51. a) We form pairs of the type (t, R) where t is the number of years since 1930 and R is the record. We have two pairs, $(0, 46.8)$ and $(40, 43.8)$. These are two points on the graph of the linear function we are seeking. First we find the slope:

$m = \dfrac{43.8 - 46.8}{40 - 0} = \dfrac{-3}{40} = -0.075$

Using the slope and the y-intercept, $(0, 46.8)$ we write the equation of the line.

$R = -0.075t + 46.8$

$R(t) = -0.075t + 46.8 \quad$ Using function notation

b) 2003 is 73 years since 1930, so to predict the record in 2003, we find $R(73)$:

$R(73) = -0.075(73) + 46.8$

$\qquad = 41.325$

The estimated record is 41.325 seconds in 2003.

2006 is 76 years since 1930, so to predict the record in 2006, we find $R(76)$:

$R(76) = -0.075(76) + 46.8$

$\qquad = 41.1$

The estimated record is 41.1 seconds in 2006.

Exercise Set 7.5

   c) Substitute 40 for $R(t)$ and solve for t:
$$40 = -0.075t + 46.8$$
$$-6.8 = -0.075t$$
$$91 \approx t$$

   The record will be 40 seconds about 91 years after 1930, or in 2021.

53. a) We form pairs of the type (t, E) where t is the number of years since 1990 and E is the life expectancy. We have two pairs, $(0, 71.8)$ and $(7, 73.6)$. These are two points on the graph of the linear function we are seeking. First we find the slope.
$$m = \frac{73.6 - 71.8}{7 - 0} = \frac{1.8}{7} = \frac{18}{70} = \frac{9}{35}$$

Using the slope and the y-intercept $(0, 71.8)$, we write the equation of the line.
$$E = \frac{9}{35}t + 71.8$$
$$E(t) = \frac{9}{35}t + 71.8 \quad \text{Using function notation}$$

  b) 2007 is 17 years since 1990, so we find $E(17)$:
$$E(17) = \frac{9}{35}(17) + 71.8 \approx 72.6$$

We predict that the life expectancy of males will be about 76.2 years in 2007.

55. Discussion and Writing Exercise

57. $\dfrac{w-t}{t-w} = \dfrac{w-t}{-(-t+w)}$
$$= \frac{w-t}{-1(w-t)}$$
$$= -1 \cdot \frac{w-t}{w-t}$$
$$= -1 \cdot 1$$
$$= -1$$

59. $\dfrac{3x^2 + 15x - 72}{6x^2 + 18x - 240} = \dfrac{3(x^2 + 5x - 24)}{6(x^2 + 3x - 40)}$
$$= \frac{3(x+8)(x-3)}{2 \cdot 3(x+8)(x-5)}$$
$$= \frac{\cancel{3}(\cancel{x+8})(x-3)}{2 \cdot \cancel{3}(\cancel{x+8})(x-5)}$$
$$= \frac{x-3}{2(x-5)}$$

Chapter 8

Systems of Equations

Exercise Set 8.1

1. Graph both lines on the same set of axes.

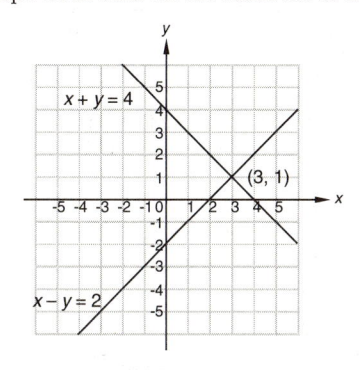

The solution (point of intersection) seems to be the point $(3, 1)$.

Check:

$x + y = 4$	$x - y = 2$
$3 + 1 \;?\; 4$	$3 - 1 \;?\; 2$
$4 \;\mid\;$ TRUE	$2 \;\mid\;$ TRUE

The solution is $(3, 1)$.

Since the system of equations has a solution it is consistent. Since there is exactly one solution, the equations are independent.

3. Graph both lines on the same set of axes.

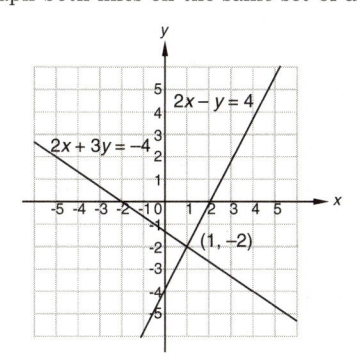

The solution (point of intersection) seems to be the point $(1, -2)$.

Check:

$2x - y = 4$	$2x + 3y = -4$
$2 \cdot 1 - (-2) \;?\; 4$	$2 \cdot 1 + 3(-2) \;?\; -4$
$2 + 2$	$2 - 6$
$4 \;\mid\;$ TRUE	$-4 \;\mid\;$ TRUE

The solution is $(1, -2)$.

Since the system of equations has a solution, it is consistent. Since there is exactly one solution, the equations are independent.

5. Graph both lines on the same set of axes.

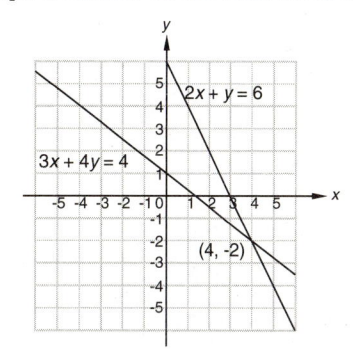

The solution (point of intersection) seems to be the point $(4, -2)$.

Check:

$2x + y = 6$	$3x + 4y = 4$
$2 \cdot 4 + (-2) \;?\; 6$	$3 \cdot 4 + 4(-2) \;?\; 4$
$8 - 2$	$12 - 8$
$6 \;\mid\;$ TRUE	$4 \;\mid\;$ TRUE

The solution is $(4, -2)$.

Since the system of equations has a solution, it is consistent. Since there is exactly one solution, the equations are independent.

7. Graph both lines on the same set of axes.

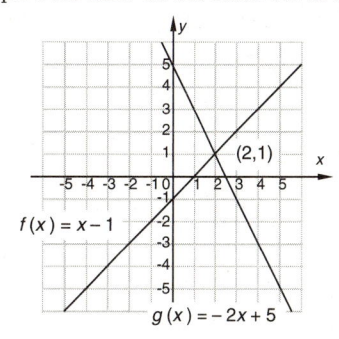

The solution seems to be the point $(2, 1)$.

Check:

$f(x) = x - 1$	$g(x) = -2x + 5$
$1 \;?\; 2 - 1$	$1 \;?\; -2 \cdot 2 + 5$
$1 \;\mid\;$ TRUE	$-4 + 5$
	$1 \;\mid\;$ TRUE

The solution is $(2, 1)$.

Since the system of equations has a solution, it is consistent. Since there is exactly one solution, the equations are independent.

9. Graph both lines on the same set of axes.

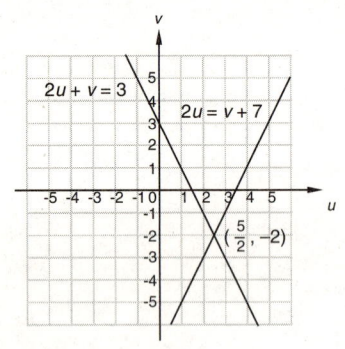

The solution seems to be $\left(\frac{5}{2}, -2\right)$.

Check:

$$\begin{array}{c|c} 2u + v = 3 & 2u = v + 7 \\ \hline 2 \cdot \frac{5}{2} + (-2) \ ? \ 3 & 2 \cdot \frac{5}{2} \ ? \ -2 + 7 \\ 5 - 2 & 5 \ \vert \ 5 \quad \text{TRUE} \\ 3 \ \vert \ \text{TRUE} & \end{array}$$

The solution is $\left(\frac{5}{2}, -2\right)$.

Since the system of equations has a solution, it is consistent. Since there is exactly one solution, the equations are independent.

11. Graph both lines on the same set of axes.

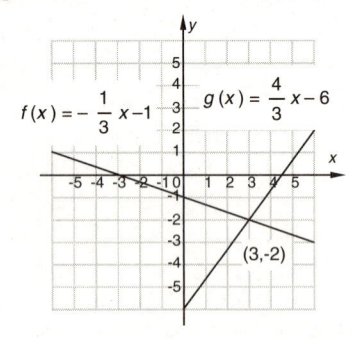

The ordered pair $(3, -2)$ checks in both equations. It is the solution.

Since the system of equations has a solution, it is consistent. Since there is exactly one solution, the equations are independent.

13. Graph both lines on the same set of axes.

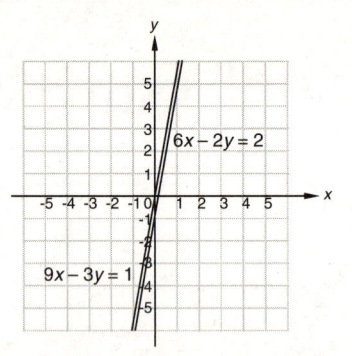

The lines are parallel. There is no solution.

Since the system of equations has no solution, it is inconsistent. Since there is no solution, the equations are independent.

15. Graph both lines on the same set of axes.

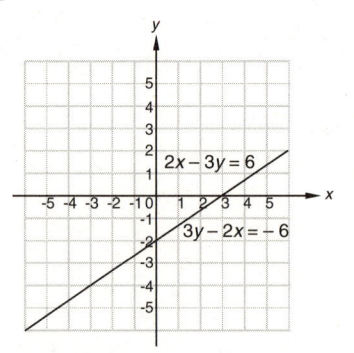

The graphs are the same. Any solution of one of the equations is also a solution of the other. Each equation has an infinite number of solutions. Thus the system of equations has an infinite number of solutions. Since the system of equations has a solution, it is consistent. Since there are infinitely many solutions, the equations are dependent.

17. Graph both lines on the same set of axes.

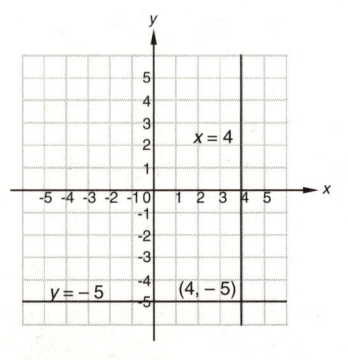

The ordered pair $(4, -5)$ checks in both equations. It is the solution.

Since the system of equations has a solution, it is consistent. Since there is exactly one solution, the equations are independent.

Exercise Set 8.2

19. Graph both lines on the same set of axes.

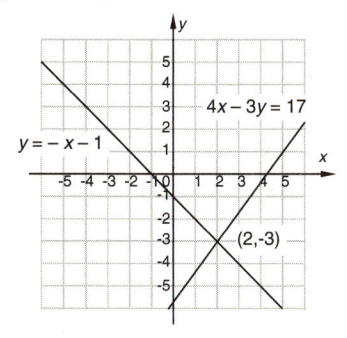

The ordered pair $(2, -3)$ checks in both equations. It is the solution.

Since the system of equations has a solution, it is consistent. Since there is exactly one solution, the equations are independent.

21. Since the system of equations has a solution, it is consistent. Since there is exactly one solution, the equations are independent. The graph of the system consists of a vertical line and a horizontal line, each passing through $(3,3)$. Thus, system **F** corresponds to this graph.

23. Since the system of equations has a solution, it is consistent. Since there are infinitely many solutions, the equations are dependent. The equations in system B are equivalent, so their graphs are the same. In addition the graph corresponds to the one shown, so system **B** corresponds to this graph.

25. Since the system of equations has no solution, it is inconsistent. Since there is no solution, the equations are independent. The equations in system **D** have the same slope and different y-intercepts and have the graphs shown, so this system corresponds to the given graph.

27. Discussion and Writing Exercise

29. First solve the equation for y and determine the slope of the given line.

$$3x = 5y - 4 \quad \text{Given line}$$
$$3x + 4 = 5y$$
$$\frac{3}{5}x + \frac{4}{5} = y$$

The slope of the given line is $\frac{3}{5}$. The line through $(-4, 2)$ must have slope $\frac{3}{5}$. We find an equation of this new line using the equation $y = mx + b$. We substitute -4 for x, 2 for y, and $\frac{3}{5}$ for m and solve for b.

$$y = mx + b$$
$$2 = \frac{3}{5}(-4) + b$$
$$2 = -\frac{12}{5} + b$$
$$\frac{22}{5} = b$$

Then we use the equation $y = mx + b$ and substitute $\frac{3}{5}$ for m and $\frac{22}{5}$ for b.

$$y = \frac{3}{5}x + \frac{22}{5}$$

31. Graph these equations, solving each equation for y first, if necessary. We get $y = \dfrac{13.78 - 2.18x}{7.81}$ and $y = \dfrac{5.79x - 8.94}{3.45}$. Using the INTERSECT feature, we find that the point of intersection is $(2.23, 1.14)$.

33. Graph both lines on the same set of axes.

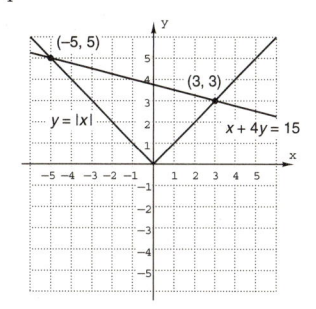

The solutions appear to be $(-5, 5)$ and $(3, 3)$.

Check:

For $(-5, 5)$:

$y = \|x\|$	$x + 4y = 15$
$5 \; ? \; \|-5\|$	$-5 + 4 \cdot 5 \; ? \; 15$
$\quad 5 \quad$ TRUE	$-5 + 20$
	$\quad 15 \quad$ TRUE

For $(3, 3)$:

$y = \|x\|$	$x + 4y = 15$
$3 \; ? \; \|3\|$	$3 + 4 \cdot 3 \; ? \; 15$
$\quad 3 \quad$ TRUE	$3 + 12$
	$\quad 15 \quad$ TRUE

Both pairs check. The solutions are $(-5, 5)$ and $(3, 3)$.

Exercise Set 8.2

1. $y = 5 - 4x, \quad (1)$
$2x - 3y = 13 \quad (2)$

We substitute $5 - 4x$ for y in the second equation and solve for x.

$$2x - 3y = 13 \quad (2)$$
$$2x - 3(5 - 4x) = 13 \quad \text{Substituting}$$
$$2x - 15 + 12x = 13$$
$$14x - 15 = 13$$
$$14x = 28$$
$$x = 2$$

Next we substitute 2 for x in either equation of the original system and solve for y.

$y = 5 - 4x$ (1)

$y = 5 - 4 \cdot 2$ Substituting

$y = 5 - 8$

$y = -3$

We check the ordered pair $(2, -3)$.

$$\begin{array}{c|c} y = 5 - 4x \\ \hline -3 \ ? \ 5 - 4 \cdot 2 \\ \quad | \ 5 - 8 \\ -3 \ | \ -3 \quad \text{TRUE} \end{array}$$

$$\begin{array}{c|c} 2x - 3y = 13 \\ \hline 2 \cdot 2 - 3(-3) \ ? \ 13 \\ 4 + 9 \ | \\ 13 \ | \ 13 \quad \text{TRUE} \end{array}$$

Since $(2, -3)$ checks, it is the solution.

3. $2y + x = 9$, (1)

$x = 3y - 3$ (2)

We substitute $3y - 3$ for x in the first equation and solve for y.

$2y + x = 9$ (1)

$2y + (3y - 3) = 9$ Substituting

$5y - 3 = 9$

$5y = 12$

$y = \dfrac{12}{5}$

Next we substitute $\dfrac{12}{5}$ for y in either equation of the original system and solve for x.

$x = 3y - 3$ (2)

$x = 3 \cdot \dfrac{12}{5} - 3 = \dfrac{36}{5} - \dfrac{15}{5} = \dfrac{21}{5}$

We check the ordered pair $\left(\dfrac{21}{5}, \dfrac{12}{5}\right)$.

$$\begin{array}{c|c} 2y + x = 9 \\ \hline 2 \cdot \dfrac{12}{5} + \dfrac{21}{15} \ ? \ 9 \\ \dfrac{24}{5} + \dfrac{21}{5} \ | \\ \dfrac{45}{5} \ | \\ 9 \ | \ 9 \quad \text{TRUE} \end{array}$$

$$\begin{array}{c|c} x = 3y - 3 \\ \hline \dfrac{21}{5} \ ? \ 3 \cdot \dfrac{12}{5} - 3 \\ \ | \ \dfrac{36}{5} - \dfrac{15}{5} \\ \dfrac{21}{5} \ | \ \dfrac{21}{5} \quad \text{TRUE} \end{array}$$

Since $\left(\dfrac{21}{5}, \dfrac{12}{5}\right)$ checks, it is the solution.

5. $3s - 4t = 14$, (1)

$5s + t = 8$ (2)

We solve the second equation for t.

$5s + t = 8$ (2)

$t = 8 - 5s$ (3)

We substitute $8 - 5s$ for t in the first equation and solve for s.

$3s - 4t = 14$ (1)

$3s - 4(8 - 5s) = 14$ Substituting

$3s - 32 + 20s = 14$

$23s - 32 = 14$

$23s = 46$

$s = 2$

Next we substitute 2 for s in Equation (1), (2), or (3). It is easiest to use Equation (3) since it is already solved for t.

$t = 8 - 5 \cdot 2 = 8 - 10 = -2$

We check the ordered pair $(2, -2)$.

$$\begin{array}{c|c} 3s - 4t = 14 \\ \hline 3 \cdot 2 - 4(-2) \ ? \ 14 \\ 6 + 8 \ | \\ 14 \ | \ 14 \quad \text{TRUE} \end{array}$$

$$\begin{array}{c|c} 5s + t = 8 \\ \hline 5 \cdot 2 + (-2) \ ? \ 8 \\ 10 - 2 \ | \\ 8 \ | \ 8 \quad \text{TRUE} \end{array}$$

Since $(2, -2)$ checks, it is the solution.

7. $9x - 2y = -6$, (1)

$7x + 8 = y$ (2)

We substitute $7x + 8$ for y in the first equation and solve for x.

$9x - 2y = -6$ (1)

$9x - 2(7x + 8) = -6$ Substituting

$9x - 14x - 16 = -6$

$-5x - 16 = -6$

$-5x = 10$

$x = -2$

Exercise Set 8.2

Next we substitute -2 for x in either equation of the original system and solve for y.
$$7x + 8 = y \quad (2)$$
$$7(-2) + 8 = y$$
$$-14 + 8 = y$$
$$-6 = y$$

We check the ordered pair $(-2, -6)$.

$$\begin{array}{c|c} 9x - 2y = -6 \\ \hline 9(-2) - 2(-6) \;?\; -6 \\ -18 + 12 \\ -6 & \text{TRUE} \end{array}$$

$$\begin{array}{c|c} 7x + 8 = y \\ \hline 7(-2) + 8 \;?\; -6 \\ -14 + 8 \\ -6 & \text{TRUE} \end{array}$$

Since $(-2, -6)$ checks, it is the solution.

9. $-5s + t = 11, \quad (1)$
$\; 4s + 12t = 4 \quad (2)$

We solve the first equation for t.
$$-5s + t = 11 \quad (1)$$
$$t = 5s + 11 \quad (3)$$

We substitute $5s + 11$ for t in the second equation and solve for s.
$$4s + 12t = 4 \quad (2)$$
$$4s + 12(5s + 11) = 4$$
$$4s + 60s + 132 = 4$$
$$64s + 132 = 4$$
$$64s = -128$$
$$s = -2$$

Next we substitute -2 for s in Equation (3).
$$t = 5s + 11 = 5(-2) + 11 = -10 + 11 = 1$$

We check the ordered pair $(-2, 1)$.

$$\begin{array}{c|c} -5s + t = 11 \\ \hline -5(-2) + 1 \;?\; 11 \\ 10 + 1 \\ 11 & 11 \quad \text{TRUE} \end{array}$$

$$\begin{array}{c|c} 4s + 12t = 4 \\ \hline 4(-2) + 12 \cdot 1 \;?\; 4 \\ -8 + 12 \\ 4 & 4 \quad \text{TRUE} \end{array}$$

Since $(-2, 1)$ checks, it is the solution.

11. $2x + 2y = 2, \quad (1)$
$\; 3x - y = 1 \quad (2)$

We solve the second equation for y.

$$3x - y = 1 \quad (2)$$
$$-y = -3x + 1$$
$$y = 3x - 1 \quad (3)$$

We substitute $3x - 1$ for y in the first equation and solve for x.
$$2x + 2y = 2 \quad (1)$$
$$2x + 2(3x - 1) = 2$$
$$2x + 6x - 2 = 2$$
$$8x - 2 = 2$$
$$8x = 4$$
$$x = \frac{1}{2}$$

Next we substitute $\frac{1}{2}$ for x in Equation (3).
$$y = 3x - 1 = 3 \cdot \frac{1}{2} - 1 = \frac{3}{2} - 1 = \frac{1}{2}$$

The ordered pair $\left(\frac{1}{2}, \frac{1}{2}\right)$ checks in both equations. It is the solution.

13. $3a - b = 7, \quad (1)$
$\; 2a + 2b = 5 \quad (2)$

We solve the first equation for b.
$$3a - b = 7 \quad (1)$$
$$-b = -3a + 7$$
$$b = 3a - 7 \quad (3)$$

We substitute $3a - 7$ for b in the second equation and solve for a.
$$2a + 2b = 5 \quad (2)$$
$$2a + 2(3a - 7) = 5$$
$$2a + 6a - 14 = 5$$
$$8a - 14 = 5$$
$$8a = 19$$
$$a = \frac{19}{8}$$

We substitute $\frac{19}{8}$ for a in Equation (3).
$$b = 3a - 7 = 3 \cdot \frac{19}{8} - 7 = \frac{57}{8} - \frac{56}{8} = \frac{1}{8}$$

The ordered pair $\left(\frac{19}{8}, \frac{1}{8}\right)$ checks in both equations. It is the solution.

15. $2x - 3 = y \quad (1)$
$\; y - 2x = 1, \quad (2)$

We substitute $2x - 3$ for y in the second equation and solve for x.
$$y - 2x = 1 \quad (2)$$
$$2x - 3 - 2x = 1 \quad \text{Substituting}$$
$$-3 = 1 \quad \text{Collecting like terms}$$

We have a false equation. Therefore, there is no solution.

17. Familiarize. Refer to the drawing in the text. Let $l=$ the length of the court and $w=$ the width. Recall that the perimeter P of a rectangle with length l and width w is given by $P = 2l + 2w$.

Translate.

The perimeter is 120 ft.
$$2l + 2w = 120$$

The length is twice the width.
$$l = 2 \cdot w$$

We have system of equations.
$$2l + 2w = 120, \quad (1)$$
$$l = 2w \quad (2)$$

Solve. We substitute $2w$ for l in Equation (1) and solve for w.
$$2l + 2w = 120$$
$$2 \cdot 2w + 2w = 120$$
$$4w + 2w = 120$$
$$6w = 120$$
$$w = 20$$

Now substitute 20 for w in Equation (2) and find l.
$$l = 2w = 2 \cdot 20 = 40$$

Check. If the length is 40 ft and the width is 20 ft, then the perimeter is $2 \cdot 40 + 2 \cdot 20$, or 120 ft. Also, the length is twice the width. The answer checks.

State. The length of the court is 40 ft, and the width is 20 ft.

19. Familiarize. Using the drawing in the text, we let x and y represent the measures of the angles.

Translate.

The sum of the measures is 180°.
$$x + y = 180$$

One angle is 3 times the other less 12°.
$$x = 3 \cdot y - 12$$

We have a system of equations.
$$x + y = 180, \quad (1)$$
$$x = 3y - 12 \quad (2)$$

Solve. Substitute $3y - 12$ for x in Equation (1) and solve for y.
$$x + y = 180$$
$$(3y - 12) + y = 180$$
$$4y - 12 = 180$$
$$4y = 192$$
$$y = 48$$

Now substitute 48 for y in Equation (2) and find x.
$$x = 3y - 12 = 3 \cdot 48 - 12 = 132$$

Check. The sum of the measures is $48° + 132°$, or 180°. Also, 132° is 12° less than three times 48°. The answer checks.

State. The measures of the angles are 48° and 132°.

21. Familiarize. Let $x =$ number of games won and $y =$ number of games tied. The total points earned in x wins is $2x$; the total points earned in y ties is $1 \cdot y$, or y.

Translate.

Points from wins plus points from ties is 60.
$$2x + y = 60$$

Number of wins is 9 more than the number of ties.
$$x = 9 + y$$

We have a system of equations:
$$2x + y = 60,$$
$$x = 9 + y$$

Solve. We solve the system of equations. We use substitution.
$$2(9 + y) + y = 60 \quad \text{Substituting } 9+y \text{ for } x \text{ in (1)}$$
$$18 + 2y + y = 60$$
$$18 + 3y = 60$$
$$3y = 42$$
$$y = 14$$
$$x = 9 + 14 \quad \text{Substituting 14 for } y \text{ in (2)}$$
$$x = 23$$

Check. The number of wins, 23, is 9 more than the number of ties, 14.

Points from wins: $\quad 23 \times 2 = 46$
Points from ties: $\quad\ \ 14 \times 1 = \underline{14}$
$\qquad\qquad\qquad$ Total $\quad\ \ 60$

The numbers check.

State. The team had 23 wins and 14 ties.

23. Discussion and Writing Exercise

25. $y = 1.3x - 7$

The equation is in slope-intercept form, $y = mx + b$. The slope is 1.3.

27. $A = \dfrac{pq}{7}$

$7A = pq \quad$ Multiplying by 7

$\dfrac{7A}{q} = p \quad$ Dividing by q

29. $2 = m + b, \quad (1) \quad$ Substituting $(1, 2)$
$\ \ \ \ 4 = -3m + b \quad (2) \quad$ Substituting $(-3, 4)$

$\ \ \ \ 2 = \ \ \ m + b \quad (1)$
$\underline{-4 = 3m - b} \quad$ Multiplying (2) by -1
$-2 = 4m$

$-\dfrac{1}{2} = m$

Substitute $-\dfrac{1}{2}$ for m in (1).

$2 = -\dfrac{1}{2} + b$

$\dfrac{5}{2} = b$

Thus, $m = -\dfrac{1}{2}$ and $b = \dfrac{5}{2}$.

31. **Familiarize**. Let l = the original length, in inches, and w = the original width, in inches. Then $w - 6$ = the width after 6 in. is cut off.

Translate.

The original perimeter is 156 in.

$2l + 2w = 156$

The length becomes 4 times the new width.

$l = 4 \cdot (w - 6)$

We have a system of equations:

$2l + 2w = 156,$ (1)

$l = 4 \cdot (w - 6)$ (2)

Solve. Substitute $4(w - 6)$ for l in Equation (1) and solve for w.

$2l + 2w = 156$

$2 \cdot 4(w - 6) + 2w = 156$

$8w - 48 + 2w = 156$

$10w - 48 = 156$

$10w = 204$

$w = 20.4$

Now substitute 20.4 for w in Equation (2) and find l.

$l = 4 \cdot (w - 6) = 4(20.4 - 6) = 4(14.4) = 57.6$

Check. The original perimeter is $2(57.6) + 2(20.4)$, or $115.2 + 40.8$, or 156 in. If 6 in. is cut off the width, then the width becomes $20.4 - 6$, or 14.4 in., and the length is 4 times the width, or 57.6 in. The answer checks.

State. The length is 57.6 in., and the width is 20.4 in.

Exercise Set 8.3

1. $x + 3y = 7$ (1)

$\underline{-x + 4y = 7}$ (2)

$0 + 7y = 14$ Adding

$7y = 14$

$y = 2$

Substitute 2 for y in one of the original equations and solve for x.

$x + 3y = 7$ Equation (1)

$x + 3 \cdot 2 = 7$ Substituting

$x + 6 = 7$

$x = 1$

Check:

$x + 3y = 7$	$-x + 4y = 7$
$1 + 3 \cdot 2\ ?\ 7$	$-1 + 4 \cdot 2\ ?\ 7$
$1 + 6$	$-1 + 8$
7 | TRUE	7 | TRUE

Since $(1, 2)$ checks, it is the solution.

3. $9x + 5y = 6$ (1)

$\underline{2x - 5y = -17}$ (2)

$11x + 0 = -11$ Adding

$11x = -11$

$x = -1$

Substitute -1 for x in one of the original equations and solve for y.

$9x + 5y = 6$ Equation (1)

$9(-1) + 5y = 6$ Substituting

$-9 + 5y = 6$

$5y = 15$

$y = 3$

We obtain $(-1, 3)$. This checks, so it is the solution.

5. $5x + 3y = 19,$ (1)

$2x - 5y = 11$ (2)

We multiply twice to make two terms become additive inverses.

From (1): $25x + 15y = 95$ Multiplying by 5

From (2): $\underline{6x - 15y = 33}$ Multiplying by 3

$31x + 0 = 128$ Adding

$31x = 128$

$x = \dfrac{128}{31}$

Substitute $\dfrac{128}{31}$ for x in one of the original equations and solve for y.

$5x + 3y = 19$ Equation (1)

$5 \cdot \dfrac{128}{31} + 3y = 19$ Substituting

$\dfrac{640}{31} + 3y = \dfrac{589}{31}$

$3y = -\dfrac{51}{31}$

$\dfrac{1}{3} \cdot 3y = \dfrac{1}{3} \cdot \left(-\dfrac{51}{31}\right)$

$y = -\dfrac{17}{31}$

We obtain $\left(\dfrac{128}{31}, -\dfrac{17}{31}\right)$. This checks, so it is the solution.

7. $5r - 3s = 24,$ (1)

$3r + 5s = 28$ (2)

We multiply twice to make two terms become additive inverses.

From (1): $25r - 15s = 120$ Multiplying by 5
From (2): $\underline{9r + 15s = 84}$ Multiplying by 3
$34r + 0 = 204$ Adding
$34r = 204$
$r = 6$

Substitute 6 for r in one of the original equations and solve for s.

$3r + 5s = 28$ Equation (2)
$3 \cdot 6 + 5s = 28$ Substituting
$18 + 5s = 28$
$5s = 10$
$s = 2$

We obtain $(6, 2)$. This checks, so it is the solution.

9. $0.3x - 0.2y = 4$,
$0.2x + 0.3y = 1$

We first multiply each equation by 10 to clear decimals.
$3x - 2y = 40$ (1)
$2x + 3y = 10$ (2)

We use the multiplication principle with both equations of the resulting system.

From (1): $9x - 6y = 120$ Multiplying by 3
From (2): $\underline{4x + 6y = 20}$ Multiplying by 2
$13x + 0 = 140$ Adding
$13x = 140$
$x = \dfrac{140}{13}$

Substitute $\dfrac{140}{13}$ for x in one of the equations in which the decimals were cleared and solve for y.

$2x + 3y = 10$ Equation (2)
$2 \cdot \dfrac{140}{13} + 3y = 10$ Substituting
$\dfrac{280}{13} + 3y = \dfrac{130}{13}$
$3y = -\dfrac{150}{13}$
$y = -\dfrac{50}{13}$

We obtain $\left(\dfrac{140}{13}, -\dfrac{50}{13}\right)$. This checks, so it is the solution.

11. $\dfrac{1}{2}x + \dfrac{1}{3}y = 4$,
$\dfrac{1}{4}x + \dfrac{1}{3}y = 3$

We first multiply each equation by the LCM of the denominators to clear fractions.
$3x + 2y = 24$ Multiplying by 6
$3x + 4y = 36$ Multiplying by 12

We multiply by -1 on both sides of the first equation and then add.

$-3x - 2y = -24$ Multiplying by -1
$\underline{3x + 4y = 36}$
$0 + 2y = 12$ Adding
$2y = 12$
$y = 6$

Substitute 6 for y in one of the equations in which the fractions were cleared and solve for x.

$3x + 2y = 24$
$3x + 2 \cdot 6 = 24$ Substituting
$3x + 12 = 24$
$3x = 12$
$x = 4$

We obtain $(4, 6)$. This checks, so it is the solution.

13. $\dfrac{2}{5}x + \dfrac{1}{2}y = 2$,
$\dfrac{1}{2}x - \dfrac{1}{6}y = 3$

We first multiply each equation by the LCM of the denominators to clear fractions.
$4x + 5y = 20$ Multiplying by 10
$3x - y = 18$ Multiplying by 6

We multiply by 5 on both sides of the second equation and then add.

$4x + 5y = 20$
$\underline{15x - 5y = 90}$ Multiplying by 5
$19x + 0 = 110$ Adding
$19x = 110$
$x = \dfrac{110}{19}$

Substitute $\dfrac{110}{19}$ for x in one of the equations in which the fractions were cleared and solve for y.

$3x - y = 18$
$3\left(\dfrac{110}{19}\right) - y = 18$ Substituting
$\dfrac{330}{19} - y = \dfrac{342}{19}$
$-y = \dfrac{12}{19}$
$y = -\dfrac{12}{19}$

We obtain $\left(\dfrac{110}{19}, -\dfrac{12}{19}\right)$. This checks, so it is the solution.

15. $2x + 3y = 1$,
$4x + 6y = 2$

Multiply the first equation by -2 and then add.

$-4x - 6y = -2$
$\underline{4x + 6y = 2}$
$0 = 0$ Adding

Exercise Set 8.3

We have an equation that is true for all numbers x and y. The system is dependent and has an infinite number of solutions.

17. $2x - 4y = 5$,
$2x - 4y = 6$

Multiply the first equation by -1 and then add.
$$-2x + 4y = -5$$
$$\underline{2x - 4y = 6}$$
$$0 = 1$$

We have a false equation. The system has no solution.

19. $5x - 9y = 7$,
$7y - 3x = -5$

We first write the second equation in the form $Ax + By = C$.

$5x - 9y = 7$ (1)
$-3x + 7y = -5$ (2)

We use the multiplication principle with both equations and then add.

$15x - 27y = 21$ Multiplying by 3
$\underline{-15x + 35y = -25}$ Multiplying by 5
$0 + 8y = -4$ Adding
$8y = -4$
$y = -\dfrac{1}{2}$

Substitute $-\dfrac{1}{2}$ for y in one of the original equations and solve for x.

$5x - 9y = 7$ Equation (1)
$5x - 9\left(-\dfrac{1}{2}\right) = 7$ Substituting
$5x + \dfrac{9}{2} = \dfrac{14}{2}$
$5x = \dfrac{5}{2}$
$x = \dfrac{1}{2}$

We obtain $\left(\dfrac{1}{2}, -\dfrac{1}{2}\right)$. This checks, so it is the solution.

21. $3(a - b) = 15$,
$4a = b + 1$

We first write each equation in the form $Ax + By = C$.

$3a - 3b = 15$ (1)
$4a - b = 1$ (2)

We multiply by -3 on both sides of the second equation and then add.

$3a - 3b = 15$
$\underline{-12a + 3b = -3}$ Multiplying by -3
$-9a + 0 = 12$
$-9a = 12$
$a = -\dfrac{12}{9}$
$a = -\dfrac{4}{3}$

Substitute $-\dfrac{4}{3}$ for a in either Equation (1) or Equation (2) and solve for b.

$4a - b = 1$ Equation (2)
$4\left(-\dfrac{4}{3}\right) - b = 1$ Substituting
$-\dfrac{16}{3} - b = \dfrac{3}{3}$
$-b = \dfrac{19}{3}$
$b = -\dfrac{19}{3}$

We obtain $\left(-\dfrac{4}{3}, -\dfrac{19}{3}\right)$. This checks, so it is the solution.

23. $x - \dfrac{1}{10}y = 100$,
$y - \dfrac{1}{10}x = -100$

We first write the second equation in the form $Ax + By = C$.

$x - \dfrac{1}{10}y = 100$
$-\dfrac{1}{10}x + y = -100$

Next we multiply each equation by 10 to clear fractions.
$10x - y = 1000$ (1)
$-x + 10y = -1000$ (2)   Equation (1)

We multiply by 10 on both sides of Equation (1) and then add.

$100x - 10y = 10,000$ Multiplying by 10
$\underline{-x + 10y = -1000}$
$99x + 0 = 9000$
$99x = 9000$
$x = \dfrac{9000}{99}$
$x = \dfrac{1000}{11}$

Substitute $\dfrac{1000}{11}$ for x in one of the equations in which the fractions were cleared and solve for y.

$10x - y = 1000$ Equation (1)

$10\left(\dfrac{1000}{11}\right) - y = 1000$ Substituting

$\dfrac{10,000}{11} - y = \dfrac{11,000}{11}$

$-y = \dfrac{1000}{11}$

$y = -\dfrac{1000}{11}$

We obtain $\left(\dfrac{1000}{11}, -\dfrac{1000}{11}\right)$. This checks, so it is the solution.

25. $0.05x + 0.25y = 22,$

$0.15x + 0.05y = 24$

We first multiply each equation by 100 to clear decimals.

$5x + 25y = 2200$ (1)

$15x + 5y = 2400$ (2)

We multiply by -5 on both sides of the second equation and add.

$5x + 25y = 2200$

$-75x - 25y = -12,000$ Multiplying by -5

$-70x + 0 = -9800$ Adding

$-70x = -9800$

$x = \dfrac{-9800}{-70}$

$x = 140$

Substitute 140 for x in one of the equations in which the decimals were cleared and solve for y.

$5x + 25y = 2200$ Equation (1)

$5 \cdot 140 + 25y = 2200$ Substituting

$700 + 25y = 2200$

$25y = 1500$

$y = 60$

We obtain $(140, 60)$. This checks, so it is the solution.

27. Familiarize. Let l = the length of the field and w = the width, in meters. Recall that the perimeter P of a rectangle with length l and width w is given by $P = 2l + 2w$.

Translate.

The perimeter is 340 m.
$2l + 2w = 340$

The length is 50 m more than the width.
$l = 50 + w$

We have system of equations.

$2l + 2w = 340,$ (1)

$l = 50 + w$ (2)

Solve. First we subtract w on both sides of Equation (2).

$l = 50 + w$

$l - w = 50$

Now we have

$2l + 2w = 340,$ (1)

$l - w = 50.$ (3)

Multiply Equation (3) by 2 and then add.

$2l + 2w = 340$

$2l - 2w = 100$

$4l = 440$

$l = 110$

Now substitute 110 for l in one of the original equations and solve for w.

$l = 50 + w$ (2)

$110 = 50 + w$

$60 = w$

Check. The perimeter is $2 \cdot 110 + 2 \cdot 60$, or $220 + 120$, or 340 m. Also, the length, 110 m, exceeds the width, 60 m, by 50 m. The answer checks.

State. The length is 110 m and the width is 60 m.

29. Familiarize. Using the drawing in the text, we let x and y represent the measures of the angles.

Translate.

The sum of the measures is 90°.
$x + y = 90$

One angle is 6° more than 5 times the other.
$y = 6 + 5 \cdot x$

We have system of equations.

$x + y = 90,$ (1)

$y = 6 + 5x$ (2)

Solve. First we subtract $5x$ on both sides of Equation (2).

$y = 6 + 5x$

$-5x + y = 6$ (3)

Now we have

$x + y = 90,$ (1)

$-5x + y = 6.$ (3)

Multiply Equation (1) by 5 and then add.

$5x + 5y = 450$

$-5x + y = 6$

$6y = 456$

$y = 76$

Substitute 76 for y in one of the original equations and solve for x.

$x + y = 90$ (1)

$x + 76 = 90$

$x = 14$

Check. The sum of the measures is $14° + 76°$, or $90°$. Also, 6° more than 5 times the measure of the 14° angle is

Exercise Set 8.3

$5 \cdot 14° + 6° = 70° + 6° = 76°$, the measure of the other angle. The answer checks.

State. The angles measure 14° and 76°.

31. **Familiarize**. Let $c =$ the number of coach seats and $f =$ the number of first-class seats.

    Translate.

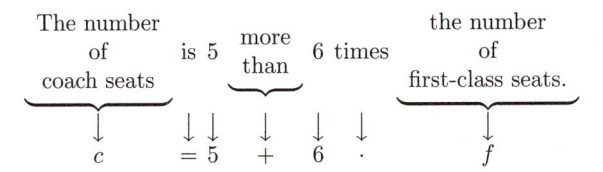

    $\underbrace{\text{The total number of seats}}_{c + f}$ is 152.
    $c + f = 152$

    $\underbrace{\text{The number of coach seats}}_{c}$ is 5 more than 6 times $\underbrace{\text{the number of first-class seats.}}_{f}$
    $c = 5 + 6 \cdot f$

    We have system of equations.
    $c + f = 152$, (1)
    $c = 5 + 6f$ (2)

    Solve. First we subtract $6f$ on both sides of Equation (2).
    $c = 5 + 6f$
    $c - 6f = 5$ (3)

    Now we have
    $c + f = 152$, (1)
    $c - 6f = 5$. (3)

    Multiply Equation (1) by 6 and then add.
    $6c + 6f = 912$
    $\underline{c - 6f = 5}$
    $7c = 917$
    $c = 131$

    Substitute 131 for c in one of the original equations and solve for f.
    $c + f = 152$ (1)
    $131 + f = 152$
    $f = 21$

    Check. The total number of seats is $131 + 21$, or 152. Five more than six times the number of first-class seats is $5 + 6 \cdot 21$, or $5 + 126$, or 131, the number of coach seats. The answer checks.

    State. There are 131 coach-class seats and 21 first-class seats.

33. Discussion and Writing Exercise

35. First solve the equation for y and determine the slope of the given line.
    $2x - 7y = 3$ Given line
    $-7y = -2x + 3$
    $y = \frac{2}{7}x - \frac{3}{7}$

    The slope of the given line is $\frac{2}{7}$. The slope of the perpendicular line is the opposite of the reciprocal of $\frac{2}{7}$. Thus, the line through $(10, 1)$ must have slope $-\frac{7}{2}$. We find an equation of this new line using the equation $y = mx + b$. We substitute 10 for x, 1 for y, and $-\frac{7}{2}$ for m and solve for b.

    $y = mx + b$
    $1 = -\frac{7}{2} \cdot 10 + b$
    $1 = -35 + b$
    $36 = b$

    Then we use the equation $y = mx + b$ and substitute $-\frac{7}{2}$ for m and 36 for b.
    $y = -\frac{7}{2}x + 36$

37. $f(x) = 3x^2 - x + 1$
    $f(0) = 3 \cdot 0^2 - 0 + 1 = 0 - 0 + 1 = 1$

39. $f(x) = 3x^2 - x + 1$
    $f(1) = 3 \cdot 1^2 - 1 + 1 = 3 - 1 + 1 = 3$

41. $f(x) = 3x^2 - x + 1$
    $f(-2) = 3(-2)^2 - (-2) + 1 = 12 + 2 + 1 = 15$

43. $f(x) = 3x^2 - x + 1$
    $f(-4) = 3(-4)^2 - (-4) + 1 = 48 + 4 + 1 = 53$

45. Graph these equations, solving each equation for y first, if necessary. We get $y = \dfrac{3.5x - 106.2}{2.1}$ and $y = \dfrac{-4.1x - 106.28}{16.7}$. Using the INTERSECT feature, we find that the point of intersection is $(23.12, -12.04)$.

47. Substitute -5 for x and -1 for y in the first equation.
    $A(-5) - 7(-1) = -3$
    $-5A + 7 = -3$
    $-5A = -10$
    $A = 2$

    Then substitute -5 for x and -1 for y in the second equation.
    $-5 - B(-1) = -1$
    $-5 + B = -1$
    $B = 4$

    We have $A = 2$, $B = 4$.

49. $(0, -3)$ and $\left(-\dfrac{3}{2}, 6\right)$ are two solutions of $px - qy = -1$. Substitute 0 for x and -3 for y.
    $p \cdot 0 - q \cdot (-3) = -1$
    $3q = -1$
    $q = -\dfrac{1}{3}$

    Substitute $-\dfrac{3}{2}$ for x and 6 for y.
    $p \cdot \left(-\dfrac{3}{2}\right) - q \cdot 6 = -1$
    $-\dfrac{3}{2}p - 6q = -1$

Substitute $-\frac{1}{3}$ for q and solve for p.

$$-\frac{3}{2}p - 6 \cdot \left(-\frac{1}{3}\right) = -1$$

$$-\frac{3}{2}p + 2 = -1$$

$$-\frac{3}{2}p = -3$$

$$-\frac{2}{3} \cdot \left(-\frac{3}{2}p\right) = -\frac{2}{3} \cdot (-3)$$

$$p = 2$$

Thus, $p = 2$ and $q = -\frac{1}{3}$.

Exercise Set 8.4

1. Familiarize. Let $x =$ the number of less expensive brushes sold and $y =$ the number of more expensive brushes sold.

Translate. We organize the information in a table.

Kind of brush	Less expensive	More expensive	Total
Number sold	x	y	45
Price	$8.50	$9.75	
Amount taken in	$8.50x$	$9.75y$	398.75

The "Number sold" row of the table gives us one equation:
$$x + y = 45$$

The "Amount taken in" row gives us a second equation:
$$8.50x + 9.75y = 398.75$$

We have a system of equations:
$$x + y = 45,$$
$$8.50x + 9.75y = 398.75$$

We can multiply the second equation on both sides by 100 to clear the decimals:
$$x + y = 45, \quad (1)$$
$$850x + 975y = 39,875 \quad (2)$$

Solve. We solve the system of equations using the elimination method. Begin by multiplying Equation (1) by -850.

$$\begin{array}{r} -850x - 850y = -38,250 \\ 850x + 975y = 39,875 \\ \hline 125y = 1625 \\ y = 13 \end{array}$$ Multiplying (1)

Substitute 13 for y in (1) and solve for x.
$$x + 13 = 45$$
$$x = 32$$

Check. The number of brushes sold is $32 + 13$, or 45. The amount taken in was $\$8.50(32) + \$9.75(13) = \$272 + \$126.75 = \$398.75$. The answer checks.

State. 32 of the less expensive brushes were sold, and 13 of the more expensive brushes were sold.

3. Familiarize. Let $h =$ the number of vials of Humulin Insulin sold and $n =$ the number of vials of Novolin Insulin sold.

Translate. We organize the information in a table.

Brand	Humulin	Novolin	Total
Number sold	h	n	65
Price	$15.75	$12.95	
Amount taken in	$15.75h$	$12.95n$	959.35

The "Number sold" row of the table gives us one equation:
$$h + n = 65$$

The "Amount taken in" row gives us a second equation:
$$15.75h + 12.95n = 959.35$$

We have a system of equations:
$$h + n = 65,$$
$$15.75h + 12.95n = 959.35$$

We can multiply the second equation on both sides by 100 to clear the decimals:
$$h + n = 65, \quad (1)$$
$$1575h + 1295n = 95,935 \quad (2)$$

Solve. We solve the system of equations using the elimination method.

$$\begin{array}{r} -1295h - 1295n = -84,175 \\ 1575h + 1295n = 95,935 \\ \hline 280h = 11,760 \\ h = 42 \end{array}$$ Multiplying (1) by -1295

Substitute 42 for h in (1) and solve for n.
$$42 + n = 65$$
$$n = 23$$

Check. A total of $42 + 23$, or 65 vials, was sold. The amount collected was $\$15.75(42) + \$12.95(23) = \$661.50 + \$297.85 = \$959.35$. The answer checks.

State. 42 vials of Humulin Insulin and 23 vials of Novolin Insulin were sold.

5. Familiarize. Let x and y represent the number of 30-sec and 60-sec commercials played, respectively. We will convert 10 min to seconds:
$$10 \text{ min} = 10 \times 1 \text{ min} = 10 \times 60 \text{ sec} = 600 \text{ sec}.$$

Translate. We organize the information in a table.

Type	30-sec	60-sec	Total
Number	x	y	12
Time	$30x$	$60y$	600

The "Number" row of the table gives us one equation:
$$x + y = 12$$

The "Time" row gives us a second equation:

Exercise Set 8.4

$30x + 60y = 600$

We have a system of equations:

$x + y = 12$, (1)
$30x + 60y = 600$ (2)

Solve. We solve the system of equations using the elimination method.

$-30x - 30y = -360$ Multiplying (1) by -30
$30x + 60y = 600$
$\overline{30y = 240}$
$y = 8$

Substitute 8 for y in Equation (1) and solve for x.

$x + 8 = 12$
$x = 4$

Check. If Rudy plays 4 30-sec and 8 60-sec commercials, then the total number of commercials played is $4 + 8$, or 12. Also, the time for 4 30-sec commercials is $4 \cdot 30$, or 120 sec, and the time for 8 60-sec commercials is $8 \cdot 60$, or 480 sec. Then the total commercial time is $120 + 480$, or 600 sec, or 10 min. The answer checks.

State. Rudy plays 4 30-sec commercials and 8 60-sec commercials.

7. **Familiarize.** Let x and y represent the number of pounds of the 40% and the 10% mixture to be used. The final mixture contains 25% (10 lb), or 0.25(10 lb), or 2.5 lb of peanuts.

Translate. We organize the information in a table.

	40% mixture	10% mixture	Wedding mixture
Number of pounds	x	y	10
Percent of peanuts	40%	10%	25%
Pounds of peanuts	$0.4x$	$0.1y$	2.5

The first row of the table gives us one equation:

$x + y = 10$

The last row gives us a second equation:

$0.4x + 0.1y = 2.5$

After clearing decimals, we have the problem translated to a system of equations:

$x + y = 10$, (1)
$4x + y = 25$ (2)

Solve. We solve the system of equations using the elimination method.

$-x - y = -10$ Multiplying (1) by -1
$4x + y = 25$
$\overline{3x = 15}$
$x = 5$

Now substitute 5 for x in Equation (1) and solve for y.

$5 + y = 10$
$y = 5$

Check. If 5 lb of each mixture is used, the total wedding mixture is $5 + 5$, or 10 lb. The amount of peanuts in the wedding mixture is $0.4(5) + 0.1(5)$, or $2 + 0.5$, or 2.5 lb. The answer checks.

State. 5 lb of each type of mixture should be used.

9. **Familiarize.** Let $x =$ the number of liters of 25% solution and $y =$ the number of liters of 50% solution to be used. The mixture contains 40%(10 L), or 0.4(10 L) = 4 L of acid.

Translate. We organize the information in a table.

	25% solution	50% solution	Mixture
Number of liters	x	y	10
Percent of acid	25%	50%	40%
Amount of acid	$0.25x$	$0.5y$	4 L

We get one equation from the "Number of liters" row of the table.

$x + y = 10$

The last row of the table yields a second equation.

$0.25x + 0.5y = 4$

After clearing decimals, we have the problem translated to a system of equations:

$x + y = 10$, (1)
$25x + 50y = 400$ (2)

Solve. We use the elimination method to solve the system of equations.

$-25x - 25y = -250$ Multiplying (1) by -25
$25x + 50y = 400$
$\overline{25y = 150}$
$y = 6$

Substitute 6 for y in (1) and solve for x.

$x + 6 = 10$
$x = 4$

Check. The total amount of the mixture is 4 lb + 6 lb, or 10 lb. The amount of acid in the mixture is $0.25(4 \text{ L}) + 0.5(6 \text{ L}) = 1 \text{ L} + 3 \text{ L} = 4 \text{ L}$. The answer checks.

State. 4 L of the 25% solution and 6 L of the 50% solution should be mixed.

11. **Familiarize.** Let $x =$ the amount of the 6% loan and $y =$ the amount of the 9% loan. Recall that the formula for simple interest is

Interest = Principal · Rate · Time.

Translate. We organize the information in a table.

	6% loan	9% loan	Total
Principal	x	y	$12,000
Interest Rate	6%	9%	
Time	1 yr	1 yr	
Interest	$0.06x$	$0.09y$	$855

The "Principal" row of the table gives us one equation:
$$x + y = 12,000$$
The last row of the table yields another equation:
$$0.06x + 0.09y = 855$$
After clearing decimals, we have the problem translated to a system of equations:
$$x + y = 12,000 \quad (1)$$
$$6x + 9y = 85,500 \quad (2)$$

Solve. We use the elimination method to solve the system of equations.
$$-6x - 6y = -72,000 \quad \text{Multiplying (1) by } -6$$
$$\underline{6x + 9y = 85,500}$$
$$3y = 13,500$$
$$y = 4500$$

Substitute 4500 for y in (1) and solve for x.
$$x + 4500 = 12,000$$
$$x = 7500$$

Check. The loans total $7500 + $4500, or $12,000. The total interest is $0.06(\$7500) + 0.09(\$4500) = \$450 + \$405 = \$855$. The answer checks.

State. The 6% loan was for $7500, and the 9% loan was for $4500.

13. *Familiarize*. From the bar graph we see that whole milk is 4% milk fat, milk for cream cheese is 8% milk fat, and cream is 30% milk fat. Let $x =$ the number of pounds of whole milk and $y =$ the number of pounds of cream to be used. The mixture contains 8%(200 lb), or 0.08(200 lb) = 16 lb of milk fat.

Translate. We organize the information in a table.

	Whole milk	Cream	Mixture
Number of pounds	x	y	200
Percent of milk fat	4%	30%	8%
Amount of milk fat	$0.04x$	$0.3y$	16 lb

We get one equation from the "Number of pounds" row of the table:
$$x + y = 200$$
The last row of the table yields a second equation:
$$0.04x + 0.3y = 16$$
After clearing decimals, we have the problem translated to a system of equations:
$$x + y = 200, \quad (1)$$
$$4x + 30y = 1600 \quad (2)$$

Solve. We use the elimination method to solve the system of equations.
$$-4x - 4y = -800 \quad \text{Multiplying (1) by } -4$$
$$\underline{4x + 30y = 1600}$$
$$26y = 800$$
$$y = \frac{400}{13}, \text{ or } 30\frac{10}{13}$$

Substitute $\frac{400}{13}$ for y in (1) and solve for x.
$$x + \frac{400}{13} = 200$$
$$x = \frac{2200}{13}, \text{ or } 169\frac{3}{13}$$

Check. The total amount of the mixture is
$\frac{2200}{13}$ lb $+ \frac{400}{13}$ lb $= \frac{2600}{13}$ lb $= 200$ lb. The amount of milk fat in the mixture is $0.04\left(\frac{2200}{13} \text{ lb}\right) + 0.3\left(\frac{400}{13} \text{ lb}\right) = \frac{88}{13}$ lb $+ \frac{120}{13}$ lb $= \frac{208}{13}$ lb $= 16$ lb.
The answer checks.

State. $169\frac{3}{13}$ lb of whole milk and $30\frac{10}{13}$ lb of cream should be mixed.

15. *Familiarize*. Let $x =$ the number of $5 bills and $y =$ the number of $1 bills. The total value of the $5 bills is $5x$, and the total value of the $1 bills is $1 \cdot y$, or y.

Translate.

The total number of bills is 22.
$$x + y = 22$$

The total value of the bills is $50.
$$5x + y = 50$$

We have a system of equations:
$$x + y = 22, \quad (1)$$
$$5x + y = 50 \quad (2)$$

Solve. We use the elimination method.
$$-x - y = -22 \quad \text{Multiplying (1) by } -1$$
$$\underline{5x + y = 50}$$
$$4x = 28$$
$$x = 7$$

$7 + y = 22$ Substituting 7 for x in (1)
$y = 15$

Check. Total number of bills: $7 + 15 = 22$

Total value of bills: $\$5 \cdot 7 + \$1 \cdot 15 = \$35 + \$15 = \$50$.

The numbers check.

State. There are 7 $5 bills and 15 $1 bills.

17. Familiarize. We first make a drawing.

```
Slow train
  d miles      75 mph      (t + 2) hr
Fast train
  d miles      125 mph     t hr
```

From the drawing we see that the distances are the same. Now complete the chart.

	Distance	Rate	Time
Slow train	d	75	t + 2
Fast train	d	125	t

→ $d = 75(t+2)$
→ $d = 125t$

Translate. Using $d = rt$ in each row of the table, we get a system of equations:

$d = 75(t + 2)$,
$d = 125t$

Solve. We solve the system of equations.

$125t = 75(t + 2)$ Using substitution
$125t = 75t + 150$
$50t = 150$
$t = 3$

Then $d = 125t = 125 \cdot 3 = 375$.

Check. At 125 mph, in 3 hr the fast train will travel $125 \cdot 3 = 375$ mi. At 75 mph, in $3 + 2$, or 5 hr the slow train will travel $75 \cdot 5 = 375$ mi. The numbers check.

State. The trains will meet 375 mi from the station.

19. Familiarize. We first make a drawing. Let $d =$ the distance and $r =$ the speed of the canoe in still water. Then when the canoe travels downstream its speed is $r + 6$, and its speed upstream is $r - 6$. From the drawing we see that the distances are the same.

```
Downstream, 6 km/h current
d km, r + 6, 4 hr

Upstream, 6 km/h current
d km, r − 6, 10 hr
```

Organize the information in a table.

	Distance	Rate	Time
With current	d	r + 6	4
Against current	d	r − 6	10

Translate. Using $d = rt$ in each row of the table, we get a system of equations:

$d = 4(r + 6)$, $d = 4r + 24$,
           or
$d = 10(r - 6)$ $d = 10r - 60$

Solve. Solve the system of equations.

$4r + 24 = 10r - 60$ Using substitution
$24 = 6r - 60$
$84 = 6r$
$14 = r$

Check. When $r = 14$, then $r + 6 = 14 + 6 = 20$, and the distance traveled in 4 hr is $4 \cdot 20 = 80$ km. Also, $r - 6 = 14 - 6 = 8$, and the distance traveled in 10 hr is $8 \cdot 10 = 80$ km. The answer checks.

State. The speed of the canoe in still water is 14 km/h.

21. Familiarize. We first make a drawing. Let $d =$ the distance and $t =$ the time at 32 mph. At 4 mph faster, the speed is 36 mph.

```
32 mph       t hr             d mi
36 mph       (t − 1/2) hr     d mi
```

From the drawing, we see that the distances are the same. List the information in a table.

	Distance	Rate	Time
Slower trip	d	32	t
Faster trip	d	36	$t - \frac{1}{2}$

→ $d = 32t$
→ $d = 36\left(t - \frac{1}{2}\right)$

Translate. Using $d = rt$ in each row of the table, we get a system of equations:

$d = 32t$,   (1)
$d = 36\left(t - \frac{1}{2}\right)$ (2)

Solve. We solve the system of equations.

$32t = 36\left(t - \frac{1}{2}\right)$ Substituting $32t$ for d in (2)
$32t = 36t - 18$
$-4t = -18$
$t = \frac{18}{4}$, or $\frac{9}{2}$

The time at 32 mph is $\frac{9}{2}$ hr, and the time at 36 mph is $\frac{9}{2} - \frac{1}{2}$, or 4 hr.

Check. At 32 mph, in $\frac{9}{2}$ hr the salesperson will travel $32 \cdot \frac{9}{2}$, or 144 mi. At 36 mph, in 4 hr she will travel $36 \cdot 4$, or 144 mi. Since the distances are the same, the numbers check.

State. The towns are 144 mi apart.

23. Familiarize. We first make a drawing. Let $t =$ the time, $d =$ the distance traveled at 190 km/h, and $780 - d =$ the distance traveled at 200 km/h.

| 190 km/h | t hr | t hr | 200 km/h |

|←————— 780 km —————→|

We list the information in a table.

	Distance	Rate	Time
Slower plane	d	190	t	→ $d = 190t$
Faster plane	$780 - d$	200	t	→ $780 - d = 200t$

Translate. Using $d = rt$ in each row of the table, we get a system of equations:

$$d = 190t, \quad (1)$$
$$780 - d = 200t \quad (2)$$

Solve. We solve the system of equations.

$780 - 190t = 200t$ Substituting $190t$ for d in (2)
$780 = 390t$
$2 = t$

Check. In 2 hr the slower plane will travel $190 \cdot 2$, or 380 km, and the faster plane will travel $200 \cdot 2$, or 400 km. The sum of the distances is $380 + 400$, or 780 km. The value checks.

State. The planes will meet in 2 hr.

25. **Familiarize.** We first make a drawing. Let $d =$ the distance traveled at 420 km/h and $t =$ the time traveled. Then $1000 - d =$ the distance traveled at 330 km/h.

| d km, 420 km/h, t hr | $1000 - d$ km, 330 km/h, t hr |

|←————— 1000 km —————→|

We list the information in a table.

	Distance	Rate	Time
Faster airplane	d	420	t	→ $d = 420t$
Slower airplane	$1000 - d$	330	t	→ $1000 - d = 330t$

Translate. Using $d = rt$ in each row of the table, we get a system of equations:

$$d = 420t, \quad (1)$$
$$1000 - d = 330t \quad (2)$$

Solve. We use substitution.

$1000 - 420t = 330t$ Substituting $420t$ for d in (2)
$1000 = 750t$
$\frac{4}{3} = t$

Check. If $t = \frac{4}{3}$, then $420 \cdot \frac{4}{3} = 560$, the distance traveled by the faster airplane. Also, $330 \cdot \frac{4}{3} = 440$, the distance traveled by the slower plane. The sum of the distances is $560 + 440$, or 1000 km. The values check.

State. The airplanes will meet after $\frac{4}{3}$ hr, or $1\frac{1}{3}$ hr.

27. **Familiarize.** We make a drawing. Note that the plane's speed traveling toward London is $360 + 50$, or 410 mph, and the speed traveling toward New York City is $360 - 50$, or 310 mph. Also, when the plane is d mi from New York City, it is $3458 - d$ mi from London.

| New York City | | | London |
| 310 mph | t hours | t hours | 410 mph |

|←————— 3458 mi —————→|

|←—— d ——|—— 3458 mi $-d$ ——→|

Organize the information in a table.

	Distance	Rate	Time
Toward NYC	d	310	t
Toward London	$3458 - d$	410	t

Translate. Using $d = rt$ in each row of the table, we get a system of equations:

$$d = 310t, \quad (1)$$
$$3458 - d = 410t \quad (2)$$

Solve. We solve the system of equations.

$3458 - 310t = 410t$ Using substitution
$3458 = 720t$
$4.8028 \approx t$

Substitute 4.8028 for t in (1).

$d \approx 310(4.8028) \approx 1489$

Check. If the plane is 1489 mi from New York City, it can return to New York City, flying at 310 mph, in $1489/310 \approx 4.8$ hr. If the plane is $3458 - 1489$, or 1969 mi from London, it can fly to London, traveling at 410 mph, in $1969/410 \approx 4.8$ hr. Since the times are the same, the answer checks.

State. The point of no return is about 1489 mi from New York City.

29. Discussion and Writing Exercise

31. $f(x) = 4x - 7$
$f(0) = 4 \cdot 0 - 7 = 0 - 7 = -7$

33. $f(x) = 4x - 7$
$f(1) = 4 \cdot 1 - 7 = 4 - 7 = -3$

35. $f(x) = 4x - 7$
$f(-2) = 4(-2) - 7 = -8 - 7 = -15$

37. $f(x) = 4x - 7$
$f(-4) = 4(-4) - 7 = -16 - 7 = -23$

39. $f(x) = 4x - 7$
$f\left(\frac{3}{4}\right) = 4 \cdot \frac{3}{4} - 7 = 3 - 7 = -4$

41. $f(x) = 4x - 7$
$f(-3h) = 4(-3h) - 7 = -12h - 7$

43. Familiarize. Let $x =$ the amount of the original solution that remains after some of the original solution is drained and replaced with pure antifreeze. Let $y =$ the amount of the original solution that is drained and replaced with pure antifreeze.

Translate. We organize the information in a table. Keep in mind that the table contains information regarding the solution *after* some of the original solution is drained and replaced with pure antifreeze.

	Original Solution	Pure Antifreeze	New Mixture
Amount of solution	x	y	16 L
Percent of antifreeze	30%	100%	50%
Amount of antifreeze in solution	$0.3x$	$1 \cdot y$, or y	$0.5(16)$, or 8

The "Amount of solution" row gives us one equation:
$x + y = 16$

The last row gives us a second equation:
$0.3x + y = 8$

After clearing the decimal we have the following system of equations:
$$x + y = 16, \quad (1)$$
$$3x + 10y = 80 \quad (2)$$

Solve. We use the elimination method.
$$\begin{aligned} -3x - 3y &= -48 \quad \text{Multiplying (1) by } -3 \\ \underline{3x + 10y} &= \underline{80} \\ 7y &= 32 \\ y &= \frac{32}{7}, \text{ or } 4\frac{4}{7} \end{aligned}$$

Although the problem only asks for the amount of pure antifreeze added, we will also find x in order to check.

$x + 4\frac{4}{7} = 16 \quad$ Substituting $4\frac{4}{7}$ for y in (1)

$x = 11\frac{3}{7}$

Check. Total amount of new mixture: $11\frac{3}{7} + 4\frac{4}{7} = 16$ L

Amount of antifreeze in new mixture:
$0.3\left(11\frac{3}{7}\right) + 4\frac{4}{7} = \frac{3}{10} \cdot \frac{80}{7} + \frac{32}{7} = \frac{56}{7} = 8$ L
The numbers check.

State. Michelle should drain $4\frac{4}{7}$ L of the original solution and replace it with pure antifreeze.

45. Familiarize. Let x and y represent the number of city miles and highway miles that were driven, respectively. Then in city driving, $\dfrac{x}{18}$ gallons of gasoline are used; in highway driving, $\dfrac{y}{24}$ gallons are used.

Translate. We organize the information in a table.

Type of driving	City	Highway	Total
Number of miles	x	y	465
Gallons of gasoline used	$\dfrac{x}{18}$	$\dfrac{y}{24}$	23

The first row of the table gives us one equation:
$x + y = 465$

The second row gives us another equation:
$\dfrac{x}{18} + \dfrac{y}{24} = 23$

After clearing fractions, we have the following system of equations:
$$x + y = 465, \quad (1)$$
$$24x + 18y = 9936 \quad (2)$$

Solve. We solve the system of equations using the elimination method.
$$\begin{aligned} -18x - 18y &= -8370 \quad \text{Multiplying (1) by } -18 \\ \underline{24x + 18y} &= \underline{9936} \\ 6x &= 1566 \\ x &= 261 \end{aligned}$$

Now substitute 261 for x in Equation (1) and solve for y.
$261 + y = 465$
$y = 204$

Check. The total mileage is $261 + 204$, or 465. In 216 city miles, $261/18$, or 14.5 gal of gasoline are used; in 204 highway miles, $204/24$, or 8.5 gal are used. Then a total of $14.5 + 8.5$ or 23 gal of gasoline are used. The answer checks.

State. 261 miles were driven in the city, and 204 miles were driven on the highway.

47. Familiarize. Let $x =$ the number of gallons of pure brown and $y =$ the number of gallons of neutral stain that should be added to the original 0.5 gal. Note that a total of 1 gal of stain needs to be added to bring the amount of stain up to 1.5 gal. The original 0.5 gal of stain contains $20\%(0.5 \text{ gal})$, or $0.2(0.5 \text{ gal}) = 0.1$ gal of brown stain. The final solution contains $60\%(1.5 \text{ gal})$, or $0.6(1.5 \text{ gal}) = 0.9$ gal of brown stain. This is composed of the original 0.1 gal and the x gal that are added.

Translate.

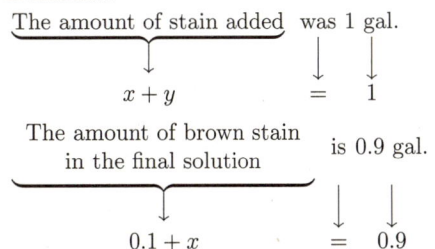

$\underbrace{\text{The amount of stain added}}$ was 1 gal.
$\qquad\qquad x + y \qquad\qquad\quad = \quad 1$

$\underbrace{\text{The amount of brown stain in the final solution}}$ is 0.9 gal.
$\qquad\qquad 0.1 + x \qquad\qquad\quad = \quad 0.9$

We have a system of equations.
$$x + y = 1, \quad (1)$$
$$0.1 + x = 0.9 \quad (2)$$

Carry out. First we solve (2) for x.
$$0.1 + x = 0.9$$
$$x = 0.8$$

Then substitute 0.8 for x in (1) and solve for y.
$$0.8 + y = 1$$
$$y = 0.2$$

Check. Total amount of stain: $0.5 + 0.8 + 0.2 = 1.5$ gal

Total amount of brown stain: $0.1 + 0.8 = 0.9$ gal

Total amount of neutral stain: $0.8(0.5) + 0.2 = 0.4 + 0.2 = 0.6$ gal $= 0.4(1.5$ gal$)$

The answer checks.

State. 0.8 gal of pure brown and 0.2 gal of neutral stain should be added.

Exercise Set 8.5

1.
$$x + y + z = 2, \quad (1)$$
$$2x - y + 5z = -5, \quad (2)$$
$$-x + 2y + 2z = 1 \quad (3)$$

Add Equations (1) and (2) to eliminate y:
$$\begin{aligned} x + y + z &= 2 \quad (1) \\ 2x - y + 5z &= -5 \quad (2) \\ \hline 3x \quad\quad + 6z &= -3 \quad (4) \text{ Adding} \end{aligned}$$

Use a different pair of equations and eliminate y:
$$\begin{aligned} 4x - 2y + 10z &= -10 \text{ Multiplying (2) by 2} \\ -x + 2y + 2z &= 1 \quad (3) \\ \hline 3x \quad\quad + 12z &= -9 \quad (5) \text{ Adding} \end{aligned}$$

Now solve the system of Equations (4) and (5).
$$3x + 6z = -3 \quad (4)$$
$$3x + 12z = -9 \quad (5)$$

$$\begin{aligned} -3x - 6z &= 3 \text{ Multiplying (4) by } -1 \\ 3x + 12z &= -9 \quad (5) \\ \hline 6z &= -6 \text{ Adding} \\ z &= -1 \end{aligned}$$

$$3x + 6(-1) = -3 \text{ Substituting } -1 \text{ for } z \text{ in (4)}$$
$$3x - 6 = -3$$
$$3x = 3$$
$$x = 1$$

$$1 + y + (-1) = 2 \text{ Substituting 1 for } x \text{ and } -1 \text{ for } z \text{ in (1)}$$
$$y = 2 \text{ Simplifying}$$

We obtain $(1, 2, -1)$. This checks, so it is the solution.

3.
$$2x - y + z = 5, \quad (1)$$
$$6x + 3y - 2z = 10, \quad (2)$$
$$x - 2y + 3z = 5 \quad (3)$$

We start by eliminating z from two different pairs of equations.

$$\begin{aligned} 4x - 2y + 2z &= 10 \text{ Multiplying (1) by 2} \\ 6x + 3y - 2z &= 10 \quad (2) \\ \hline 10x + y \quad\quad &= 20 \quad (4) \text{ Adding} \end{aligned}$$

$$\begin{aligned} -6x + 3y - 3z &= -15 \text{ Multiplying (1) by } -3 \\ x - 2y + 3z &= 5 \quad (3) \\ \hline -5x + y \quad\quad &= -10 \quad (5) \text{ Adding} \end{aligned}$$

Now solve the system of Equations (4) and (5).
$$10x + y = 20 \quad (4)$$
$$\begin{aligned} 5x - y &= 10 \text{ Multiplying (5) by } -1 \\ \hline 15x &= 30 \text{ Adding} \\ x &= 2 \end{aligned}$$

$$10 \cdot 2 + y = 20 \text{ Substituting 2 for } x \text{ in (4)}$$
$$20 + y = 20$$
$$y = 0$$

$$2 \cdot 2 - 0 + z = 5 \text{ Substituting 2 for } x \text{ and 0 for } y \text{ in (1)}$$
$$4 + z = 5$$
$$z = 1$$

We obtain $(2, 0, 1)$. This checks, so it is the solution.

5.
$$2x - 3y + z = 5, \quad (1)$$
$$x + 3y + 8z = 22, \quad (2)$$
$$3x - y + 2z = 12 \quad (3)$$

We start by eliminating y from two different pairs of equations.

$$\begin{aligned} 2x - 3y + z &= 5 \quad (1) \\ x + 3y + 8z &= 22 \quad (2) \\ \hline 3x \quad\quad + 9z &= 27 \quad (4) \text{ Adding} \end{aligned}$$

$$\begin{aligned} x + 3y + 8z &= 22 \quad (2) \\ 9x - 3y + 6z &= 36 \text{ Multiplying (3) by 3} \\ \hline 10x \quad\quad + 14z &= 58 \quad (5) \text{ Adding} \end{aligned}$$

Solve the system of Equations (4) and (5).
$$3x + 9z = 27 \quad (4)$$
$$10x + 14z = 58 \quad (5)$$

$$\begin{aligned} 30x + 90z &= 270 \text{ Multiplying (4) by 10} \\ -30x - 42z &= -174 \text{ Multiplying (5) by } -3 \\ \hline 48z &= 96 \text{ Adding} \\ z &= 2 \end{aligned}$$

$$3x + 9 \cdot 2 = 27 \text{ Substituting 2 for } z \text{ in (4)}$$
$$3x + 18 = 27$$
$$3x = 9$$
$$x = 3$$

Exercise Set 8.5

$2 \cdot 3 - 3y + 2 = 5$ Substituting 3 for x and 2 for z in (1)

$-3y + 8 = 5$

$-3y = -3$

$y = 1$

We obtain $(3, 1, 2)$. This checks, so it is the solution.

7. $3a - 2b + 7c = 13,$ (1)

$a + 8b - 6c = -47,$ (2)

$7a - 9b - 9c = -3$ (3)

We start by eliminating a from two different pairs of equations.

$3a - 2b + 7c = 13$ (1)

$\underline{-3a - 24b + 18c = 141}$ Multiplying (2) by -3

$-26b + 25c = 154$ (4)   Adding

$-7a - 56b + 42c = 329$ Multiplying (2) by -7

$\underline{7a - 9b - 9c = -3}$ (3)

$-65b + 33c = 326$ (5)   Adding

Now solve the system of Equations (4) and (5).

$-26b + 25c = 154$ (4)

$-65b + 33c = 326$ (5)

$-130b + 125c = 770$ Multiplying (4) by 5

$\underline{130b - 66c = -652}$ Multiplying (5) by -2

$59c = 118$

$c = 2$

$-26b + 25 \cdot 2 = 154$ Substituting 2 for c in (4)

$-26b + 50 = 154$

$-26b = 104$

$b = -4$

$a + 8(-4) - 6(2) = -47$ Substituting -4 for b and 2 for c in (2)

$a - 32 - 12 = -47$

$a - 44 = -47$

$a = -3$

We obtain $(-3, -4, 2)$. This checks, so it is the solution.

9. $2x + 3y + z = 17,$ (1)

$x - 3y + 2z = -8,$ (2)

$5x - 2y + 3z = 5$ (3)

We start by eliminating y from two different pairs of equations.

$2x + 3y + z = 17$ (1)

$\underline{x - 3y + 2z = -8}$ (2)

$3x + 3z = 9$ (4)   Adding

$4x + 6y + 2z = 34$ Multiplying (1) by 2

$\underline{15x - 6y + 9z = 15}$ Multiplying (3) by 3

$19x + 11z = 49$ (5)   Adding

Now solve the system of Equations (4) and (5).

$3x + 3z = 9$ (4)

$19x + 11z = 49$ (5)

$33x + 33z = 99$ Multiplying (4) by 11

$\underline{-57x - 33z = -147}$ Multiplying (5) by -3

$-24x = -48$

$x = 2$

$3 \cdot 2 + 3z = 9$ Substituting 2 for x in (4)

$6 + 3z = 9$

$3z = 3$

$z = 1$

$2 \cdot 2 + 3y + 1 = 17$ Substituting 2 for x and 1 for z in (1)

$3y + 5 = 17$

$3y = 12$

$y = 4$

We obtain $(2, 4, 1)$. This checks, so it is the solution.

11. $2x + y + z = -2,$ (1)

$2x - y + 3z = 6,$ (2)

$3x - 5y + 4z = 7$ (3)

We start by eliminating y from two different pairs of equations.

$2x + y + z = -2$ (1)

$\underline{2x - y + 3z = 6}$ (2)

$4x + 4z = 4$ (4)   Adding

$10x + 5y + 5z = -10$ Multiplying (1) by 5

$\underline{3x - 5y + 4z = 7}$ (3)

$13x + 9z = -3$ (5)   Adding

Now solve the system of Equations (4) and (5).

$4x + 4z = 4$ (4)

$13x + 9z = -3$ (5)

$36x + 36z = 36$ Multiplying (4) by 9

$\underline{-52x - 36z = 12}$ Multiplying (5) by -4

$-16x = 48$ Adding

$x = -3$

$4(-3) + 4z = 4$ Substituting -3 for x in (4)

$-12 + 4z = 4$

$4z = 16$

$z = 4$

$2(-3) + y + 4 = -2$ Substituting -3 for x and 4 for z in (1)

$y - 2 = -2$

$y = 0$

We obtain $(-3, 0, 4)$. This checks, so it is the solution.

13. $x - y + z = 4,$ (1)
$5x + 2y - 3z = 2,$ (2)
$3x - 7y + 4z = 8$ (3)

We start by eliminating z from two different pairs of equations.

$3x - 3y + 3z = 12$ Multiplying (1) by 3
$\underline{5x + 2y - 3z = 2}$ (2)
$8x - y = 14$ (4) Adding

$-4x + 4y - 4z = -16$ Multiplying (1) by -4
$\underline{3x - 7y + 4z = 8}$ (3)
$-x - 3y = -8$ (5) Adding

Now solve the system of Equations (4) and (5).

$8x - y = 14$ (4)
$-x - 3y = -8$ (5)

$8x - y = 14$ (4)
$\underline{-8x - 24y = -64}$ Multiplying (5) by 8
$ - 25y = -50$
$y = 2$

$8x - 2 = 14$ Substituting 2 for y in (4)
$8x = 16$
$x = 2$

$2 - 2 + z = 4$ Substituting 2 for x and 2 for y in (1)
$z = 4$

We obtain $(2, 2, 4)$. This checks, so it is the solution.

15. $4x - y - z = 4,$ (1)
$2x + y + z = -1,$ (2)
$6x - 3y - 2z = 3$ (3)

We start by eliminating y from two different pairs of equations.

$4x - y - z = 4$ (1)
$\underline{2x + y + z = -1}$ (2)
$6x = 3$ (4) Adding

At this point we can either continue by eliminating y from a second pair of equations or we can solve (4) for x and substitute that value in a different pair of the original equations to obtain a system of two equations in two variables. We take the second option.

$6x = 3$ (4)
$x = \frac{1}{2}$

Substitute $\frac{1}{2}$ for x in (1):

$4\left(\frac{1}{2}\right) - y - z = 4$
$2 - y - z = 4$
$-y - z = 2$ (5)

Substitute $\frac{1}{2}$ for x in (3):

$6\left(\frac{1}{2}\right) - 3y - 2z = 3$
$3 - 3y - 2z = 3$
$-3y - 2z = 0$ (6)

Solve the system of Equations (5) and (6).

$2y + 2z = -4$ Multiplying (5) by -2
$\underline{-3y - 2z = 0}$ (6)
$-y = -4$
$y = 4$

$-4 - z = 2$ Substituting 4 for y in (5)
$-z = 6$
$z = -6$

We obtain $\left(\frac{1}{2}, 4, -6\right)$. This checks, so it is the solution.

17. $2r + 3s + 12t = 4,$ (1)
$4r - 6s + 6t = 1,$ (2)
$r + s + t = 1$ (3)

We start by eliminating s from two different pairs of equations.

$4r + 6s + 24t = 8$ Multiplying (1) by 2
$\underline{4r - 6s + 6t = 1}$ (2)
$8r + 30t = 9$ (4) Adding

$4r - 6s + 6t = 1$ (2)
$\underline{6r + 6s + 6t = 6}$ Multiplying (3) by 6
$10r + 12t = 7$ (5) Adding

Solve the system of Equations (4) and (5).

$40r + 150t = 45$ Multiplying (4) by 5
$\underline{-40r - 48t = -28}$ Multiplying (5) by -4
$102t = 17$
$t = \frac{17}{102}$
$t = \frac{1}{6}$

$8r + 30\left(\frac{1}{6}\right) = 9$ Substituting $\frac{1}{6}$ for t in (4)
$8r + 5 = 9$
$8r = 4$
$r = \frac{1}{2}$

$\frac{1}{2} + s + \frac{1}{6} = 1$ Substituting $\frac{1}{2}$ for r and $\frac{1}{6}$ for t in (3)
$s + \frac{2}{3} = 1$
$s = \frac{1}{3}$

We obtain $\left(\frac{1}{2}, \frac{1}{3}, \frac{1}{6}\right)$. This checks, so it is the solution.

Exercise Set 8.5

19. $4a + 9b = 8,\quad (1)$
$8a + 6c = -1,\quad (2)$
$6b + 6c = -1\quad (3)$

We will use the elimination method. Note that there is no c in Equation (1). We will use equations (2) and (3) to obtain another equation with no c terms.

$8a + 6c = -1\quad (2)$
$\underline{-6b - 6c = 1}\quad \text{Multiplying (3) by } -1$
$8a - 6b = 0\quad (4)\quad \text{Adding}$

Now solve the system of Equations (1) and (4).

$-8a - 18b = -16\quad \text{Multiplying (1) by } -2$
$\underline{8a - 6b = 0}$
$- 24b = -16$
$b = \dfrac{2}{3}$

$8a - 6\left(\dfrac{2}{3}\right) = 0\quad \text{Substituting } \dfrac{2}{3} \text{ for } b \text{ in } (4)$
$8a - 4 = 0$
$8a = 4$
$a = \dfrac{1}{2}$

$8\left(\dfrac{1}{2}\right) + 6c = -1\quad \text{Substituting } \dfrac{1}{2} \text{ for } a \text{ in } (2)$
$4 + 6c = -1$
$6c = -5$
$c = -\dfrac{5}{6}$

We obtain $\left(\dfrac{1}{2}, \dfrac{2}{3}, -\dfrac{5}{6}\right)$. This checks, so it is the solution.

21. $x + y + z = 57,\quad (1)$
$-2x + y = 3,\quad (2)$
$x - z = 6\quad (3)$

We will use the substitution method. Solve Equations (2) and (3) for y and z, respectively. Then substitute in Equation (1) to solve for x.

$-2x + y = 3\qquad \text{Solving (2) for } y$
$y = 2x + 3$
$x - z = 6\qquad \text{Solving (3) for } z$
$-z = -x + 6$
$z = x - 6$

$x + (2x + 3) + (x - 6) = 57\quad \text{Substituting in (1)}$
$4x - 3 = 57$
$4x = 60$
$x = 15$

To find y, substitute 15 for x in $y = 2x + 3$:
$y = 2 \cdot 15 + 3 = 33$

To find z, substitute 15 for x in $z = x - 6$:
$z = 15 - 6 = 9$

We obtain $(15, 33, 9)$. This checks, so it is the solution.

23. $r + s = 5,\quad (1)$
$3s + 2t = -1,\quad (2)$
$4r + t = 14\quad (3)$

We will use the elimination method. Note that there is no t in Equation (1). We will use Equations (2) and (3) to obtain another equation with no t terms.

$3s + 2t = -1\quad (2)$
$\underline{-8r - 2t = -28}\quad \text{Multiplying (3) by } -2$
$-8r + 3s = -29\quad (4)\quad \text{Adding}$

Now solve the system of Equations (1) and (4).

$r + s = 5\quad (1)$
$-8r + 3s = -29\quad (4)$

$8r + 8s = 40\quad \text{Multiplying (1) by 8}$
$\underline{-8r + 3s = -29}\quad (4)$
$11s = 11\quad \text{Adding}$
$s = 1$

$r + 1 = 5\quad \text{Substituting 1 for } s \text{ in (1)}$
$r = 4$

$4 \cdot 4 + t = 14\quad \text{Substituting 4 for } r \text{ in (3)}$
$16 + t = 14$
$t = -2$

We obtain $(4, 1, -2)$. This checks, so it is the solution.

25. Discussion and Writing Exercise

27. $F = 3ab$
$\dfrac{F}{3b} = a\quad \text{Dividing by } 3b$

29. $F = \dfrac{1}{2}t(c - d)$
$2F = t(c - d)\quad \text{Multiplying by 2}$
$2F = tc - td\quad \text{Removing parentheses}$
$2F + td = tc\quad \text{Adding } td$
$\dfrac{2F + td}{t} = c,\text{ or}\quad \text{Dividing by } t$
$\dfrac{2F}{t} + d = c$

31. $Ax - By = c$
$Ax = By + c\quad \text{Adding } By$
$Ax - c = By\quad \text{Subtracting } c$
$\dfrac{Ax - c}{B} = y\quad \text{Dividing by } B$

33. $y = -\dfrac{2}{3}x - \dfrac{5}{4}$

The equation is in slope-intercept form, $y = mx + b$. The slope is $-\dfrac{2}{3}$, and the y-intercept is $\left(0, -\dfrac{5}{4}\right)$.

35.
$$2x - 5y = 10$$
$$-5y = -2x + 10$$
$$-\frac{1}{5}(-5y) = -\frac{1}{5}(-2x + 10)$$
$$y = \frac{2}{5}x - 2$$

The equation is now in slope-intercept form, $y = mx + b$. The slope is $\frac{2}{5}$, and the y-intercept is $(0, -2)$.

37.
$$w + x + y + z = 2, \quad (1)$$
$$w + 2x + 2y + 4z = 1, \quad (3)$$
$$w - x + y + z = 6, \quad (3)$$
$$w - 3x - y + z = 2 \quad (4)$$

Start by eliminating w from three different pairs of equations.

$$\begin{array}{r}w + x + y + z = 2 \quad (1) \\ -w - 2x - 2y - 4z = -1 \quad \text{Multiplying (2) by } -1 \\ \hline -x - y - 3z = 1 \quad (5) \quad \text{Adding}\end{array}$$

$$\begin{array}{r}w + x + y + z = 2 \quad (1) \\ -w + x - y - z = -6 \quad \text{Multiplying (3) by } -1 \\ \hline 2x = -4 \quad (6) \quad \text{Adding}\end{array}$$

$$\begin{array}{r}w + x + y + z = 2 \quad (1) \\ -w + 3x + y - z = -2 \quad \text{Multiplying (4) by } -1 \\ \hline 4x + 2y = 0 \quad (7) \quad \text{Adding}\end{array}$$

We can solve (6) for x:
$$2x = -4$$
$$x = -2$$

Substitute -2 for x in (7):
$$4(-2) + 2y = 0$$
$$-8 + 2y = 0$$
$$2y = 8$$
$$y = 4$$

Substitute -2 for x and 4 for y in (5):
$$-(-2) - 4 - 3z = 1$$
$$-2 - 3z = 1$$
$$-3z = 3$$
$$z = -1$$

Substitute -2 for x, 4 for y, and -1 for z in (1):
$$w - 2 + 4 - 1 = 2$$
$$w + 1 = 2$$
$$w = 1$$

We obtain $(1, -2, 4, -1)$. This checks, so it is the solution.

Exercise Set 8.6

1. Familiarize. Let $x =$ the number of 10-oz cups, $y =$ the number of 14-oz cups, and $z =$ the number of 20-oz cups that Reggie filled. Note that five 96-oz pots contain $5 \cdot 96$ oz, or 480 oz of coffee. Also, x 10-oz cups contain a total of $10x$ oz of coffee and bring in $\$1.09x$, y 14-oz cups contain $14y$ oz and bring in $\$1.29y$, and z 20-oz cups contain $20z$ oz and bring in $\$1.49z$.

Translate.

The total number of coffees served was 34.
$$x + y + z = 34$$

The total amount of coffee served was 480 oz.
$$10x + 14y + 20z = 480$$

The total amount collected was $44.06.
$$1.09x + 1.29y + 1.49z = 44.06$$

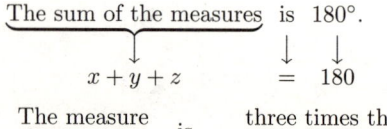

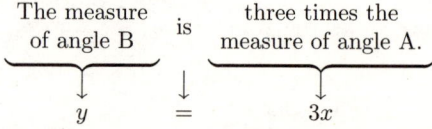

Now we have a system of equations.
$$x + y + z = 34,$$
$$10x + 14y + 20z = 480,$$
$$1.09x + 1.29y + 1.49z = 43.46$$

Solve. Solving the system we get $(8, 20, 6)$.

Check. The total number of coffees served was $8 + 20 + 6$, or 34. The total amount of coffee served was $10 \cdot 8 + 14 \cdot 20 + 20 \cdot 6 = 80 + 280 + 120 = 480$ oz. The total amount collected was $\$1.09(8) + \$1.29(20) + \$1.49(6) = \$8.72 + \$25.80 + \$8.94 = \$43.46$. The numbers check.

State. Reggie filled 8 10-oz cups, 20 14-oz cups, and 6 20-oz cups.

3. Familiarize. We first make a drawing.

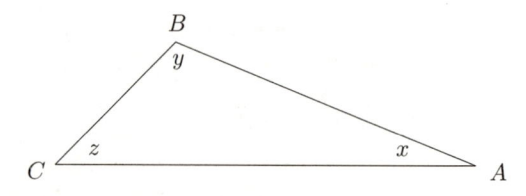

We let x, y, and z represent the measures of angles A, B, and C, respectively. The measures of the angles of a triangle add up to 180°.

Translate.

The sum of the measures is 180°.
$$x + y + z = 180$$

The measure of angle B is three times the measure of angle A.
$$y = 3x$$

Exercise Set 8.6

$\underbrace{\text{The measure of angle C}}_{z}$ $\underset{\downarrow}{\text{is}}$ $\underbrace{20° \text{ more than the measure of angle A.}}_{x+20}$
$z = x + 20$

We now have a system of equations.

$x + y + z = 180,$

$y = 3x,$

$z = x + 20$

Solve. Solving the system we get $(32, 96, 52)$.

Check. The sum of the measures is $32° + 96° + 52°$, or $180°$. Three times the measure of angle A is $3 \cdot 32°$, or $96°$, the measure of angle B. $20°$ more than the measure of angle A is $32° + 20°$, or $52°$, the measure of angle C. The numbers check.

State. The measures of angles A, B, and C are $32°$, $96°$, and $52°$, respectively.

5. **Familiarize.** Let $x =$ the cost of automatic transmission, $y =$ the cost of power door locks, and $z =$ the cost of air conditioning. The prices of the options are added to the basic price of $12,685.

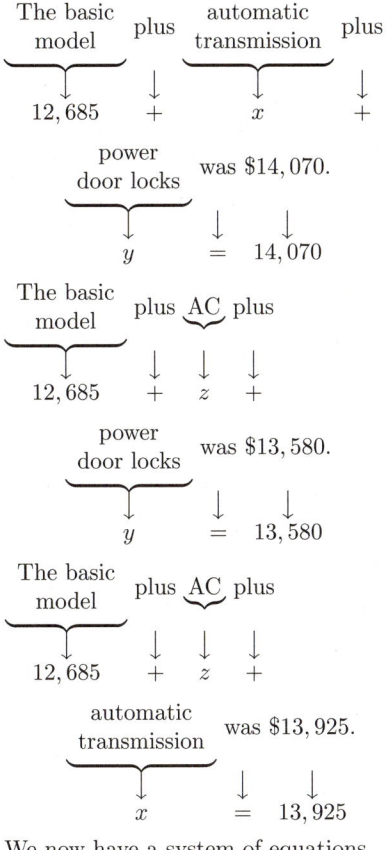

Translate.

$\underbrace{\text{The basic model}}_{12,685}$ plus $\underbrace{\text{automatic transmission}}_{x}$ plus
$\underbrace{\text{power door locks}}_{y}$ was $14,070.
$= 14,070$

$\underbrace{\text{The basic model}}_{12,685}$ plus $\underbrace{\text{AC}}_{z}$ plus
$\underbrace{\text{power door locks}}_{y}$ was $13,580.
$= 13,580$

$\underbrace{\text{The basic model}}_{12,685}$ plus $\underbrace{\text{AC}}_{z}$ plus
$\underbrace{\text{automatic transmission}}_{x}$ was $13,925.
$= 13,925$

We now have a system of equations.

$12,685 + x + y = 14,070,$

$12,685 + z + y = 13,580,$

$12,685 + z + x = 13,925$

Solve. Solving the system we get $(865, 520, 375)$.

Check. The basic model with automatic transmission and power door locks costs $12,685 + $865 + $520, or $14,070. The basic model with AC and power door locks costs $12,685 + $375 + $520, or $13,580. The basic model with AC and automatic transmission costs $12,685 + $375 + $865, or $13,925. The numbers check.

State. Automatic transmission costs $865, power door locks cost $520, and AC costs $375.

7. **Familiarize.** It helps to organize the information in a table. We let x, y, and z represent the weekly productions of the individual machines.

Machines Working	A	B	C
Weekly Production	x	y	z

Machines Working	A & B	B & C	A, B, & C
Weekly Production	3400	4200	5700

Translate. From the table, we obtain three equations.

$x + y + z = 5700$ (All three machines working)

$x + y \phantom{{}+z} = 3400$ (A and B working)

$\phantom{x+{}} y + z = 4200$ (B and C working)

Solve. Solving the system we get $(1500, 1900, 2300)$.

Check. The sum of the weekly productions of machines A, B & C is $1500 + 1900 + 2300$, or 5700. The sum of the weekly productions of machines A and B is $1500 + 1900$, or 3400. The sum of the weekly productions of machines B and C is $1900 + 2300$, or 4200. The numbers check.

State. In a week Machine A can polish 1500 lenses, Machine B can polish 1900 lenses, and Machine C can polish 2300 lenses.

9. **Familiarize.** Let $x =$ the amount invested in the first fund, $y =$ the amount invested in the second fund, and $z =$ the amount invested in the third fund. Then the earnings from the investments were $0.1x$, $0.06y$, and $0.15z$.

Translate.

$\underbrace{\text{The total amount invested}}_{x+y+z}$ was $80,000.
$= 80,000$

$\underbrace{\text{The total earnings}}_{0.1x + 0.06y + 0.15z}$ were $8850.
$= 8850$

$\underbrace{\text{The earnings from the first fund}}_{0.1x}$ were $750 $\underbrace{\text{more than}}_{}$ $\underbrace{\text{the earnings from the third fund.}}_{0.15z}$
$= 750 + 0.15z$

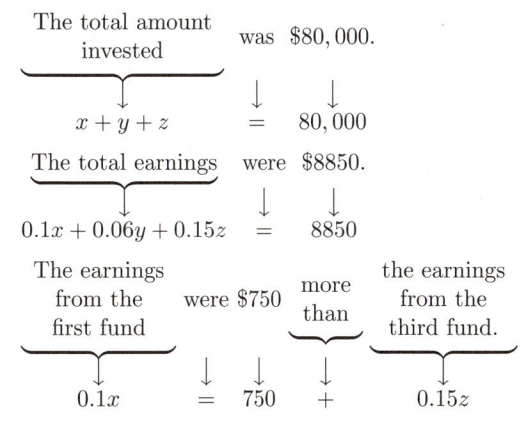

205

Now we have a system of equations.
$$x + y + z = 80{,}000$$
$$0.1x + 0.06y + 0.15z = 8850,$$
$$0.1x = 750 + 0.15z$$

Solve. Solving the system we get $(45{,}000, 10{,}000, 25{,}000)$.

Check. The total investment was $\$45{,}000 + \$10{,}000 + \$25{,}000$, or $\$80{,}000$. The total earnings were $0.1(\$45{,}000) + 0.06(10{,}000) + 0.15(25{,}000) = \$4500 + \$600 + \$3750 = \$8850$. The earnings from the first fund, $\$4500$, were $\$750$ more than the earnings from the second fund, $\$3750$.

State. $\$45{,}000$ was invested in the first fund, $\$10{,}000$ in the second fund, and $\$25{,}000$ in the third fund.

11. **Familiarize.** Let x, y, and z represent the number of fraternal twin births for Asian-Americans, African-Americans, and Caucasians in the U.S., respectively, out of every 15,400 births.

    Translate. Out of every 15,400 births, we have the following statistics:

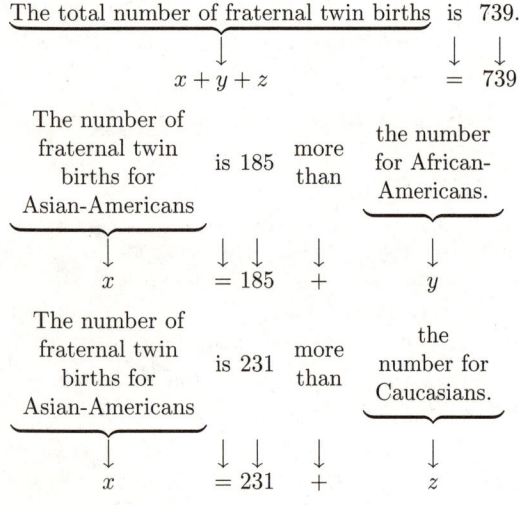

    The total number of fraternal twin births is 739.
    $$x + y + z = 739$$

    The number of fraternal twin births for Asian-Americans is 185 more than the number for African-Americans.
    $$x = 185 + y$$

    The number of fraternal twin births for Asian-Americans is 231 more than the number for Caucasians.
    $$x = 231 + z$$

    We have a system of equations.
    $$x + y + z = 739,$$
    $$x = 185 + y,$$
    $$x = 231 + y,$$

    Solve. Solving the system we get $(385, 200, 154)$.

    Check. The total of the numbers is 739. Also 385 is 185 more than 200, and it is 231 more than 154.

    State. Out of every 15,400 births, there are 385 births of fraternal twins for Asian-Americans, 200 for African-Americans, and 154 for Caucasians.

13. **Familiarize.** Let $r =$ the number of servings of roast beef, $p =$ the number of baked potatoes, and $b =$ the number of servings of broccoli. Then r servings of roast beef contain $300r$ Calories, $20r$ g of protein, and no vitamin C. In p baked potatoes there are $100p$ Calories, $5p$ g of protein, and $20p$ mg of vitamin C. And b servings of broccoli contain $50b$ Calories, $5b$ g of protein, and $100b$ mg of vitamin C. The patient requires 800 Calories, 55 g of protein, and 220 mg of vitamin C.

    Translate. Write equations for the total number of calories, the total amount of protein, and the total amount of vitamin C.
    $$300r + 100p + 50b = 800 \quad \text{(Calories)}$$
    $$20r + 5p + 5b = 55 \quad \text{(protein)}$$
    $$20p + 100b = 220 \quad \text{(vitamin C)}$$

    We now have a system of equations.

    Solve. Solving the system we get $(2, 1, 2)$.

    Check. Two servings of roast beef provide 600 Calories, 40 g of protein, and no vitamin C. One baked potato provides 100 Calories, 5 g of protein, and 20 mg of vitamin C. And 2 servings of broccoli provide 100 Calories, 10 g of protein, and 200 mg of vitamin C. Together, then, they provide 800 Calories, 55 g of protein, and 220 mg of vitamin C. The values check.

    State. The dietician should prepare 2 servings of roast beef, 1 baked potato, and 2 servings of broccoli.

15. **Familiarize.** Let x, y, and z represent the number of par-3, par-4, and par-5 holes, respectively. Then a par golfer shoots $3x$ on the par-3 holes, $4x$ on the par-4 holes, and $5x$ on the par-5 holes.

    Translate.

    The total number of holes is 18.
    $$x + y + z = 18$$

    A par golfer's score is 70.
    $$3x + 4y + 5z = 70$$

    The number of par-4 holes is 2 times the number of par-5 holes.
    $$y = 2 \cdot z$$

    We have a system of equations.
    $$x + y + z = 18,$$
    $$3x + 4y + 5z = 70,$$
    $$y = 2z$$

    Solve. Solving the system we get $(6, 8, 4)$.

    Check. The numbers add up to 18. A par golfer would shoot $3 \cdot 6 + 4 \cdot 8 + 5 \cdot 4$, or 70. The number of par-4 holes, 8, is twice the number of par-5 holes, 4. The numbers check.

    State. There are 6 par-3 holes, 8 par-4 holes, and 4 par-5 holes.

17. **Familiarize.** Let x, y, and z represent the number of 2-point field goals, 3-point field goals, and 1-point foul shots made, respectively. The total number of points scored from each of these types of goals is $2x$, $3y$, and z.

    Translate.

    The total number of points was 92.
    $$2x + 3y + z = 92$$

Exercise Set 8.6

The total number of baskets was 50.
$$x + y + z = 50$$

The number of 2-pointers was 19 more than the number of foul shots.
$$x = 19 + z$$

Now we have a system of equations.

$2x + 3y + z = 92,$

$x + y + z = 50,$

$x = 19 + z$

Solve. Solving the system we get $(32, 5, 13)$.

Check. The total number of points was $2 \cdot 32 + 3 \cdot 5 + 13 = 64 + 15 + 13 = 92$. The number of baskets was $32 + 5 + 13$, or 50. The number of 2-pointers, 32, was 19 more than the number of foul shots, 13. The numbers check.

State. The Knicks made 32 two-point field goals, 5 three-point field goals, and 13 foul shots.

19. Discussion and Writing Exercise

21. Discussion and Writing Exercise

23. The correspondence is not a function, because an input (in fact, both inputs) corresponds to more than one output.

25. We cannot calculate $f(x)$ when the denominator is 0. We set the denominator equal to 0 and solve for x.

$x + 7 = 0$

$x = -7$

The domain is $\{x | x \text{ is a real number } and \ x \neq -7\}$, or $(-\infty, -7) \cup (-7, \infty)$.

27. Substitute $-\dfrac{3}{5}$ for m and -7 for b in the slope-intercept equation.

$y = mx + b$

$y = -\dfrac{3}{5}x - 7$

29. Familiarize. We first make a drawing with additional labels.

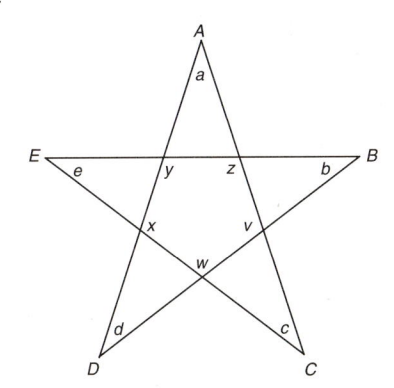

We let a, b, c, d, and e represent the angle measures at the tips of the star. We also label the interior angles of the pentagon v, w, x, y, and z. We recall the following geometric fact:

The sum of the measures of the interior angles of a polygon of n sides is given by $(n-2)180°$.

Using this fact we know:

1. The sum of the angle measures of a triangle is $(3-2)180°$, or $180°$.

2. The sum of the angle measures of a pentagon is $(5-2)180°$, or $3(180°)$.

Translate. Using fact (1) listed above we obtain a system of 5 equations.

$a + v + d = 180$

$b + w + e = 180$

$c + x + a = 180$

$d + y + b = 180$

$e + z + c = 180$

Solve. Adding we obtain

$2a + 2b + 2c + 2d + 2e + v + w + x + y + z = 5(180)$

$2(a + b + c + d + e) + (v + w + x + y + z) = 5(180)$

Using fact (2) listed above we substitute $3(180)$ for $(v + w + x + y + z)$ and solve for $(a + b + c + d + e)$.

$2(a + b + c + d + e) + 3(180) = 5(180)$

$2(a + b + c + d + e) = 2(180)$

$a + b + c + d + e = 180$

Check. We should repeat the above calculations.

State. The sum of the angle measures at the tips of the star is $180°$.

31. Familiarize. Let $x =$ the one's digit, $y =$ the ten's digit, and $z =$ the hundred's digit. Then the number is represented by $100z + 10y + x$. When the digits are reversed, the resulting number is represented by $100x + 10y + z$.

Translate.

The sum of the digits is 14.
$$x + y + z = 14$$

The ten's digit is 2 more than the one's digit.
$$y = 2 + x$$

The number is the same as the number with the digits reversed.
$$100z + 10y + x = 100x + 10y + z$$

Now we have a system of equations.

$x + y + z = 14,$

$y = 2 + x,$

$100z + 10y + x = 100x + 10y + z$

Solve. Solving the system we get $(4, 6, 4)$.

Check. If the number is 464, then the sum of the digits is $4+6+4$, or 14. The ten's digit, 6, is 2 more than the one's digit, 4. If the digits are reversed the number is unchanged. The result checks.

State. The number is 464.

Exercise Set 8.7

1. $C(x) = 25x + 270,000 \qquad R(x) = 70x$

   a) $P(x) = R(x) - C(x)$
   $= 70x - (25x + 270,000)$
   $= 70x - 25x - 270,000$
   $= 45x - 270,000$

   b) To find the break-even point we solve the system
   $R(x) = 70x,$
   $C(x) = 25x + 270,000.$

   Since both $R(x)$ and $C(x)$ are in dollars and they are equal at the break-even point, we can rewrite the system:
   $d = 70x, \qquad (1)$
   $d = 25x + 270,000 \qquad (2)$

   We solve using substitution.
   $70x = 25x + 270,000$ Substituting $65x$
   for d in (2)
   $45x = 270,000$
   $x = 6000$

   Thus, 6000 units must be produced and sold in order to break even.

   The amount taken in is $R(6000) = 70 \cdot 6000 = \$420,000$. Thus, the break-even point is $(6000, \$420,000)$.

3. $C(x) = 10x + 120,000 \qquad R(x) = 60x$

   a) $P(x) = R(x) - C(x)$
   $= 60x - (10x + 120,000)$
   $= 60x - 10x - 120,000$
   $= 50x - 120,000$

   b) Solve the system
   $R(x) = 60x,$
   $C(x) = 10x + 120,000.$

   Since both $R(x)$ and $C(x)$ are in dollars and they are equal at the break-even point, we can rewrite the system:
   $d = 60x, \qquad (1)$
   $d = 10x + 120,000 \qquad (2)$

   We solve using substitution.
   $60x = 10x + 120,000$ Substituting $60x$
   for d in (2)
   $50x = 120,000$
   $x = 2400$

   Thus, 2400 units must be produced and sold in order to break even.

   The amount taken in is $R(2400) = 60 \cdot 2400 = \$144,000$. Thus, the break-even point is $(2400, \$144,000)$.

5. $C(x) = 20x + 10,000 \qquad R(x) = 100x$

   a) $P(x) = R(x) - C(x)$
   $= 100x - (20x + 10,000)$
   $= 100x - 20x - 10,000$
   $= 80x - 10,000$

   b) Solve the system
   $R(x) = 100x,$
   $C(x) = 20x + 10,000.$

   Since both $R(x)$ and $C(x)$ are in dollars and they are equal at the break-even point, we can rewrite the system:
   $d = 100x, \qquad (1)$
   $d = 20x + 10,000 \qquad (2)$

   We solve using substitution.
   $100x = 20x + 10,000$ Substituting $100x$
   for d in (2)
   $80x = 10,000$
   $x = 125$

   Thus, 125 units must be produced and sold in order to break even.

   The amount taken in is $R(125) = 100 \cdot 125 = \$12,500$. Thus, the break-even point is $(125, \$12,500)$.

7. $C(x) = 22x + 16,000 \qquad R(x) = 40x$

   a) $P(x) = R(x) - C(x)$
   $= 40x - (22x + 16,000)$
   $= 40x - 22x - 16,000$
   $= 18x - 16,000$

   b) Solve the system
   $R(x) = 40x,$
   $C(x) = 22x + 16,000.$

   Since both $R(x)$ and $C(x)$ are in dollars and they are equal at the break-even point, we can rewrite the system:
   $d = 40x, \qquad (1)$
   $d = 22x + 16,000 \qquad (2)$

   We solve using substitution.
   $40x = 22x + 16,000$ Substituting $40x$ for
   d in (2)
   $18x = 16,000$
   $x \approx 889$ units

   Thus, 889 units must be produced and sold in order to break even.

   The amount taken in is $R(889) = 40 \cdot 889 = \$35,560$. Thus, the break-even point is $(889, \$35,560)$.

Exercise Set 8.7

9. $C(x) = 50x + 195{,}000 \quad R(x) = 125x$

a) $P(x) = R(x) - C(x)$
$= 125x - (50x + 195{,}000)$
$= 125x - 50x - 195{,}000$
$= 75x - 195{,}000$

b) Solve the system
$R(x) = 125x,$
$C(x) = 50x + 195{,}000.$

Since $R(x) = C(x)$ at the break-even point, we can rewrite the system:
$R(x) = 125x, \quad (1)$
$R(x) = 50x + 195{,}000 \quad (2)$

We solve using substitution.
$125x = 50x + 195{,}000$ Substituting $125x$ for $R(x)$ in (2)
$75x = 195{,}000$
$x = 2600$

To break even 2600 units must be produced and sold.

The amount taken in is $R(2600) = 125 \cdot 2600 = \$325{,}000$. Thus, the break-even point is $(2600, \$325{,}000)$.

11. a) $C(x) = $ Fixed costs + Variable costs
$C(x) = 22{,}500 + 40x,$
where x is the number of lamps produced.

b) Each lamp sells for \$85. The total revenue is 85 times the number of lamps sold. We assume that all lamps produced are sold.
$R(x) = 85x$

c) $P(x) = R(x) - C(x)$
$P(x) = 85x - (22{,}500 + 40x)$
$= 85x - 22{,}500 - 40x$
$= 45x - 22{,}500$

d) $P(3000) = 45(3000) - 22{,}500$
$= 135{,}000 - 22{,}500$
$= 112{,}500$

The company will realize a profit of \$112,500 when 3000 lamps are produced and sold.

$P(400) = 45(400) - 22{,}500$
$= 18{,}000 - 22{,}500$
$= -4500$

The company will realize a \$4500 loss when 400 lamps are produced and sold.

e) Solve the system
$R(x) = 85x,$
$C(x) = 22{,}500 + 40x.$

Since both $R(x)$ and $C(x)$ are in dollars and they are equal at the break-even point, we can rewrite the system:
$d = 85x, \quad (1)$
$d = 22{,}500 + 40x \quad (2)$

We solve using substitution.
$85x = 22{,}500 + 40x$ Substituting $85x$ for d in (2)
$45x = 22{,}500$
$x = 500$

The firm will break even if it produces and sells 500 lamps and takes in a total of $R(500) = 85 \cdot 500 = \$42{,}500$ in revenue. Thus, the break-even point is $(500, \$42{,}500)$.

13. a) $C(x) = $ Fixed costs + Variable costs
$C(x) = 16{,}404 + 6x,$
where x is the number of caps produced, in dozens.

b) Each dozen caps sell for \$18. The total revenue is 18 times the number of caps sold, in dozens. We assume that all caps produced are sold.
$R(x) = 18x$

c) $P(x) = R(x) - C(x)$
$P(x) = 18x - (16{,}404 + 6x)$
$= 18x - 16{,}404 - 6x$
$= 12x - 16{,}404$

d) $P(3000) = 12(3000) - 16{,}404$
$= 36{,}000 - 16{,}404$
$= 19{,}596$

The company will realize a profit of \$19,596 when 3000 dozen caps are produced and sold.

$P(1000) = 12(1000) - 16{,}404$
$= 12{,}000 - 16{,}404$
$= -4404$

The company will realize a \$4404 loss when 1000 dozen caps are produced and sold.

e) Solve the system
$R(x) = 18x,$
$C(x) = 16{,}404 + 6x.$

Since both $R(x)$ and $C(x)$ are in dollars and they are equal at the break-even point, we can rewrite the system:
$d = 18x, \quad (1)$
$d = 16{,}404 + 6x \quad (2)$

We solve using substitution.
$18x = 16{,}404 + 6x$ Substituting $18x$ for d in (2)
$12x = 16{,}404$
$x = 1367$

The firm will break even if it produces and sells 1367 dozen caps and takes in a total of $R(1367) = 18 \cdot 1367 = \$24,606$ in revenue. Thus, the break-even point is (1367, \$24,606).

15. $D(p) = 1000 - 10p,$
    $S(p) = 230 + p$

    Since both demand and supply are quantities, the system can be rewritten:

    $q = 1000 - 10p, \quad (1)$
    $q = 230 + p \quad (2)$

    Substitute $1000 - 10p$ for q in (2) and solve.

    $1000 - 10p = 230 + p$
    $770 = 11p$
    $70 = p$

    The equilibrium price is \$70 per unit. To find the equilibrium quantity we substitute \$70 into either $D(p)$ or $S(p)$.

    $D(70) = 1000 - 10 \cdot 70 = 1000 - 700 = 300$

    The equilibrium quantity is 300 units.

    The equilibrium point is (\$70, 300).

17. $D(p) = 760 - 13p,$
    $S(p) = 430 + 2p$

    Rewrite the system:

    $q = 760 - 13p, \quad (1)$
    $q = 430 + 2p \quad (2)$

    Substitute $760 - 13p$ for q in (2) and solve.

    $760 - 13p = 430 + 2p$
    $330 = 15p$
    $22 = p$

    The equilibrium price is \$22 per unit.

    To find the equilibrium quantity we substitute \$22 into either $D(p)$ or $S(p)$.

    $S(22) = 430 + 2(22) = 430 + 44 = 474$

    The equilibrium quantity is 474 units.

    The equilibrium point is (\$22, 474).

19. $D(p) = 7500 - 25p,$
    $S(p) = 6000 + 5p$

    Rewrite the system:

    $q = 7500 - 25p, \quad (1)$
    $q = 6000 + 5p \quad (2)$

    Substitute $7500 - 25p$ for q in (2) and solve.

    $7500 - 25p = 6000 + 5p$
    $1500 = 30p$
    $50 = p$

    The equilibrium price is \$50 per unit.

    To find the equilibrium quantity we substitute \$50 into either $D(p)$ or $S(p)$.

    $D(50) = 7500 - 25(50) = 7500 - 1250 = 6250$

    The equilibrium quantity is 6250 units.

    The equilibrium point is (\$50, 6250).

21. $D(p) = 1600 - 53p,$
    $S(p) = 320 + 75p$

    Rewrite the system:

    $q = 1600 - 53p, \quad (1)$
    $q = 320 + 75p \quad (2)$

    Substitute $1600 - 53p$ for q in (2) and solve.

    $1600 - 53p = 320 + 75p$
    $1280 = 128p$
    $10 = p$

    The equilibrium price is \$10 per unit.

    To find the equilibrium quantity we substitute \$10 into either $D(p)$ or $S(p)$.

    $S(10) = 320 + 75(10) = 320 + 750 = 1070$

    The equilibrium quantity is 1070 units.

    The equilibrium point is (\$10, 1070).

23. Discussion and Writing Exercise

25. $5y - 3x = 8$
    $5y = 3x + 8$
    $\frac{1}{5} \cdot 5y = \frac{1}{5}(3x + 8)$
    $y = \frac{3}{5}x + \frac{8}{5}$

    The equation is now in slope-intercept form, $y = mx + b$. The slope is $\frac{3}{5}$, and the y-intercept is $\left(0, \frac{8}{5}\right)$.

27. $2y = 3.4x + 98$
    $\frac{1}{2} \cdot 2y = \frac{1}{2}(3.4x + 98)$
    $y = 1.7x + 49$

    The equation is now in slope-intercept form, $y = mx + b$. The slope is 1.7, and the y-intercept is (0, 49).

Chapter 9

More on Inequalities

Exercise Set 9.1

1. $x - 2 \geq 6$

    -4: We substitute and get $-4 - 2 \geq 6$, or $-6 \geq 6$, a false sentence. Therefore, -4 is not a solution.

    0: We substitute and get $0 - 2 \geq 6$, or $-2 \geq 6$, a false sentence. Therefore, 0 is not a solution.

    4: We substitute and get $4 - 2 \geq 6$, or $2 \geq 6$, a false sentence. Therefore, 4 is not a solution.

    8: We substitute and get $8 - 2 \geq 6$, or $6 \geq 6$, a true sentence. Therefore, 8 is a solution.

3. $t - 8 > 2t - 3$

    0: We substitute and get $0 - 8 > 2 \cdot 0 - 3$, or $-8 > -3$, a false sentence. Therefore, 0 is not a solution.

    -8: We substitute and get $-8 - 8 > 2(-8) - 3$, or $-16 > -19$, a true sentence. Therefore, -8 is a solution.

    -9: We substitute and get $-9 - 8 > 2(-9) - 3$, or $-17 > -21$, a true sentence. Therefore, -9 is a solution.

    -3: We substitute and get $-3 - 8 > 2(-3) - 3$, or $-11 > -9$, a false sentence. Therefore, -3 is not a solution.

    $-\dfrac{7}{8}$: We substitute and get $-\dfrac{7}{8} - 8 > 2\left(-\dfrac{7}{8}\right) - 3$, or $-\dfrac{71}{8} > -\dfrac{38}{8}$, a false sentence. Therefore, $-\dfrac{7}{8}$ is not a solution.

5. Interval notation for $\{x | x < 5\}$ is $(-\infty, 5)$.

7. Interval notation for $\{x | -3 \leq x \leq 3\}$ is $[-3, 3]$.

9. Interval notation for the given graph is $(-2, 5)$.

11. Interval notation for the given graph is $(-\sqrt{2}, \infty)$.

13. $x + 2 > 1$

    $x + 2 - 2 > 1 - 2$ Subtracting 2

    $x > -1$

    The solution set is $\{x | x > -1\}$, or $(-1, \infty)$.

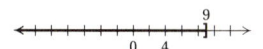

15. $y + 3 < 9$

    $y + 3 - 3 < 9 - 3$ Subtracting 3

    $y < 6$

    The solution set is $\{y | y < 6\}$, or $(-\infty, 6)$.

17. $a - 9 \leq -31$

    $a - 9 + 9 \leq -31 + 9$ Adding 9

    $a \leq -22$

    The solution set is $\{a | a \leq -22\}$, or $(-\infty, -22]$.

19. $t + 13 \geq 9$

    $t + 13 - 13 \geq 9 - 13$ Subtracting 13

    $t \geq -4$

    The solution set is $\{t | t \geq -4\}$, or $[-4, \infty)$.

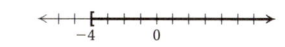

21. $y - 8 > -14$

    $y - 8 + 8 > -14 + 8$ Adding 8

    $y > -6$

    The solution set is $\{y | y > -6\}$, or $(-6, \infty)$.

23. $x - 11 \leq -2$

    $x - 11 + 11 \leq -2 + 11$ Adding 11

    $x \leq 9$

    The solution set is $\{x | x \leq 9\}$, or $(-\infty, 9]$.

25. $8x \geq 24$

    $\dfrac{8x}{8} \geq \dfrac{24}{8}$ Dividing by 8

    $x \geq 3$

    The solution set is $\{x | x \geq 3\}$, or $[3, \infty)$.

27. $0.3x < -18$

    $\dfrac{0.3x}{0.3} < \dfrac{-18}{0.3}$ Dividing by 0.3

    $x < -60$

    The solution set is $\{x | x < -60\}$, or $(-\infty, -60)$.

29. $-9x \geq -8.1$

    $\dfrac{-9x}{-9} \leq \dfrac{-8.1}{-9}$ Dividing by -9 and reversing the inequality symbol

    $x \leq 0.9$

    The solution set is $\{x | x \leq 0.9\}$, or $(-\infty, 0.9]$.

31.
$$-\frac{3}{4}x \geq -\frac{5}{8}$$
$$-\frac{4}{3}\left(-\frac{3}{4}x\right) \leq -\frac{4}{3}\left(-\frac{5}{8}\right) \quad \text{Multiplying by } -\frac{4}{3} \text{ and reversing the inequality symbol}$$
$$x \leq \frac{20}{24}$$
$$x \leq \frac{5}{6}$$

The solution set is $\left\{x \mid x \leq \frac{5}{6}\right\}$, or $\left(-\infty, \frac{5}{6}\right]$.

33.
$$2x + 7 < 19$$
$$2x + 7 - 7 < 19 - 7 \quad \text{Subtracting 7}$$
$$2x < 12$$
$$\frac{2x}{2} < \frac{12}{2} \quad \text{Dividing by 2}$$
$$x < 6$$

The solution set is $\{x \mid x < 6\}$, or $(-\infty, 6)$.

35.
$$5y + 2y \leq -21$$
$$7y \leq -21 \quad \text{Collecting like terms}$$
$$\frac{7y}{7} \leq \frac{-21}{7} \quad \text{Dividing by 7}$$
$$y \leq -3$$

The solution set is $\{y \mid y \leq -3\}$, or $(-\infty, -3]$.

37.
$$2y - 7 < 5y - 9$$
$$-5y + 2y - 7 < -5y + 5y - 9 \quad \text{Adding } -5y$$
$$-3y - 7 < -9$$
$$-3y - 7 + 7 < -9 + 7 \quad \text{Adding 7}$$
$$-3y < -2$$
$$\frac{-3y}{-3} > \frac{-2}{-3} \quad \text{Dividing by } -3 \text{ and reversing the inequality symbol}$$
$$y > \frac{2}{3}$$

The solution set is $\left\{y \mid y > \frac{2}{3}\right\}$, or $\left(\frac{2}{3}, \infty\right)$.

39.
$$0.4x + 5 \leq 1.2x - 4$$
$$-1.2x + 0.4x + 5 \leq -1.2x + 1.2x - 4 \quad \text{Adding } -1.2x$$
$$-0.8x + 5 \leq -4$$
$$-0.8x + 5 - 5 \leq -4 - 5 \quad \text{Subtracting 5}$$
$$-0.8x \leq -9$$
$$\frac{-0.8x}{-0.8} \geq \frac{-9}{-0.8} \quad \text{Dividing by } -0.8 \text{ and reversing the inequality symbol}$$
$$x \geq 11.25$$

The solution set is $\{x \mid x \geq 11.25\}$, or $[11.25, \infty)$.

41.
$$5x - \frac{1}{12} \leq \frac{5}{12} + 4x$$
$$12\left(5x - \frac{1}{12}\right) \leq 12\left(\frac{5}{12} + 4x\right) \quad \text{Clearing fractions}$$
$$60x - 1 \leq 5 + 48x$$
$$60x - 1 - 48x \leq 5 + 48x - 48x \quad \text{Subtracting } 48x$$
$$12x - 1 \leq 5$$
$$12x - 1 + 1 \leq 5 + 1 \quad \text{Adding 1}$$
$$12x \leq 6$$
$$\frac{12x}{12} \leq \frac{6}{12} \quad \text{Dividing by 12}$$
$$x \leq \frac{1}{2}$$

The solution set is $\left\{x \mid x \leq \frac{1}{2}\right\}$, or $\left(-\infty, \frac{1}{2}\right]$.

43.
$$4(4y - 3) \geq 9(2y + 7)$$
$$16y - 12 \geq 18y + 63 \quad \text{Removing parentheses}$$
$$16y - 12 - 18y \geq 18y + 63 - 18y \quad \text{Subtracting } 18y$$
$$-2y - 12 \geq 63$$
$$-2y - 12 + 12 \geq 63 + 12 \quad \text{Adding 12}$$
$$-2y \geq 75$$
$$\frac{-2y}{-2} \leq \frac{75}{-2} \quad \text{Dividing by } -2 \text{ and reversing the inequality symbol}$$
$$y \leq -\frac{75}{2}$$

The solution set is $\left\{y \mid y \leq -\frac{75}{2}\right\}$, or $\left(-\infty, -\frac{75}{2}\right]$.

45.
$$3(2 - 5x) + 2x < 2(4 + 2x)$$
$$6 - 15x + 2x < 8 + 4x$$
$$6 - 13x < 8 + 4x \quad \text{Collecting like terms}$$
$$6 - 17x < 8 \quad \text{Subtracting } 4x$$
$$-17x < 2 \quad \text{Subtracting 6}$$
$$x > -\frac{2}{17} \quad \text{Dividing by } -17 \text{ and reversing the inequality symbol}$$

The solution set is $\left\{x \mid x > -\frac{2}{17}\right\}$, or $\left(-\frac{2}{17}, \infty\right)$.

47.
$$5[3m - (m + 4)] > -2(m - 4)$$
$$5(3m - m - 4) > -2(m - 4)$$
$$5(2m - 4) > -2(m - 4)$$
$$10m - 20 > -2m + 8$$
$$12m - 20 > 8 \quad \text{Adding } 2m$$
$$12m > 28 \quad \text{Adding 20}$$
$$m > \frac{28}{12}$$
$$m > \frac{7}{3}$$

The solution set is $\left\{m \mid m > \frac{7}{3}\right\}$, or $\left(\frac{7}{3}, \infty\right)$.

Exercise Set 9.1 213

49. $3(r-6) + 2 > 4(r+2) - 21$
$3r - 18 + 2 > 4r + 8 - 21$
$3r - 16 > 4r - 13$ Collecting like terms
$-r - 16 > -13$ Subtracting $4r$
$-r > 3$ Adding 16
$r < -3$ Multiplying by -1 and reversing the inequality symbol

The solution set is $\{r|r < -3\}$, or $(-\infty, -3)$.

51. $19 - (2x+3) \leq 2(x+3) + x$
$19 - 2x - 3 \leq 2x + 6 + x$
$16 - 2x \leq 3x + 6$ Collecting like terms
$16 - 5x \leq 6$ Subtracting $3x$
$-5x \leq -10$ Subtracting 16
$x \geq 2$ Dividing by -5 and reversing the inequality symbol

The solution set is $\{x|x \geq 2\}$, or $[2, \infty)$.

53. $\frac{1}{4}(8y+4) - 17 < -\frac{1}{2}(4y-8)$
$2y + 1 - 17 < -2y + 4$
$2y - 16 < -2y + 4$ Collecting like terms
$4y - 16 < 4$ Adding $2y$
$4y < 20$ Adding 16
$y < 5$

The solution set is $\{y|y < 5\}$, or $(-\infty, 5)$.

55. $2[4 - 2(3-x)] - 1 \geq 4[2(4x-3) + 7] - 25$
$2[4 - 6 + 2x] - 1 \geq 4[8x - 6 + 7] - 25$
$2[-2 + 2x] - 1 \geq 4[8x + 1] - 25$
$-4 + 4x - 1 \geq 32x + 4 - 25$
$4x - 5 \geq 32x - 21$
$-28x - 5 \geq -21$
$-28x \geq -16$
$x \leq \frac{-16}{-28}$ Dividing by -28 and reversing the inequality symbol
$x \leq \frac{4}{7}$

The solution set is $\left\{x\big|x \leq \frac{4}{7}\right\}$, or $\left(-\infty, \frac{4}{7}\right]$.

57. $\frac{4}{5}(7x - 6) < 40$
$5 \cdot \frac{4}{5}(7x-6) < 5 \cdot 40$ Clearing the fraction
$4(7x - 6) < 200$
$28x - 24 < 200$
$28x < 224$
$x < 8$

The solution set is $\{x|x < 8\}$, or $(-\infty, 8)$.

59. $\frac{3}{4}(3 + 2x) + 1 \geq 13$
$4\left[\frac{3}{4}(3+2x) + 1\right] \geq 4 \cdot 13$ Clearing the fraction
$3(3 + 2x) + 4 \geq 52$
$9 + 6x + 4 \geq 52$
$6x + 13 \geq 52$
$6x \geq 39$
$x \geq \frac{39}{6}$, or $\frac{13}{2}$

The solution set is $\left\{x\big|x \geq \frac{13}{2}\right\}$, or $\left[\frac{13}{2}, \infty\right)$.

61. $\frac{3}{4}\left(3x - \frac{1}{2}\right) - \frac{2}{3} < \frac{1}{3}$
$\frac{9x}{4} - \frac{3}{8} - \frac{2}{3} < \frac{1}{3}$
$24\left(\frac{9x}{4} - \frac{3}{8} - \frac{2}{3}\right) < 24 \cdot \frac{1}{3}$ Clearing fractions
$54x - 9 - 16 < 8$
$54x - 25 < 8$
$54x < 33$
$x < \frac{33}{54}$, or $\frac{11}{18}$

The solution set is $\left\{x\big|x < \frac{11}{18}\right\}$, or $\left(-\infty, \frac{11}{18}\right)$.

63. $0.7(3x+6) \geq 1.1 - (x+2)$
$10[0.7(3x+6)] \geq 10[1.1 - (x+2)]$ Clearing decimals
$7(3x + 6) \geq 11 - 10(x+2)$
$21x + 42 \geq 11 - 10x - 20$
$21x + 42 \geq -9 - 10x$
$31x + 42 \geq -9$
$31x \geq -51$
$x \geq -\frac{51}{31}$

The solution set is $\left\{x\big|x \geq -\frac{51}{31}\right\}$, or $\left[-\frac{51}{31}, \infty\right)$.

65. $a + (a-3) \leq (a+2) - (a+1)$
$a + a - 3 \leq a + 2 - a - 1$
$2a - 3 \leq 1$
$2a \leq 4$
$a \leq 2$

The solution set is $\{a|a \leq 2\}$, or $(-\infty, 2]$.

67. a) We substitute 214 for W and 73 for H and calculate I.
$I = \frac{704.5W}{H^2}$
$I = \frac{704.5(192)}{73^2}$
$I \approx 25.38$

b) **Familiarize.** We will use the formula $I = \dfrac{704.5W}{H^2}$. Recall that $H = 73$ in.

Translate. An index I less than 25 indicates the lowest risk category, so we have
$$I < 25, \text{ or } \dfrac{704.5W}{H^2} < 25.$$
Now we replace H with 73.
$$\dfrac{704.5W}{73^2} < 25$$

Solve. We solve the inequality.
$$\dfrac{704.5W}{73^2} < 25$$
$$\dfrac{704.5W}{5329} < 25$$
$$704.5W < 133,225$$
$$W < 189.1 \quad \text{Rounding}$$

Check. As a partial check we can substitute a value of W less than 189.1 and a value greater than 189.1 in the formula.

For $W = 189$: $I = \dfrac{704.5(189)}{73^2} \approx 24.97$

For $W = 190$: $I = \dfrac{704.5(190)}{73^2} \approx 25.12$

Since a value of W less than 189.1 gives a body mass index less than 25 and a value of W greater than 189.1 gives an index greater than 25, we have a partial check.

State. Weights of approximately 189.1 lb or less will keep Marv in the lowest risk category. In terms of an inequality we write
$\{W | W < \text{(approximately) } 189.1 \text{ lb}\}$.

69. **Familiarize.** List the information in a table. Let $x =$ the score on the fourth test.

Test	Score
Test 1	89
Test 2	92
Test 3	95
Test 4	x
Total	360 or more

Translate. We can easily get an inequality from the table.
$$89 + 92 + 95 + x \geq 360$$

Solve.
$$276 + x \geq 360 \quad \text{Collecting like terms}$$
$$x \geq 84 \quad \text{Adding } -276$$

Check. If you get 84 on the fourth test, your total score will be $89 + 92 + 95 + 84$, or 360. Any higher score will also give you an A.

State. A score of 84 or better will give you an A. In terms of an inequality we write $\{x | x \geq 84\}$.

71. **Familiarize.** Let $v =$ the blue book value of the car. Since the car was not replaced, we know that $9200 does not exceed 80% of the blue book value.

Translate. We write an inequality stating that $9200 does not exceed 80% of the blue book value.
$$9200 \leq 0.8v$$

Solve.
$$9200 \leq 0.8v$$
$$11,500 \leq v \quad \text{Multiplying by } \dfrac{1}{0.8}$$

Check. We can do a partial check by substituting a value for v greater than 11,500. When $v = 11,525$, then 80% of v is $0.8(11,525)$, or $9220. This is greater than $9200; that is, $9200 does not exceed this amount. We cannot check all possible values for v, so we stop here.

State. The blue book value of the car is $11,500 or more. In terms of an inequality we write $\{v | v \geq \$11,500\}$.

73. **Familiarize.** We make a table of information.

Plan A: Monthly Income	Plan B: Monthly Income
$400 salary	$610 salary
8% of sales	5% of sales
Total: $400 + 8%$ of sales	Total: $610 + 5%$ of sales

Translate. We write an inequality stating that the income from Plan A is greater than the income from Plan B. We let $S =$ gross sales.
$$400 + 8\%S > 610 + 5\%S$$

Solve.
$$400 + 0.08S > 610 + 0.05S$$
$$400 + 0.03S > 610$$
$$0.03S > 210$$
$$S > 7000$$

Check. We calculate for $S = \$7000$ and for some amount greater than $7000 and some amount less than $7000.

Plan A: Plan B:

$400 + 8\%(7000)$ $610 + 5\%(7000)$

$400 + 0.08(7000)$ $610 + 0.05(7000)$

$400 + 560$ $610 + 350$

$960 $960

When $S = \$7000$, the income from Plan A is equal to the income from Plan B.

Plan A: Plan B:

$400 + 8\%(8000)$ $610 + 5\%(8000)$

$400 + 0.08(8000)$ $610 + 0.05(8000)$

$400 + 640$ $610 + 400$

$1040 $1010

When $S = \$8000$, the income from Plan A is greater than the income from Plan B.

Plan A: Plan B:
$400 + 8\%(6000)$ $610 + 5\%(6000)$
$400 + 0.08(6000)$ $610 + 0.05(6000)$
$400 + 480$ $610 + 300$
$\$880$ $\$910$

When $S = \$6000$, the income from Plan A is less than the income from Plan B.

State. Plan A is better than Plan B when gross sales are greater than $\$7000$. In terms of an inequality we write $\{S|S > \$7000\}$.

75. *Familiarize*. Let $c =$ the number of checks per month. Then the Anywhere plan will cost $\$0.20c$ per month and the Acu-checking plan will cost $\$2 + \$0.12c$ per month.

Translate. We write an inequality stating that the Acu-checking plan costs less than the Anywhere plan.

$$2 + 0.12c < 0.20c$$

Solve.
$$2 + 0.12c < 0.20c$$
$$2 < 0.08c$$
$$25 < c$$

Check. We can do a partial check by substituting a value for c less than 25 and a value for c greater than 25. When $c = 24$, the Acu-checking plan costs $\$2 + \$0.12(24)$, or $\$4.88$, and the Anywhere plan costs $\$0.20(24)$, or $\$4.80$, so the Anywhere plan is less expensive. When $c = 26$, the Acu-checking plan costs $\$2 + \$0.12(26)$, or $\$5.12$, and the Anywhere plan costs $\$0.20(26)$, or $\$5.20$, so Acu-checking is less expensive. We cannot check all possible values for c, so we stop here.

State. The Acu-checking plan costs less for more than 25 checks per month. In terms of an inequality we write $\{c|c > 25\}$.

77. *Familiarize*. Let $p =$ the number of guests at the wedding party. Then the number of guests in excess of 25 is $p - 25$. The cost under plan A is $30p$, and the cost under plan B is $1300 + 20(p - 25)$.

Translate. We write an inequality stating that plan B costs less than plan A.

$$1300 + 20(p - 25) < 30p$$

Solve. We solve the inequality.
$$1300 + 20(p - 25) < 30p$$
$$1300 + 20p - 500 < 30p$$
$$800 + 20p < 30p$$
$$800 < 10p$$
$$80 < p$$

Check. We calculate for $p = 80$ and for some number less than 80 and some number greater than 80.

Plan A: Plan B:
$30 \cdot 80$ $1300 + 20(80 - 25)$
$\$2400$ $\$2400$

When 80 people attend, plan B costs the same as plan A.

Plan A: Plan B:
$30 \cdot 79$ $1300 + 20(79 - 25)$
$\$2370$ $\$2380$

When fewer than 80 people attend, plan B costs more than plan A.

Plan A: Plan B:
$30 \cdot 81$ $1300 + 20(81 - 25)$
$\$2430$ $\$2420$

When more than 80 people attend, plan B costs less than plan A.

State. For parties of more than 80 people, plan B will cost less. In terms of an inequality we write $\{p|p > 80\}$.

79. *Familiarize*. We want to find the values of s for which $I > 36$.

Translate. $2(s + 10) > 36$

Solve.
$$2s + 20 > 36$$
$$2s > 16$$
$$s > 8$$

Check. For $s = 8$, $I = 2(8 + 10) = 2 \cdot 18 = 36$. Then any U.S. size larger than 8 will give a size larger than 36 in Italy.

State. For U.S. dress sizes larger than 8, dress sizes in Italy will be larger than 36. In terms of an inequality we write $\{s|s > 8\}$.

81. a) Substitute 0 for t and carry out the calculation.
$$N = 0.733(0) + 8.398$$
$$N = 0 + 8.398$$
$$N = 8.398$$

Each person drank 8.398 gal of bottled water in 1990.

Substitute 5 for t and carry out the calculation.
$$N = 0.733(5) + 8.398$$
$$N = 3.665 + 8.398$$
$$N = 12.063$$

Each person drank 12.063 gal of bottled water in 1995.

In 2000, $t = 2000 - 1990 = 10$. Substitute 10 for t and carry out the calculation.
$$N = 0.733(10) + 8.398$$
$$N = 7.33 + 8.398$$
$$N = 15.728$$

Each person will drink 15.728 gal of bottled water in 2000.

b) *Familiarize*. The amount of bottled water that each person drinks t years after 1990 is given by $0.733t + 8.398$.

Translate.

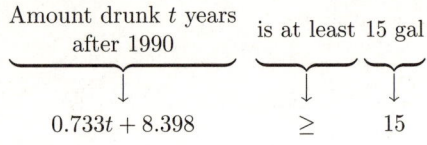

$$\underbrace{\text{Amount drunk } t \text{ years after 1990}}_{0.733t + 8.398} \quad \underbrace{\text{is at least}}_{\geq} \quad \underbrace{15 \text{ gal}}_{15}$$

Solve. We solve the inequality.

$$0.733t + 8.398 \geq 15$$
$$0.733t \geq 6.602 \quad \text{Subtracting } 8.398$$
$$t \geq 9 \quad \text{Rounding}$$

Check. We calculate for 9, for some number less than 9, and for some number greater than 9.

For $t = 9$: $0.733(9) + 8.398 \approx 15$.

For $t = 8$: $0.733(8) + 8.398 \approx 14.3$.

For $t = 10$: $0.733(10) + 8.398 \approx 15.7$.

For a value of t greater than or equal to 9, the number of gallons of bottled water each person drinks is at 15. We cannot check all the possible values of t, so we stop here.

State. Each person will drink at least 15 gal of bottled water 9 or more years after 1990, or for all years after 1999.

83. Discussion and Writing Exercise

85. $(3x - 4)(x + 8) = 3x \cdot x + 3x \cdot 8 - 4 \cdot x - 4 \cdot 8$
$$= 3x^2 + 24x - 4x - 32$$
$$= 3x^2 + 20x - 32$$

87. $(2a - 5)(3a + 11) = 2a \cdot 3a + 2a \cdot 11 - 5 \cdot 3a - 5 \cdot 11$
$$= 6a^2 + 22a - 15a - 55$$
$$= 6a^2 + 7a - 55$$

89. $f(x) = \dfrac{-3}{x + 8}$

Since $\dfrac{-3}{x + 8}$ cannot be calculated when the denominator is 0, we find the x-value that causes $x + 8$ to be 0:

$$x + 8 = 0$$
$$x = -8 \quad \text{Subtracting 8 on both sides}$$

Thus, -8 is not in the domain of f, while all other real numbers are. The domain of f is
$\{x | x \text{ is a real number } and \ x \neq -8\}$, or
$(-\infty, -8) \cup (-8, \infty)$.

91. $f(x) = |x| - 4$

Since we can calculate $|x| - 4$ for any real number x, the domain is the set of all real numbers.

93. a) ***Familiarize***. We will use
$$S = 460 + 94p \quad \text{and} \quad D = 2000 - 60p.$$

Translate. Supply is to exceed demand, so we have
$$S > D, \text{ or}$$
$$460 + 94p > 2000 - 60p.$$

Solve. We solve the inequality.

$$460 + 94p > 2000 - 60p$$
$$460 + 154p > 2000 \quad \text{Adding } 60p$$
$$154p > 1540 \quad \text{Subtracting } 460$$
$$p > 10 \quad \text{Dividing by } 154$$

Check. We calculate for $p = 10$, for some value of p less than 10, and for some value of p greater than 10.

For $p = 10$: $S = 460 + 94 \cdot 10 = 1400$
$D = 2000 - 60 \cdot 10 = 1400$

For $p = 9$: $S = 460 + 94 \cdot 9 = 1306$
$D = 2000 - 60 \cdot 9 = 1460$

For $p = 11$: $S = 460 + 94 \cdot 11 = 1494$
$D = 2000 - 60 \cdot 11 = 1340$

For a value of p greater than 10, supply exceeds demand. We cannot check all possible values of p, so we stop here.

State. Supply exceeds demand for values of p greater than 10. In terms of an inequality we write $\{p | p > 10\}$.

b) We have seen in part (a) that $D = S$ for $p = 10$, $S < D$ for a value of p less than 10, and $S > D$ for a value of p greater than 10. Since we cannot check all possible values of p, we stop here. Supply is less than demand for values of p less than 10. In terms of an inequality we write $\{p | p < 10\}$.

95. True

97. $x + 5 \leq 5 + x$
$5 \leq 5 \quad \text{Subtracting } x$

We get a true inequality, so all real numbers are solutions.

99. $x^2 + 1 > 0$

$x^2 \geq 0$ for all real numbers, so $x^2 + 1 \geq 1 > 0$ for all real numbers.

Exercise Set 9.2

1. $\{9, 10, 11\} \cap \{9, 11, 13\}$

The numbers 9 and 11 are common to the two sets, so the intersection is $\{9, 11\}$.

3. $\{a, b, c, d\} \cap \{b, f, g\}$

Only the letter b is common to the two sets. The intersection is $\{b\}$.

5. $\{9, 10, 11\} \cup \{9, 11, 13\}$

The numbers in either or both sets are 9, 10, 11, and 13, so the union is $\{9, 10, 11, 13\}$.

7. $\{a, b, c, d\} \cup \{b, f, g\}$

The letters in either or both sets are a, b, c, d, f, and g, so the union is $\{a, b, c, d, f, g\}$.

Exercise Set 9.2 217

9. $\{2,5,7,9\} \cap \{1,3,4\}$

There are no numbers common to the two sets. The intersection is the empty set, $\emptyset$.

11. $\{3,5,7\} \cup \emptyset$

The numbers in either or both sets are 3, 5, and 7, so the union is $\{3,5,7\}$.

13. $-4 < a$ and $a \leq 1$ can be written $-4 < a \leq 1$. In interval notation we have $(-4, 1]$.

The graph is the intersection of the graphs of $a > -4$ and $a \leq 1$.

15. We can write $1 < x < 6$ in interval notation as $(1, 6)$.

The graph is the intersection of the graphs of $x > 1$ and $x < 6$.

17. $-10 \leq 3x + 2$ and $3x + 2 < 17$
$-12 \leq 3x$ and $3x < 15$
$-4 \leq x$ and $x < 5$

The solution set is the intersection of the solution sets of the individual inequalities. The numbers common to both sets are those that are greater than or equal to -4 *and* less than 5. Thus the solution set is $\{x | -4 \leq x < 5\}$, or $[-4, 5)$.

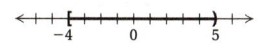

19. $3x + 7 \geq 4$ and $2x - 5 \geq -1$
$3x \geq -3$ and $2x \geq 4$
$x \geq -1$ and $x \geq 2$

The solution set is $\{x | x \geq -1\} \cap \{x | x \geq 2\} = \{x | x \geq 2\}$, or $[2, \infty)$.

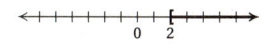

21. $4 - 3x \geq 10$ and $5x - 2 > 13$
$-3x \geq 6$ and $5x > 15$
$x \leq -2$ and $x > 3$

The solution set is $\{x | x \leq -2\} \cap \{x | x > 3\} = \emptyset$.

23. $\phantom{-4 - 4\ <\ }-4 < x + 4 < 10$
$-4 - 4 < x + 4 - 4 < 10 - 4$ Subtracting 4
$\phantom{-4 - 4\ <\ }-8 < x < 6$

The solution set is $\{x | -8 < x < 6\}$, or $(-8, 6)$.

25. $6 > -x \geq -2$
$-6 < x \leq 2$ Multiplying by -1

The solution set is $\{x | -6 < x \leq 2\}$, or $(-6, 2]$.

27. $\phantom{1 - 4\ <\ }1 < 3y + 4 \leq 19$
$1 - 4 < 3y + 4 - 4 \leq 19 - 4$ Subtracting 4
$\phantom{1 - 4\ <\ }-3 < 3y \leq 15$
$\phantom{1 - 4\ <\ }\dfrac{-3}{3} < \dfrac{3y}{3} \leq \dfrac{15}{3}$ Dividing by 3
$\phantom{1 - 4\ <\ }-1 < y \leq 5$

The solution set is $\{y | -1 < y \leq 5\}$, or $(-1, 5]$.

29. $-10 \leq 3x - 5 \leq -1$
$-10 + 5 \leq 3x - 5 + 5 \leq -1 + 5$ Adding 5
$-5 \leq 3x \leq 4$
$\dfrac{-5}{3} \leq \dfrac{3x}{3} \leq \dfrac{4}{3}$ Dividing by 3
$-\dfrac{5}{3} \leq x \leq \dfrac{4}{3}$

The solution set is $\left\{x \mid -\dfrac{5}{3} \leq x \leq \dfrac{4}{3}\right\}$, or $\left[-\dfrac{5}{3}, \dfrac{4}{3}\right]$.

31. $\phantom{2 - 3\ <\ }2 < x + 3 \leq 9$
$2 - 3 < x + 3 - 3 \leq 9 - 3$ Subtracting 3
$\phantom{2 - 3\ <\ }-1 < x \leq 6$

The solution set is $\{x | -1 < x \leq 6\}$, or $(-1, 6]$.

33. $-6 \leq 2x - 3 < 6$
$-6 + 3 \leq 2x - 3 + 3 < 6 + 3$
$-3 \leq 2x < 9$
$\dfrac{-3}{2} \leq \dfrac{2x}{2} < \dfrac{9}{2}$
$-\dfrac{3}{2} \leq x < \dfrac{9}{2}$

The solution set is $\left\{x \mid -\dfrac{3}{2} \leq x < \dfrac{9}{2}\right\}$, or $\left[-\dfrac{3}{2}, \dfrac{9}{2}\right)$.

35. $\phantom{-\dfrac{1}{2} + 3\ <\ }-\dfrac{1}{2} < \dfrac{1}{4}x - 3 \leq \dfrac{1}{2}$
$-\dfrac{1}{2} + 3 < \dfrac{1}{4}x - 3 + 3 \leq \dfrac{1}{2} + 3$
$\phantom{-\dfrac{1}{2} + 3\ <\ }\dfrac{5}{2} < \dfrac{1}{4}x \leq \dfrac{7}{2}$
$4 \cdot \dfrac{5}{2} < 4 \cdot \dfrac{1}{4}x \leq 4 \cdot \dfrac{7}{2}$
$\phantom{4 \cdot \dfrac{5}{2}\ <\ }10 < x \leq 14$

The solution set is $\{x | 10 < x \leq 14\}$, or $(10, 14]$.

37. $-3 < \dfrac{2x-5}{4} < 8$

$4(-3) < 4\left(\dfrac{2x-5}{4}\right) < 4 \cdot 8$

$-12 < 2x - 5 < 32$

$-12 + 5 < 2x - 5 + 5 < 32 + 5$

$-7 < 2x < 37$

$\dfrac{-7}{2} < \dfrac{2x}{2} < \dfrac{37}{2}$

$-\dfrac{7}{2} < x < \dfrac{37}{2}$

The solution set is $\left\{x \middle| -\dfrac{7}{2} < x < \dfrac{37}{2}\right\}$, or $\left(-\dfrac{7}{2}, \dfrac{37}{2}\right)$.

39. $x < -2$ or $x > 1$ can be written in interval notation as $(-\infty, -2) \cup (1, \infty)$.

The graph is the union of the graphs of $x < -2$ and $x > 1$.

41. $x \leq -3$ or $x > 1$ can be written in interval notation as $(-\infty, -3] \cup (1, \infty)$.

The graph is the union of the graphs of $x \leq -3$ and $x > 1$.

43. $x + 3 < -2$ or $x + 3 > 2$

$x + 3 - 3 < -2 - 3$ or $x + 3 - 3 > 2 - 3$

$x < -5$ or $x > -1$

The solution set is $\{x | x < -5 \text{ or } x > -1\}$, or $(-\infty, -5) \cup (-1, \infty)$.

45. $2x - 8 \leq -3$ or $x - 1 \geq 3$

$2x - 8 + 8 \leq -3 + 8$ or $x - 1 + 1 \geq 3 + 1$

$2x \leq 5$ or $x \geq 4$

$\dfrac{2x}{2} \leq \dfrac{5}{2}$ or $x \geq 4$

$x \leq \dfrac{5}{2}$ or $x \geq 4$

The solution set is $\left\{x \middle| x \leq \dfrac{5}{2} \text{ or } x \geq 4\right\}$, or $\left(-\infty, \dfrac{5}{2}\right] \cup [4, \infty)$.

47. $7x + 4 \geq -17$ or $6x + 5 \geq -7$

$7x \geq -21$ or $6x \geq -12$

$x \geq -3$ or $x \geq -2$

The solution set is $\{x | x \geq -3\}$, or $[-3, \infty)$.

49. $7 > -4x + 5$ or $10 \leq -4x + 5$

$7 - 5 > -4x + 5 - 5$ or $10 - 5 \leq -4x + 5 - 5$

$2 > -4x$ or $5 \leq -4x$

$\dfrac{2}{-4} < \dfrac{-4x}{-4}$ or $\dfrac{5}{-4} \geq \dfrac{-4x}{-4}$

$-\dfrac{1}{2} < x$ or $-\dfrac{5}{4} \geq x$

The solution set is $\left\{x \middle| x \leq -\dfrac{5}{4} \text{ or } x > -\dfrac{1}{2}\right\}$, or $\left(-\infty, -\dfrac{5}{4}\right] \cup \left(-\dfrac{1}{2}, \infty\right)$.

51. $3x - 7 > -10$ or $5x + 2 \leq 22$

$3x > -3$ or $5x \leq 20$

$x > -1$ or $x \leq 4$

All real numbers are solutions. In interval notation, the solution set is $(-\infty, \infty)$.

53. $-2x - 2 < -6$ or $-2x - 2 > 6$

$-2x - 2 + 2 < -6 + 2$ or $-2x - 2 + 2 > 6 + 2$

$-2x < -4$ or $-2x > 8$

$\dfrac{-2x}{-2} > \dfrac{-4}{-2}$ or $\dfrac{-2x}{-2} < \dfrac{8}{-2}$

$x > 2$ or $x < -4$

The solution set is $\{x | x < -4 \text{ or } x > 2\}$, or $(-\infty, -4) \cup (2, \infty)$.

55. $\dfrac{2}{3}x - 14 < -\dfrac{5}{6}$ or $\dfrac{2}{3}x - 14 > \dfrac{5}{6}$

$6\left(\dfrac{2}{3}x - 14\right) < 6\left(-\dfrac{5}{6}\right)$ or $6\left(\dfrac{2}{3}x - 14\right) > 6 \cdot \dfrac{5}{6}$

$4x - 84 < -5$ or $4x - 84 > 5$

$4x - 84 + 84 < -5 + 84$ or $4x - 84 + 84 > 5 + 84$

$4x < 79$ or $4x > 89$

$\dfrac{4x}{4} < \dfrac{79}{4}$ or $\dfrac{4x}{4} > \dfrac{89}{4}$

$x < \dfrac{79}{4}$ or $x > \dfrac{89}{4}$

The solution set is $\left\{x \middle| x < \dfrac{79}{4} \text{ or } x > \dfrac{89}{4}\right\}$, or $\left(-\infty, \dfrac{79}{4}\right) \cup \left(\dfrac{89}{4}, \infty\right)$.

Exercise Set 9.2

57. $$\frac{2x-5}{6} \leq -3 \quad \text{or} \quad \frac{2x-5}{6} \geq 4$$
$$6\left(\frac{2x-5}{6}\right) \leq 6(-3) \quad \text{or} \quad 6\left(\frac{2x-5}{6}\right) \geq 6 \cdot 4$$
$$2x - 5 \leq -18 \quad \text{or} \quad 2x - 5 \geq 24$$
$$2x - 5 + 5 \leq -18 + 5 \quad \text{or} \quad 2x - 5 + 5 \geq 24 + 5$$
$$2x \leq -13 \quad \text{or} \quad 2x \geq 29$$
$$\frac{2x}{2} \leq \frac{-13}{2} \quad \text{or} \quad \frac{2x}{2} \geq \frac{29}{2}$$
$$x \leq -\frac{13}{2} \quad \text{or} \quad x \geq \frac{29}{2}$$

The solution set is $\left\{x \,\middle|\, x \leq -\frac{13}{2} \text{ or } x \geq \frac{29}{2}\right\}$, or $\left(-\infty, -\frac{13}{2}\right] \cup \left[\frac{29}{2}, \infty\right)$.

59. Familiarize. We will use the formula $P = 1 + \frac{d}{33}$.

Translate. We want to find those values of d for which
$$1 \leq P \leq 7$$
or
$$1 \leq 1 + \frac{d}{33} \leq 7.$$

Solve. We solve the inequality.
$$1 \leq 1 + \frac{d}{33} \leq 7$$
$$0 \leq \frac{d}{33} \leq 6$$
$$0 \leq d \leq 198$$

Check. We could do a partial check by substituting some values for d in the formula. The result checks.

State. The pressure is at least 1 atm and at most 7 atm for depths d in the set $\{d | 0 \text{ ft} \leq d \leq 198 \text{ ft}\}$.

61. Familiarize. Let $b =$ the number of beats per minute. Note that $10 \text{ sec} = 10 \text{ sec} \times \frac{1 \text{ min}}{60 \text{ sec}} = \frac{10}{60} \times \frac{\text{sec}}{\text{sec}} \times 1 \text{ min} = \frac{1}{6}$ min. Then in 10 sec, or $\frac{1}{6}$ min, the woman should have between $\frac{1}{6} \cdot 138$ and $\frac{1}{6} \cdot 162$ beats.

Translate. We want to find the value of b for which
$$\frac{1}{6} \cdot 138 < b < \frac{1}{6} \cdot 162$$

Solve. We solve the inequality.
$$\frac{1}{6} \cdot 138 < b < \frac{1}{6} \cdot 162$$
$$23 < b < 27$$

Check. If the number of beats in 10 sec, or $\frac{1}{6}$ min, is between 23 and 27, then the number of beats per minute is between $6 \cdot 23$ and $6 \cdot 27$, or between 138 and 162. The answer checks.

State. The number of beats should be between 23 and 27.

63. Familiarize. We will use the equation $y = 14.57x + 62.91$.

Translate. Since y is given in millions, we want to find the values of x for which

$$106.62 < y < 194.04$$
or
$$106.62 < 14.57x + 62.91 < 194.04.$$

Solve. We solve the inequality.
$$106.62 < 14.57x + 62.91 < 194.04$$
$$43.71 < 14.57x < 131.13$$
$$3 < x < 9$$

Check. We could do a partial check by substituting some values for x in the equation. The result checks.

State. The number of online shoppers will be between 106,620,000 and 194,040,000 for years x in the set $\{x | 3 \text{ yr} < x < 9 \text{ yr}\}$, where x is the number of years after 2000, or between 2003 and 2009.

65. Discussion and Writing Exercise

67. $3x - 2y = -7$, (1)
$2x + 5y = 8$ (2)

We multiply twice to make two terms become additive inverses.

From (1): $15x - 10y = -35$ Multiplying by 5
From (2): $\underline{4x + 10y = 16}$ Multiplying by 2
$19x + 0 = -19$ Adding
$19x = -19$
$x = -1$

Substitute -1 for x in one of the original equations and solve for y.
$2x + 5y = 8$ Equation (2)
$2(-1) + 5y = 8$ Substituting
$-2 + 5y = 8$
$5y = 10$
$y = 2$

We obtain $(-1, 2)$. This checks, so it is the solution.

69. $x + y = 0$ (1)
$\underline{x - y = 8}$ (2)
$2x + 0 = 8$ Adding
$2x = 8$
$x = 4$

Substitute 4 for x in one of the original equations and solve for y.
$x + y = 0$ Equation (1)
$4 + y = 0$ Substituting
$y = -4$

We obtain $(4, -4)$. This checks, so it is the solution.

71. First find the slope of the line:
$$m = \frac{7 - (-1)}{0 - 2} = \frac{8}{-2} = -4$$

Now we use the slope and one of the given points to find b. We choose $(0, 7)$ and substitute 0 for x, 7 for y, and -4 for m in $y = mx + b$. Then we solve for b.

$y = mx + b$
$7 = -4 \cdot 0 + b$
$7 = b$

Finally, we use the equation $y = mx + b$ and substitute -4 for m and 7 for b.
$y = -4x + 7$

73. $(2a - b)(3a + 5b) = 2a \cdot 3a + 2a \cdot 5b - b \cdot 3a - b \cdot 5b$
$= 6a^2 + 10ab - 3ab - 5b^2$
$= 6a^2 + 7ab - 5b^2$

75. $(7x - 8)(3x - 5) = 7x \cdot 3x - 7x \cdot 5 - 8 \cdot 3x + 8 \cdot 5$
$= 21x^2 - 35x - 24x + 40$
$= 21x^2 - 59x + 40$

77. $-\dfrac{2}{15} \le \dfrac{2}{3}x - \dfrac{2}{5} \le \dfrac{2}{15}$

$-\dfrac{2}{15} \le \dfrac{2}{3}x - \dfrac{6}{15} \le \dfrac{2}{15}$

$\dfrac{4}{15} \le \dfrac{2}{3}x \le \dfrac{8}{15}$

$\dfrac{3}{2} \cdot \dfrac{4}{15} \le \dfrac{3}{2} \cdot \dfrac{2}{3}x \le \dfrac{3}{2} \cdot \dfrac{8}{15}$

$\dfrac{2}{5} \le x \le \dfrac{4}{5}$

The solution set is $\left\{x \middle| \dfrac{2}{5} \le x \le \dfrac{4}{5}\right\}$, or $\left[\dfrac{2}{5}, \dfrac{4}{5}\right]$.

79. $3x < 4 - 5x < 5 + 3x$
$0 < 4 - 8x < 5$ Subtracting $3x$
$-4 < -8x < 1$
$\dfrac{1}{2} > x > -\dfrac{1}{8}$

The solution set is $\left\{x \middle| -\dfrac{1}{8} < x < \dfrac{1}{2}\right\}$, or $\left(-\dfrac{1}{8}, \dfrac{1}{2}\right)$.

81. $x + 4 < 2x - 6 \le x + 12$
$4 < x - 6 \le 12$ Subtracting x
$10 < x \le 18$

The solution set is $\{x | 10 < x \le 18\}$, or $(10, 18]$.

83. If $-b < -a$, then $-1(-b) > -1(-a)$, or $b > a$, or $a < b$. The statement is true.

85. Let $a = 5$, $c = 12$, and $b = 2$. Then $a < c$ and $b < c$, but $a \not< b$. The given statement is false.

87. The numbers in either the set of all rational numbers or the set of all irrational numbers are all real numbers, so the union is all real numbers.

There are no numbers common to the set of all rational numbers and the set of all irrational numbers, so the intersection is $\emptyset$.

Exercise Set 9.3

1. $|9x| = |9| \cdot |x| = 9|x|$

3. $|2x^2| = |2| \cdot |x^2|$
$= 2|x^2|$
$= 2x^2$ Since x^2 is never negative

5. $|-2x^2| = |-2| \cdot |x^2|$
$= 2|x^2|$
$= 2x^2$ Since x^2 is never negative

7. $|-6y| = |-6| \cdot |y| = 6|y|$

9. $\left|\dfrac{-2}{x}\right| = \dfrac{|-2|}{|x|} = \dfrac{2}{|x|}$

11. $\left|\dfrac{x^2}{-y}\right| = \dfrac{|x^2|}{|-y|}$
$= \dfrac{x^2}{|-y|}$
$= \dfrac{x^2}{|y|}$ The absolute value of the opposite of a number is the same as the absolute value of the number.

13. $\left|\dfrac{-8x^2}{2x}\right| = |-4x| = |-4| \cdot |x| = 4|x|$

15. $|-8 - (-46)| = |38| = 38$, or
$|-46 - (-8)| = |-38| = 38$

17. $|36 - 17| = |19| = 19$, or
$|17 - 36| = |-19| = 19$

19. $|-3.9 - 2.4| = |-6.3| = 6.3$, or
$|2.4 - (-3.9)| = |6.3| = 6.3$

21. $|-5 - 0| = |-5| = 5$, or
$|0 - (-5)| = |5| = 5$

23. $|x| = 3$
$x = -3$ or $x = 3$ Absolute-value principle
The solution set is $\{-3, 3\}$.

25. $|x| = -3$
The absolute value of a number is always nonnegative. Therefore, the solution set is $\emptyset$.

27. $|q| = 0$
The only number whose absolute value is 0 is 0. The solution set is $\{0\}$.

29. $|x - 3| = 12$
$x - 3 = -12$ or $x - 3 = 12$ Absolute-value principle
$x = -9$ or $x = 15$

The solution set is $\{-9, 15\}$.

Exercise Set 9.3

31. $|2x - 3| = 4$

$2x - 3 = -4$ or $2x - 3 = 4$ Absolute-value principle

$2x = -1$ or $2x = 7$

$x = -\dfrac{1}{2}$ or $x = \dfrac{7}{2}$

The solution set is $\left\{-\dfrac{1}{2}, \dfrac{7}{2}\right\}$.

33. $|4x - 9| = 14$

$4x - 9 = -14$ or $4x - 9 = 14$

$4x = -5$ or $4x = 23$

$x = -\dfrac{5}{4}$ or $x = \dfrac{23}{4}$

The solution set is $\left\{-\dfrac{5}{4}, \dfrac{23}{4}\right\}$.

35. $|x| + 7 = 18$

$|x| + 7 - 7 = 18 - 7$ Subtracting 7

$|x| = 11$

$x = -11$ or $x = 11$ Absolute-value principle

The solution set is $\{-11, 11\}$.

37. $574 = 283 + |t|$

$291 = |t|$ Subtracting 283

$t = -291$ or $t = 291$ Absolute-value principle

The solution set is $\{-291, 291\}$.

39. $|5x| = 40$

$5x = -40$ or $5x = 40$

$x = -8$ or $x = 8$

The solution set is $\{-8, 8\}$.

41. $|3x| - 4 = 17$

$|3x| = 21$ Adding 4

$3x = -21$ or $3x = 21$

$x = -7$ or $x = 7$

The solution set is $\{-7, 7\}$.

43. $7|w| - 3 = 11$

$7|w| = 14$ Adding 3

$|w| = 2$ Dividing by 7

$w = -2$ or $w = 2$ Absolute-value principle

The solution set is $\{-2, 2\}$.

45. $\left|\dfrac{2x - 1}{3}\right| = 5$

$\dfrac{2x - 1}{3} = -5$ or $\dfrac{2x - 1}{3} = 5$

$2x - 1 = -15$ or $2x - 1 = 15$

$2x = -14$ or $2x = 16$

$x = -7$ or $x = 8$

The solution set is $\{-7, 8\}$.

47. $|m + 5| + 9 = 16$

$|m + 5| = 7$ Subtracting 9

$m + 5 = -7$ or $m + 5 = 7$

$m = -12$ or $m = 2$

The solution set is $\{-12, 2\}$.

49. $10 - |2x - 1| = 4$

$-|2x - 1| = -6$ Subtracting 10

$|2x - 1| = 6$ Multiplying by -1

$2x - 1 = -6$ or $2x - 1 = 6$

$2x = -5$ or $2x = 7$

$x = -\dfrac{5}{2}$ or $x = \dfrac{7}{2}$

The solution set is $\left\{-\dfrac{5}{2}, \dfrac{7}{2}\right\}$.

51. $|3x - 4| = -2$

The absolute value of a number is always nonnegative. The solution set is $\emptyset$.

53. $\left|\dfrac{5}{9} + 3x\right| = \dfrac{1}{6}$

$\dfrac{5}{9} + 3x = -\dfrac{1}{6}$ or $\dfrac{5}{9} + 3x = \dfrac{1}{6}$

$3x = -\dfrac{13}{18}$ or $3x = -\dfrac{7}{18}$

$x = -\dfrac{13}{54}$ or $x = -\dfrac{7}{54}$

The solution set is $\left\{-\dfrac{13}{54}, -\dfrac{7}{54}\right\}$.

55. $|3x + 4| = |x - 7|$

$3x + 4 = x - 7$ or $3x + 4 = -(x - 7)$

$2x + 4 = -7$ or $3x + 4 = -x + 7$

$2x = -11$ or $4x + 4 = 7$

$x = -\dfrac{11}{2}$ or $4x = 3$

$x = -\dfrac{11}{2}$ or $x = \dfrac{3}{4}$

The solution set is $\left\{-\dfrac{11}{2}, \dfrac{3}{4}\right\}$.

57. $|x + 3| = |x - 6|$

$x + 3 = x - 6$ or $x + 3 = -(x - 6)$

$3 = -6$ or $x + 3 = -x + 6$

$3 = -6$ or $2x = 3$

$3 = -6$ or $x = \dfrac{3}{2}$

The first equation has no solution. The second equation has $\dfrac{3}{2}$ as a solution. There is only one solution of the original equation. The solution set is $\left\{\dfrac{3}{2}\right\}$.

59. $|2a+4| = |3a-1|$

$\begin{aligned} 2a+4 &= 3a-1 & \text{or} \quad 2a+4 &= -(3a-1) \\ -a+4 &= -1 & \text{or} \quad 2a+4 &= -3a+1 \\ -a &= -5 & \text{or} \quad 5a+4 &= 1 \\ a &= 5 & \text{or} \quad 5a &= -3 \\ a &= 5 & \text{or} \quad a &= -\frac{3}{5} \end{aligned}$

The solution set is $\left\{5, -\frac{3}{5}\right\}$.

61. $|y-3| = |3-y|$

$\begin{aligned} y-3 &= 3-y & \text{or} \quad y-3 &= -(3-y) \\ 2y-3 &= 3 & \text{or} \quad y-3 &= -3+y \\ 2y &= 6 & \text{or} \quad -3 &= -3 \\ y &= 3 & & \text{True for all real values of } y \end{aligned}$

All real numbers are solutions.

63. $|5-p| = |p+8|$

$\begin{aligned} 5-p &= p+8 & \text{or} \quad 5-p &= -(p+8) \\ 5-2p &= 8 & \text{or} \quad 5-p &= -p-8 \\ -2p &= 3 & \text{or} \quad 5 &= -8 \\ p &= -\frac{3}{2} & & \text{False} \end{aligned}$

The solution set is $\left\{-\frac{3}{2}\right\}$.

65. $\left|\dfrac{2x-3}{6}\right| = \left|\dfrac{4-5x}{8}\right|$

$\begin{aligned} \frac{2x-3}{6} &= \frac{4-5x}{8} & \text{or} \quad \frac{2x-3}{6} &= -\left(\frac{4-5x}{8}\right) \\ 24\left(\frac{2x-3}{6}\right) &= 24\left(\frac{4-5x}{8}\right) & \text{or} \quad \frac{2x-3}{6} &= \frac{-4+5x}{8} \\ 8x-12 &= 12-15x & \text{or} \quad 24\left(\frac{2x-3}{6}\right) &= 24\left(\frac{-4+5x}{8}\right) \\ 23x-12 &= 12 & \text{or} \quad 8x-12 &= -12+15x \\ 23x &= 24 & \text{or} \quad -7x-12 &= -12 \\ x &= \frac{24}{23} & \text{or} \quad -7x &= 0 \\ & & x &= 0 \end{aligned}$

The solution set is $\left\{\dfrac{24}{23}, 0\right\}$.

67. $\left|\dfrac{1}{2}x - 5\right| = \left|\dfrac{1}{4}x + 3\right|$

$\begin{aligned} \frac{1}{2}x - 5 &= \frac{1}{4}x + 3 & \text{or} \quad \frac{1}{2}x - 5 &= -\left(\frac{1}{4}x + 3\right) \\ \frac{1}{4}x - 5 &= 3 & \text{or} \quad \frac{1}{2}x - 5 &= -\frac{1}{4}x - 3 \\ \frac{1}{4}x &= 8 & \text{or} \quad \frac{3}{4}x - 5 &= -3 \\ x &= 32 & \text{or} \quad \frac{3}{4}x &= 2 \\ x &= 32 & \text{or} \quad x &= \frac{8}{3} \end{aligned}$

The solution set is $\left\{32, \dfrac{8}{3}\right\}$.

69. $|x| < 3$

$-3 < x < 3$

The solution set is $\{x | -3 < x < 3\}$, or $(-3, 3)$.

71. $|x| \geq 2$

$x \leq -2 \text{ or } x \geq 2$

The solution set is $\{x | x \leq -2 \text{ or } x \geq 2\}$, or $(-\infty, -2] \cup [2, \infty)$.

73. $|x-1| < 1$

$-1 < x - 1 < 1$

$0 < x < 2$

The solution set is $\{x | 0 < x < 2\}$, or $(0, 2)$.

75. $5|x+4| \leq 10$

$\begin{aligned} |x+4| &\leq 2 & & \text{Dividing by 5} \\ -2 &\leq x+4 \leq 2 \\ -6 &\leq x \leq -2 & & \text{Subtracting 4} \end{aligned}$

The solution set is $\{x | -6 \leq x \leq -2\}$, or $[-6, -2]$.

77. $|2x-3| \leq 4$

$\begin{aligned} -4 &\leq 2x - 3 \leq 4 \\ -1 &\leq 2x \leq 7 & & \text{Adding 3} \\ -\frac{1}{2} &\leq x \leq \frac{7}{2} & & \text{Dividing by 2} \end{aligned}$

The solution set is $\left\{x \,\middle|\, -\dfrac{1}{2} \leq x \leq \dfrac{7}{2}\right\}$, or $\left[-\dfrac{1}{2}, \dfrac{7}{2}\right]$.

79. $|2y-7| > 10$

$\begin{aligned} 2y - 7 &< -10 & \text{or} \quad 2y - 7 &> 10 \\ 2y &< -3 & \text{or} \quad 2y &> 17 & \text{Adding 7} \\ y &< -\frac{3}{2} & \text{or} \quad y &> \frac{17}{2} & \text{Dividing by 2} \end{aligned}$

The solution set is $\left\{y \,\middle|\, y < -\dfrac{3}{2} \text{ or } y > \dfrac{17}{2}\right\}$, or $\left(-\infty, -\dfrac{3}{2}\right) \cup \left(\dfrac{17}{2}, \infty\right)$.

Exercise Set 9.3   223

81. $|4x - 9| \geq 14$

$\qquad 4x - 9 \leq -14 \quad \text{or} \quad 4x - 9 \geq 14$

$\qquad\quad 4x \leq -5 \quad \text{or} \qquad 4x \geq 23$

$\qquad\quad\; x \leq -\dfrac{5}{4} \quad \text{or} \qquad x \geq \dfrac{23}{4}$

The solution set is $\left\{x \middle| x \leq -\dfrac{5}{4} \text{ or } x \geq \dfrac{23}{4}\right\}$, or

$\left(-\infty, -\dfrac{5}{4}\right] \cup \left[\dfrac{23}{4}, \infty\right)$.

83. $|y - 3| < 12$

$\qquad -12 < y - 3 < 12$

$\qquad -9 < y < 15 \qquad \text{Adding 3}$

The solution set is $\{y| -9 < y < 15\}$, or $(-9, 15)$.

85. $|2x + 3| \leq 4$

$\qquad -4 \leq 2x + 3 \leq 4$

$\qquad -7 \leq 2x \leq 1 \qquad \text{Subtracting 3}$

$\qquad -\dfrac{7}{2} \leq x \leq \dfrac{1}{2} \qquad \text{Dividing by 2}$

The solution set is $\left\{x \middle| -\dfrac{7}{2} \leq x \leq \dfrac{1}{2}\right\}$, or $\left[-\dfrac{7}{2}, \dfrac{1}{2}\right]$.

87. $|4 - 3y| > 8$

$\qquad 4 - 3y < -8 \quad \text{or} \quad 4 - 3y > 8$

$\qquad\;\; -3y < -12 \quad \text{or} \quad\;\; -3y > 4 \qquad \text{Subtracting 4}$

$\qquad\qquad y > 4 \quad \text{or} \qquad y < -\dfrac{4}{3} \qquad \text{Dividing by } -3$

The solution set is $\left\{y \middle| y < -\dfrac{4}{3} \text{ or } y > 4\right\}$, or

$\left(-\infty, -\dfrac{4}{3}\right) \cup (4, \infty)$.

89. $|9 - 4x| \geq 14$

$\qquad 9 - 4x \leq -14 \quad \text{or} \quad 9 - 4x \geq 14$

$\qquad\;\; -4x \leq -23 \quad \text{or} \qquad -4x \geq 5 \qquad \text{Subtracting 9}$

$\qquad\quad x \geq \dfrac{23}{4} \quad \text{or} \qquad x \leq -\dfrac{5}{4} \qquad \text{Dividing by } -4$

The solution set is $\left\{x \middle| x \leq -\dfrac{5}{4} \text{ or } x \geq \dfrac{23}{4}\right\}$, or

$\left(-\infty, -\dfrac{5}{4}\right] \cup \left[\dfrac{23}{4}, \infty\right)$.

91. $|3 - 4x| < 21$

$\qquad -21 < 3 - 4x < 21$

$\qquad -24 < -4x < 18 \qquad \text{Subtracting 3}$

$\qquad\;\; 6 > x > -\dfrac{9}{2} \qquad \text{Dividing by } -4 \text{ and simplifying}$

The solution set is $\left\{x \middle| 6 > x > -\dfrac{9}{2}\right\}$, or

$\left\{x \middle| -\dfrac{9}{2} < x < 6\right\}$, or $\left(-\dfrac{9}{2}, 6\right)$.

93. $\left|\dfrac{1}{2} + 3x\right| \geq 12$

$\qquad \dfrac{1}{2} + 3x \leq -12 \quad \text{or} \quad \dfrac{1}{2} + 3x \geq 12$

$\qquad\qquad 3x \leq -\dfrac{25}{2} \quad \text{or} \qquad 3x \geq \dfrac{23}{2} \qquad \text{Subtracting } \dfrac{1}{2}$

$\qquad\qquad\;\; x \leq -\dfrac{25}{6} \quad \text{or} \qquad\; x \geq \dfrac{23}{6} \qquad \text{Dividing by 3}$

The solution set is $\left\{x \middle| x \leq -\dfrac{25}{6} \text{ or } x \geq \dfrac{23}{6}\right\}$, or

$\left(-\infty, -\dfrac{25}{6}\right] \cup \left[\dfrac{23}{6}, \infty\right)$.

95. $\left|\dfrac{x - 7}{3}\right| < 4$

$\qquad -4 < \dfrac{x - 7}{3} < 4$

$\qquad -12 < x - 7 < 12 \qquad \text{Multiplying by 3}$

$\qquad\;\; -5 < x < 19 \qquad \text{Adding 7}$

The solution set is $\{x| -5 < x < 19\}$, or $(-5, 19)$.

97. $\left|\dfrac{2 - 5x}{4}\right| \geq \dfrac{2}{3}$

$\qquad \dfrac{2 - 5x}{4} \leq -\dfrac{2}{3} \quad \text{or} \quad \dfrac{2 - 5x}{4} \geq \dfrac{2}{3}$

$\qquad 2 - 5x \leq -\dfrac{8}{3} \quad \text{or} \quad 2 - 5x \geq \dfrac{8}{3} \qquad \text{Multiplying by 4}$

$\qquad\;\; -5x \leq -\dfrac{14}{3} \quad \text{or} \qquad -5x \geq \dfrac{2}{3} \qquad \text{Subtracting 2}$

$\qquad\qquad x \geq \dfrac{14}{15} \quad \text{or} \qquad x \leq -\dfrac{2}{15} \qquad \text{Dividing by } -5$

The solution set is $\left\{x \middle| x \leq -\dfrac{2}{15} \text{ or } x \geq \dfrac{14}{15}\right\}$, or

$\left(-\infty, -\dfrac{2}{15}\right] \cup \left[\dfrac{14}{15}, \infty\right)$.

99. $|m + 5| + 9 \leq 16$

$\qquad |m + 5| \leq 7 \qquad \text{Subtracting 9}$

$\qquad -7 \leq m + 5 \leq 7$

$\qquad -12 \leq m \leq 2$

The solution set is $\{m| -12 \leq m \leq 2\}$, or $[-12, 2]$.

101. $7 - |3 - 2x| \geq 5$

$\qquad -|3 - 2x| \geq -2 \qquad \text{Subtracting 7}$

$\qquad\;\; |3 - 2x| \leq 2 \qquad \text{Multiplying by } -1$

$\qquad -2 \leq 3 - 2x \leq 2$

$\qquad -5 \leq -2x \leq -1 \qquad \text{Subtracting 3}$

$\qquad\;\; \dfrac{5}{2} \geq x \geq \dfrac{1}{2} \qquad \text{Dividing by } -2$

The solution set is $\left\{x \middle| \dfrac{5}{2} \geq x \geq \dfrac{1}{2}\right\}$, or $\left\{x \middle| \dfrac{1}{2} \leq x \leq \dfrac{5}{2}\right\}$, or

$\left[\dfrac{1}{2}, \dfrac{5}{2}\right]$.

103. $\left|\dfrac{2x-1}{0.0059}\right| \leq 1$

$$-1 \leq \dfrac{2x-1}{0.0059} \leq 1$$
$$-0.0059 \leq 2x - 1 \leq 0.0059$$
$$0.9941 \leq 2x \leq 1.0059$$
$$0.49705 \leq x \leq 0.50295$$

The solution set is $\{x|0.49705 \leq x \leq 0.50295\}$, or $[0.49705, 0.50295]$.

105. Discussion and Writing Exercise

107. $f(x) = |x - 2|$

Since we can calculate $|x - 2|$ for any real number x, the domain is the set of all real numbers.

109. $f(x) = \dfrac{1}{x-5}$

Since $\dfrac{1}{x-5}$ cannot be calculated when the denominator is 0, we find the x-value that causes $x - 5$ to be 0.
$$x - 5 = 0$$
$$x = 5 \quad \text{Adding 5 on both sides}$$

Thus, 5 is not in the domain of f, while all other real numbers are. The domain of f is
$\{x|x \text{ is a real number } and\ x \neq 5\}$, or $(-\infty, 5) \cup (5, \infty)$.

111. $2x + y = 7$, (1)
$x - 3y = 21$ (2)

Multiply the first equation by 3 and then add.
$$\begin{aligned}6x + 3y &= 21 \quad \text{Multiplying by 3}\\ \underline{x - 3y} &= \underline{21}\\ 7x + 0 &= 42 \quad \text{Adding}\\ 7x &= 42\\ x &= 6\end{aligned}$$

Substitute 6 for x in one of the original equations and solve for y.
$$\begin{aligned}2x + y &= 7 \quad \text{Equation (1)}\\ 2(6) + y &= 7 \quad \text{Substituting}\\ 12 + y &= 7\\ y &= -5\end{aligned}$$

We obtain $(6, -5)$. This checks, so it is the solution.

113. $|d - 6\text{ ft}| \leq \dfrac{1}{2}\text{ ft}$

$$-\dfrac{1}{2}\text{ ft} \leq d - 6\text{ ft} \leq \dfrac{1}{2}\text{ ft}$$
$$5\dfrac{1}{2}\text{ ft} \leq d \leq 6\dfrac{1}{2}\text{ ft}$$

The solution set is $\left\{d\bigg|5\dfrac{1}{2}\text{ ft} \leq d \leq 6\dfrac{1}{2}\text{ ft}\right\}$.

115. $|x + 5| = x + 5$

From the definition of absolute value, $|x + 5| = x + 5$ only when $x + 5 \geq 0$, or $x \geq -5$. The solution set is $\{x|x \geq -5\}$, or $[-5, \infty)$.

117. $|7x - 2| = x + 4$

From the definition of absolute value, we know $x + 4 \geq 0$, or $x \geq -4$. So we have $x \geq -4$ and
$$\begin{aligned}7x - 2 &= x + 4 \quad or \quad 7x - 2 = -(x+4)\\ 6x &= 6 \quad \quad\ or \quad 7x - 2 = -x - 4\\ x &= 1 \quad \quad\ or \quad 8x = -2\\ x &= 1 \quad \quad\ or \quad x = -\dfrac{1}{4}\end{aligned}$$

The solution set is $\left\{x\bigg|x \geq -4\ and\ x = 1\ or\ x = -\dfrac{1}{4}\right\}$, or $\left\{1, -\dfrac{1}{4}\right\}$.

119. $|x - 6| \leq -8$

From the definition of absolute value we know that $|x - 6| \geq 0$. Thus $|x - 6| \leq -8$ is false for all x. The solution set is $\emptyset$.

121. $|x + 5| > x$

The inequality is true for all $x < 0$ (because absolute value must be nonnegative). The solution set in this case is $\{x|x < 0\}$. If $x = 0$, we have $|0 + 5| > 0$, which is true. The solution set in this case is $\{0\}$. If $x > 0$, we have the following:
$$\begin{aligned}x + 5 < -x \quad &or \quad x + 5 > x\\ 2x < -5 \quad &or \quad\quad 5 > 0\\ x < -\dfrac{5}{2} \quad &or \quad\quad 5 > 0\end{aligned}$$

Although $x > 0$ and $x < -\dfrac{5}{2}$ yields no solution, $x > 0$ and $5 > 0$ (true for all x) yield the solution set $\{x|x > 0\}$ in this case. The solution set for the inequality is $\{x|x < 0\} \cup \{0\} \cup \{x|x > 0\}$, or all real numbers.

123. $|x| \geq 0$

Since the absolute value of a number is always nonnegative, all real numbers are solutions.

125. $-3 < x < 3$ is equivalent to $|x| < 3$.

127. $x \leq -6$ or or $x \geq 6$ is equivalent to $|x| \geq 6$.

129. $\quad x < -8 \quad or \quad\quad x > 2$
$x + 3 < -5 \quad or \quad x + 3 > 5 \quad \text{Adding 3}$
$|x + 3| > 5$

Exercise Set 9.4

1. We use alphabetical order to replace x by -3 and y by 3.
$$\begin{array}{c|c}3x + y < -5\\ \hline 3(-3) + 3\ ?\ -5\\ -9 + 3\\ -6 & \text{TRUE}\end{array}$$

Since $-6 < -5$ is true, $(-3, 3)$ is a solution.

Exercise Set 9.4

3. We use alphabetical order to replace x by 5 and y by 9.

$$\begin{array}{c|c} 2x - y > -1 \\ \hline 2 \cdot 5 - 9 \ ? \ -1 \\ 10 - 9 \\ 1 & \text{TRUE} \end{array}$$

Since $1 > -1$ is true, $(5, 9)$ is a solution.

5. Graph: $y > 2x$

We first graph the line $y = 2x$. We draw the line dashed since the inequality symbol is $>$. To determine which half-plane to shade, test a point not on the line. We try $(1, 1)$ and substitute:

$$\begin{array}{c|c} y > 2x \\ \hline 1 \ ? \ 2 \cdot 1 \\ 2 & \text{FALSE} \end{array}$$

Since $1 > 2$ is false, $(1, 1)$ is not a solution, nor are any points in the half-plane containing $(1, 1)$. The points in the opposite half-plane are solutions, so we shade that half-plane and obtain the graph.

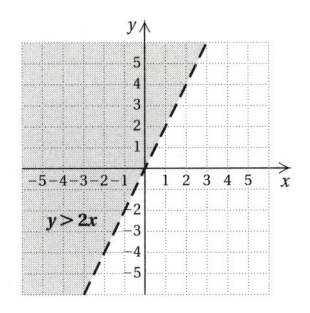

7. Graph: $y < x + 1$

First graph the line $y = x + 1$. Draw it dashed since the inequality symbol is $<$. Test the point $(0, 0)$ to determine if it is a solution.

$$\begin{array}{c|c} y < x + 1 \\ \hline 0 \ ? \ 0 + 1 \\ 1 & \text{TRUE} \end{array}$$

Since $0 < 1$ is true, we shade the half-plane containing $(0, 0)$ and obtain the graph.

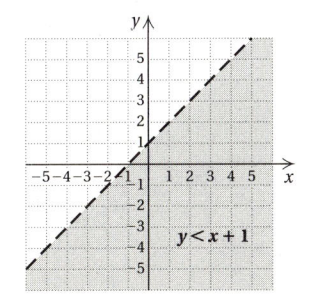

9. Graph: $y > x - 2$

First graph the line $y = x - 2$. Draw a dashed line since the inequality symbol is $>$. Test the point $(0, 0)$ to determine if it is a solution.

$$\begin{array}{c|c} y > x - 2 \\ \hline 0 \ ? \ 0 - 2 \\ -2 & \text{TRUE} \end{array}$$

Since $0 > -2$ is true, we shade the half-plane containing $(0, 0)$ and obtain the graph.

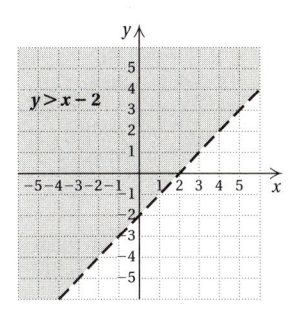

11. Graph: $x + y < 4$

First graph $x + y = 4$. Draw the line dashed since the inequality symbol is $<$. Test the point $(0, 0)$ to determine if it is a solution.

$$\begin{array}{c|c} x + y < 4 \\ \hline 0 + 0 \ ? \ 4 \\ 0 & \text{TRUE} \end{array}$$

Since $0 < 4$ is true, we shade the half-plane containing $(0, 0)$ and obtain the graph.

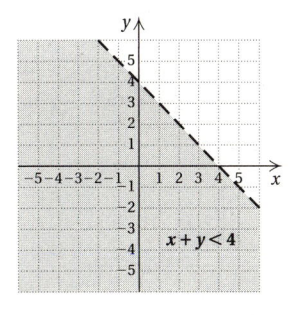

13. Graph: $3x + 4y \leq 12$

We first graph $3x + 4y = 12$. Draw the line solid since the inequality symbol is $\leq$. Test the point $(0, 0)$ to determine if it is a solution.

$$\begin{array}{c|c} 3x + 4y \leq 12 \\ \hline 3 \cdot 0 + 4 \cdot 0 \ ? \ 12 \\ 0 & \text{TRUE} \end{array}$$

Since $0 \leq 12$ is true, we shade the half-plane containing $(0, 0)$ and obtain the graph.

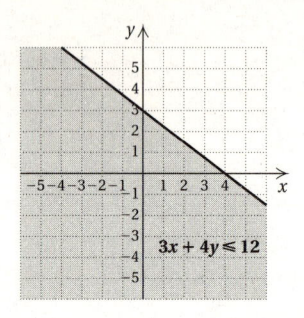

[Graph: $3x + 4y \leq 12$]

15. Graph: $2y - 3x > 6$

We first graph $2y - 3x = 6$. Draw the line dashed since the inequality symbol is $>$. Test the point $(0,0)$ to determine if it is a solution.

$$\frac{2y - 3x > 6}{2 \cdot 0 - 3 \cdot 0 \;?\; 6}$$
$$0 \;\big|\; \text{FALSE}$$

Since $0 > 6$ is false, we shade the half-plane that does not contain $(0,0)$ and obtain the graph.

[Graph: $2y - 3x > 6$]

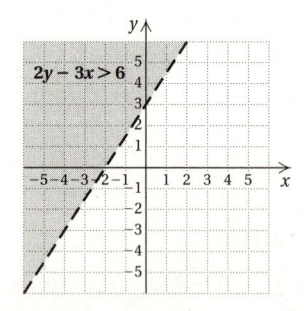

17. Graph: $3x - 2 \leq 5x + y$
$$-2 \leq 2x + y$$

We first graph $-2 = 2x + y$. Draw the line solid since the inequality symbol is $\leq$. Test the point $(0,0)$ to determine if it is a solution.

$$\frac{-2 \leq 2x + y}{-2 \;?\; 2 \cdot 0 + 0}$$
$$\big|\; 0 \quad \text{TRUE}$$

Since $-2 \leq 0$ is true, we shade the half-plane containing $(0,0)$ and obtain the graph.

[Graph: $3x - 2 \leq 5x + y$]

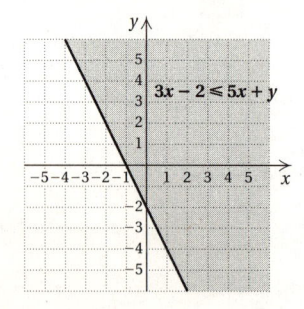

19. Graph: $x < 5$

We first graph $x = 5$. Draw the line dashed since the inequality symbol is $<$. Test the point $(0,0)$ to determine if it is a solution.

$$\frac{x < 5}{0 \;?\; 5 \quad \text{TRUE}}$$

Since $0 < 5$ is true, we shade the half-plane containing $(0,0)$ and obtain the graph.

[Graph: $x < 5$]

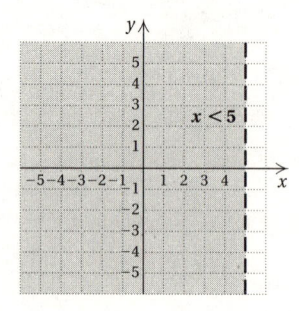

21. Graph: $y > 2$

We first graph $y = 2$. We draw the line dashed since the inequality symbol is $>$. Test the point $(0,0)$ to determine if it is a solution.

$$\frac{y > 2}{0 \;?\; 2 \quad \text{FALSE}}$$

Since $0 > 2$ is false, we shade the half-plane that does not contain $(0,0)$ and obtain the graph.

[Graph: $y > 2$]

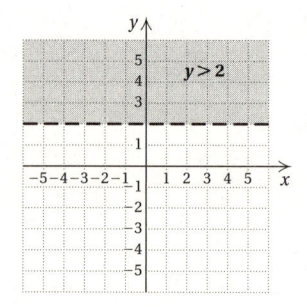

23. Graph: $2x + 3y \leq 6$

We first graph $2x + 3y = 6$. We draw the line solid since the inequality symbol is $\leq$. Test the point $(0,0)$ to determine if it is a solution.

$$\frac{2x + 3y \leq 6}{2 \cdot 0 + 3 \cdot 0 \;?\; 6}$$
$$0 \;\big|\; \text{TRUE}$$

Since $0 \leq 6$ is true, we shade the half-plane containing $(0,0)$ and obtain the graph.

[Graph: $2x + 3y \leq 6$]

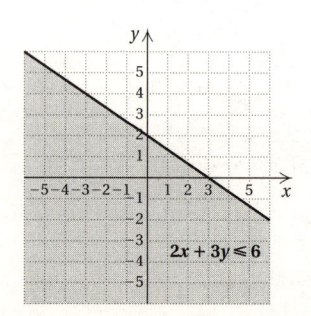

Exercise Set 9.4   227

25. The intercepts of the graph of the related equation are $(0, -2)$ and $(3, 0)$, so inequality **F** could be the correct one. Since $(0, 0)$ is in the solution set of this inequality and the half-plane containing $(0, 0)$ is shaded, we know that inequality **F** corresponds to this graph.

27. The intercepts of the graph of the related equation are $(-5, 0)$ and $(0, 3)$, so inequality **B** could be the correct one. Since $(0, 0)$ is in the solution set of this inequality and the half-plane containing $(0, 0)$ is shaded, we know that inequality **B** corresponds to this graph.

29. The intercepts of the graph of the related equation are $(-3, 0)$ and $(0, -3)$, so inequality **C** could be the correct one. Since $(0, 0)$ is not in the solution set of the inequality and the half-plane that does not contain $(0, 0)$ is shaded, we know that inequality **C** corresponds to this graph.

31. Graph: $y \geq x$,
$\quad\quad\quad\quad y \leq -x + 2$

We graph the lines $y = x$ and $y = -x + 2$, using solid lines. We indicate the region for each inequality by the arrows at the ends of the lines. Note where the regions overlap, and shade the region of solutions.

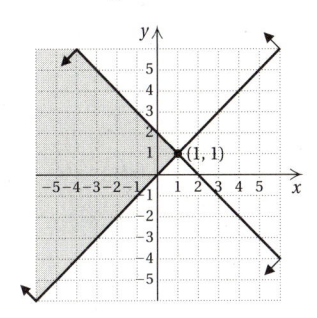

To find the vertex we solve the system of related equations:
$y = x$,
$y = -x + 2$

Solving, we obtain the vertex $(1, 1)$.

33. Graph: $y > x$,
$\quad\quad\quad\quad y < -x + 1$

We graph the lines $y = x$ and $y = -x + 1$, using dashed lines. We indicate the region for each inequality by arrows at the ends of the lines. Note where the regions overlap, and shade the region of solutions.

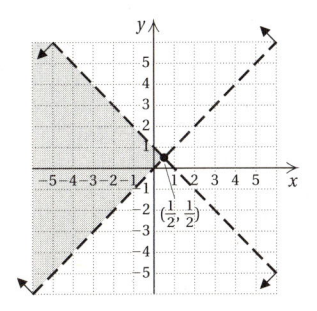

To find the vertex we solve the system of related equations:
$y = x$,
$y = -x + 1$

Solving, we obtain the vertex $\left(\dfrac{1}{2}, \dfrac{1}{2}\right)$.

35. Graph: $y \geq -2$,
$\quad\quad\quad\quad x \geq 1$

We graph the lines $y = -2$ and $x = 1$, using solid lines. We indicate the region for each inequality by arrows. Shade the region where they overlap.

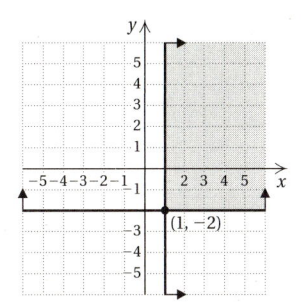

To find the vertex, we solve the system of related equations:
$y = -2$,
$x = 1$

Solving, we obtain the vertex $(1, -2)$.

37. Graph: $x \leq 3$,
$\quad\quad\quad\quad y \geq -3x + 2$

Graph the lines $x = 3$ and $y = -3x + 2$, using solid lines. Indicate the region for each inequality by arrows, and shade the region where they overlap.

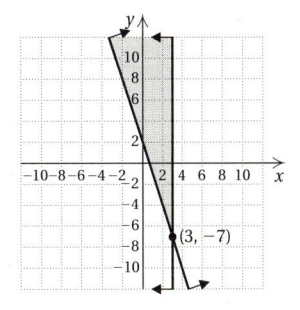

To find the vertex we solve the system of related equations:
$x = 3$,
$y = -3x + 2$

Solving, we obtain the vertex $(3, -7)$.

39. Graph: $y \leq 2x + 1$,   (1)
$\quad\quad\quad\quad y \geq -2x + 1$,   (2)
$\quad\quad\quad\quad x \leq 2$ (3)

Shade the intersection of the graphs of $y \leq 2x + 1$, $y \geq -2x + 1$, and $x \leq 2$.

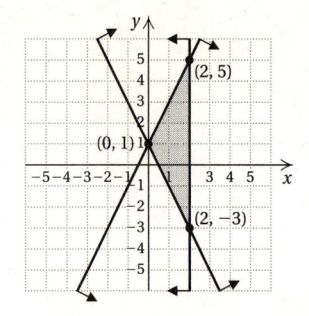

To find the vertices we solve three different systems of equations. From (1) and (2) we obtain the vertex $(0,1)$. From (1) and (3) we obtain the vertex $(2,5)$. From (2) and (3) we obtain the vertex $(2,-3)$.

41. Graph: $x + y \leq 1$,
    $x - y \leq 2$

    Graph the lines $x + y = 1$ and $x - y = 2$, using solid lines. Indicate the region for each inequality by arrows, and shade the region where they overlap.

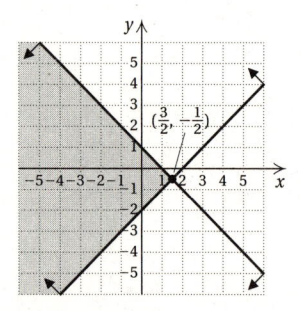

    To find the vertex we solve the system of related equations:
    $x + y = 1$,
    $x - y = 2$
    The vertex is $\left(\dfrac{3}{2}, -\dfrac{1}{2}\right)$.

43. Graph: $x + 2y \leq 12$, (1)
    $2x + y \leq 12$, (2)
    $x \geq 0$, (3)
    $y \geq 0$ (4)

    Shade the intersection of the graphs of the four inequalities above.

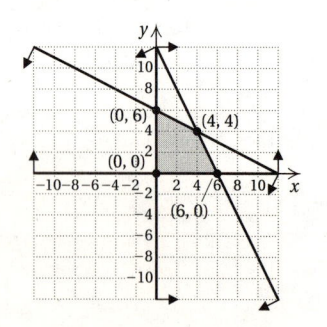

    To find the vertices we solve four different systems of equations, as follows:

System of equations	Vertex
From (1) and (2)	$(4, 4)$
From (1) and (3)	$(0, 6)$
From (2) and (4)	$(6, 0)$
From (3) and (4)	$(0, 0)$

45. Graph: $8x + 5y \leq 40$, (1)
    $x + 2y \leq 8$, (2)
    $x \geq 0$, (3)
    $y \geq 0$ (4)

    Shade the intersection of the graphs of the four inequalities above.

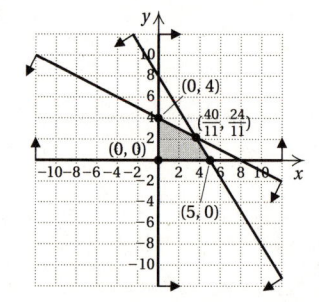

    To find the vertices we solve four different systems of equations, as follows:

System of equations	Vertex
From (1) and (2)	$\left(\dfrac{40}{11}, \dfrac{24}{11}\right)$
From (1) and (4)	$(5, 0)$
From (2) and (3)	$(0, 4)$
From (3) and (4)	$(0, 0)$

47. Discussion and Writing Exercise

49. First find the slope of the line:
    $$m = \dfrac{-6 - (-2)}{-5 - 3} = \dfrac{-4}{-8} = \dfrac{1}{2}$$
    Now we use the slope and one of the given points to find b. We choose $(3, -2)$ and substitute 3 for x, -2 for y, and $\dfrac{1}{2}$ for m in $y = mx + b$. Then we solve for b.
    $$y = mx + b$$
    $$-2 = \dfrac{1}{2} \cdot 3 + b$$
    $$-2 = \dfrac{3}{2} + b$$
    $$-\dfrac{7}{2} = b$$
    Finally, we use the equation $y = mx + b$ and substitute $\dfrac{1}{2}$ for m and $-\dfrac{7}{2}$ for b.
    $$y = \dfrac{1}{2}x - \dfrac{7}{2}$$

Exercise Set 9.4

51. First find the slope of the line:
$$m = \frac{6-12}{3-1} = \frac{-6}{2} = -3$$

Now we use the slope and one of the given points to find b. We choose $(3,6)$ and substitute 3 for x, 6 for y, and -3 for m in $y = mx + b$. Then we solve for b.
$$y = mx + b$$
$$6 = -3 \cdot 3 + b$$
$$6 = -9 + b$$
$$15 = b$$

Finally, we use the equation $y = mx + b$ and substitute -3 for m and 15 for b.
$$y = -3x + 15$$

53. First find the slope of the line:
$$m = \frac{5-(-1)}{-5-(-1)} = \frac{6}{-4} = -\frac{3}{2}$$

Now we use the slope and one of the given points to find b. We choose $(-5, 5)$ and substitute -5 for x, 5 for y, and $-\frac{3}{2}$ for m in $y = mx + b$. Then we solve for b.
$$y = mx + b$$
$$5 = -\frac{3}{2}(-5) + b$$
$$5 = \frac{15}{2} + b$$
$$-\frac{5}{2} = b$$

Finally, we use the equation $y = mx + b$ and substitute $-\frac{3}{2}$ for m and $-\frac{5}{2}$ for b.
$$y = -\frac{3}{2}x - \frac{5}{2}$$

55. $f(x) = |2 - x|$
$f(0) = |2 - 0| = |2| = 2$

57. $f(x) = |2 - x|$
$f(1) = |2 - 1| = |1| = 1$

59. $f(x) = |2 - x|$
$f(-2) = |2 - (-2)| = |4| = 4$

61. $f(x) = |2 - x|$
$f(-4) = |2 - (-4)| = |6| = 6$

63. Both the width and the height must be positive, but they must be less than 62 in. in order to be checked as luggage, so we have:
$$0 < w \leq 62,$$
$$0 < h \leq 62$$

The girth is represented by $2w + 2h$ and the length is 62 in. In order to meet postal regulations the sum of the girth and the length cannot exceed 108 in., so we have:
$$62 + 2w + 2h \leq 108, \text{ or}$$
$$2w + 2h \leq 46, \text{ or}$$
$$w + h \leq 23$$

Thus, have a system of inequalities:
$$0 < w \leq 62,$$
$$0 < h \leq 62,$$
$$w + h \leq 23$$

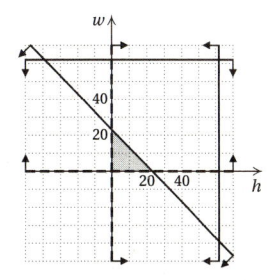

65.

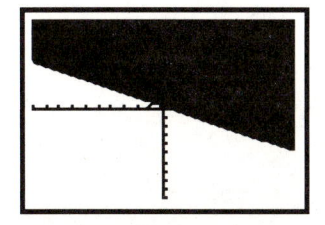

67.

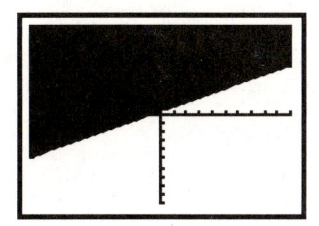

69. Left to the student

Chapter 10

Radical Expressions, Equations, and Functions

Exercise Set 10.1

1. The square roots of 16 are 4 and -4, because $4^2 = 16$ and $(-4)^2 = 16$.

3. The square roots of 144 are 12 and -12, because $12^2 = 144$ and $(-12)^2 = 144$.

5. The square roots of 400 are 20 and -20, because $20^2 = 400$ and $(-20)^2 = 400$.

7. $-\sqrt{\dfrac{49}{36}} = -\dfrac{7}{6}$ Since $\sqrt{\dfrac{49}{36}} = \dfrac{7}{6}$, $-\sqrt{\dfrac{49}{36}} = -\dfrac{7}{6}$.

9. $\sqrt{196} = 14$ Remember, $\sqrt{}$ indicates the principle square root.

11. $\sqrt{0.0036} = 0.06$

13. $\sqrt{347} \approx 18.628$

15. $\sqrt{\dfrac{285}{74}} \approx 1.962$

17. $9\sqrt{y^2 + 16}$

    The radicand is the expression written under the radical sign, $y^2 + 16$.

19. $x^4 y^5 \sqrt{\dfrac{x}{y-1}}$

    The radicand is the expression written under the radical sign, $\dfrac{x}{y-1}$.

21. $f(x) = \sqrt{5x - 10}$

    $f(6) = \sqrt{5 \cdot 6 - 10} = \sqrt{20} \approx 4.472$

    $f(2) = \sqrt{5 \cdot 2 - 10} = \sqrt{0} = 0$

    $f(1) = \sqrt{5 \cdot 1 - 10} = \sqrt{-5}$

    Since negative numbers do not have real-number square roots, $f(1)$ does not exist as a real number.

    $f(-1) = \sqrt{5(-1) - 10} = \sqrt{-15}$

    Since negative numbers do not have real-number square roots, $f(-1)$ does not exist as a real number.

23. $g(x) = \sqrt{x^2 - 25}$

    $g(-6) = \sqrt{(-6)^2 - 25} = \sqrt{11} \approx 3.317$

    $g(3) = \sqrt{3^2 - 25} = \sqrt{-16}$

    Since negative numbers do not have real-number square roots, $g(3)$ does not exist as a real number.

    $g(6) = \sqrt{6^2 - 25} = \sqrt{11} \approx 3.317$

    $g(13) = \sqrt{13^2 - 25} = \sqrt{144} = 12$

25. The domain of $f(x) = \sqrt{5x - 10}$ is the set of all x-values for which $5x - 10 \geq 0$.

    $5x - 10 \geq 0$

    $5x \geq 10$

    $x \geq 2$

    The domain is $\{x | x \geq 2\}$, or $[2, \infty)$.

27. $S(x) = 2\sqrt{5x}$

    $S(30) = 2\sqrt{5 \cdot 30} = 2\sqrt{150} \approx 24.5$

    The speed of a car that left skid marks of length 30 ft was about 24.5 mph.

    $S(150) = 2\sqrt{5 \cdot 150} = 2\sqrt{750} \approx 54.8$

    The speed of a car that left skid marks of length 150 ft was about 54.8 mph.

29. Graph: $f(x) = 2\sqrt{x}$.

    We find some ordered pairs, plot points, and draw the curve.

x	$f(x)$	$(x, f(x))$
0	0	$(0, 0)$
1	2	$(1, 2)$
2	2.8	$(2, 2.8)$
3	3.5	$(3, 3.5)$
4	4	$(4, 4)$
5	4.5	$(5, 4.5)$

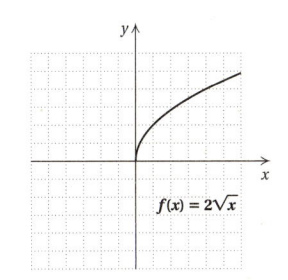

31. Graph: $F(x) = -3\sqrt{x}$.

We find some ordered pairs, plot points, and draw the curve.

x	$f(x)$	$(x, f(x))$
0	0	$(0, 0)$
1	-3	$(1, -3)$
2	-4.2	$(2, -4.2)$
3	-5.2	$(3, -5.2)$
4	-6	$(4, -6)$
5	-6.7	$(5, -6.7)$

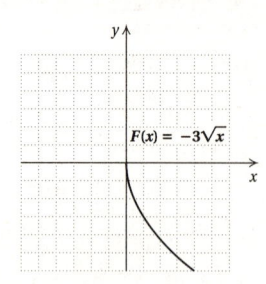

33. Graph: $f(x) = \sqrt{x}$.

We find some ordered pairs, plot points, and draw the curve.

x	$f(x)$	$(x, f(x))$
0	0	$(0, 0)$
1	1	$(1, 1)$
2	1.4	$(2, 1.4)$
3	1.7	$(3, 1.7)$
4	2	$(4, 2)$
5	2.2	$(5, 2.2)$

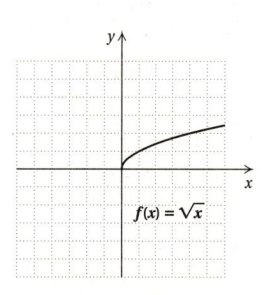

35. Graph: $f(x) = \sqrt{x - 2}$.

We find some ordered pairs, plot points, and draw the curve.

x	$f(x)$	$(x, f(x))$
2	0	$(2, 0)$
3	1	$(3, 1)$
4	1.4	$(4, 1.4)$
5	1.7	$(5, 1.7)$
7	2.2	$(7, 2.2)$
9	2.6	$(9, 2.6)$

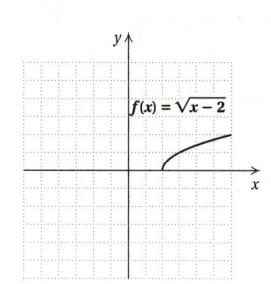

37. Graph: $f(x) = \sqrt{12 - 3x}$.

We find some ordered pairs, plot points, and draw the curve.

x	$f(x)$	$(x, f(x))$
-5	5.2	$(-5, 5.2)$
-3	4.6	$(-3, 4.6)$
-1	3.9	$(-1, 3.9)$
0	3.5	$(0, 3.5)$
2	2.4	$(2, 2.4)$
4	0	$(4, 0)$

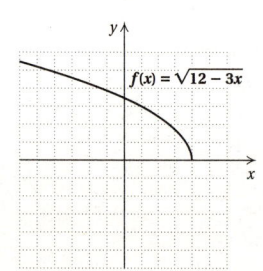

39. Graph: $g(x) = \sqrt{3x + 9}$.

We find some ordered pairs, plot points, and draw the curve.

x	$f(x)$	$(x, f(x))$
-3	0	$(-3, 0)$
-1	2.4	$(-1, 2.4)$
0	3	$(0, 3)$
1	3.5	$(1, 3.5)$
3	4.2	$(3, 4.2)$
5	4.9	$(5, 4.9)$

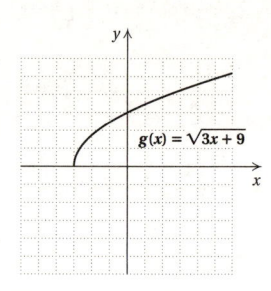

41. $\sqrt{16x^2} = \sqrt{(4x)^2} = |4x| = 4|x|$

(The absolute value is used to ensure that the principal square root is nonnegative.)

43. $\sqrt{(-12c)^2} = |-12c| = |-12| \cdot |c| = 12|c|$

(The absolute value is used to ensure that the principal square root is nonnegative.)

45. $\sqrt{(p+3)^2} = |p + 3|$

(The absolute value is used to ensure that the principal square root is nonnegative.)

47. $\sqrt{x^2 - 4x + 4} = \sqrt{(x-2)^2} = |x - 2|$

(The absolute value is used to ensure that the principal square root is nonnegative.)

49. $\sqrt[3]{27} = 3 \quad [3^3 = 27]$

51. $\sqrt[3]{-64x^3} = -4x \quad [(-4x)^3 = -64x^3]$

53. $\sqrt[3]{-216} = -6 \quad [(-6)^3 = -216]$

55. $\sqrt[3]{0.343(x+1)^3} = 0.7(x+1)$
$\qquad [(0.7(x+1))^3 = 0.343(x+1)^3]$

57. $\quad f(x) = \sqrt[3]{x + 1}$
$\quad f(7) = \sqrt[3]{7 + 1} = \sqrt[3]{8} = 2$
$\quad f(26) = \sqrt[3]{26 + 1} = \sqrt[3]{27} = 3$
$\quad f(-9) = \sqrt[3]{-9 + 1} = \sqrt[3]{-8} = -2$
$\quad f(-65) = \sqrt[3]{-65 + 1} = \sqrt[3]{-64} = -4$

59. $\quad f(x) = -\sqrt[3]{3x + 1}$
$\quad f(0) = -\sqrt[3]{3 \cdot 0 + 1} = -\sqrt[3]{1} = -1$
$\quad f(-7) = -\sqrt[3]{3(-7) + 1} = -\sqrt[3]{-20}, \text{ or } \sqrt[3]{20} \approx 2.7144$
$\quad f(21) = -\sqrt[3]{3 \cdot 21 + 1} = -\sqrt[3]{64} = -4$
$\quad f(333) = -\sqrt[3]{3 \cdot 333 + 1} = -\sqrt[3]{1000} = -10$

61. $-\sqrt[4]{625} = -5$ Since $5^4 = 625$, then $\sqrt[4]{625} = 5$ and $-\sqrt[4]{625} = -5$.

63. $\sqrt[5]{-1} = -1$ Since $(-1)^5 = -1$

65. $\sqrt[5]{-\dfrac{32}{243}} = -\dfrac{2}{3}$ Since $\left(-\dfrac{2}{3}\right)^5 = -\dfrac{32}{243}$

67. $\sqrt[6]{x^6} = |x|$

The index is even so we use absolute-value notation.

69. $\sqrt[4]{(5a)^4} = |5a| = 5|a|$

The index is even so we use absolute-value notation.

Exercise Set 10.2

71. $\sqrt[10]{(-6)^{10}} = |-6| = 6$

73. $\sqrt[414]{(a+b)^{414}} = |a+b|$

    The index is even so we use absolute-value notation.

75. $\sqrt[7]{y^7} = y$

    We do not use absolute-value notation when the index is odd.

77. $\sqrt[5]{(x-2)^5} = x - 2$

    We do not use absolute-value notation when the index is odd.

79. Discussion and Writing Exercise

81. $x^2 + x - 2 = 0$
    $(x+2)(x-1) = 0$ Factoring
    $x + 2 = 0$ or $x - 1 = 0$ Principle of zero products
    $x = -2$ or $x = 1$

    The solutions are -2 and 1.

83. $4x^2 - 49 = 0$
    $(2x+7)(2x-7) = 0$ Factoring
    $2x + 7 = 0$ or $2x - 7 = 0$ Principle of zero products
    $2x = -7$ or $2x = 7$
    $x = -\dfrac{7}{2}$ or $x = \dfrac{7}{2}$

    The solutions are $-\dfrac{7}{2}$ and $\dfrac{7}{2}$.

85. $3x^2 + x = 10$
    $3x^2 + x - 10 = 0$
    $(3x-5)(x+2) = 0$
    $3x - 5 = 0$ or $x + 2 = 0$
    $3x = 5$ or $x = -2$
    $x = \dfrac{5}{3}$ or $x = -2$

    The solutions are $\dfrac{5}{3}$ and -2.

87. $4x^3 - 20x^2 + 25x = 0$
    $x(4x^2 - 20x + 25) = 0$
    $x(2x-5)(2x-5) = 0$
    $x = 0$ or $2x - 5 = 0$ or $2x - 5 = 0$
    $x = 0$ or $2x = 5$ or $2x = 5$
    $x = 0$ or $x = \dfrac{5}{2}$ or $x = \dfrac{5}{2}$

    The solutions are 0 and $\dfrac{5}{2}$.

89. $(a^3 b^2 c^5)^3 = a^{3\cdot 3} b^{2\cdot 3} c^{5\cdot 3} = a^9 b^6 c^{15}$

91. $f(x) = \dfrac{\sqrt{x+3}}{\sqrt{2-x}}$

    In the numerator we must have $x + 3 \geq 0$, or $x \geq -3$, and in the denominator we must have $2 - x > 0$, or $x < 2$. Thus, we have $x \geq -3$ and $x < 2$, so

    Domain of $f = \{x \mid -3 \leq x < 2\}$, or $[-3, 2)$.

93. From 3 on the x-axis, go up to the graph and across to the y-axis to find $f(3) = \sqrt{3} \approx 1.7$.

    From 5 on the x-axis, go up to the graph and across to the y-axis to find $f(5) = \sqrt{5} \approx 2.2$.

    From 10 on the x-axis, go up to the graph and across to the y-axis to find $f(10) = \sqrt{10} \approx 3.2$.

95. a) $f(x) = \sqrt[3]{x}$

    Domain $= (-\infty, \infty)$; range $= (-\infty, \infty)$

    b) $g(x) = \sqrt[3]{4x - 5}$

    Domain $= (-\infty, \infty)$; range $= (-\infty, \infty)$

    c) $q(x) = 2 - \sqrt{x+3}$

    Domain $= [-3, \infty)$; range $= (-\infty, 2]$

    d) $h(x) = \sqrt[4]{x}$

    Domain $= [0, \infty)$; range $= [0, \infty)$

    e) $t(x) = \sqrt[4]{x-3}$

    Domain $= [3, \infty)$; range $= [0, \infty)$

Exercise Set 10.2

1. $y^{1/7} = \sqrt[7]{y}$

3. $(8)^{1/3} = \sqrt[3]{8} = 2$

5. $(a^3 b^3)^{1/5} = \sqrt[5]{a^3 b^3}$

7. $16^{3/4} = \sqrt[4]{16^3} = (\sqrt[4]{16})^3 = 2^3 = 8$

9. $49^{3/2} = \sqrt{49^3} = (\sqrt{49})^3 = 7^3 = 343$

11. $\sqrt{17} = 17^{1/2}$

13. $\sqrt[3]{18} = 18^{1/3}$

15. $\sqrt[5]{xy^2 z} = (xy^2 z)^{1/5}$

17. $(\sqrt{3mn})^3 = (3mn)^{3/2}$

19. $(\sqrt[7]{8x^2 y})^5 = (8x^2 y)^{5/7}$

21. $27^{-1/3} = \dfrac{1}{27^{1/3}} = \dfrac{1}{\sqrt[3]{27}} = \dfrac{1}{3}$

23. $100^{-3/2} = \dfrac{1}{100^{3/2}} = \dfrac{1}{(\sqrt{100})^3} = \dfrac{1}{10^3} = \dfrac{1}{1000}$

25. $x^{-1/4} = \dfrac{1}{x^{1/4}}$

27. $(2rs)^{-3/4} = \dfrac{1}{(2rs)^{3/4}}$

29. $2a^{3/4} b^{-1/2} c^{2/3} = 2 \cdot a^{3/4} \cdot \dfrac{1}{b^{1/2}} \cdot c^{2/3} = \dfrac{2a^{3/4} c^{2/3}}{b^{1/2}}$

31. $\left(\dfrac{7x}{8yz}\right)^{-3/5} = \left(\dfrac{8yz}{7x}\right)^{3/5}$ (Since $\left(\dfrac{a}{b}\right)^{-n} = \left(\dfrac{b}{a}\right)^n$)

33. $\dfrac{1}{x^{-2/3}} = x^{2/3}$

35. $2^{-1/3}x^4y^{-2/7} = \dfrac{1}{2^{1/3}} \cdot x^4 \cdot \dfrac{1}{y^{2/7}} = \dfrac{x^4}{2^{1/3}y^{2/7}}$

37. $\dfrac{7x}{\sqrt[3]{z}} = \dfrac{7x}{z^{1/3}}$

39. $\dfrac{5a}{3c^{-1/2}} = \dfrac{5a}{3} \cdot c^{1/2} = \dfrac{5ac^{1/2}}{3}$

41. $5^{3/4} \cdot 5^{1/8} = 5^{3/4+1/8} = 5^{6/8+1/8} = 5^{7/8}$

43. $\dfrac{7^{5/8}}{7^{3/8}} = 7^{5/8-3/8} = 7^{2/8} = 7^{1/4}$

45. $\dfrac{4.9^{-1/6}}{4.9^{-2/3}} = 4.9^{-1/6-(-2/3)} = 4.9^{-1/6+4/6} = 4.9^{3/6} = 4.9^{1/2}$

47. $(6^{3/8})^{2/7} = 6^{3/8 \cdot 2/7} = 6^{6/56} = 6^{3/28}$

49. $a^{2/3} \cdot a^{5/4} = a^{2/3+5/4} = a^{8/12+15/12} = a^{23/12}$

51. $(a^{2/3} \cdot b^{5/8})^4 = (a^{2/3})^4(b^{5/8})^4 = a^{8/3}b^{20/8} = a^{8/3}b^{5/2}$

53. $(x^{2/3})^{-3/7} = x^{2/3(-3/7)} = x^{-2/7} = \dfrac{1}{x^{2/7}}$

55. $\sqrt[6]{a^2} = a^{2/6}$ Converting to exponential notation
$\phantom{\sqrt[6]{a^2}} = a^{1/3}$ Simplifying the exponent
$\phantom{\sqrt[6]{a^2}} = \sqrt[3]{a}$ Returning to radical notation

57. $\sqrt[3]{x^{15}} = x^{15/3}$ Converting to exponential notation
$\phantom{\sqrt[3]{x^{15}}} = x^5$ Simplifying

59. $\sqrt[6]{x^{-18}} = x^{-18/6}$ Converting to exponential notation
$\phantom{\sqrt[6]{x^{-18}}} = x^{-3}$ Simplifying
$\phantom{\sqrt[6]{x^{-18}}} = \dfrac{1}{x^3}$

61. $(\sqrt[3]{ab})^{15} = (ab)^{15/3}$ Converting to exponential notation
$\phantom{(\sqrt[3]{ab})^{15}} = (ab)^5$ Simplifying the exponent
$\phantom{(\sqrt[3]{ab})^{15}} = a^5b^5$ Using the law of exponents

63. $\sqrt[14]{128} = \sqrt[14]{2^7} = 2^{7/14} = 2^{1/2} = \sqrt{2}$

65. $\sqrt[6]{4x^2} = (2^2x^2)^{1/6} = 2^{2/6}x^{2/6}$
$\phantom{\sqrt[6]{4x^2}} = 2^{1/3}x^{1/3} = (2x)^{1/3} = \sqrt[3]{2x}$

67. $\sqrt{x^4y^6} = (x^4y^6)^{1/2} = x^{4/2}y^{6/2} = x^2y^3$

69. $\sqrt[5]{32c^{10}d^{15}} = (2^5c^{10}d^{15})^{1/5} = 2^{5/5}c^{10/5}d^{15/5}$
$\phantom{\sqrt[5]{32c^{10}d^{15}}} = 2c^2d^3$

71. $\sqrt[3]{7} \cdot \sqrt[4]{5} = 7^{1/3} \cdot 5^{1/4} = 7^{4/12} \cdot 5^{3/12} = $
$(7^4 \cdot 5^3)^{1/12} = \sqrt[12]{7^4 \cdot 5^3}$

73. $\sqrt[4]{5} \cdot \sqrt[5]{7} = 5^{1/4} \cdot 7^{1/5} = 5^{5/20} \cdot 7^{4/20} = (5^5 \cdot 7^4)^{1/20} = $
$\sqrt[20]{5^5 \cdot 7^4}$

75. $\sqrt{x}\sqrt[3]{2x} = x^{1/2} \cdot (2x)^{1/3} = x^{3/6} \cdot (2x)^{2/6} = $
$[x^3(2x)^2]^{1/6} = (x^3 \cdot 4x^2)^{1/6} = (4x^5)^{1/6} = \sqrt[6]{4x^5}$

77. $(\sqrt[5]{a^2b^4})^{15} = (a^2b^4)^{15/5} = (a^2b^4)^3 = a^6b^{12}$

79. $\sqrt[3]{\sqrt[6]{m}} = \sqrt[3]{m^{1/6}} = (m^{1/6})^{1/3} = m^{1/18} = \sqrt[18]{m}$

81. $x^{1/3} \cdot y^{1/4} \cdot z^{1/6} = x^{4/12} \cdot y^{3/12} \cdot z^{2/12} = $
$(x^4y^3z^2)^{1/12} = \sqrt[12]{x^4y^3z^2}$

83. $\left(\dfrac{c^{-4/5}d^{5/9}}{c^{3/10}d^{1/6}}\right)^3 = (c^{-4/5-3/10}d^{5/9-1/6})^3 = $
$(c^{-8/10-3/10}d^{10/18-3/18})^3 = (c^{-11/10}d^{7/18})^3 = $
$c^{-33/10}d^{7/6} = c^{-99/30}d^{35/30} = (c^{-99}d^{35})^{1/30} = $
$\left(\dfrac{d^{35}}{c^{99}}\right)^{1/30} = \sqrt[30]{\dfrac{d^{35}}{c^{99}}}$

85. Discussion and Writing Exercise

87. $|7x - 5| = 9$
$7x - 5 = -9 \quad \text{or} \quad 7x - 5 = 9$
$7x = -4 \quad \text{or} \quad 7x = 14$
$x = -\dfrac{4}{7} \quad \text{or} \quad x = 2$

The solution set is $\left\{-\dfrac{4}{7}, 2\right\}$.

89. $8 - |2x + 5| = -2$
$-|2x + 5| = -10$
$|2x + 5| = 10 \quad \text{Multiplying by } -1$
$2x + 5 = -10 \quad \text{or} \quad 2x + 5 = 10$
$2x = -15 \quad \text{or} \quad 2x = 5$
$x = -\dfrac{15}{2} \quad \text{or} \quad x = \dfrac{5}{2}$

The solution set is $\left\{-\dfrac{15}{2}, \dfrac{5}{2}\right\}$.

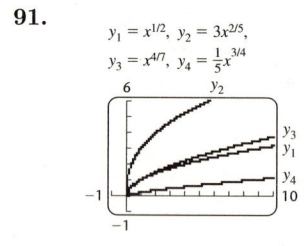

91. $y_1 = x^{1/2}, \ y_2 = 3x^{2/5},$
$y_3 = x^{4/7}, \ y_4 = \dfrac{1}{5}x^{3/4}$

Exercise Set 10.3

1. $\sqrt{24} = \sqrt{4 \cdot 6} = \sqrt{4}\sqrt{6} = 2\sqrt{6}$

3. $\sqrt{90} = \sqrt{9 \cdot 10} = \sqrt{9}\sqrt{10} = 3\sqrt{10}$

5. $\sqrt[3]{250} = \sqrt[3]{125 \cdot 2} = \sqrt[3]{125}\sqrt[3]{2} = 5\sqrt[3]{2}$

7. $\sqrt{180x^4} = \sqrt{36 \cdot 5 \cdot x^4} = \sqrt{36x^4}\sqrt{5} = 6x^2\sqrt{5}$

9. $\sqrt[3]{54x^8} = \sqrt[3]{27 \cdot 2 \cdot x^6 \cdot x^2} = \sqrt[3]{27x^6}\sqrt[3]{2x^2} = 3x^2\sqrt[3]{2x^2}$

11. $\sqrt[3]{80t^8} = \sqrt[3]{8 \cdot 10 \cdot t^6 \cdot t^2} = \sqrt[3]{8t^6}\sqrt[3]{10t^2} = 2t^2\sqrt[3]{10t^2}$

13. $\sqrt[4]{80} = \sqrt[4]{16 \cdot 5} = \sqrt[4]{16}\sqrt[4]{5} = 2\sqrt[4]{5}$

15. $\sqrt{32a^2b} = \sqrt{16a^2 \cdot 2b} = \sqrt{16a^2} \cdot \sqrt{2b} = 4a\sqrt{2b}$

Exercise Set 10.3

17. $\sqrt[4]{243x^8y^{10}} = \sqrt[4]{81x^8y^8 \cdot 3y^2} = \sqrt[4]{81x^8y^8} \sqrt[4]{3y^2} = 3x^2y^2 \sqrt[4]{3y^2}$

19. $\sqrt[5]{96x^7y^{15}} = \sqrt[5]{32x^5y^{15} \cdot 3x^2} = \sqrt[5]{32x^5y^{15}} \sqrt[5]{3x^2} = 2xy^3 \sqrt[5]{3x^2}$

21. $\sqrt{10}\sqrt{5} = \sqrt{10 \cdot 5} = \sqrt{50} = \sqrt{25 \cdot 2} = 5\sqrt{2}$

23. $\sqrt{15}\sqrt{6} = \sqrt{15 \cdot 6} = \sqrt{90}$
    $= \sqrt{9 \cdot 10} = \sqrt{9}\sqrt{10} = 3\sqrt{10}$

25. $\sqrt[3]{2}\sqrt[3]{4} = \sqrt[3]{2 \cdot 4} = \sqrt[3]{8} = 2$

27. $\sqrt{45}\sqrt{60} = \sqrt{45 \cdot 60} = \sqrt{2700}$
    $= \sqrt{900 \cdot 3} = \sqrt{900}\sqrt{3} = 30\sqrt{3}$

29. $\sqrt{3x^3}\sqrt{6x^5} = \sqrt{18x^8} = \sqrt{9x^8 \cdot 2} = 3x^4\sqrt{2}$

31. $\sqrt{5b^3}\sqrt{10c^4} = \sqrt{5b^3 \cdot 10c^4}$
    $= \sqrt{50b^3c^4}$
    $= \sqrt{25 \cdot 2 \cdot b^2 \cdot b \cdot c^4}$
    $= \sqrt{25b^2c^4}\sqrt{2b}$
    $= 5bc^2\sqrt{2b}$

33. $\sqrt[3]{5a^2}\sqrt[3]{2a} = \sqrt[3]{5a^2 \cdot 2a} = \sqrt[3]{10a^3} = \sqrt[3]{a^3 \cdot 10} = a\sqrt[3]{10}$

35. $\sqrt[3]{y^4}\sqrt[3]{16y^5} = \sqrt[3]{y^4 \cdot 16y^5}$
    $= \sqrt[3]{16y^9}$
    $= \sqrt[3]{8 \cdot 2 \cdot y^9}$
    $= \sqrt[3]{8y^9}\sqrt[3]{2}$
    $= 2y^3\sqrt[3]{2}$

37. $\sqrt[4]{16}\sqrt[4]{64} = \sqrt[4]{16 \cdot 64} = \sqrt[4]{1024} = \sqrt[4]{256 \cdot 4} = \sqrt[4]{256}\sqrt[4]{4} = 4\sqrt[4]{4}$

39. $\sqrt{12a^3b}\sqrt{8a^4b^2} = \sqrt{12a^3b \cdot 8a^4b^2} = \sqrt{96a^7b^3} = \sqrt{16a^6b^2 \cdot 6ab} = \sqrt{16a^6b^2}\sqrt{6ab} = 4a^3b\sqrt{6ab}$

41. $\sqrt{2}\sqrt[3]{5}$
    $= 2^{1/2} \cdot 5^{1/3}$ Converting to exponential notation
    $= 2^{3/6} \cdot 5^{2/6}$ Rewriting so that exponents have a common denominator
    $= (2^3 \cdot 5^2)^{1/6}$ Using $a^n b^n = (ab)^n$
    $= \sqrt[6]{2^3 \cdot 5^2}$ Converting to radical notation
    $= \sqrt[6]{8 \cdot 25}$ Simplifying
    $= \sqrt[6]{200}$ Multiplying

43. $\sqrt[4]{3}\sqrt{2}$
    $= 3^{1/4} \cdot 2^{1/2}$ Converting to exponential notation
    $= 3^{1/4} \cdot 2^{2/4}$ Rewriting so that exponents have a common denominator
    $= (3 \cdot 2^2)^{1/4}$ Using $a^n b^n = (ab)^n$
    $= \sqrt[4]{3 \cdot 2^2}$ Converting to radical notation
    $= \sqrt[4]{3 \cdot 4}$ Squaring 2
    $= \sqrt[4]{12}$ Multiplying

45. $\sqrt{a}\sqrt[4]{a^3}$
    $= a^{1/2} \cdot a^{3/4}$ Converting to exponential notation
    $= a^{5/4}$ Adding exponents
    $= a^{1+1/4}$ Writing 5/4 as a mixed number
    $= a \cdot a^{1/4}$ Factoring
    $= a\sqrt[4]{a}$ Returning to radical notation

47. $\sqrt[5]{b^2}\sqrt{b^3}$
    $= b^{2/5} \cdot b^{3/2}$ Converting to exponential notation
    $= b^{19/10}$ Adding exponents
    $= b^{1+9/10}$ Writing 19/10 as a mixed number
    $= b \cdot b^{9/10}$ Factoring
    $= b\sqrt[10]{b^9}$ Returning to radical notation

49. $\sqrt{xy^3}\sqrt[3]{x^2y} = (xy^3)^{1/2}(x^2y)^{1/3}$
    $= (xy^3)^{3/6}(x^2y)^{2/6}$
    $= [(xy^3)^3(x^2y)^2]^{1/6}$
    $= \sqrt[6]{x^3y^9 \cdot x^4y^2}$
    $= \sqrt[6]{x^7y^{11}}$
    $= \sqrt[6]{x^6y^6 \cdot xy^5}$
    $= xy\sqrt[6]{xy^5}$

51. $\dfrac{\sqrt{90}}{\sqrt{5}} = \sqrt{\dfrac{90}{5}} = \sqrt{18} = \sqrt{9 \cdot 2} = \sqrt{9}\sqrt{2} = 3\sqrt{2}$

53. $\dfrac{\sqrt{35q}}{\sqrt{7q}} = \sqrt{\dfrac{35q}{7q}} = \sqrt{5}$

55. $\dfrac{\sqrt[3]{54}}{\sqrt[3]{2}} = \sqrt[3]{\dfrac{54}{2}} = \sqrt[3]{27} = 3$

57. $\dfrac{\sqrt{56xy^3}}{\sqrt{8x}} = \sqrt{\dfrac{56xy^3}{8x}} = \sqrt{7y^3} = \sqrt{y^2 \cdot 7y} = \sqrt{y^2}\sqrt{7y} = y\sqrt{7y}$

59. $\dfrac{\sqrt[3]{96a^4b^2}}{\sqrt[3]{12a^2b}} = \sqrt[3]{\dfrac{96a^4b^2}{12a^2b}} = \sqrt[3]{8a^2b} = \sqrt[3]{8}\sqrt[3]{a^2b} = 2\sqrt[3]{a^2b}$

61. $\dfrac{\sqrt{128xy}}{2\sqrt{2}} = \dfrac{1}{2}\dfrac{\sqrt{128xy}}{\sqrt{2}} = \dfrac{1}{2}\sqrt{\dfrac{128xy}{2}} = \dfrac{1}{2}\sqrt{64xy} = \dfrac{1}{2}\sqrt{64}\sqrt{xy} = \dfrac{1}{2} \cdot 8\sqrt{xy} = 4\sqrt{xy}$

63. $\dfrac{\sqrt[4]{48x^9y^{13}}}{\sqrt[4]{3xy^5}} = \sqrt[4]{\dfrac{48x^9y^{13}}{3xy^5}} = \sqrt[4]{16x^8y^8} = 2x^2y^2$

65. $\dfrac{\sqrt[3]{a}}{\sqrt{a}}$
$= \dfrac{a^{1/3}}{a^{1/2}}$ Converting to exponential notation
$= a^{1/3 - 1/2}$ Subtracting exponents
$= a^{2/6 - 3/6}$
$= a^{-1/6}$
$= \dfrac{1}{a^{1/6}}$
$= \dfrac{1}{\sqrt[6]{a}}$ Converting to radical notation

67. $\dfrac{\sqrt[3]{a^2}}{\sqrt[4]{a}}$
$= \dfrac{a^{2/3}}{a^{1/4}}$ Converting to exponential notation
$= a^{2/3 - 1/4}$ Subtracting exponents
$= a^{5/12}$ Converting back
$= \sqrt[12]{a^5}$ to radical notation

69. $\dfrac{\sqrt[4]{x^2y^3}}{\sqrt[3]{xy}}$
$= \dfrac{(x^2y^3)^{1/4}}{(xy)^{1/3}}$ Converting to exponential notation
$= \dfrac{x^{2/4}y^{3/4}}{x^{1/3}y^{1/3}}$ Using the power and product rules
$= x^{2/4 - 1/3}y^{3/4 - 1/3}$ Subtracting exponents
$= x^{2/12}y^{5/12}$
$= (x^2y^5)^{1/12}$ Converting back to
$= \sqrt[12]{x^2y^5}$ radical notation

71. $\sqrt{\dfrac{25}{36}} = \dfrac{\sqrt{25}}{\sqrt{36}} = \dfrac{5}{6}$

73. $\sqrt{\dfrac{16}{49}} = \dfrac{\sqrt{16}}{\sqrt{49}} = \dfrac{4}{7}$

75. $\sqrt[3]{\dfrac{125}{27}} = \dfrac{\sqrt[3]{125}}{\sqrt[3]{27}} = \dfrac{5}{3}$

77. $\sqrt{\dfrac{49}{y^2}} = \dfrac{\sqrt{49}}{\sqrt{y^2}} = \dfrac{7}{y}$

79. $\sqrt{\dfrac{25y^3}{x^4}} = \dfrac{\sqrt{25y^3}}{\sqrt{x^4}} = \dfrac{\sqrt{25y^2 \cdot y}}{\sqrt{x^4}} = \dfrac{\sqrt{25y^2}\sqrt{y}}{\sqrt{x^4}} = \dfrac{5y\sqrt{y}}{x^2}$

81. $\sqrt[3]{\dfrac{27a^4}{8b^3}} = \dfrac{\sqrt[3]{27a^4}}{\sqrt[3]{8b^3}} = \dfrac{\sqrt[3]{27a^3 \cdot a}}{\sqrt[3]{8b^3}} = \dfrac{\sqrt[3]{27a^3}\sqrt[3]{a}}{\sqrt[3]{8b^3}} = \dfrac{3a\sqrt[3]{a}}{2b}$

83. $\sqrt[4]{\dfrac{81x^4}{16}} = \dfrac{\sqrt[4]{81x^4}}{\sqrt[4]{16}} = \dfrac{3x}{2}$

85. $\sqrt[5]{\dfrac{32x^8}{y^{10}}} = \dfrac{\sqrt[5]{32x^8}}{\sqrt[5]{y^{10}}} = \dfrac{\sqrt[5]{32 \cdot x^5 \cdot x^3}}{\sqrt[5]{y^{10}}} = \dfrac{\sqrt[5]{32x^5}\sqrt[5]{x^3}}{\sqrt[5]{y^{10}}} = \dfrac{2x\sqrt[5]{x^3}}{y^2}$

87. $\sqrt[6]{\dfrac{x^{13}}{y^6z^{12}}} = \dfrac{\sqrt[6]{x^{13}}}{\sqrt[6]{y^6z^{12}}} = \dfrac{\sqrt[6]{x^{12} \cdot x}}{\sqrt[6]{y^6z^{12}}} = \dfrac{\sqrt[6]{x^{12}}\sqrt[6]{x}}{\sqrt[6]{y^6z^{12}}} = \dfrac{x^2\sqrt[6]{x}}{yz^2}$

89. Discussion and Writing Exercise

91. **Familiarize.** We will use the formula $d = rt$. When the boat travels downstream, its rate is $14 + 7$, or 21 mph. Its rate traveling upstream is $14 - 7$ or 7 mph.

Translate. We substitute in the formula.

Downstream: $56 = 21t$

Upstream: $56 = 7t$

Solve. We solve the equation.

Downstream: $56 = 21t$
$\dfrac{56}{21} = t$
$\dfrac{8}{3} = t$, or
$2\dfrac{2}{3} = t$

Upstream: $56 = 7t$
$8 = t$

Check. At a rate of 21 mph, in $\dfrac{8}{3}$ hr the boat would travel $21 \cdot \dfrac{8}{3}$, or 56 mi. At a rate of 7 mph, in 8 hr the boat would travel $7 \cdot 8$, or 56 mi. The answer checks.

State. It will take the boat $2\dfrac{2}{3}$ hr to travel 56 mi downstream and 8 hr to travel 56 mi upstream.

93.
$\dfrac{12x}{x-4} - \dfrac{3x^2}{x+4} = \dfrac{384}{x^2-16}$

$\dfrac{12x}{x-4} - \dfrac{3x^2}{x+4} = \dfrac{384}{(x+4)(x-4)}$,
     LCM is $(x+4)(x-4)$.

$(x+4)(x-4)\left[\dfrac{12x}{x-4} - \dfrac{3x^2}{x+4}\right] = (x+4)(x-4) \cdot \dfrac{384}{(x+4)(x-4)}$

$12x(x+4) - 3x^2(x-4) = 384$
$12x^2 + 48x - 3x^3 + 12x^2 = 384$
$-3x^3 + 24x^2 + 48x - 384 = 0$
$-3(x^3 - 8x^2 - 16x + 128) = 0$
$-3[x^2(x-8) - 16(x-8)] = 0$
$-3(x-8)(x^2 - 16) = 0$
$-3(x-8)(x+4)(x-4) = 0$

$x - 8 = 0$ or $x + 4 = 0$ or $x - 4 = 0$
$x = 8$ or $x = -4$ or $x = 4$

Check: For 8:
$$\frac{12x}{x-4} - \frac{3x^2}{x+4} = \frac{384}{x^2-16}$$
$$\frac{12 \cdot 8}{8-4} - \frac{3 \cdot 8^2}{8+4} \;?\; \frac{384}{8^2-16}$$
$$\frac{96}{4} - \frac{192}{12} \;\bigg|\; \frac{384}{48}$$
$$24 - 16 \;\bigg|\; 8$$
$$8 \qquad\qquad \text{TRUE}$$

8 is a solution.

For -4:
$$\frac{12x}{x-4} - \frac{3x^2}{x+4} = \frac{384}{x^2-16}$$
$$\frac{12(-4)}{-4-4} - \frac{3(-4)^2}{-4+4} \;?\; \frac{384}{(-4)^2-16}$$
$$\frac{-48}{-8} - \frac{48}{0} \;\bigg|\; \frac{384}{16-16} \quad \text{UNDEFINED}$$

-4 is not a solution.

For 4:
$$\frac{12x}{x-4} - \frac{3x^2}{x+4} = \frac{384}{x^2-16}$$
$$\frac{12 \cdot 4}{4-4} - \frac{3 \cdot 4^2}{4+4} \;?\; \frac{384}{4^2-16}$$
$$\frac{48}{0} - \frac{48}{8} \;\bigg|\; \frac{384}{16-16} \quad \text{UNDEFINED}$$

4 is not a solution.
The solution is 8.

95.
$$\frac{18}{x^2-3x} = \frac{2x}{x-3} - \frac{6}{x}$$
$$\frac{18}{x(x-3)} = \frac{2x}{x-3} - \frac{6}{x},$$
$$\text{LCM is } x(x-3)$$
$$x(x-3) \cdot \frac{18}{x(x-3)} = x(x-3)\left(\frac{2x}{x-3} - \frac{6}{x}\right)$$
$$18 = x(x-3) \cdot \frac{2x}{x-3} - x(x-3) \cdot \frac{6}{x}$$
$$18 = 2x^2 - 6x + 18$$
$$0 = 2x^2 - 6x$$
$$0 = 2x(x-3)$$
$$2x = 0 \;\text{ or }\; x - 3 = 0$$
$$x = 0 \;\text{ or }\; x = 3$$

Each value makes a denominator 0. There is no solution.

97. a) $T = 2\pi\sqrt{\dfrac{65}{980}} \approx 1.62$ sec

b) $T = 2\pi\sqrt{\dfrac{98}{980}} \approx 1.99$ sec

c) $T = 2\pi\sqrt{\dfrac{120}{980}} \approx 2.20$ sec

99. $\dfrac{\sqrt{44x^2y^9z}\sqrt{22y^9z^6}}{(\sqrt{11xy^8z^2})^2} = \dfrac{\sqrt{44 \cdot 22x^2y^{18}z^7}}{\sqrt{11 \cdot 11x^2y^{16}z^4}} =$

$\sqrt{\dfrac{44 \cdot 22x^2y^{18}z^7}{11 \cdot 11x^2y^{16}z^4}} = \sqrt{4 \cdot 2y^2z^3} = \sqrt{4y^2z^2 \cdot 2z} = 2yz\sqrt{2z}$

Exercise Set 10.4

1. $7\sqrt{5} + 4\sqrt{5} = (7+4)\sqrt{5}$ Factoring out $\sqrt{5}$
$= 11\sqrt{5}$

3. $6\sqrt[3]{7} - 5\sqrt[3]{7} = (6-5)\sqrt[3]{7}$ Factoring out $\sqrt[3]{7}$
$= \sqrt[3]{7}$

5. $4\sqrt[3]{y} + 9\sqrt[3]{y} = (4+9)\sqrt[3]{y} = 13\sqrt[3]{y}$

7. $5\sqrt{6} - 9\sqrt{6} - 4\sqrt{6} = (5-9-4)\sqrt{6} = -8\sqrt{6}$

9. $4\sqrt[3]{3} - \sqrt{5} + 2\sqrt[3]{3} + \sqrt{5} =$
$(4+2)\sqrt[3]{3} + (-1+1)\sqrt{5} = 6\sqrt[3]{3}$

11. $8\sqrt{27} - 3\sqrt{3} = 8\sqrt{9 \cdot 3} - 3\sqrt{3}$ ⎫ Factoring the
$= 8\sqrt{9} \cdot \sqrt{3} - 3\sqrt{3}$ ⎭ first radical
$= 8 \cdot 3\sqrt{3} - 3\sqrt{3}$ Taking the square root
$= 24\sqrt{3} - 3\sqrt{3}$
$= (24-3)\sqrt{3}$ Factoring out $\sqrt{3}$
$= 21\sqrt{3}$

13. $8\sqrt{45} + 7\sqrt{20} = 8\sqrt{9 \cdot 5} + 7\sqrt{4 \cdot 5}$ ⎫ Factoring
$= 8\sqrt{9} \cdot \sqrt{5} + 7\sqrt{4} \cdot \sqrt{5}$ ⎬ the
$= 8 \cdot 3\sqrt{5} + 7 \cdot 2\sqrt{5}$ ⎭ radicals
   Taking the square roots
$= 24\sqrt{5} + 14\sqrt{5}$
$= (24+14)\sqrt{5}$ Factoring out $\sqrt{5}$
$= 38\sqrt{5}$

15. $18\sqrt{72} + 2\sqrt{98} = 18\sqrt{36 \cdot 2} + 2\sqrt{49 \cdot 2} =$
$18\sqrt{36} \cdot \sqrt{2} + 2\sqrt{49} \cdot \sqrt{2} = 18 \cdot 6\sqrt{2} + 2 \cdot 7\sqrt{2} =$
$108\sqrt{2} + 14\sqrt{2} = (108+14)\sqrt{2} = 122\sqrt{2}$

17. $3\sqrt[3]{16} + \sqrt[3]{54} = 3\sqrt[3]{8 \cdot 2} + \sqrt[3]{27 \cdot 2} =$
$3\sqrt[3]{8} \cdot \sqrt[3]{2} + \sqrt[3]{27} \cdot \sqrt[3]{2} = 3 \cdot 2\sqrt[3]{2} + 3\sqrt[3]{2} =$
$6\sqrt[3]{2} + 3\sqrt[3]{2} = (6+3)\sqrt[3]{2} = 9\sqrt[3]{2}$

19. $2\sqrt{128} - \sqrt{18} + 4\sqrt{32} =$
$2\sqrt{64 \cdot 2} - \sqrt{9 \cdot 2} + 4\sqrt{16 \cdot 2} =$
$2\sqrt{64} \cdot \sqrt{2} - \sqrt{9} \cdot \sqrt{2} + 4\sqrt{16} \cdot \sqrt{2} =$
$2 \cdot 8\sqrt{2} - 3\sqrt{2} + 4 \cdot 4\sqrt{2} = 16\sqrt{2} - 3\sqrt{2} + 16\sqrt{2} =$
$(16 - 3 + 16)\sqrt{2} = 29\sqrt{2}$

21. $\sqrt{5a} + 2\sqrt{45a^3} = \sqrt{5a} + 2\sqrt{9a^2 \cdot 5a} =$
$\sqrt{5a} + 2\sqrt{9a^2} \cdot \sqrt{5a} = \sqrt{5a} + 2 \cdot 3a\sqrt{5a} =$
$\sqrt{5a} + 6a\sqrt{5a} = (1+6a)\sqrt{5a}$

23. $\sqrt[3]{24x} - \sqrt[3]{3x^4} = \sqrt[3]{8 \cdot 3x} - \sqrt[3]{x^3 \cdot 3x} =$
$\sqrt[3]{8} \cdot \sqrt[3]{3x} - \sqrt[3]{x^3} \cdot \sqrt[3]{3x} = 2\sqrt[3]{3x} - x\sqrt[3]{3x} =$
$(2-x)\sqrt[3]{3x}$

25. $5\sqrt[3]{32} - \sqrt[3]{108} + 2\sqrt[3]{256} =$
$5\sqrt[3]{8 \cdot 4} - \sqrt[3]{27 \cdot 4} + 2\sqrt[3]{64 \cdot 4} =$
$5\sqrt[3]{8} \cdot \sqrt[3]{4} - \sqrt[3]{27} \cdot \sqrt[3]{4} + 2\sqrt[3]{64} \cdot \sqrt[3]{4} =$
$5 \cdot 2\sqrt[3]{4} - 3\sqrt[3]{4} + 2 \cdot 4\sqrt[3]{4} = 10\sqrt[3]{4} - 3\sqrt[3]{4} + 8\sqrt[3]{4} =$
$(10 - 3 + 8)\sqrt[3]{4} = 15\sqrt[3]{4}$

27. $\sqrt[3]{6x^4} + \sqrt[3]{48x} - \sqrt[3]{6x}$
$= \sqrt[3]{x^3 \cdot 6x} + \sqrt[3]{8 \cdot 6x} - \sqrt[3]{6x}$
$= \sqrt[3]{x^3} \cdot \sqrt[3]{6x} + \sqrt[3]{8} \cdot \sqrt[3]{6x} - \sqrt[3]{6x}$
$= x\sqrt[3]{6x} + 2\sqrt[3]{6x} - \sqrt[3]{6x}$
$= (x + 2 - 1)\sqrt[3]{6x}$
$= (x + 1)\sqrt[3]{6x}$

29. $\sqrt{4a-4} + \sqrt{a-1} = \sqrt{4(a-1)} + \sqrt{a-1}$
$= \sqrt{4} \cdot \sqrt{a-1} + \sqrt{a-1}$
$= 2\sqrt{a-1} + \sqrt{a-1}$
$= (2+1)\sqrt{a-1}$
$= 3\sqrt{a-1}$

31. $\sqrt{x^3 - x^2} + \sqrt{9x - 9} = \sqrt{x^2(x-1)} + \sqrt{9(x-1)}$
$= \sqrt{x^2}\sqrt{x-1} + \sqrt{9}\sqrt{x-1}$
$= x\sqrt{x-1} + 3\sqrt{x-1}$
$= (x+3)\sqrt{x-1}$

33. $\sqrt{5}(4 - 2\sqrt{5}) = \sqrt{5} \cdot 4 - 2(\sqrt{5})^2$ Distributive law
$= 4\sqrt{5} - 2 \cdot 5$
$= 4\sqrt{5} - 10$

35. $\sqrt{3}(\sqrt{2} - \sqrt{7}) = \sqrt{3}\sqrt{2} - \sqrt{3}\sqrt{7}$ Distributive law
$= \sqrt{6} - \sqrt{21}$

37. $\sqrt{3}(2\sqrt{5} - 3\sqrt{4}) = \sqrt{3}(2\sqrt{5} - 3 \cdot 2) =$
$\sqrt{3} \cdot 2\sqrt{5} - \sqrt{3} \cdot 6 = 2\sqrt{15} - 6\sqrt{3}$

39. $\sqrt[3]{2}(\sqrt[3]{4} - 2\sqrt[3]{32}) = \sqrt[3]{2} \cdot \sqrt[3]{4} - \sqrt[3]{2} \cdot 2\sqrt[3]{32} =$
$\sqrt[3]{8} - 2\sqrt[3]{64} = 2 - 2 \cdot 4 = 2 - 8 = -6$

41. $\sqrt[3]{a}(\sqrt[3]{2a^2} + \sqrt[3]{16a^2}) = \sqrt[3]{a} \cdot \sqrt[3]{2a^2} + \sqrt[3]{a} \cdot \sqrt[3]{16a^2} =$
$\sqrt[3]{2a^3} + \sqrt[3]{16a^3} = \sqrt[3]{a^3 \cdot 2} + \sqrt[3]{8a^3 \cdot 2} = a\sqrt[3]{2} + 2a\sqrt[3]{2} =$
$3a\sqrt[3]{2}$

43. $(\sqrt{3} - \sqrt{2})(\sqrt{3} + \sqrt{2}) = (\sqrt{3})^2 - (\sqrt{2})^2 = 3 - 2 = 1$

45. $(\sqrt{8} + 2\sqrt{5})(\sqrt{8} - 2\sqrt{5}) = (\sqrt{8})^2 - (2\sqrt{5})^2 =$
$8 - 4 \cdot 5 = 8 - 20 = -12$

47. $(7 + \sqrt{5})(7 - \sqrt{5}) = 7^2 - (\sqrt{5})^2 = 49 - 5 = 44$

49. $(2 - \sqrt{3})(2 + \sqrt{3}) = 2^2 - (\sqrt{3})^2 = 4 - 3 = 1$

51. $(\sqrt{8} + \sqrt{5})(\sqrt{8} - \sqrt{5}) = (\sqrt{8})^2 - (\sqrt{5})^2 = 8 - 5 = 3$

53. $(3 + 2\sqrt{7})(3 - 2\sqrt{7}) = 3^2 - (2\sqrt{7})^2 =$
$9 - 4 \cdot 7 = 9 - 28 = -19$

55. $(\sqrt{a} + \sqrt{b})(\sqrt{a} - \sqrt{b}) = (\sqrt{a})^2 - (\sqrt{b})^2 = a - b$

57. $(3 - \sqrt{5})(2 + \sqrt{5})$
$= 3 \cdot 2 + 3\sqrt{5} - 2\sqrt{5} - (\sqrt{5})^2$ Using FOIL
$= 6 + 3\sqrt{5} - 2\sqrt{5} - 5$
$= 1 + \sqrt{5}$ Simplifying

59. $(\sqrt{3} + 1)(2\sqrt{3} + 1)$
$= \sqrt{3} \cdot 2\sqrt{3} + \sqrt{3} \cdot 1 + 1 \cdot 2\sqrt{3} + 1^2$ Using FOIL
$= 2 \cdot 3 + \sqrt{3} + 2\sqrt{3} + 1$
$= 7 + 3\sqrt{3}$ Simplifying

61. $(2\sqrt{7} - 4\sqrt{2})(3\sqrt{7} + 6\sqrt{2}) =$
$2\sqrt{7} \cdot 3\sqrt{7} + 2\sqrt{7} \cdot 6\sqrt{2} - 4\sqrt{2} \cdot 3\sqrt{7} - 4\sqrt{2} \cdot 6\sqrt{2} =$
$6 \cdot 7 + 12\sqrt{14} - 12\sqrt{14} - 24 \cdot 2 =$
$42 + 12\sqrt{14} - 12\sqrt{14} - 48 = -6$

63. $(\sqrt{a} + \sqrt{2})(\sqrt{a} + \sqrt{3}) =$
$(\sqrt{a})^2 + \sqrt{a} \cdot \sqrt{3} + \sqrt{2} \cdot \sqrt{a} + \sqrt{2} \cdot \sqrt{3} =$
$a + \sqrt{3a} + \sqrt{2a} + \sqrt{6}$

65. $(2\sqrt[3]{3} + \sqrt[3]{2})(\sqrt[3]{3} - 2\sqrt[3]{2}) =$
$2\sqrt[3]{3} \cdot \sqrt[3]{3} - 2\sqrt[3]{3} \cdot 2\sqrt[3]{2} + \sqrt[3]{2} \cdot \sqrt[3]{3} - \sqrt[3]{2} \cdot 2\sqrt[3]{2} =$
$2\sqrt[3]{9} - 4\sqrt[3]{6} + \sqrt[3]{6} - 2\sqrt[3]{4} = 2\sqrt[3]{9} - 3\sqrt[3]{6} - 2\sqrt[3]{4}$

67. $(2 + \sqrt{3})^2 = 2^2 + 4\sqrt{3} + (\sqrt{3})^2$ Squaring a binomial
$\phantom{(2 + \sqrt{3})^2} = 4 + 4\sqrt{3} + 3$
$\phantom{(2 + \sqrt{3})^2} = 7 + 4\sqrt{3}$

69. $(\sqrt[5]{9} - \sqrt[5]{3})(\sqrt[5]{8} + \sqrt[5]{27})$
$= \sqrt[5]{9} \cdot \sqrt[5]{8} + \sqrt[5]{9} \cdot \sqrt[5]{27} - \sqrt[5]{3} \cdot \sqrt[5]{8} - \sqrt[5]{3} \cdot \sqrt[5]{27}$
$\phantom{=\sqrt[5]{9} \cdot \sqrt[5]{8} + \sqrt[5]{9} \cdot \sqrt[5]{27} - \sqrt[5]{3} \cdot \sqrt[5]{8}}$ Using FOIL
$= \sqrt[5]{72} + \sqrt[5]{243} - \sqrt[5]{24} - \sqrt[5]{81}$
$= \sqrt[5]{72} + 3 - \sqrt[5]{24} - \sqrt[5]{81}$

71. Discussion and Writing Exercise

73. $\dfrac{x^3 + 4x}{x^2 - 16} \div \dfrac{x^2 + 8x + 15}{x^2 + x - 20}$
$= \dfrac{x^3 + 4x}{x^2 - 16} \cdot \dfrac{x^2 + x - 20}{x^2 + 8x + 15}$
$= \dfrac{(x^3 + 4x)(x^2 + x - 20)}{(x^2 - 16)(x^2 + 8x + 15)}$
$= \dfrac{x(x^2 + 4)(x+5)(x-4)}{(x+4)(x-4)(x+3)(x+5)}$
$= \dfrac{x(x^2 + 4)\cancel{(x+5)}\cancel{(x-4)}}{(x+4)\cancel{(x-4)}(x+3)\cancel{(x+5)}}$
$= \dfrac{x(x^2 + 4)}{(x+4)(x+3)}$

Exercise Set 10.5

75. $\dfrac{a^3+8}{a^2-4} \cdot \dfrac{a^2-4a+4}{a^2-2a+4}$

$= \dfrac{(a^3+8)(a^2-4a+4)}{(a^2-4)(a^2-2a+4)}$

$= \dfrac{(a+2)(a^2-2a+4)(a-2)(a-2)}{(a+2)(a-2)(a^2-2a+4)(1)}$

$= \dfrac{(a+2)(a^2-2a+4)(a-2)}{(a+2)(a^2-2a+4)(a-2)} \cdot \dfrac{a-2}{1}$

$= a-2$

77. $\dfrac{x-\frac{1}{3}}{x+\frac{1}{4}} = \dfrac{x-\frac{1}{3}}{x+\frac{1}{4}} \cdot \dfrac{12}{12}$

$= \dfrac{\left(x-\frac{1}{3}\right)(12)}{\left(x+\frac{1}{4}\right)(12)}$

$= \dfrac{12x-4}{12x+3}, \text{ or } \dfrac{4(3x-1)}{3(4x+1)}$

79. $\dfrac{\frac{1}{p}-\frac{1}{q}}{\frac{1}{p^2}-\frac{1}{q^2}} = \dfrac{\frac{1}{p}-\frac{1}{q}}{\frac{1}{p^2}-\frac{1}{q^2}} \cdot \dfrac{p^2q^2}{p^2q^2}$

$= \dfrac{\left(\frac{1}{p}-\frac{1}{q}\right)(p^2q^2)}{\left(\frac{1}{p^2}-\frac{1}{q^2}\right)(p^2q^2)}$

$= \dfrac{p^2q^2 \cdot \frac{1}{p} - p^2q^2 \cdot \frac{1}{q}}{p^2q^2 \cdot \frac{1}{p^2} - p^2q^2 \cdot \frac{1}{q^2}}$

$= \dfrac{pq^2 - p^2q}{q^2 - p^2}$

$= \dfrac{pq(q-p)}{(q+p)(q-p)}$

$= \dfrac{pq(q-p)}{(q+p)(q-p)}$

$= \dfrac{pq}{q+p}$

81.

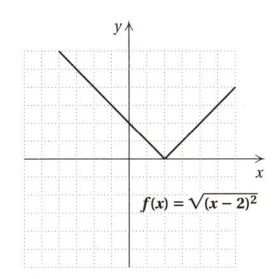

$f(x) = \sqrt{(x-2)^2}$

Since $(x-2)^2$ is nonnegative for all values of x, the domain of f is $\{x|x \text{ is a real number}\}$, or $(-\infty, \infty)$.

83. $\sqrt{9+3\sqrt{5}}\sqrt{9-3\sqrt{5}} = \sqrt{(9+3\sqrt{5})(9-3\sqrt{5})} =$

$\sqrt{9^2 - (3\sqrt{5})^2} = \sqrt{81 - 9 \cdot 5} = \sqrt{81-45} = \sqrt{36} = 6$

85. $(\sqrt{3}+\sqrt{5}-\sqrt{6})^2 = [(\sqrt{3}+\sqrt{5})-\sqrt{6}]^2 =$

$(\sqrt{3}+\sqrt{5})^2 - 2(\sqrt{3}+\sqrt{5})(\sqrt{6}) + (\sqrt{6})^2 =$

$3 + 2\sqrt{15} + 5 - 2\sqrt{18} - 2\sqrt{30} + 6 =$

$14 + 2\sqrt{15} - 2\sqrt{9\cdot 2} - 2\sqrt{30} =$

$14 + 2\sqrt{15} - 6\sqrt{2} - 2\sqrt{30}$

87. $(\sqrt[3]{9}-2)(\sqrt[3]{9}+4)$

$= \sqrt[3]{9}\sqrt[3]{9} + 4\sqrt[3]{9} - 2\sqrt[3]{9} - 2\cdot 4$

$= \sqrt[3]{81} + 2\sqrt[3]{9} - 8$

$= \sqrt[3]{27\cdot 3} + 2\sqrt[3]{9} - 8$

$= 3\sqrt[3]{3} + 2\sqrt[3]{9} - 8$

Exercise Set 10.5

1. $\sqrt{\dfrac{5}{3}} = \sqrt{\dfrac{5}{3}\cdot\dfrac{3}{3}} = \sqrt{\dfrac{15}{9}} = \dfrac{\sqrt{15}}{\sqrt{9}} = \dfrac{\sqrt{15}}{3}$

3. $\sqrt{\dfrac{11}{2}} = \sqrt{\dfrac{11}{2}\cdot\dfrac{2}{2}} = \sqrt{\dfrac{22}{4}} = \dfrac{\sqrt{22}}{\sqrt{4}} = \dfrac{\sqrt{22}}{2}$

5. $\dfrac{2\sqrt{3}}{7\sqrt{5}} = \dfrac{2\sqrt{3}}{7\sqrt{5}}\cdot\dfrac{\sqrt{5}}{\sqrt{5}} = \dfrac{2\sqrt{15}}{7\sqrt{5^2}} = \dfrac{2\sqrt{15}}{7\cdot 5} = \dfrac{2\sqrt{15}}{35}$

7. $\sqrt[3]{\dfrac{16}{9}} = \sqrt[3]{\dfrac{16}{9}\cdot\dfrac{3}{3}} = \sqrt[3]{\dfrac{48}{27}} = \dfrac{\sqrt[3]{8\cdot 6}}{\sqrt[3]{27}} = \dfrac{2\sqrt[3]{6}}{3}$

9. $\dfrac{\sqrt[3]{3a}}{\sqrt[3]{5c}} = \dfrac{\sqrt[3]{3a}}{\sqrt[3]{5c}}\cdot\dfrac{\sqrt[3]{5^2c^2}}{\sqrt[3]{5^2c^2}} = \dfrac{\sqrt[3]{75ac^2}}{\sqrt[3]{5^3c^3}} = \dfrac{\sqrt[3]{75ac^2}}{5c}$

11. $\dfrac{\sqrt[3]{2y^4}}{\sqrt[3]{6x^4}} = \dfrac{\sqrt[3]{2y^4}}{\sqrt[3]{6x^4}}\cdot\dfrac{\sqrt[3]{6^2x^2}}{\sqrt[3]{6^2x^2}} = \dfrac{\sqrt[3]{72x^2y^4}}{\sqrt[3]{6^3x^6}} = \dfrac{\sqrt[3]{8y^3\cdot 9x^2y}}{6x^2} =$

$\dfrac{2y\sqrt[3]{9x^2y}}{6x^2} = \dfrac{y\sqrt[3]{9x^2y}}{3x^2}$

13. $\dfrac{1}{\sqrt[4]{st}} = \dfrac{1}{\sqrt[4]{st}}\cdot\dfrac{\sqrt[4]{s^3t^3}}{\sqrt[4]{s^3t^3}} = \dfrac{\sqrt[4]{s^3t^3}}{\sqrt[4]{s^4t^4}} = \dfrac{\sqrt[4]{s^3t^3}}{st}$

15. $\sqrt{\dfrac{3x}{20}} = \sqrt{\dfrac{3x}{20}\cdot\dfrac{5}{5}} = \sqrt{\dfrac{15x}{100}} = \dfrac{\sqrt{15x}}{\sqrt{100}} = \dfrac{\sqrt{15x}}{10}$

17. $\sqrt[3]{\dfrac{4}{5x^5y^2}} = \sqrt[3]{\dfrac{4}{5x^5y^2}\cdot\dfrac{25xy}{5^2xy}} = \sqrt[3]{\dfrac{100xy}{5^3x^6y^3}} =$

$\dfrac{\sqrt[3]{100xy}}{\sqrt[3]{5^3x^6y^3}} = \dfrac{\sqrt[3]{100xy}}{5x^2y}$

19. $\sqrt[4]{\dfrac{1}{8x^7y^3}} = \sqrt[4]{\dfrac{1}{2^3x^7y^3}\cdot\dfrac{2xy}{2xy}} = \sqrt[4]{\dfrac{2xy}{2^4x^8y^4}} = \dfrac{\sqrt[4]{2xy}}{\sqrt[4]{2^4x^8y^4}} =$

$\dfrac{\sqrt[4]{2xy}}{2x^2y}$

21. $\dfrac{9}{6-\sqrt{10}} = \dfrac{9}{6-\sqrt{10}}\cdot\dfrac{6+\sqrt{10}}{6+\sqrt{10}} = \dfrac{9(6+\sqrt{10})}{6^2-(\sqrt{10})^2} =$

$\dfrac{9(6+\sqrt{10})}{36-10} = \dfrac{54+9\sqrt{10}}{26}$

23. $\dfrac{-4\sqrt{7}}{\sqrt{5}-\sqrt{3}} = \dfrac{-4\sqrt{7}}{\sqrt{5}-\sqrt{3}} \cdot \dfrac{\sqrt{5}+\sqrt{3}}{\sqrt{5}+\sqrt{3}} =$
$\dfrac{-4\sqrt{7}(\sqrt{5}+\sqrt{3})}{(\sqrt{5})^2-(\sqrt{3})^2} = \dfrac{-4\sqrt{7}(\sqrt{5}+\sqrt{3})}{5-3} =$
$\dfrac{-4\sqrt{7}(\sqrt{5}+\sqrt{3})}{2} = -2\sqrt{7}(\sqrt{5}+\sqrt{3}) = -2\sqrt{35} - 2\sqrt{21}$

25. $\dfrac{\sqrt{5}-2\sqrt{6}}{\sqrt{3}-4\sqrt{5}} = \dfrac{\sqrt{5}-2\sqrt{6}}{\sqrt{3}-4\sqrt{5}} \cdot \dfrac{\sqrt{3}+4\sqrt{5}}{\sqrt{3}+4\sqrt{5}} =$
$\dfrac{\sqrt{15}+4\cdot 5 - 2\sqrt{18} - 8\sqrt{30}}{(\sqrt{3})^2 - (4\sqrt{5})^2} =$
$\dfrac{\sqrt{15}+20 - 2\sqrt{9\cdot 2} - 8\sqrt{30}}{3 - 16\cdot 5} =$
$\dfrac{\sqrt{15}+20 - 6\sqrt{2} - 8\sqrt{30}}{-77}$, or $-\dfrac{\sqrt{15}+20 - 6\sqrt{2} - 8\sqrt{30}}{77}$

27. $\dfrac{2-\sqrt{a}}{3+\sqrt{a}} = \dfrac{2-\sqrt{a}}{3+\sqrt{a}} \cdot \dfrac{3-\sqrt{a}}{3-\sqrt{a}} = \dfrac{6 - 2\sqrt{a} - 3\sqrt{a} + a}{9-a} =$
$\dfrac{6 - 5\sqrt{a} + a}{9-a}$

29. $\dfrac{5\sqrt{3}-3\sqrt{2}}{3\sqrt{2}-2\sqrt{3}} = \dfrac{5\sqrt{3}-3\sqrt{2}}{3\sqrt{2}-2\sqrt{3}} \cdot \dfrac{3\sqrt{2}+2\sqrt{3}}{3\sqrt{2}+2\sqrt{3}} =$
$\dfrac{15\sqrt{6}+10\cdot 3 - 9\cdot 2 - 6\sqrt{6}}{9\cdot 2 - 4\cdot 3} = \dfrac{12+9\sqrt{6}}{6} =$
$\dfrac{3(4+3\sqrt{6})}{3\cdot 2} = \dfrac{4+3\sqrt{6}}{2}$

31. $\dfrac{\sqrt{x}-\sqrt{y}}{\sqrt{x}+\sqrt{y}} = \dfrac{\sqrt{x}-\sqrt{y}}{\sqrt{x}+\sqrt{y}} \cdot \dfrac{\sqrt{x}-\sqrt{y}}{\sqrt{x}-\sqrt{y}} =$
$\dfrac{x - \sqrt{xy} - \sqrt{xy} + y}{x-y} = \dfrac{x - 2\sqrt{xy} + y}{x-y}$

33. Discussion and Writing Exercise

35. $\dfrac{1}{2} - \dfrac{1}{3} = \dfrac{5}{t}$, LCM is $6t$
$6t\left(\dfrac{1}{2} - \dfrac{1}{3}\right) = 6t\left(\dfrac{5}{t}\right)$
$3t - 2t = 30$
$t = 30$
Check:
$\dfrac{1}{2} - \dfrac{1}{3} = \dfrac{1}{t}$
$\dfrac{1}{2} - \dfrac{1}{3}\ ?\ \dfrac{5}{30}$
$\dfrac{3}{6} - \dfrac{2}{6}\ \bigg|\ \dfrac{1}{6}$
$\dfrac{1}{6}\ \bigg|$ TRUE

The solution is 30.

37. $\dfrac{1}{x^3-y^2} \div \dfrac{1}{(x-y)(x^2+xy+y^2)}$
$= \dfrac{1}{(x-y)(x^2+xy+y^2)} \cdot \dfrac{(x-y)(x^2+xy+y^2)}{1}$
$= \dfrac{(x-y)(x^2+xy+y^2)}{(x-y)(x^2+xy+y^2)}$
$= 1$

39. Left to the student

41. $\sqrt{a^2-3} - \dfrac{a^2}{\sqrt{a^2-3}}$
$= \sqrt{a^2-3} - \dfrac{a^2}{\sqrt{a^2-3}} \cdot \dfrac{\sqrt{a^2-3}}{\sqrt{a^2-3}}$
$= \sqrt{a^2-3} - \dfrac{a^2\sqrt{a^2-3}}{a^2-3}$
$= \sqrt{a^2-3} \cdot \dfrac{a^2-3}{a^2-3} - \dfrac{a^2\sqrt{a^2-3}}{a^2-3}$
$= \dfrac{a^2\sqrt{a^2-3} - 3\sqrt{a^2-3} - a^2\sqrt{a^2-3}}{a^2-3}$
$= \dfrac{-3\sqrt{a^2-3}}{a^2-3}$, or $-\dfrac{3\sqrt{a^2-3}}{a^2-3}$

Exercise Set 10.6

1. $\sqrt{2x-3} = 4$
$(\sqrt{2x-3})^2 = 4^2$ Principle of powers
$2x - 3 = 16$
$2x = 19$
$x = \dfrac{19}{2}$

Check: $\sqrt{2x-3} = 4$
$\sqrt{2\cdot\dfrac{19}{2} - 3}\ ?\ 4$
$\sqrt{19-3}\ \bigg|$
$\sqrt{16}\ \bigg|$
$4\ \bigg|$ TRUE

The solution is $\dfrac{19}{2}$.

3. $\sqrt{6x} + 1 = 8$
$\sqrt{6x} = 7$ Subtracting to isolate the radical
$(\sqrt{6x})^2 = 7^2$ Principle of powers
$6x = 49$
$x = \dfrac{49}{6}$

Check: $\sqrt{6x} + 1 = 8$
$\sqrt{6\cdot\dfrac{49}{6}} + 1\ ?\ 8$
$\sqrt{49} + 1\ \bigg|$
$7 + 1\ \bigg|$
$8\ \bigg|$ TRUE

The solution is $\dfrac{49}{6}$.

Exercise Set 10.6 241

5. $\sqrt{y+7} - 4 = 4$
$\sqrt{y+7} = 8$ Adding to isolate the radical
$(\sqrt{y+7})^2 = 8^2$ Principle of powers
$y + 7 = 64$
$y = 57$

Check: $\dfrac{\sqrt{y+7} - 4 = 4}{\sqrt{57+7} - 4 \;?\; 4}$
$\sqrt{64} - 4$
$8 - 4$
$4 \;\big|\; \text{TRUE}$

The solution is 57.

7. $\sqrt{5y+8} = 10$
$(\sqrt{5y+8})^2 = 10^2$ Principle of powers
$5y + 8 = 100$
$5y = 92$
$y = \dfrac{92}{5}$

Check: $\dfrac{\sqrt{5y+8} = 10}{\sqrt{5 \cdot \dfrac{92}{5} + 8} \;?\; 10}$
$\sqrt{92+8}$
$\sqrt{100}$
$10 \;\big|\; \text{TRUE}$

The solution is $\dfrac{92}{5}$.

9. $\sqrt[3]{x} = -1$
$(\sqrt[3]{x})^3 = (-1)^3$ Principle of powers
$x = -1$

Check: $\dfrac{\sqrt[3]{x} = -1}{\sqrt[3]{-1} \;?\; -1}$
$-1 \;\big|\; \text{TRUE}$

The solution is -1.

11. $\sqrt{x+2} = -4$
$(\sqrt{x+2})^2 = (-4)^2$
$x + 2 = 16$
$x = 14$

Check: $\dfrac{\sqrt{x+2} = -4}{\sqrt{14+2} \;?\; -4}$
$\sqrt{16}$
$4 \;\big|\; \text{FALSE}$

The number 14 does not check. The equation has no solution. We might have observed at the outset that this equation has no solution because the principle square root of a number is never negative.

13. $\sqrt[3]{x+5} = 2$
$(\sqrt[3]{x+5})^3 = 2^3$
$x + 5 = 8$
$x = 3$

Check: $\dfrac{\sqrt[3]{x+5} = 2}{\sqrt[3]{3+5} \;?\; 2}$
$\sqrt[3]{8}$
$2 \;\big|\; \text{TRUE}$

The solution is 3.

15. $\sqrt[4]{y-3} = 2$
$(\sqrt[4]{y-3})^4 = 2^4$
$y - 3 = 16$
$y = 19$

Check: $\dfrac{\sqrt[4]{y-3} = 2}{\sqrt[4]{19-3} \;?\; 2}$
$\sqrt[4]{16}$
$2 \;\big|\; \text{TRUE}$

The solution is 19.

17. $\sqrt[3]{6x+9} + 8 = 5$
$\sqrt[3]{6x+9} = -3$
$(\sqrt[3]{6x+9})^3 = (-3)^3$
$6x + 9 = -27$
$6x = -36$
$x = -6$

Check: $\dfrac{\sqrt[3]{6x+9} + 8 = 5}{\sqrt[3]{6(-6)+9} + 8 \;?\; 5}$
$\sqrt[3]{-27} + 8$
$-3 + 8$
$5 \;\big|\; \text{TRUE}$

The solution is -6.

19. $8 = \dfrac{1}{\sqrt{x}}$
$8 \cdot \sqrt{x} = \dfrac{1}{\sqrt{x}} \cdot \sqrt{x}$
$8\sqrt{x} = 1$
$(8\sqrt{x})^2 = 1^2$
$64x = 1$
$x = \dfrac{1}{64}$

Check: $8 = \dfrac{1}{\sqrt{x}}$

$8 \ ? \ \dfrac{1}{\sqrt{\dfrac{1}{64}}}$

$\quad\quad \Big| \ \dfrac{1}{\frac{1}{8}}$

$\quad\quad 8 \quad$ TRUE

The solution is $\dfrac{1}{64}$.

21. $\quad x - 7 = \sqrt{x - 5}$
$(x - 7)^2 = (\sqrt{x - 5})^2$
$x^2 - 14x + 49 = x - 5$
$x^2 - 15x + 54 = 0$
$(x - 6)(x - 9) = 0$
$x - 6 = 0 \ \text{or} \ x - 9 = 0$
$x = 6 \ \text{or} \ \quad x = 9$

Check: For 6: $\quad x - 7 = \sqrt{x - 5}$

$\quad\quad 6 - 7 \ ? \ \sqrt{6 - 5}$

$\quad\quad -1 \ \Big| \ \sqrt{1}$

$\quad\quad -1 \ \Big| \ 1 \quad$ FALSE

Check: For 9: $\quad x - 7 = \sqrt{x - 5}$

$\quad\quad 9 - 7 \ ? \ \sqrt{9 - 5}$

$\quad\quad 2 \ \Big| \ \sqrt{4}$

$\quad\quad 2 \ \Big| \ 2 \quad$ TRUE

The number 6 does not check, but 9 does. The solution is 9.

23. $\quad 2\sqrt{x + 1} + 7 = x$
$2\sqrt{x + 1} = x - 7$
$(2\sqrt{x + 1})^2 = (x - 7)^2$
$4(x + 1) = x^2 - 14x + 49$
$4x + 4 = x^2 - 14x + 49$
$0 = x^2 - 18x + 45$
$0 = (x - 3)(x - 15)$
$x - 3 = 0 \ \text{or} \ x - 15 = 0$
$x = 3 \ \text{or} \ \quad x = 15$

Check: For 3: $\quad 2\sqrt{x + 1} + 7 = x$

$\quad\quad 2\sqrt{3 + 1} + 7 \ ? \ 3$

$\quad\quad 2\sqrt{4} + 7 \ \Big|$

$\quad\quad 2 \cdot 2 + 7 \ \Big|$

$\quad\quad 4 + 7 \ \Big|$

$\quad\quad 11 \ \Big| \ 3 \quad$ FALSE

Check: For 15: $\quad 2\sqrt{x + 1} + 7 = x$

$\quad\quad 2\sqrt{15 + 1} + 7 \ ? \ 15$

$\quad\quad 2\sqrt{16} + 7 \ \Big|$

$\quad\quad 2 \cdot 4 + 7 \ \Big|$

$\quad\quad 8 + 7 \ \Big|$

$\quad\quad 15 \ \Big| \ 15 \quad$ TRUE

The number 3 does not check, but 15 does. The solution is 15.

25. $\quad 3\sqrt{x - 1} - 1 = x$
$3\sqrt{x - 1} = x + 1$
$(3\sqrt{x - 1})^2 = (x + 1)^2$
$9(x - 1) = x^2 + 2x + 1$
$9x - 9 = x^2 + 2x + 1$
$0 = x^2 - 7x + 10$
$0 = (x - 2)(x - 5)$
$x - 2 = 0 \ \text{or} \ x - 5 = 0$
$x = 2 \ \text{or} \ \quad x = 5$

Check: For 2: $\quad 3\sqrt{x - 1} - 1 = x$

$\quad\quad 3\sqrt{2 - 1} - 1 \ ? \ 2$

$\quad\quad 3\sqrt{1} - 1 \ \Big|$

$\quad\quad 3 \cdot 1 - 1 \ \Big|$

$\quad\quad 3 - 1 \ \Big|$

$\quad\quad 2 \ \Big| \ 2 \quad$ TRUE

Check: For 5: $\quad 3\sqrt{x - 1} - 1 = x$

$\quad\quad 3\sqrt{5 - 1} - 1 \ ? \ 5$

$\quad\quad 3\sqrt{4} - 1 \ \Big|$

$\quad\quad 3 \cdot 2 - 1 \ \Big|$

$\quad\quad 6 - 1 \ \Big|$

$\quad\quad 5 \ \Big| \ 5 \quad$ TRUE

Both numbers check. The solutions are 2 and 5.

27. $\quad x - 3 = \sqrt{27 - 3x}$
$(x - 3)^2 = (\sqrt{27 - 3x})^2$
$x^2 - 6x + 9 = 27 - 3x$
$x^2 - 3x - 18 = 0$
$(x - 6)(x + 3) = 0$
$x - 6 = 0 \ \text{or} \ x + 3 = 0$
$x = 6 \ \text{or} \ \quad x = -3$

Check: For 6: $\quad x - 3 = \sqrt{27 - 3x}$

$\quad\quad 6 - 3 \ ? \ \sqrt{27 - 3 \cdot 6}$

$\quad\quad 3 \ \Big| \ \sqrt{27 - 18}$

$\quad\quad \ \Big| \ \sqrt{9}$

$\quad\quad 3 \ \Big| \ 3 \quad$ TRUE

Exercise Set 10.6

Check: For -3:

$$\begin{array}{c|c} x-3 = & \sqrt{27-3x} \\ \hline -3-3 \;?\; & \sqrt{27-3(-3)} \\ -6 & \sqrt{27+9} \\ & \sqrt{36} \\ -6 & 6 \quad \text{FALSE} \end{array}$$

The number 6 checks but -3 does not. The solution is 6.

29. $\sqrt{3y+1} = \sqrt{2y+6}$
$(\sqrt{3y+1})^2 = (\sqrt{2y+6})^2$
$3y+1 = 2y+6$
$y = 5$

Check:
$$\begin{array}{c|c} \sqrt{3y+1} = & \sqrt{2y+6} \\ \hline \sqrt{3 \cdot 5+1} \;?\; & \sqrt{2 \cdot 5+6} \\ \sqrt{16} & \sqrt{16} \quad \text{TRUE} \end{array}$$

The solution is 5.

31. $\sqrt{y-5} + \sqrt{y} = 5$
$\sqrt{y-5} = 5 - \sqrt{y}$ Isolating one radical
$(\sqrt{y-5})^2 = (5-\sqrt{y})^2$
$y - 5 = 25 - 10\sqrt{y} + y$
$10\sqrt{y} = 30$ Isolating the remaining radical
$\sqrt{y} = 3$ Dividing by 10
$(\sqrt{y})^2 = 3^2$
$y = 9$

The number 9 checks, so it is the solution.

33. $3 + \sqrt{z-6} = \sqrt{z+9}$
$(3+\sqrt{z-6})^2 = (\sqrt{z+9})^2$
$9 + 6\sqrt{z-6} + z - 6 = z+9$
$6\sqrt{z-6} = 6$
$\sqrt{z-6} = 1$ Dividing by 6
$(\sqrt{z-6})^2 = 1^2$
$z - 6 = 1$
$z = 7$

The number 7 checks, so it is the solution.

35. $\sqrt{20-x} + 8 = \sqrt{9-x} + 11$
$\sqrt{20-x} = \sqrt{9-x} + 3$ Isolating one radical
$(\sqrt{20-x})^2 = (\sqrt{9-x}+3)^2$
$20 - x = 9 - x + 6\sqrt{9-x} + 9$
$2 = 6\sqrt{9-x}$ Isolating the remaining radical
$1 = 3\sqrt{9-x}$ Dividing by 2
$1^2 = (3\sqrt{9-x})^2$
$1 = 9(9-x)$
$1 = 81 - 9x$
$9x = 80$
$x = \dfrac{80}{9}$

The number $\dfrac{80}{9}$ checks, so it is the solution.

37. $\sqrt{4y+1} - \sqrt{y-2} = 3$
$\sqrt{4y+1} = 3 + \sqrt{y-2}$ Isolating one radical
$(\sqrt{4y+1})^2 = (3+\sqrt{y-2})^2$
$4y + 1 = 9 + 6\sqrt{y-2} + y - 2$
$3y - 6 = 6\sqrt{y-2}$ Isolating the remaining radical
$y - 2 = 2\sqrt{y-2}$ Multiplying by $\dfrac{1}{3}$
$(y-2)^2 = (2\sqrt{y-2})^2$
$y^2 - 4y + 4 = 4(y-2)$
$y^2 - 4y + 4 = 4y - 8$
$y^2 - 8y + 12 = 0$
$(y-6)(y-2) = 0$
$y - 6 = 0 \text{ or } y - 2 = 0$
$y = 6 \text{ or } \quad y = 2$

The numbers 6 and 2 check, so they are the solutions.

39. $\sqrt{x+2} + \sqrt{3x+4} = 2$
$\sqrt{x+2} = 2 - \sqrt{3x+4}$ Isolating one radical
$(\sqrt{x+2})^2 = (2-\sqrt{3x-4})^2$
$x + 2 = 4 - 4\sqrt{3x+4} + 3x + 4$
$-2x - 6 = -4\sqrt{3x+4}$ Isolating the remaining radical
$x + 3 = 2\sqrt{3x+4}$ Dividing by 2
$(x+3)^2 = (2\sqrt{3x+4})^2$
$x^2 + 6x + 9 = 4(3x+4)$
$x^2 + 6x + 9 = 12x + 16$
$x^2 - 6x - 7 = 0$
$(x-7)(x+1) = 0$
$x - 7 = 0 \text{ or } x + 1 = 0$
$x = 7 \text{ or } \quad x = -1$

Check: For 7:
$$\sqrt{x+2}+\sqrt{3x+4}=2$$
$$\sqrt{7+2}+\sqrt{3\cdot 7+4} \;?\; 2$$
$$\sqrt{9}+\sqrt{25}$$
$$8 \;|\; \text{FALSE}$$

Check: For -1:
$$\sqrt{x+2}+\sqrt{3x+4}=2$$
$$\sqrt{-1+2}+\sqrt{3(-1)+4} \;?\; 2$$
$$\sqrt{1}+\sqrt{1}$$
$$2 \;|\; \text{TRUE}$$

Since -1 checks but 7 does not, the solution is -1.

41. $\sqrt{3x-5}+\sqrt{2x+3}+1=0$
$$\sqrt{3x-5}+1 = -\sqrt{2x+3}$$
$$(\sqrt{3x-5}+1)^2 = (-\sqrt{2x+3})^2$$
$$3x-5+2\sqrt{3x-5}+1 = 2x+3$$
$$2\sqrt{3x-5} = -x+7$$
$$(2\sqrt{3x-5})^2 = (-x+7)^2$$
$$4(3x-5) = x^2-14x+49$$
$$12x-20 = x^2-14x+49$$
$$0 = x^2-26x+69$$
$$0 = (x-23)(x-3)$$
$$x-23=0 \quad \text{or} \quad x-3=0$$
$$x=23 \quad \text{or} \quad x=3$$

Neither number checks. There is no solution. (At the outset we might have observed that there is no solution since the sum on the left side of the equation must be at least 1.)

43. $2\sqrt{t-1}-\sqrt{3t-1}=0$
$$2\sqrt{t-1} = \sqrt{3t-1}$$
$$(2\sqrt{t-1})^2 = (\sqrt{3t-1})^2$$
$$4(t-1) = 3t-1$$
$$4t-4 = 3t-1$$
$$t=3$$

Since 3 checks, it is the solution.

45. $V = 3.5\sqrt{h}$
$$V = 3.5\sqrt{9000}$$
$$V \approx 332$$

You can see about 332 km to the horizon at an altitude of 9000 m.

47. $V = 3.5\sqrt{h}$
$$50.4 = 3.5\sqrt{h}$$
$$(50.4)^2 = (3.5\sqrt{h})^2$$
$$2540.16 = 12.25h$$
$$207.36 = h$$

The altitude is 207.36 m.

49. $V = 3.5\sqrt{h}$
$$21 = 3.5\sqrt{h}$$
$$21^2 = (3.5\sqrt{h})^2$$
$$441 = 12.25h$$
$$36 = h$$

The height of the steeplejack's eyes is 36 m.

51. $V = 3.5\sqrt{h}$
$$V = 3.5\sqrt{37}$$
$$V \approx 21$$

The sailor can see about 21 km to the horizon.

53. At 55 mph: $r = 2\sqrt{5L}$
$$55 = 2\sqrt{5L}$$
$$27.5 = \sqrt{5L}$$
$$(27.5)^2 = (\sqrt{5L})^2$$
$$756.25 = 5L$$
$$151.25 = L$$

At 55 mph, a car will skid 151.25 ft.

At 75 mph: $r = 2\sqrt{5L}$
$$75 = 2\sqrt{5L}$$
$$37.5 = \sqrt{5L}$$
$$(37.5)^2 = (\sqrt{5L})^2$$
$$1406.25 = 5L$$
$$281.25 = L$$

At 75 mph, a car will skid 281.25 ft.

55. $S = 21.9\sqrt{5t+2457}$
$$1113 = 21.9\sqrt{5t+2457}$$
$$\frac{1113}{21.9} = \sqrt{5t+2457}$$
$$\left(\frac{1113}{21.9}\right)^2 = (\sqrt{5t+2457})^2$$
$$2583 \approx 5t+2457$$
$$126 \approx 5t$$
$$25.2 \approx t$$

The temperature was approximately 25°F.

Exercise Set 10.6

57.
$$T = 2\pi\sqrt{\frac{L}{32}}$$
$$1 = 2(3.14)\sqrt{\frac{L}{32}}$$
$$1 = 6.28\sqrt{\frac{L}{32}}$$
$$1^2 = \left(6.28\sqrt{\frac{L}{32}}\right)^2$$
$$1 = 39.4384 \cdot \frac{L}{32}$$
$$\frac{32}{39.4384} = L$$
$$0.81 \approx L$$

The pendulum is about 0.81 ft long.

59. Discussion and Writing Exercise

61. *Familiarize.* Let t = the time it will take Julia and George to paint the room working together.

Translate. We use the work principle.
$$\frac{t}{a} + \frac{t}{b} = 1$$
$$\frac{t}{8} + \frac{t}{10} = 1$$

Solve. We first multiply by 40 to clear fractions.
$$40\left(\frac{t}{8} + \frac{t}{10}\right) = 40 \cdot 1$$
$$40 \cdot \frac{t}{8} + 40 \cdot \frac{t}{10} = 40$$
$$5t + 4t = 40$$
$$9t = 40$$
$$t = \frac{40}{9}, \text{ or } 4\frac{4}{9}$$

Check. In $\frac{40}{9}$ hr, Julia does $\frac{40}{9}\left(\frac{1}{8}\right)$, or $\frac{5}{9}$, of the job and George does $\frac{40}{9}\left(\frac{1}{10}\right)$, or $\frac{4}{9}$, of the job. Together they do $\frac{5}{9} + \frac{4}{9}$, or 1 entire job. The answer checks.

State. It will take them $4\frac{4}{9}$ hr to paint the room, working together.

63. *Familiarize.* Let d = the distance the cyclist would travel in 56 days at the same rate.

Translate. We translate to a proportion.

Distance → $\dfrac{702}{14} = \dfrac{d}{56}$ ← Distance
Days →   ← Days

Solve. We equate cross products.
$$\frac{702}{14} = \frac{d}{56}$$
$$702 \cdot 56 = 14 \cdot d$$
$$\frac{702 \cdot 56}{14} = d$$
$$2808 = d$$

Check. We substitute in the proportion and check cross products.
$$\frac{702}{14} = \frac{2808}{56}; \quad 702 \cdot 56 = 39,312; \quad 14 \cdot 2808 = 39,312$$

The cross products are the same, so the answer checks.

State. The cyclist would have traveled 2808 mi in 56 days.

65.
$$x^2 + 2.8x = 0$$
$$x(x + 2.8) = 0$$
$$x = 0 \text{ or } x + 2.8 = 0$$
$$x = 0 \text{ or } \quad x = -2.8$$

The solutions are 0 and -2.8.

67.
$$x^2 - 64 = 0$$
$$(x+8)(x-8) = 0$$
$$x + 8 = 0 \text{ or } x - 8 = 0$$
$$x = -8 \text{ or } \quad x = 8$$

The solutions are -8 and 8.

69. Left to the student

71.
$$\sqrt[3]{\frac{z}{4}} - 10 = 2$$
$$\sqrt[3]{\frac{z}{4}} = 12$$
$$\left(\sqrt[3]{\frac{z}{4}}\right)^3 = 12^3$$
$$\frac{z}{4} = 1728$$
$$z = 6912$$

The number 6912 checks, so it is the solution.

73.
$$\sqrt{\sqrt{y+49} - \sqrt{y}} = \sqrt{7}$$
$$\left(\sqrt{\sqrt{y+49} - \sqrt{y}}\right)^2 = (\sqrt{7})^2$$
$$\sqrt{y+49} - \sqrt{y} = 7$$
$$\sqrt{y+49} = 7 + \sqrt{y}$$
$$(\sqrt{y+49})^2 = (7 + \sqrt{y})^2$$
$$y + 49 = 49 + 14\sqrt{y} + y$$
$$0 = 14\sqrt{y}$$
$$0 = \sqrt{y}$$
$$0^2 = (\sqrt{y})^2$$
$$0 = y$$

The number 0 checks and is the solution.

75.
$$\sqrt{\sqrt{x^2+9x+34}} = 2$$
$$\left(\sqrt{\sqrt{x^2+9x+34}}\right)^2 = 2^2$$
$$\sqrt{x^2+9x+34} = 4$$
$$(\sqrt{x^2+9x+34})^2 = 4^2$$
$$x^2+9x+34 = 16$$
$$x^2+9x+18 = 0$$
$$(x+6)(x+3) = 0$$
$$x+6 = 0 \quad \text{or} \quad x+3 = 0$$
$$x = -6 \quad \text{or} \quad x = -3$$

Both values check. The solutions are -6 and -3.

77.
$$\sqrt{x-2} - \sqrt{x+2} + 2 = 0$$
$$\sqrt{x-2} + 2 = \sqrt{x+2}$$
$$(\sqrt{x-2}+2)^2 = (\sqrt{x+2})^2$$
$$(x-2) + 4\sqrt{x-2} + 4 = x+2$$
$$4\sqrt{x-2} = 0$$
$$\sqrt{x-2} = 0$$
$$(\sqrt{x-2})^2 = 0^2$$
$$x-2 = 0$$
$$x = 2$$

The number 2 checks, so it is the solution.

79.
$$\sqrt{a^2+30a} = a + \sqrt{5a}$$
$$(\sqrt{a^2+30a})^2 = (a+\sqrt{5a})^2$$
$$a^2+30a = a^2 + 2a\sqrt{5a} + 5a$$
$$25a = 2a\sqrt{5a}$$
$$(25a)^2 = (2a\sqrt{5a})^2$$
$$625a^2 = 4a^2 \cdot 5a$$
$$625a^2 = 20a^3$$
$$0 = 20a^3 - 625a^2$$
$$0 = 5a^2(4a - 125)$$
$$5a^2 = 0 \quad \text{or} \quad 4a - 125 = 0$$
$$a^2 = 0 \quad \text{or} \quad 4a = 125$$
$$a = 0 \quad \text{or} \quad a = \frac{125}{4}$$

Both values check. The solutions are 0 and $\frac{125}{4}$.

81.
$$\frac{x-1}{\sqrt{x^2+3x+6}} = \frac{1}{4},$$
$$\text{LCM} = 4\sqrt{x^2+3x+6}$$
$$4\sqrt{x^2+3x+6} \cdot \frac{x-1}{\sqrt{x^2+3x+6}} = 4\sqrt{x^2+3x+6} \cdot \frac{1}{4}$$
$$4x - 4 = \sqrt{x^2+3x+6}$$
$$16x^2 - 32x + 16 = x^2 + 3x + 6$$
$$\text{Squaring both sides}$$
$$15x^2 - 35x + 10 = 0$$
$$3x^2 - 7x + 2 = 0 \quad \text{Dividing by 5}$$
$$(3x-1)(x-2) = 0$$
$$3x-1 = 0 \quad \text{or} \quad x-2 = 0$$
$$3x = 1 \quad \text{or} \quad x = 2$$
$$x = \frac{1}{3} \quad \text{or} \quad x = 2$$

The number 2 checks but $\frac{1}{3}$ does not. The solution is 2.

83.
$$\sqrt{y^2+6} + y - 3 = 0$$
$$\sqrt{y^2+6} = 3 - y$$
$$(\sqrt{y^2+6})^2 = (3-y)^2$$
$$y^2 + 6 = 9 - 6y + y^2$$
$$-3 = -6y$$
$$\frac{1}{2} = y$$

The number $\frac{1}{2}$ checks and is the solution.

85.
$$\sqrt{y+1} - \sqrt{2y-5} = \sqrt{y-2}$$
$$(\sqrt{y+1} - \sqrt{2y-5})^2 = (\sqrt{y-2})^2$$
$$y+1 - 2\sqrt{(y+1)(2y-5)} + 2y - 5 = y - 2$$
$$-2\sqrt{2y^2 - 3y - 5} = -2y + 2$$
$$\sqrt{2y^2 - 3y - 5} = y - 1$$
$$\text{Dividing by } -2$$
$$(\sqrt{2y^2-3y-5})^2 = (y-1)^2$$
$$2y^2 - 3y - 5 = y^2 - 2y + 1$$
$$y^2 - y - 6 = 0$$
$$(y-3)(y+2) = 0$$
$$y - 3 = 0 \quad \text{or} \quad y + 2 = 0$$
$$y = 3 \quad \text{or} \quad y = -2$$

The number 3 checks but -2 does not. The solution is 3.

Exercise Set 10.7

1. $a = 3, \quad b = 5$

Find c.
$$c^2 = a^2 + b^2 \quad \text{Pythagorean equation}$$
$$c^2 = 3^2 + 5^2 \quad \text{Substituting}$$
$$c^2 = 9 + 25$$
$$c^2 = 34$$
$$c = \sqrt{34} \quad \text{Exact answer}$$
$$c \approx 5.831 \quad \text{Approximation}$$

Exercise Set 10.7

3. $a = 15$, $b = 15$
Find c.
$c^2 = a^2 + b^2$ Pythagorean equation
$c^2 = 15^2 + 15^2$ Substituting
$c^2 = 225 + 225$
$c^2 = 450$
$c = \sqrt{450}$ Exact answer
$c \approx 21.213$ Approximation

5. $b = 12$, $c = 13$
Find a.
$a^2 + b^2 = c^2$ Pythagorean equation
$a^2 + 12^2 = 13^2$ Substituting
$a^2 + 144 = 169$
$a^2 = 25$
$a = 5$

7. $c = 7$, $a = \sqrt{6}$
Find b.
$c^2 = a^2 + b^2$ Pythagorean equation
$7^2 = (\sqrt{6})^2 + b^2$ Substituting
$49 = 6 + b^2$
$43 = b^2$
$\sqrt{43} = b$ Exact answer
$6.557 \approx b$ Approximation

9. $b = 1$, $c = \sqrt{13}$
Find a.
$a^2 + b^2 = c^2$ Pythagorean equation
$a^2 + 1^2 = (\sqrt{13})^2$ Substituting
$a^2 + 1 = 13$
$a^2 = 12$
$a = \sqrt{12}$ Exact answer
$a \approx 3.464$ Approximation

11. $a = 1$, $c = \sqrt{n}$
Find b.
$a^2 + b^2 = c^2$
$1^2 + b^2 = (\sqrt{n})^2$
$1 + b^2 = n$
$b^2 = n - 1$
$b = \sqrt{n-1}$

13. We add labels to the drawing in the text. We let $h =$ the height of the bulge.

Note that 1 mi = 5280 ft, so 1 mi + 1 ft = 5280 + 1, or 5281 ft.
We use the Pythagorean equation to find h.
$5281^2 = 5280^2 + h^2$
$27{,}888{,}961 = 27{,}878{,}400 + h^2$
$10{,}561 = h^2$
$\sqrt{10{,}561} = h$
$102.767 \approx h$
The bulge is $\sqrt{10{,}561}$ ft, or about 102.767 ft high.

15. We add some labels to the drawing in the text.

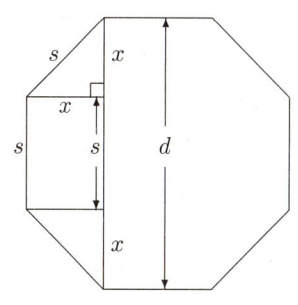

Note that $d = s + 2x$. We use the Pythagorean equation to find x.
$x^2 + x^2 = s^2$
$2x^2 = s^2$
$x^2 = \dfrac{s^2}{2}$
$x = \sqrt{\dfrac{s^2}{2}}$
$x = \dfrac{s}{\sqrt{2}}$
$x = \dfrac{s\sqrt{2}}{2}$ Rationalizing the denominator
Then $d = s + 2x = s + 2\left(\dfrac{s\sqrt{2}}{2}\right) = s + s\sqrt{2}$.

17. We make a drawing and let $d =$ the length of the guy wire.

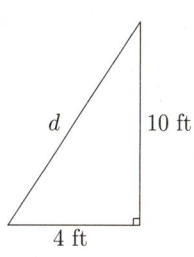

We use the Pythagorean equation to find d.
$d^2 = 4^2 + 10^2$
$d^2 = 16 + 100$
$d^2 = 116$
$d = \sqrt{116}$
$d \approx 10.770$

The wire is $\sqrt{116}$ ft, or about 10.770 ft long.

19. $L = \dfrac{0.000169 d^{2.27}}{h}$

$L = \dfrac{0.000169(200)^{2.27}}{4}$

≈ 7.1

The length of the letters should be about 7.1 ft.

21. We make a drawing. Let $x =$ the width of the rectangle. Then $x + 1 =$ the length.

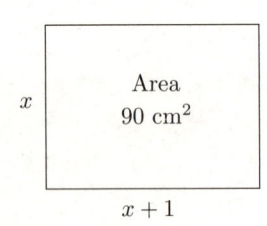

[Rectangle with width x, length $x+1$, Area 90 cm²]

We first find the length and width of the rectangle. Recall the formula for the area of a rectangle, $A = lw$. We substitute 90 for A, $x + 1$ for l, and x for w in this formula and solve for x.

$90 = (x+1)x$

$90 = x^2 + x$

$0 = x^2 + x - 90$

$0 = (x+10)(x-9)$

$x + 10 = 0$ or $x - 9 = 0$

$x = -10$ or $x = 9$

Since the width cannot be negative, we know that the width is 9 cm. Thus the length is 10 cm. (These numbers check since 9 and 10 are consecutive integers and the area of a rectangle with width 9 cm and length 10 cm is $10 \cdot 9$, or 90 cm².)

Now we find the length of the diagonal of the rectangle. We make another drawing, letting $d =$ the length of the diagonal.

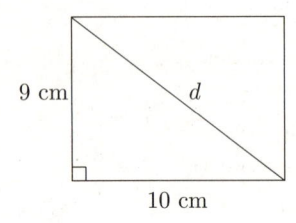

[Rectangle 9 cm by 10 cm with diagonal d]

We use the Pythagorean equation to find d.

$d^2 = 9^2 + 10^2$

$d^2 = 81 + 100$

$d^2 = 181$

$d = \sqrt{181}$

$d \approx 13.454$

The length of the diagonal is $\sqrt{181}$ cm, or about 13.454 cm.

23. We use the drawing in the text, replacing w with 16 in.

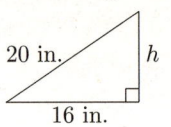

[Right triangle with hypotenuse 20 in., legs 16 in. and h]

We use the Pythagorean equation to find h.

$h^2 + 16^2 = 20^2$

$h^2 + 256 = 400$

$h^2 = 144$

$h = 12$

The height is 12 in.

25. We first make a drawing. A point on the x-axis has coordinates $(x, 0)$ and is $|x|$ units from the origin.

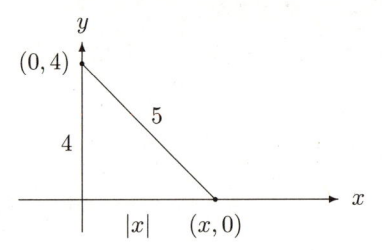

[Graph showing point (0,4), hypotenuse 5, horizontal leg $|x|$ to point $(x,0)$, vertical leg 4]

We use the Pythagorean equation to find x.

$4^2 + |x|^2 = 5^2$

$16 + x^2 = 25$ $|x|^2 = x^2$

$x^2 - 9 = 0$ Subtracting 25

$(x+3)(x-3) = 0$

$x - 3 = 0$ or $x + 3 = 0$

$x = 3$ or $x = -3$

The points are $(3, 0)$ and $(-3, 0)$.

27. We make a drawing, letting $d =$ the distance the wire will run diagonally, disregarding the slack.

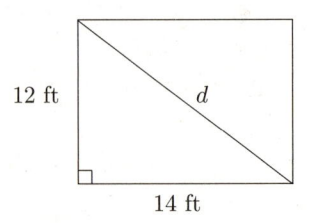

[Rectangle 12 ft by 14 ft with diagonal d]

We use the Pythagorean equation to find d.

$d^2 = 12^2 + 14^2$

$d^2 = 144 + 196$

$d^2 = 340$

$d = \sqrt{340}$

Adding 4 ft of slack on each end, we find that a wire of length $\sqrt{340} + 4 + 4$, or $\sqrt{340} + 8$ ft should be purchased. This is approximately 26.439 ft of wire.

29. Referring to the drawing in the text, we let t = the travel. Then we use the Pythagorean equation.
$$t^2 = (17.75)^2 + (10.25)^2$$
$$t^2 = 315.0625 + 105.0625$$
$$t^2 = 420.125$$
$$t = \sqrt{420.125}$$
$$t \approx 20.497$$

The travel is $\sqrt{420.125}$ in., or about 20.497 in.

31. Discussion and Writing Exercise

33. *Familiarize.* Let r = the speed of the Carmel Crawler. Then $r + 14$ = the speed of the Zionsville Flash. We organize the information in a table.

	Distance	Speed	Time
Crawler	230	r	t
Flash	290	$r + 14$	t

Translate. Using the formula $t = d/r$ and noting that the times are the same, we have
$$\frac{230}{r} = \frac{290}{r+14}.$$

Solve. We first clear fractions by multiplying by the LCM of the denominators, $r(r+14)$.
$$r(r+14) \cdot \frac{230}{r} = r(r+14) \cdot \frac{290}{r+14}$$
$$230(r+14) = 290r$$
$$230r + 3220 = 290r$$
$$3220 = 60r$$
$$\frac{161}{3} = r, \text{ or}$$
$$53\frac{2}{3} = r$$

If $r = 53\frac{2}{3}$, then $r + 14 = 67\frac{2}{3}$.

Check. At $53\frac{2}{3}$, or $\frac{161}{3}$ mph, the Crawler travels 230 mi in $\frac{230}{161/3}$, or about 4.3 hr. At $67\frac{2}{3}$, or $\frac{203}{3}$ mph, the Flash travels 290 mi in $\frac{290}{203/3}$, or about 4.3 hr. Since the times are the same, the answer checks.

State. The Carmel Crawler's speed is $53\frac{2}{3}$ mph, and the Zionsville Flash's speed is $67\frac{2}{3}$ mph.

35. $2x^2 + 11x - 21 = 0$
$(2x - 3)(x + 7) = 0$
$2x - 3 = 0 \quad or \quad x + 7 = 0$
$x = \frac{3}{2} \quad or \quad x = -7$

The solutions are $\frac{3}{2}$ and -7.

37. $\frac{x+2}{x+3} = \frac{x-4}{x-5},$
LCM is $(x+3)(x-5)$
$(x+3)(x-5) \cdot \frac{x+2}{x+3} = (x+3)(x-5) \cdot \frac{x-4}{x-5}$
$(x-5)(x+2) = (x+3)(x-4)$
$x^2 - 3x - 10 = x^2 - x - 12$
$-2x = -2$
$x = 1$

The number 1 checks, so it is the solution.

39. $\frac{x-5}{x-7} = \frac{4}{3},$ LCM is $3(x-7)$
$3(x-7) \cdot \frac{x-5}{x-7} = 3(x-7) \cdot \frac{4}{3}$
$3(x-5) = 4(x-7)$
$3x - 15 = 4x - 28$
$13 = x$

The number 13 checks and is the solution.

41.

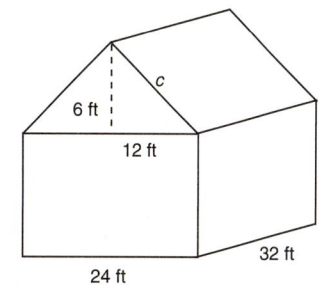

$c^2 = 6^2 + 12^2 = 36 + 144 = 180$
$c = \sqrt{180}$ ft
Area of the roof $= 2 \cdot \sqrt{180} \cdot 32 = 64\sqrt{180}$ ft^2
Number of packets $= \dfrac{64\sqrt{180}}{33\frac{1}{3}} \approx 26$

Kit should buy 26 packets of shingles.

43.

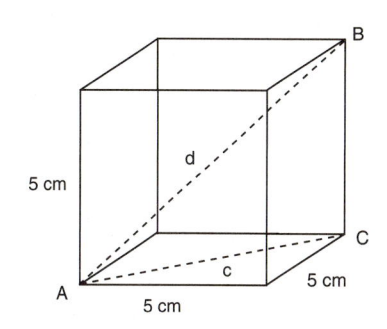

First find the length of a diagonal of the base of the cube. It is the hypotenuse of an isosceles right triangle with $a = 5$ cm. Then $c = a\sqrt{2} = 5\sqrt{2}$ cm.

Triangle ABC is a right triangle with legs of $5\sqrt{2}$ cm and 5 cm and hypotenuse d. Use the Pythagorean equation to find d, the length of the diagonal that connects two opposite corners of the cube.

$d^2 = (5\sqrt{2})^2 + 5^2$
$d^2 = 25 \cdot 2 + 25$
$d^2 = 50 + 25$
$d^2 = 75$
$d = \sqrt{75}$

Exact answer: $d = \sqrt{75}$ cm

Exercise Set 10.8

1. $\sqrt{-35} = \sqrt{-1 \cdot 35} = \sqrt{-1} \cdot \sqrt{35} = i\sqrt{35}$, or $\sqrt{35}i$

3. $\sqrt{-16} = \sqrt{-1 \cdot 16} = \sqrt{-1} \cdot \sqrt{16} = i \cdot 4 = 4i$

5. $-\sqrt{-12} = -\sqrt{-1 \cdot 12} = -\sqrt{-1} \cdot \sqrt{12} = -i \cdot 2\sqrt{3} = -2\sqrt{3}i$, or $-2i\sqrt{3}$

7. $\sqrt{-3} = \sqrt{-1 \cdot 3} = \sqrt{-1} \cdot \sqrt{3} = i\sqrt{3}$, or $\sqrt{3}i$

9. $\sqrt{-81} = \sqrt{-1 \cdot 81} = \sqrt{-1} \cdot \sqrt{81} = i \cdot 9 = 9i$

11. $\sqrt{-98} = \sqrt{-1 \cdot 98} = \sqrt{-1} \cdot \sqrt{98} = i \cdot 7\sqrt{2} = 7\sqrt{2}i$, or $7i\sqrt{2}$

13. $-\sqrt{-49} = -\sqrt{-1 \cdot 49} = -\sqrt{-1} \cdot \sqrt{49} = -i \cdot 7 = -7i$

15. $4 - \sqrt{-60} = 4 - \sqrt{-1 \cdot 60} = 4 - \sqrt{-1} \cdot \sqrt{60} = 4 - i \cdot 2\sqrt{15} = 4 - 2\sqrt{15}i$, or $4 - 2i\sqrt{15}$

17. $\sqrt{-4} + \sqrt{-12} = \sqrt{-1 \cdot 4} + \sqrt{-1 \cdot 12} = \sqrt{-1} \cdot \sqrt{4} + \sqrt{-1} \cdot \sqrt{12} = i \cdot 2 + i \cdot 2\sqrt{3} = (2 + 2\sqrt{3})i$

19. $(7 + 2i) + (5 - 6i)$
    $= (7 + 5) + (2 - 6)i$ Collecting like terms
    $= 12 - 4i$

21. $(4 - 3i) + (5 - 2i)$
    $= (4 + 5) + (-3 - 2)i$ Collecting like terms
    $= 9 - 5i$

23. $(9 - i) + (-2 + 5i) = (9 - 2) + (-1 + 5)i$
    $= 7 + 4i$

25. $(6 - i) - (10 + 3i) = (6 - 10) + (-1 - 3)i$
    $= -4 - 4i$

27. $(4 - 2i) - (5 - 3i) = (4 - 5) + [-2 - (-3)]i$
    $= -1 + i$

29. $(9 + 5i) - (-2 - i) = [9 - (-2)] + [5 - (-1)]i$
    $= 11 + 6i$

31. $\sqrt{-36} \cdot \sqrt{-9} = \sqrt{-1} \cdot \sqrt{36} \cdot \sqrt{-1} \cdot \sqrt{9}$
    $= i \cdot 6 \cdot i \cdot 3$
    $= i^2 \cdot 18$
    $= -1 \cdot 18$ $i^2 = -1$
    $= -18$

33. $\sqrt{-7} \cdot \sqrt{-2} = \sqrt{-1} \cdot \sqrt{7} \cdot \sqrt{-1} \cdot \sqrt{2}$
    $= i \cdot \sqrt{7} \cdot i \cdot \sqrt{2}$
    $= i^2(\sqrt{14})$
    $= -1(\sqrt{14})$ $i^2 = -1$
    $= -\sqrt{14}$

35. $-3i \cdot 7i = -21 \cdot i^2$
    $= -21(-1)$ $i^2 = -1$
    $= 21$

37. $-3i(-8 - 2i) = -3i(-8) - 3i(-2i)$
    $= 24i + 6i^2$
    $= 24i + 6(-1)$ $i^2 = -1$
    $= 24i - 6$
    $= -6 + 24i$

39. $(3 + 2i)(1 + i)$
    $= 3 + 3i + 2i + 2i^2$ Using FOIL
    $= 3 + 3i + 2i - 2$ $i^2 = -1$
    $= 1 + 5i$

41. $(2 + 3i)(6 - 2i)$
    $= 12 - 4i + 18i - 6i^2$ Using FOIL
    $= 12 - 4i + 18i + 6$ $i^2 = -1$
    $= 18 + 14i$

43. $(6 - 5i)(3 + 4i) = 18 + 24i - 15i - 20i^2$
    $= 18 + 24i - 15i + 20$
    $= 38 + 9i$

45. $(7 - 2i)(2 - 6i) = 14 - 42i - 4i + 12i^2$
    $= 14 - 42i - 4i - 12$
    $= 2 - 46i$

47. $(3 - 2i)^2 = 3^2 - 2 \cdot 3 \cdot 2i + (2i)^2$ Squaring a binomial
    $= 9 - 12i + 4i^2$
    $= 9 - 12i - 4$ $i^2 = -1$
    $= 5 - 12i$

49. $(1 + 5i)^2$
    $= 1^2 + 2 \cdot 1 \cdot 5i + (5i)^2$ Squaring a binomial
    $= 1 + 10i + 25i^2$
    $= 1 + 10i - 25$ $i^2 = -1$
    $= -24 + 10i$

51. $(-2 + 3i)^2 = 4 - 12i + 9i^2 = 4 - 12i - 9 = -5 - 12i$

53. $i^7 = i^6 \cdot i = (i^2)^3 \cdot i = (-1)^3 \cdot i = -1 \cdot i = -i$

55. $i^{24} = (i^2)^{12} = (-1)^{12} = 1$

57. $i^{42} = (i^2)^{21} = (-1)^{21} = -1$

59. $i^9 = (i^2)^4 \cdot i = (-1)^4 \cdot i = 1 \cdot i = i$

Exercise Set 10.8

61. $i^6 = (i^2)^3 = (-1)^3 = -1$

63. $(5i)^3 = 5^3 \cdot i^3 = 125 \cdot i^2 \cdot i = 125(-1)(i) = -125i$

65. $7 + i^4 = 7 + (i^2)^2 = 7 + (-1)^2 = 7 + 1 = 8$

67. $i^{28} - 23i = (i^2)^{14} - 23i = (-1)^{14} - 23i = 1 - 23i$

69. $i^2 + i^4 = -1 + (i^2)^2 = -1 + (-1)^2 = -1 + 1 = 0$

71. $i^5 + i^7 = i^4 \cdot i + i^6 \cdot i = (i^2)^2 \cdot i + (i^2)^3 \cdot i =$
$(-1)^2 \cdot i + (-1)^3 \cdot i = 1 \cdot i + (-1)i = i - i = 0$

73. $1 + i + i^2 + i^3 + i^4 = 1 + i + i^2 + i^2 \cdot i + (i^2)^2$
$= 1 + i + (-1) + (-1) \cdot i + (-1)^2$
$= 1 + i - 1 - i + 1$
$= 1$

75. $5 - \sqrt{-64} = 5 - \sqrt{-1} \cdot \sqrt{64} = 5 - i \cdot 8 = 5 - 8i$

77. $\dfrac{8 - \sqrt{-24}}{4} = \dfrac{8 - \sqrt{-1} \cdot \sqrt{24}}{4} = \dfrac{8 - i \cdot 2\sqrt{6}}{4} =$
$\dfrac{2(4 - i\sqrt{6})}{2 \cdot 2} = \dfrac{2(4 - i\sqrt{6})}{2 \cdot 2} = \dfrac{4 - i\sqrt{6}}{2} = 2 - \dfrac{\sqrt{6}}{2}i$

79. $\dfrac{4 + 3i}{3 - i} = \dfrac{4 + 3i}{3 - i} \cdot \dfrac{3 + i}{3 + i}$
$= \dfrac{(4 + 3i)(3 + i)}{(3 - i)(3 + i)}$
$= \dfrac{12 + 4i + 9i + 3i^2}{9 - i^2}$
$= \dfrac{12 + 13i - 3}{9 - (-1)}$
$= \dfrac{9 + 13i}{10}$
$= \dfrac{9}{10} + \dfrac{13}{10}i$

81. $\dfrac{3 - 2i}{2 + 3i} = \dfrac{3 - 2i}{2 + 3i} \cdot \dfrac{2 - 3i}{2 - 3i}$
$= \dfrac{(3 - 2i)(2 - 3i)}{(2 + 3i)(2 - 3i)}$
$= \dfrac{6 - 9i - 4i + 6i^2}{4 - 9i^2}$
$= \dfrac{6 - 13i - 6}{4 - 9(-1)}$
$= \dfrac{-13i}{13}$
$= -i$

83. $\dfrac{8 - 3i}{7i} = \dfrac{8 - 3i}{7i} \cdot \dfrac{-7i}{-7i}$
$= \dfrac{-56i + 21i^2}{-49i^2}$
$= \dfrac{-21 - 56i}{49}$
$= -\dfrac{21}{49} - \dfrac{56}{49}i$
$= -\dfrac{3}{7} - \dfrac{8}{7}i$

85. $\dfrac{4}{3 + i} = \dfrac{4}{3 + i} \cdot \dfrac{3 - i}{3 - i}$
$= \dfrac{12 - 4i}{9 - i^2}$
$= \dfrac{12 - 4i}{9 - (-1)}$
$= \dfrac{12 - 4i}{10}$
$= \dfrac{12}{10} - \dfrac{4}{10}i$
$= \dfrac{6}{5} - \dfrac{2}{5}i$

87. $\dfrac{2i}{5 - 4i} = \dfrac{2i}{5 - 4i} \cdot \dfrac{5 + 4i}{5 + 4i}$
$= \dfrac{10i + 8i^2}{25 - 16i^2}$
$= \dfrac{10i + 8(-1)}{25 - 16(-1)}$
$= \dfrac{-8 + 10i}{41}$
$= -\dfrac{8}{41} + \dfrac{10}{41}i$

89. $\dfrac{4}{3i} = \dfrac{4}{3i} \cdot \dfrac{-3i}{-3i}$
$= \dfrac{-12i}{-9i^2}$
$= \dfrac{-12i}{-9(-1)}$
$= \dfrac{-12i}{9}$
$= -\dfrac{4}{3}i$

91. $\dfrac{9 - 4i}{8i} = \dfrac{2 - 4i}{8i} \cdot \dfrac{-8i}{-8i}$
$= \dfrac{-16i + 32i^2}{-64i^2}$
$= \dfrac{-16i + 32(-1)}{-64(-1)}$
$= \dfrac{-32 - 16i}{64}$
$= -\dfrac{32}{64} - \dfrac{16}{64}i$
$= -\dfrac{1}{2} - \dfrac{1}{4}i$

93. $\dfrac{6+3i}{6-3i} = \dfrac{6+3i}{6-3i} \cdot \dfrac{6+3i}{6+3i}$

$= \dfrac{36+18i+18i+9i^2}{36-9i^2}$

$= \dfrac{36+36i-9}{36-9(-1)}$

$= \dfrac{27+36i}{45}$

$= \dfrac{27}{45} + \dfrac{36}{45}i$

$= \dfrac{3}{5} + \dfrac{4}{5}i$

95. Substitute $1-2i$ for x in the equation.

$\begin{array}{c|c} x^2 - 2x + 5 = 0 \\ \hline (1-2i)^2 - 2(1-2i) + 5 \; ? \; 0 \\ 1 - 4i + 4i^2 - 2 + 4i + 5 \\ 1 - 4i - 4 - 2 + 4i + 5 \\ 0 & \text{TRUE} \end{array}$

$1 - 2i$ is a solution.

97. Substitute $2+i$ for x in the equation.

$\begin{array}{c|c} x^2 - 4x - 5 = 0 \\ \hline (2+i)^2 - 4(2+i) - 5 \; ? \; 0 \\ 4 + 4i + i^2 - 8 - 4i - 5 \\ 4 + 4i - 1 - 8 - 4i - 5 \\ -10 & \text{FALSE} \end{array}$

$2 + i$ is not a solution.

99. Discussion and Writing Exercise

101. $\dfrac{196}{x^2 - 7x + 49} - \dfrac{2x}{x+7} = \dfrac{2058}{x^3 + 343}$

Note: $x^3 + 343 = (x+7)(x^2 - 7x + 49)$.

The LCM $= (x+7)(x^2 - 7x + 49)$.

$(x+7)(x^2 - 7x + 49)\left(\dfrac{196}{x^2 - 7x + 49} - \dfrac{2x}{x+7}\right) =$

$(x+7)(x^2 - 7x + 49) \cdot \dfrac{2058}{x^3 + 343}$

$196(x+7) - 2x(x^2 - 7x + 49) = 2058$

$196x + 1372 - 2x^3 + 14x^2 - 98x = 2058$

$98x - 686 - 2x^3 + 14x^2 = 0$

$49x - 343 - x^3 + 7x^2 = 0 \quad$ Dividing by 2

$49(x-7) - x^2(x-7) = 0$

$(49 - x^2)(x-7) = 0$

$(7-x)(7+x)(x-7) = 0$

$7 - x = 0 \quad \text{or} \quad 7 + x = 0 \quad \text{or} \quad x - 7 = 0$

$7 = x \quad \text{or} \quad x = -7 \quad \text{or} \quad x = 7$

Only 7 checks. It is the solution.

103. $|3x + 7| = 22$

$3x + 7 = -22 \quad \text{or} \quad 3x + 7 = 22$

$3x = -29 \quad \text{or} \quad 3x = 15$

$x = -\dfrac{29}{3} \quad \text{or} \quad x = 5$

The solutions are $-\dfrac{29}{3}$ and 5.

105. $|3x + 7| \geq 22$

$3x + 7 \leq -22 \quad \text{or} \quad 3x + 7 \geq 22$

$3x \leq -29 \quad \text{or} \quad 3x \geq 15$

$x \leq -\dfrac{29}{3} \quad \text{or} \quad x \geq 5$

The solution set is $\left\{x \Big| x \leq -\dfrac{29}{3} \text{ or } x \geq 5\right\}$, or $\left(-\infty, -\dfrac{29}{3}\right] \cup [5, \infty)$.

107. $g(2i) = \dfrac{(2i)^4 - (2i)^2}{2i - 1} = \dfrac{16i^4 - 4i^2}{-1 + 2i} = \dfrac{20}{-1 + 2i} =$

$\dfrac{20}{-1 + 2i} \cdot \dfrac{-1 - 2i}{-1 - 2i} = \dfrac{-20 - 40i}{5} = -4 - 8i;$

$g(i+1) = \dfrac{(i+1)^4 - (i+1)^2}{(i+1) - 1} =$

$\dfrac{(i+1)^2[(i+1)^2 - 1]}{i} = \dfrac{2i(2i-1)}{i} = 2(2i-1) =$

$-2 + 4i;$

$g(2i - 1) = \dfrac{(2i-1)^4 - (2i-1)^2}{(2i-1) - 1} =$

$\dfrac{(2i-1)^2[(2i-1)^2 - 1]}{2i - 2} = \dfrac{(-3-4i)(-4-4i)}{-2 + 2i} =$

$\dfrac{(-3-4i)(-2-2i)}{-1 + i} = \dfrac{-2 + 14i}{-1 + i} =$

$\dfrac{-2 + 14i}{-1 + i} \cdot \dfrac{-1 - i}{-1 - i} = \dfrac{16 - 12i}{2} = 8 - 6i$

109. $\dfrac{1}{8}\left(-24 - \sqrt{-1024}\right) = \dfrac{1}{8}(-24 - 32i) = -3 - 4i$

111. $7\sqrt{-64} - 9\sqrt{-256} = 7 \cdot 8i - 9 \cdot 16i = 56i - 144i = -88i$

113. $(1-i)^3(1+i)^3 =$

$(1-i)(1+i) \cdot (1-i)(1+i) \cdot (1-i)(1+i) =$

$(1 - i^2)(1 - i^2)(1 - i^2) = (1+1)(1+1)(1+1) =$

$2 \cdot 2 \cdot 2 = 8$

115. $\dfrac{6}{1 + \dfrac{3}{i}} = \dfrac{6}{\dfrac{i+3}{i}} = \dfrac{6i}{i+3} = \dfrac{6i}{i+3} \cdot \dfrac{-i+3}{-i+3} =$

$\dfrac{-6i^2 + 18i}{-i^2 + 9} = \dfrac{6 + 18i}{10} = \dfrac{6}{10} + \dfrac{18}{10}i = \dfrac{3}{5} + \dfrac{9}{5}i$

117. $\dfrac{i - i^{38}}{1 + i} = \dfrac{i - (i^2)^{19}}{1 + i} = \dfrac{i - (-1)^{19}}{1 + i} = \dfrac{i - (-1)}{1 + i} =$

$\dfrac{i + 1}{1 + i} = 1$

Chapter 11

Quadratic Equations and Functions

Exercise Set 11.1

1. a) $6x^2 = 30$

    $x^2 = 5$ Dividing by 6

    $x = \sqrt{5}$ or $x = -\sqrt{5}$ Principle of square roots

    Check: $\quad 6x^2 = 30$
    $$6(\pm\sqrt{5})^2 \;?\; 30$$
    $$6 \cdot 5$$
    $$30 \;|\; \text{TRUE}$$

    The solutions are $\sqrt{5}$ and $-\sqrt{5}$, or $\pm\sqrt{5}$.

    b) The real-number solutions of the equation $6x^2 = 30$ are the first coordinates of the x-intercepts of the graph of $f(x) = 6x^2 - 30$. Thus, the x-intercepts are $(-\sqrt{5}, 0)$ and $(\sqrt{5}, 0)$.

3. a) $9x^2 + 25 = 0$

    $9x^2 = -25$ Subtracting 25

    $x^2 = -\dfrac{25}{9}$ Dividing by 9

    $x = \sqrt{-\dfrac{25}{9}}$ or $x = -\sqrt{-\dfrac{25}{9}}$ Principle of square roots

    $x = \dfrac{5}{3}i$ or $x = -\dfrac{5}{3}i$ Simplifying

    Check: $\quad 9x^2 + 25 = 0$
    $$9\left(\pm\dfrac{5}{3}i\right)^2 + 25 \;?\; 0$$
    $$9\left(-\dfrac{25}{9}\right) + 25$$
    $$-25 + 25$$
    $$0 \;|\; \text{TRUE}$$

    The solutions are $\dfrac{5}{3}i$ and $-\dfrac{5}{3}i$, or $\pm\dfrac{5}{3}i$.

    b) Since the equation $9x^2 + 25 = 0$ has no real-number solutions, the graph of $f(x) = 9x^2 + 25$ has no x-intercepts.

5. $2x^2 - 3 = 0$

    $2x^2 = 3$

    $x^2 = \dfrac{3}{2}$

    $x = \sqrt{\dfrac{3}{2}}$ or $x = -\sqrt{\dfrac{3}{2}}$ Principle of square roots

    $x = \sqrt{\dfrac{3}{2} \cdot \dfrac{2}{2}}$ or $x = -\sqrt{\dfrac{3}{2} \cdot \dfrac{2}{2}}$ Rationalizing denominators

    $x = \dfrac{\sqrt{6}}{2}$ or $x = -\dfrac{\sqrt{6}}{2}$

    Check: $\quad 2x^2 - 3 = 0$
    $$2\left(\pm\dfrac{\sqrt{6}}{2}\right)^2 - 3 \;?\; 0$$
    $$2 \cdot \dfrac{6}{4} - 3$$
    $$3 - 3$$
    $$0 \;|\; \text{TRUE}$$

    The solutions are $\dfrac{\sqrt{6}}{2}$ and $-\dfrac{\sqrt{6}}{2}$, or $\pm\dfrac{\sqrt{6}}{2}$. Using a calculator, we find that the solutions are approximately ± 1.225.

7. $(x + 2)^2 = 49$

    $x + 2 = 7$ or $x + 2 = -7$ Principle of square roots

    $x = 5$ or $x = -9$

    The solutions are 5 and -9.

9. $(x - 4)^2 = 16$

    $x - 4 = 4$ or $x - 4 = -4$ Principle of square roots

    $x = 8$ or $x = 0$

    The solutions are 8 and 0.

11. $(x - 11)^2 = 7$

    $x - 11 = \sqrt{7}$ or $x - 11 = -\sqrt{7}$

    $x = 11 + \sqrt{7}$ or $x = 11 - \sqrt{7}$

    The solutions are $11 + \sqrt{7}$ and $11 - \sqrt{7}$, or $11 \pm \sqrt{7}$. Using a calculator, we find that the solutions are approximately 8.354 and 13.646.

13. $(x - 7)^2 = -4$

    $x - 7 = \sqrt{-4}$ or $x - 7 = -\sqrt{-4}$

    $x - 7 = 2i$ or $x - 7 = -2i$

    $x = 7 + 2i$ or $x = 7 - 2i$

    The solutions are $7 + 2i$ and $7 - 2i$, or $7 \pm 2i$.

15. $(x - 9)^2 = 81$

    $x - 9 = 9$ or $x - 9 = -9$

    $x = 18$ or $x = 0$

    The solutions are 18 and 0.

17. $\left(x - \frac{3}{2}\right)^2 = \frac{7}{2}$

$x - \frac{3}{2} = \sqrt{\frac{7}{2}}$ or $x - \frac{3}{2} = -\sqrt{\frac{7}{2}}$

$x - \frac{3}{2} = \sqrt{\frac{7}{2} \cdot \frac{2}{2}}$ or $x - \frac{3}{2} = -\sqrt{\frac{7}{2} \cdot \frac{2}{2}}$

$x - \frac{3}{2} = \frac{\sqrt{14}}{2}$ or $x - \frac{3}{2} = -\frac{\sqrt{14}}{2}$

$x = \frac{3}{2} + \frac{\sqrt{14}}{2}$ or $x = \frac{3}{2} - \frac{\sqrt{14}}{2}$

$x = \frac{3 + \sqrt{14}}{2}$ or $x = \frac{3 - \sqrt{14}}{2}$

The solutions are $\frac{3 + \sqrt{14}}{2}$ and $\frac{3 - \sqrt{14}}{2}$, or $\frac{3 \pm \sqrt{14}}{2}$. Using a calculator, we find that the solutions are approximately -0.371 and 3.371.

19. $x^2 + 6x + 9 = 64$

$(x + 3)^2 = 64$

$x + 3 = 8$ or $x + 3 = -8$

$x = 5$ or $x = -11$

The solutions are 5 and -11.

21. $y^2 - 14y + 49 = 4$

$(y - 7)^2 = 4$

$y - 7 = 2$ or $y - 7 = -2$

$y = 9$ or $y = 5$

The solutions are 9 and 5.

23. $x^2 + 4x = 2$ Original equation

$x^2 + 4x + 4 = 2 + 4$ Adding 4: $\left(\frac{4}{2}\right)^2 = 2^2 = 4$

$(x + 2)^2 = 6$

$x + 2 = \sqrt{6}$ or $x + 2 = -\sqrt{6}$ Principle of square roots

$x = -2 + \sqrt{6}$ or $x = -2 - \sqrt{6}$

The solutions are $-2 \pm \sqrt{6}$.

25. $x^2 - 22x = 11$ Original equation

$x^2 - 22x + 121 = 11 + 121$ Adding 121: $\left(\frac{-22}{2}\right)^2 = (-11)^2 = 121$

$(x - 11)^2 = 132$

$x - 11 = \sqrt{132}$ or $x - 11 = -\sqrt{132}$

$x - 11 = 2\sqrt{33}$ or $x - 11 = -2\sqrt{33}$

$x = 11 + 2\sqrt{33}$ or $x = 11 - 2\sqrt{33}$

The solutions are $11 \pm 2\sqrt{33}$.

27. $x^2 + x = 1$

$x^2 + x + \frac{1}{4} = 1 + \frac{1}{4}$ Adding $\frac{1}{4}$: $\left(\frac{1}{2}\right)^2 = \frac{1}{4}$

$\left(x + \frac{1}{2}\right)^2 = \frac{5}{4}$

$x + \frac{1}{2} = \frac{\sqrt{5}}{2}$ or $x + \frac{1}{2} = -\frac{\sqrt{5}}{2}$

$x = \frac{-1 + \sqrt{5}}{2}$ or $x = \frac{-1 - \sqrt{5}}{2}$

The solutions are $\frac{-1 \pm \sqrt{5}}{2}$.

29. $t^2 - 5t = 7$

$t^2 - 5t + \frac{25}{4} = 7 + \frac{25}{4}$ Adding $\frac{25}{4}$: $\left(\frac{-5}{2}\right)^2 = \frac{25}{4}$

$\left(t - \frac{5}{2}\right)^2 = \frac{53}{4}$

$t - \frac{5}{2} = \frac{\sqrt{53}}{2}$ or $t - \frac{5}{2} = -\frac{\sqrt{53}}{2}$

$t = \frac{5 + \sqrt{53}}{2}$ or $t = \frac{5 - \sqrt{53}}{2}$

The solutions are $\frac{5 \pm \sqrt{53}}{2}$.

31. $x^2 + \frac{3}{2}x = 3$

$x^2 + \frac{3}{2}x + \frac{9}{16} = 3 + \frac{9}{16}$ $\left(\frac{1}{2} \cdot \frac{3}{2}\right)^2 = \left(\frac{3}{4}\right)^2 = \frac{9}{16}$

$\left(x + \frac{3}{4}\right)^2 = \frac{57}{16}$

$x + \frac{3}{4} = \frac{\sqrt{57}}{4}$ or $x + \frac{3}{4} = -\frac{\sqrt{57}}{4}$

$x = \frac{-3 + \sqrt{57}}{4}$ or $x = \frac{-3 - \sqrt{57}}{4}$

The solutions are $\frac{-3 \pm \sqrt{57}}{4}$.

33. $m^2 - \frac{9}{2}m = \frac{3}{2}$ Original equation

$m^2 - \frac{9}{2}m + \frac{81}{16} = \frac{3}{2} + \frac{81}{16}$ $\left[\frac{1}{2}\left(-\frac{9}{2}\right)\right]^2 = \left(-\frac{9}{4}\right)^2 = \frac{81}{16}$

$\left(m - \frac{9}{4}\right)^2 = \frac{105}{16}$

$m - \frac{9}{4} = \frac{\sqrt{105}}{4}$ or $m - \frac{9}{4} = -\frac{\sqrt{105}}{4}$

$m = \frac{9 + \sqrt{105}}{4}$ or $m = \frac{9 - \sqrt{105}}{4}$

The solutions are $\frac{9 \pm \sqrt{105}}{4}$.

35. $x^2 + 6x - 16 = 0$

$x^2 + 6x = 16$ Adding 16

$x^2 + 6x + 9 = 16 + 9$ $\left(\frac{6}{2}\right)^2 = 3^2 = 9$

$(x + 3)^2 = 25$

$x + 3 = 5$ or $x + 3 = -5$

$x = 2$ or $x = -8$

Exercise Set 11.1

The solutions are 2 and -8.

37. $x^2 + 22x + 102 = 0$

$x^2 + 22x = -102$ Subtracting 102

$x^2 + 22x + 121 = -102 + 121$ $\left(\dfrac{22}{2}\right)^2 = 11^2 = 121$

$(x + 11)^2 = 19$

$x + 11 = \sqrt{19}$ or $x + 11 = -\sqrt{19}$

$x = -11 + \sqrt{19}$ or $x = -11 - \sqrt{19}$

The solutions are $-11 \pm \sqrt{19}$.

39. $x^2 - 10x - 4 = 0$

$x^2 - 10x = 4$ Adding 4

$x^2 - 10x + 25 = 4 + 25$ $\left(\dfrac{-10}{2}\right)^2 = (-5)^2 = 25$

$(x - 5)^2 = 29$

$x - 5 = \sqrt{29}$ or $x - 5 = -\sqrt{29}$

$x = 5 + \sqrt{29}$ or $x = 5 - \sqrt{29}$

The solutions are $5 \pm \sqrt{29}$.

41. a) $x^2 + 7x - 2 = 0$

$x^2 + 7x = 2$ Adding 2

$x^2 + 7x + \dfrac{49}{4} = 2 + \dfrac{49}{4}$ $\left(\dfrac{7}{2}\right)^2 = \dfrac{49}{4}$

$\left(x + \dfrac{7}{2}\right)^2 = \dfrac{57}{4}$

$x + \dfrac{7}{2} = \dfrac{\sqrt{57}}{2}$ or $x + \dfrac{7}{2} = -\dfrac{\sqrt{57}}{2}$

$x = \dfrac{-7 + \sqrt{57}}{2}$ or $x = \dfrac{-7 - \sqrt{57}}{2}$

The solutions are $\dfrac{-7 \pm \sqrt{57}}{2}$.

b) The real-number solutions of the equation $x^2 + 7x - 2 = 0$ are the first coordinates of the x-intercepts of the graph of $f(x) = x^2 + 7x - 2$. Thus, the x-intercepts are $\left(\dfrac{-7 - \sqrt{57}}{2}, 0\right)$ and $\left(\dfrac{-7 + \sqrt{57}}{2}, 0\right)$.

43. a) $2x^2 - 5x + 8 = 0$

$\dfrac{1}{2}(2x^2 - 5x + 8) = \dfrac{1}{2} \cdot 0$ Multiplying by $\dfrac{1}{2}$ to make the x^2-coefficient 1

$x^2 - \dfrac{5}{2}x + 4 = 0$

$x^2 - \dfrac{5}{2}x = -4$ Subtracting 4

$x^2 - \dfrac{5}{2}x + \dfrac{25}{16} = -4 + \dfrac{25}{16}$

$\left[\dfrac{1}{2}\left(-\dfrac{5}{2}\right)\right]^2 = \left(-\dfrac{5}{4}\right)^2 = \dfrac{25}{16}$

$\left(x - \dfrac{5}{4}\right)^2 = -\dfrac{64}{16} + \dfrac{25}{16}$

$\left(x - \dfrac{5}{4}\right)^2 = -\dfrac{39}{16}$

$x - \dfrac{5}{4} = \sqrt{-\dfrac{39}{16}}$ or $x - \dfrac{5}{4} = -\sqrt{-\dfrac{39}{16}}$

$x - \dfrac{5}{4} = i\sqrt{\dfrac{39}{16}}$ or $x - \dfrac{5}{4} = -i\sqrt{\dfrac{39}{16}}$

$x = \dfrac{5}{4} + i\dfrac{\sqrt{39}}{4}$ or $x = \dfrac{5}{4} - i\dfrac{\sqrt{39}}{4}$

The solutions are $\dfrac{5}{4} \pm i\dfrac{\sqrt{39}}{4}$.

b) Since the equation $2x^2 - 5x + 8 = 0$ has no real-number solutions, the graph of $f(x) = 2x^2 - 5x + 8$ has no x-intercepts.

45. $x^2 - \dfrac{3}{2}x - \dfrac{1}{2} = 0$

$x^2 - \dfrac{3}{2}x = \dfrac{1}{2}$

$x^2 - \dfrac{3}{2}x + \dfrac{9}{16} = \dfrac{1}{2} + \dfrac{9}{16}$ $\left[\dfrac{1}{2}\left(-\dfrac{3}{2}\right)\right]^2 = \left(-\dfrac{3}{4}\right)^2 = \dfrac{9}{16}$

$\left(x - \dfrac{3}{4}\right)^2 = \dfrac{17}{16}$

$x - \dfrac{3}{4} = \dfrac{\sqrt{17}}{4}$ or $x - \dfrac{3}{4} = -\dfrac{\sqrt{17}}{4}$

$x = \dfrac{3 + \sqrt{17}}{4}$ or $x = \dfrac{3 - \sqrt{17}}{4}$

The solutions are $\dfrac{3 \pm \sqrt{17}}{4}$.

47. $2x^2 - 3x - 17 = 0$

$\frac{1}{2}(2x^2 - 3x - 17) = \frac{1}{2} \cdot 0$ Multiplying by $\frac{1}{2}$ to make the x^2-coefficient 1

$x^2 - \frac{3}{2}x - \frac{17}{2} = 0$

$x^2 - \frac{3}{2}x = \frac{17}{2}$ Adding $\frac{17}{2}$

$x^2 - \frac{3}{2}x + \frac{9}{16} = \frac{17}{2} + \frac{9}{16}$ $\left[\frac{1}{2}\left(-\frac{3}{2}\right)\right]^2 = \left(-\frac{3}{4}\right)^2 = \frac{9}{16}$

$\left(x - \frac{3}{4}\right)^2 = \frac{145}{16}$

$x - \frac{3}{4} = \frac{\sqrt{145}}{4}$ or $x - \frac{3}{4} = -\frac{\sqrt{145}}{4}$

$x = \frac{3 + \sqrt{145}}{4}$ or $x = \frac{3 - \sqrt{145}}{4}$

The solutions are $\frac{3 \pm \sqrt{145}}{4}$.

49. $3x^2 - 4x - 1 = 0$

$\frac{1}{3}(3x^2 - 4x - 1) = \frac{1}{3} \cdot 0$ Multiplying to make the x^2-coefficient 1

$x^2 - \frac{4}{3}x - \frac{1}{3} = 0$

$x^2 - \frac{4}{3}x = \frac{1}{3}$ Adding $\frac{1}{3}$

$x^2 - \frac{4}{3}x + \frac{4}{9} = \frac{1}{3} + \frac{4}{9}$ $\left[\frac{1}{2}\left(-\frac{4}{3}\right)\right]^2 = \left(-\frac{2}{3}\right)^2 = \frac{4}{9}$

$\left(x - \frac{2}{3}\right)^2 = \frac{7}{9}$

$x - \frac{2}{3} = \frac{\sqrt{7}}{3}$ or $x - \frac{2}{3} = -\frac{\sqrt{7}}{3}$

$x = \frac{2 + \sqrt{7}}{3}$ or $x = \frac{2 - \sqrt{7}}{3}$

The solutions are $\frac{2 \pm \sqrt{7}}{3}$.

51. $x^2 + x + 2 = 0$

$x^2 + x = -2$ Subtracting 2

$x^2 + x + \frac{1}{4} = -2 + \frac{1}{4}$ $\left(\frac{1}{2}\right)^2 = \frac{1}{4}$

$\left(x + \frac{1}{2}\right)^2 = -\frac{7}{4}$

$x + \frac{1}{2} = \sqrt{-\frac{7}{4}}$ or $x + \frac{1}{2} = -\sqrt{-\frac{7}{4}}$

$x + \frac{1}{2} = i\sqrt{\frac{7}{4}}$ or $x + \frac{1}{2} = -i\sqrt{\frac{7}{4}}$

$x = -\frac{1}{2} + i\frac{\sqrt{7}}{2}$ or $x = -\frac{1}{2} - i\frac{\sqrt{7}}{2}$

The solutions are $-\frac{1}{2} \pm i\frac{\sqrt{7}}{2}$.

53. $x^2 - 4x + 13 = 0$

$x^2 - 4x = -13$ Subtracting 13

$x^2 - 4x + 4 = -13 + 4$ $\left(\frac{-4}{2}\right)^2 = (-2)^2 = 4$

$(x - 2)^2 = -9$

$x - 2 = \sqrt{-9}$ or $x - 2 = -\sqrt{-9}$

$x - 2 = 3i$ or $x - 2 = -3i$

$x = 2 + 3i$ or $x = 2 - 3i$

The solutions are $2 \pm 3i$.

55. $V = 48T^2$

$36 = 48T^2$ Substituting 36 for V

$\frac{36}{48} = T^2$ Solving for T^2

$0.75 = T^2$

$\sqrt{0.75} = T$

$0.866 \approx T$

The hang time is 0.866 sec.

57. $s(t) = 16t^2$

$850 = 16t^2$ Substituting 850 for $s(t)$

$\frac{850}{16} = t^2$ Solving for t^2

$53.125 = t^2$

$\sqrt{53.125} = t$

$7.3 \approx t$

It will take about 7.3 sec for an object to fall from the top.

59. $s(t) = 16t^2$

$745 = 16t^2$ Substituting 745 for $s(t)$

$\frac{745}{16} = t^2$ Solving for t^2

$46.5625 = t^2$

$\sqrt{46.5625} = t$

$6.8 \approx t$

It will take about 6.8 sec for an object to fall from the top.

61. Discussion and Writing Exercise

63. a) First find the slope.

$m = \frac{410 - 275}{5 - 1} = \frac{135}{4} = 33.75$

Now find the function. We substitute in the point-slope equation, using the point $(1, 275)$.

$T - T_1 = m(t - t_1)$

$T - 275 = 33.75(t - 1)$

$T - 275 = 33.75t - 33.75$

$T = 33.75t + 241.25$

Using the function notation we have $T(t) = 33.75t + 241.25$.

b) Since $2005 - 1995 = 10$, the year 2005 is 10 years since 1995. We find $T(10)$.
$$T(t) = 33.75t + 241.25$$
$$T(10) = 33.75(10) + 241.25$$
$$= 337.5 + 241.25$$
$$= 578.75$$

In 2005 about 578.75 thousand, or 578,750 people, will visit a doctor for tattoo removal.

c) Since $1,085,000 = 1085$ thousand, we substitute 1085 for $T(t)$ and solve for t.
$$T(t) = 33.75t + 241.25$$
$$1085 = 33.75t + 241.25$$
$$843.75 = 33.75t$$
$$25 = t$$

1,085,000 people will visit a doctor for tattoo removal 25 yr after 1995, or in 2020.

65. Graph $f(x) = 5 - 2x$

We find some ordered pairs $(x, f(x))$, plot them, and draw the graph.

x	$f(x)$	$(x, f(x))$
0	5	$(0, 5)$
1	3	$(1, 3)$
3	-1	$(3, -1)$
5	-5	$(5, -5)$

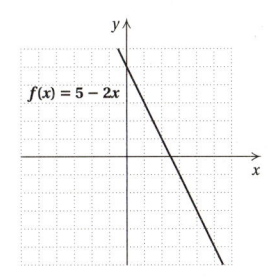

67. Graph $f(x) = |5 - 2x|$

We find some ordered pairs $(x, f(x))$, plot them, and draw the graph.

x	$f(x)$	$(x, f(x))$
0	5	$(0, 5)$
1	3	$(1, 3)$
$\frac{5}{2}$	0	$\left(\frac{5}{2}, 0\right)$
3	1	$(3, 1)$
5	5	$(5, 5)$

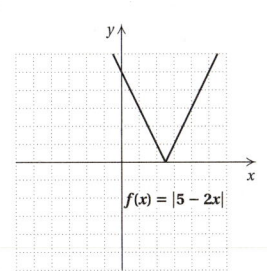

69. $\sqrt{\dfrac{2}{5}} = \sqrt{\dfrac{2}{5} \cdot \dfrac{5}{5}} = \sqrt{\dfrac{10}{25}} = \dfrac{\sqrt{10}}{\sqrt{25}} = \dfrac{\sqrt{10}}{5}$

71. Left to the student

73. In order for $x^2 + bx + 64$ to be a trinomial square, the following must be true:
$$\left(\dfrac{b}{2}\right)^2 = 64$$
$$\dfrac{b^2}{4} = 64$$
$$b^2 = 256$$
$$b = 16 \quad or \quad b = -16$$

75. $x(2x^2 + 9x - 56)(3x + 10) = 0$
$x(2x - 7)(x + 8)(3x + 10) = 0$
$x=0$ or $2x-7=0$ or $x+8=0$ or $3x+10=0$
$x=0$ or $x=\dfrac{7}{2}$ or $x=-8$ or $x=-\dfrac{10}{3}$

The solutions are -8, $-\dfrac{10}{3}$, 0, and $\dfrac{7}{2}$.

Exercise Set 11.2

1. $x^2 + 6x + 4 = 0$
$a = 1, \; b = 6, \; c = 4$
$$x = \dfrac{-b \pm \sqrt{b^2 - 4ac}}{2a}$$
$$x = \dfrac{-6 \pm \sqrt{6^2 - 4 \cdot 1 \cdot 4}}{2 \cdot 1} = \dfrac{-6 \pm \sqrt{36 - 16}}{2}$$
$$x = \dfrac{-6 \pm \sqrt{20}}{2} = \dfrac{-6 \pm 2\sqrt{5}}{2}$$
$$x = \dfrac{2(-3 \pm \sqrt{5})}{2} = -3 \pm \sqrt{5}$$

The solutions are $-3 + \sqrt{5}$ and $-3 - \sqrt{5}$.

3. $\quad 3p^2 = -8p - 1$
$3p^2 + 8p + 1 = 0 \qquad$ Finding standard form
$a = 3, \; b = 8, \; c = 1$
$$p = \dfrac{-b \pm \sqrt{b^2 - 4ac}}{2a}$$
$$p = \dfrac{-8 \pm \sqrt{8^2 - 4 \cdot 3 \cdot 1}}{2 \cdot 3} = \dfrac{-8 \pm \sqrt{64 - 12}}{6}$$
$$x = \dfrac{-8 \pm \sqrt{52}}{6} = \dfrac{-8 \pm 2\sqrt{13}}{6}$$
$$x = \dfrac{2(-4 \pm \sqrt{13})}{2 \cdot 3} = \dfrac{-4 \pm \sqrt{13}}{3}$$

The solutions are $\dfrac{-4 + \sqrt{13}}{3}$ and $\dfrac{-4 - \sqrt{13}}{3}$.

5. $x^2 - x + 1 = 0$
$a = 1, \; b = -1, \; c = 1$
$$x = \dfrac{-(-1) \pm \sqrt{(-1)^2 - 4 \cdot 1 \cdot 1}}{2 \cdot 1} = \dfrac{1 \pm \sqrt{1 - 4}}{2}$$
$$x = \dfrac{1 \pm \sqrt{-3}}{2} = \dfrac{1 \pm i\sqrt{3}}{2} = \dfrac{1}{2} \pm i\dfrac{\sqrt{3}}{2}$$

The solutions are $\dfrac{1}{2} + i\dfrac{\sqrt{3}}{2}$ and $\dfrac{1}{2} - i\dfrac{\sqrt{3}}{2}$.

7. $\quad x^2 + 13 = 4x$
$x^2 - 4x + 13 = 0 \qquad$ Finding standard form
$a = 1, \; b = -4, \; c = 13$
$$x = \dfrac{-(-4) \pm \sqrt{(-4)^2 - 4 \cdot 1 \cdot 13}}{2 \cdot 1} = \dfrac{4 \pm \sqrt{16 - 52}}{2}$$
$$x = \dfrac{4 \pm \sqrt{-36}}{2} = \dfrac{4 \pm 6i}{2} = 2 \pm 3i$$

The solutions are $2 + 3i$ and $2 - 3i$.

9. $r^2 + 3r = 8$
$r^2 + 3r - 8 = 0$ Finding standard form
$a = 1$, $b = 3$, $c = -8$
$r = \dfrac{-3 \pm \sqrt{3^2 - 4 \cdot 1 \cdot (-8)}}{2 \cdot 1} = \dfrac{-3 \pm \sqrt{9 + 32}}{2}$
$r = \dfrac{-3 \pm \sqrt{41}}{2}$

The solutions are $\dfrac{-3 + \sqrt{41}}{2}$ and $\dfrac{-3 - \sqrt{41}}{2}$.

11. $1 + \dfrac{2}{x} + \dfrac{5}{x^2} = 0$

$x^2 + 2x + 5 = 0$ Multiplying by x^2, the LCM of the denominators

$a = 1$, $b = 2$, $c = 5$
$x = \dfrac{-2 \pm \sqrt{2^2 - 4 \cdot 1 \cdot 5}}{2 \cdot 1} = \dfrac{-2 \pm \sqrt{4 - 20}}{2}$
$x = \dfrac{-2 \pm \sqrt{-16}}{2} = \dfrac{-2 \pm 4i}{2} = -1 \pm 2i$

The solutions are $-1 + 2i$ and $-1 - 2i$.

13. a) $3x + x(x - 2) = 0$
$3x + x^2 - 2x = 0$
$x^2 + x = 0$
$x(x + 1) = 0$
$x = 0$ or $x + 1 = 0$
$x = 0$ or $x = -1$

The solutions are 0 and -1.

b) The solutions of the equation $3x + x(x - 2) = 0$ are the first coordinates of the x-intercepts of the graph of $f(x) = 3x + x(x - 2)$. Thus, the x-intercepts are $(-1, 0)$ and $(0, 0)$.

15. a) $11x^2 - 3x - 5 = 0$
$a = 11$, $b = -3$, $c = -5$
$x = \dfrac{-(-3) \pm \sqrt{(-3)^2 - 4 \cdot 11 \cdot (-5)}}{2 \cdot 11}$
$x = \dfrac{3 \pm \sqrt{9 + 220}}{22} = \dfrac{3 \pm \sqrt{229}}{22}$

The solutions are $\dfrac{3 + \sqrt{229}}{22}$ and $\dfrac{3 - \sqrt{229}}{22}$.

b) The solutions of the equation $11x^2 - 3x - 5 = 0$ are the first coordinates of the x-intercepts of the graph of $f(x) = 11x^2 - 3x - 5$. Thus, the x-intercepts are $\left(\dfrac{3 - \sqrt{229}}{22}, 0\right)$ and $\left(\dfrac{3 + \sqrt{229}}{22}, 0\right)$.

17. a) $25x^2 - 20x + 4 = 0$
$(5x - 2)(5x - 2) = 0$
$5x - 2 = 0$ or $5x - 2 = 0$
$5x = 2$ or $5x = 2$
$x = \dfrac{2}{5}$ or $x = \dfrac{2}{5}$

The solution is $\dfrac{2}{5}$.

b) The solution of the equation $25x^2 - 20x + 4 = 0$ is the first coordinate of the x-intercept of $f(x) = 25x^2 - 20x + 4$. Thus, the x-intercept is $\left(\dfrac{2}{5}, 0\right)$.

19. $4x(x - 2) - 5x(x - 1) = 2$
$4x^2 - 8x - 5x^2 + 5x = 2$ Removing parentheses
$-x^2 - 3x = 2$
$-x^2 - 3x - 2 = 0$
$x^2 + 3x + 2 = 0$ Multiplying by -1
$(x + 2)(x + 1) = 0$
$x + 2 = 0$ or $x + 1 = 0$
$x = -2$ or $x = -1$

The solutions are -2 and -1.

21. $14(x - 4) - (x + 2) = (x + 2)(x - 4)$
$14x - 56 - x - 2 = x^2 - 2x - 8$
$13x - 58 = x^2 - 2x - 8$
$0 = x^2 - 15x + 50$
$0 = (x - 10)(x - 5)$
$x - 10 = 0$ or $x - 5 = 0$
$x = 10$ or $x = 5$

The solutions are 10 and 5.

23. $5x^2 = 17x - 2$
$5x^2 - 17x + 2 = 0$
$a = 5$, $b = -17$, $c = 2$
$x = \dfrac{-(-17) \pm \sqrt{(-17)^2 - 4 \cdot 5 \cdot 2}}{2 \cdot 5}$
$x = \dfrac{17 \pm \sqrt{289 - 40}}{10} = \dfrac{17 \pm \sqrt{249}}{10}$

The solutions are $\dfrac{17 + \sqrt{249}}{10}$ and $\dfrac{17 - \sqrt{249}}{10}$.

25. $x^2 + 5 = 4x$
$x^2 - 4x + 5 = 0$
$a = 1$, $b = -4$, $c = 5$
$x = \dfrac{-(-4) \pm \sqrt{(-4)^2 - 4 \cdot 1 \cdot 5}}{2 \cdot 1} = \dfrac{4 \pm \sqrt{16 - 20}}{2}$
$x = \dfrac{4 \pm \sqrt{-4}}{2} = \dfrac{4 \pm 2i}{2} = 2 \pm i$

The solutions are $2 + i$ and $2 - i$.

Exercise Set 11.2

27. $x + \dfrac{1}{x} = \dfrac{13}{6}$, LCM is $6x$

$$6x\left(x + \dfrac{1}{x}\right) = 6x \cdot \dfrac{13}{6}$$

$$6x^2 + 6 = 13x$$

$$6x^2 - 13x + 6 = 0$$

$$(2x-3)(3x-2) = 0$$

$2x - 3 = 0 \quad or \quad 3x - 2 = 0$

$2x = 3 \quad or \quad 3x = 2$

$x = \dfrac{3}{2} \quad or \quad x = \dfrac{2}{3}$

The solutions are $\dfrac{3}{2}$ and $\dfrac{2}{3}$.

29. $\dfrac{1}{y} + \dfrac{1}{y+2} = \dfrac{1}{3}$, LCM is $3y(y+2)$

$$3y(y+2)\left(\dfrac{1}{y} + \dfrac{1}{y+2}\right) = 3y(y+2) \cdot \dfrac{1}{3}$$

$$3(y+2) + 3y = y(y+2)$$

$$3y + 6 + 3y = y^2 + 2y$$

$$6y + 6 = y^2 + 2y$$

$$0 = y^2 - 4y - 6$$

$a = 1,\ b = -4,\ c = -6$

$$y = \dfrac{-(-4) \pm \sqrt{(-4)^2 - 4 \cdot 1 \cdot (-6)}}{2 \cdot 1} = \dfrac{4 \pm \sqrt{16 + 24}}{2}$$

$$y = \dfrac{4 \pm \sqrt{40}}{2} = \dfrac{4 \pm 2\sqrt{10}}{2}$$

$$y = \dfrac{2(2 \pm \sqrt{10})}{2 \cdot 1} = 2 \pm \sqrt{10}$$

The solutions are $2 + \sqrt{10}$ and $2 - \sqrt{10}$.

31. $(2t-3)^2 + 17t = 15$

$$4t^2 - 12t + 9 + 17t = 15$$

$$4t^2 + 5t - 6 = 0$$

$$(4t - 3)(t + 2) = 0$$

$4t - 3 = 0 \quad or \quad t + 2 = 0$

$t = \dfrac{3}{4} \quad or \quad t = -2$

The solutions are $\dfrac{3}{4}$ and -2.

33. $(x-2)^2 + (x+1)^2 = 0$

$$x^2 - 4x + 4 + x^2 + 2x + 1 = 0$$

$$2x^2 - 2x + 5 = 0$$

$a = 2,\ b = -2,\ c = 5$

$$x = \dfrac{-(-2) \pm \sqrt{(-2)^2 - 4 \cdot 2 \cdot 5}}{2 \cdot 2} = \dfrac{2 \pm \sqrt{4 - 40}}{4}$$

$$x = \dfrac{2 \pm \sqrt{-36}}{4} = \dfrac{2 \pm 6i}{4}$$

$$x = \dfrac{2(1 \pm 3i)}{2 \cdot 2} = \dfrac{1 \pm 3i}{2} = \dfrac{1}{2} \pm \dfrac{3}{2}i$$

The solutions are $\dfrac{1}{2} + \dfrac{3}{2}i$ and $\dfrac{1}{2} - \dfrac{3}{2}i$.

35. $x^3 - 1 = 0$

$$(x-1)(x^2 + x + 1) = 0$$

$x - 1 = 0 \quad or \quad x^2 + x + 1 = 0$

$x = 1 \quad or \quad x = \dfrac{-1 \pm \sqrt{1^2 - 4 \cdot 1 \cdot 1}}{2 \cdot 1}$

$x = 1 \quad or \quad x = \dfrac{-1 \pm \sqrt{-3}}{2}$

$x = 1 \quad or \quad x = \dfrac{-1 \pm i\sqrt{3}}{2} = -\dfrac{1}{2} \pm i\dfrac{\sqrt{3}}{2}$

The solutions are 1, $-\dfrac{1}{2} + i\dfrac{\sqrt{3}}{2}$, and $-\dfrac{1}{2} - i\dfrac{\sqrt{3}}{2}$.

37. $x^2 + 6x + 4 = 0$

$a = 1,\ b = 6,\ c = 4$

$$x = \dfrac{-6 \pm \sqrt{6^2 - 4 \cdot 1 \cdot 4}}{2 \cdot 1} = \dfrac{-6 \pm \sqrt{36 - 16}}{2}$$

$$x = \dfrac{-6 \pm \sqrt{20}}{2} = \dfrac{-6 \pm \sqrt{4 \cdot 5}}{2}$$

$$x = \dfrac{-6 \pm 2\sqrt{5}}{2} = \dfrac{2(-3 \pm \sqrt{5})}{2}$$

$$x = -3 \pm \sqrt{5}$$

We can use a calculator to approximate the solutions:

$-3 + \sqrt{5} \approx -0.764;\ -3 - \sqrt{5} \approx -5.236$

The solutions are $-3 + \sqrt{5}$ and $-3 - \sqrt{5}$, or approximately -0.764 and -5.236.

39. $x^2 - 6x + 4 = 0$

$a = 1,\ b = -6,\ c = 4$

$$x = \dfrac{-(-6) \pm \sqrt{(-6)^2 - 4 \cdot 1 \cdot 4}}{2 \cdot 1} = \dfrac{6 \pm \sqrt{36 - 16}}{2}$$

$$x = \dfrac{6 \pm \sqrt{20}}{2} = \dfrac{6 \pm \sqrt{4 \cdot 5}}{2}$$

$$x = \dfrac{6 \pm 2\sqrt{5}}{2} = \dfrac{2(3 \pm \sqrt{5})}{2}$$

$$x = 3 \pm \sqrt{5}$$

We can use a calculator to approximate the solutions:

$3 + \sqrt{5} \approx 5.236;\ 3 - \sqrt{5} \approx 0.764$

The solutions are $3 + \sqrt{5}$ and $3 - \sqrt{5}$, or approximately 5.236 and 0.764.

41. $2x^2 - 3x - 7 = 0$

$a = 2,\ b = -3,\ c = -7$

$$x = \dfrac{-(-3) \pm \sqrt{(-3)^2 - 4 \cdot 2 \cdot (-7)}}{2 \cdot 2} = \dfrac{3 \pm \sqrt{9 + 56}}{4}$$

$$x = \dfrac{3 \pm \sqrt{65}}{4}$$

We can use a calculator to approximate the solutions:

$\dfrac{3 + \sqrt{65}}{4} \approx 2.766;\ \dfrac{3 - \sqrt{65}}{4} \approx -1.266$

The solutions are $\dfrac{3 + \sqrt{65}}{4}$ and $\dfrac{3 - \sqrt{65}}{4}$, or approximately 2.766 and -1.266.

43. $5x^2 = 3 + 8x$

$5x^2 - 8x - 3 = 0$

$a = 5,\ b = -8,\ c = -3$

$$x = \frac{-(-8) \pm \sqrt{(-8)^2 - 4 \cdot 5 \cdot (-3)}}{2 \cdot 5} = \frac{8 \pm \sqrt{64 + 60}}{10}$$

$$x = \frac{8 \pm \sqrt{124}}{10} = \frac{8 \pm \sqrt{4 \cdot 31}}{10}$$

$$x = \frac{8 \pm 2\sqrt{31}}{10} = \frac{2(4 \pm \sqrt{31})}{2 \cdot 5}$$

$$x = \frac{4 \pm \sqrt{31}}{5}$$

We can use a calculator to approximate the solutions:

$\frac{4 + \sqrt{31}}{5} \approx 1.914;\ \frac{4 - \sqrt{31}}{5} \approx -0.314$

The solutions are $\frac{4 + \sqrt{31}}{5}$ and $\frac{4 - \sqrt{31}}{5}$, or approximately 1.914 and -0.314.

45. Discussion and Writing Exercise

47. $x = \sqrt{x + 2}$

$x^2 = (\sqrt{x + 2})^2$ Principle of powers

$x^2 = x + 2$

$x^2 - x - 2 = 0$

$(x - 2)(x + 1) = 0$

$x - 2 = 0$ *or* $x + 1 = 0$

$x = 2$ *or* $x = -1$

The number 2 checks but -1 does not, so the solution is 2.

49. $\sqrt{x + 2} = \sqrt{2x - 8}$

$(\sqrt{x + 2})^2 = (\sqrt{2x + 8})^2$ Principle of powers

$x + 2 = 2x - 8$

$2 = x - 8$

$10 = x$

The number 10 checks, so it is the solution.

51. $\sqrt{x + 5} = -7$

Since the square root of a number must be nonnegative, this equation has no solution.

53. $\sqrt[3]{4x - 7} = 2$

$(\sqrt[3]{4x - 7})^3 = 2^3$ Principle of powers

$4x - 7 = 8$

$4x = 15$

$x = \frac{15}{4}$

The number $\frac{15}{4}$ checks, so it is the solution.

55. The solutions of $2.2x^2 + 0.5x - 1 = 0$ are approximately -0.797 and 0.570.

57. $2x^2 - x - \sqrt{5} = 0$

$a = 2,\ b = -1,\ c = -\sqrt{5}$

$$x = \frac{-(-1) \pm \sqrt{(-1)^2 - 4 \cdot 2 \cdot (-\sqrt{5})}}{2 \cdot 2} = \frac{1 \pm \sqrt{1 + 8\sqrt{5}}}{4}$$

The solutions are $\frac{1 + \sqrt{1 + 8\sqrt{5}}}{4}$ and $\frac{1 - \sqrt{1 + 8\sqrt{5}}}{4}$.

59. $ix^2 - x - 1 = 0$

$a = i,\ b = -1,\ c = -1$

$$x = \frac{-(-1) \pm \sqrt{(-1)^2 - 4 \cdot i \cdot (-1)}}{2 \cdot i} = \frac{1 \pm \sqrt{1 + 4i}}{2i}$$

$$x = \frac{1 \pm \sqrt{1 + 4i}}{2i} \cdot \frac{i}{i} = \frac{i \pm i\sqrt{1 + 4i}}{2i^2} = \frac{i \pm i\sqrt{1 + 4i}}{-2}$$

$$x = \frac{-i \pm i\sqrt{1 + 4i}}{2}$$

The solutions are $\frac{-i + i\sqrt{1 + 4i}}{2}$ and $\frac{-i - i\sqrt{1 + 4i}}{2}$.

61. $\frac{x}{x + 1} = 4 + \frac{1}{3x^2 - 3}$

$\frac{x}{x + 1} = 4 + \frac{1}{3(x + 1)(x - 1)}$,

LCM is $3x(x + 1)(x - 1)$

$3(x + 1)(x - 1) \cdot \frac{x}{x + 1} =$

$3(x + 1)(x - 1)\left(4 + \frac{1}{3(x + 1)(x - 1)}\right)$

$3x(x - 1) = 12(x + 1)(x - 1) + 1$

$3x^2 - 3x = 12x^2 - 12 + 1$

$0 = 9x^2 + 3x - 11$

$a = 9,\ b = 3,\ c = -11$

$$x = \frac{-3 \pm \sqrt{3^2 - 4 \cdot 9 \cdot (-11)}}{2 \cdot 9} = \frac{-3 \pm \sqrt{9 + 396}}{18}$$

$$x = \frac{-3 \pm \sqrt{405}}{18} = \frac{-3 \pm 9\sqrt{5}}{18}$$

$$x = \frac{3(-1 \pm 3\sqrt{5})}{3 \cdot 6} = \frac{-1 \pm 3\sqrt{5}}{6}$$

The solutions are $\frac{-1 + 3\sqrt{5}}{6}$ and $\frac{-1 - 3\sqrt{5}}{6}$.

63. Replace $f(x)$ with 13.

$13 = (x - 3)^2$

$\pm\sqrt{13} = x - 3$

$3 \pm \sqrt{13} = x$

The solutions are $3 + \sqrt{13}$ and $3 - \sqrt{13}$.

Exercise Set 11.3

1. ***Familiarize.*** We make a drawing and label it. We let $x =$ the length of the rectangle. Then $x - 7 =$ the width.

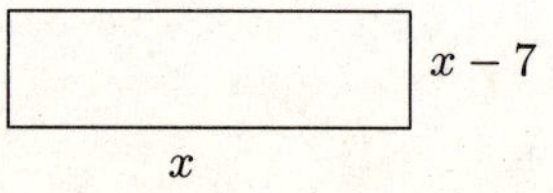

Exercise Set 11.3 261

Translate. We use the formula for the area of a rectangle.
$$A = lw$$
$$18 = x(x-7) \quad \text{Substituting}$$
Solve. We solve the equation.
$$18 = x^2 - 7x$$
$$0 = x^2 - 7x - 18$$
$$0 = (x-9)(x+2)$$
$$x - 9 = 0 \quad \text{or} \quad x + 2 = 0$$
$$x = 9 \quad \text{or} \quad x = -2$$
Check. We only check 9 since the length cannot be negative. If $x = 9$, then $x - 7 = 9 - 7$, or 2, and the area is $9 \cdot 2$, or 18 ft^2. The value checks.

State. The length of 9 ft, and the width is 2 ft.

3. *Familiarize.* We make a drawing and label it. We let $x =$ the width of the rectangle. Then $2x =$ the length.

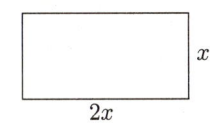

Translate.
$$A = lw$$
$$162 = 2x \cdot x \quad \text{Substituting}$$
Solve. We solve the equation.
$$162 = 2x^2$$
$$81 = x^2$$
$$\pm 9 = x$$
Check. We only check 9 since the width cannot be negative. If $x = 9$, then $2x = 2 \cdot 9$, or 18, and the area is $18 \cdot 9$, or 162 yd^2. The value checks.

State. The length is 18 yd, and the width is 9 yd.

5. *Familiarize.* Let h represent the height of the sail. Then $h - 9$ represents the base. Recall that the formula for the area of a triangle is $A = \dfrac{1}{2} \times$ base $\times$ height.

Translate. The area is 56 m^2. We substitute in the formula.
$$\frac{1}{2}(h-9)h = 56$$
Solve. We solve the equation:
$$\frac{1}{2}(h-9)h = 56$$
$$(h-9)h = 112 \quad \text{Multiplying by 2}$$
$$h^2 - 9h = 112$$
$$h^2 - 9h - 112 = 0$$
$$(h-16)(h+7) = 0$$
$$h - 16 = 0 \quad \text{or} \quad h + 7 = 0$$
$$h = 16 \quad \text{or} \quad h = -7$$
Check. We check only 16, since height cannot be negative. If the height is 16 m, the base is $16 - 9$, or 7 m, and the area is $\dfrac{1}{2} \cdot 7 \cdot 16$, or 56 m^2. We have a solution.

State. The height is 16 m, and the base is 7 m.

7. *Familiarize.* Let h represent the height of the sail. Then $h - 8$ represents the base. Recall that the formula for the area of a triangle is $A = \dfrac{1}{2} \times$ base $\times$ height.

Translate. The area 56 ft^2. We substitute n the formula.
$$\frac{1}{2}(h-8)h = 56$$
Solve. We solve the equation.
$$\frac{1}{2}(h-8)h = 56$$
$$(h-8)h = 112 \quad \text{Multiplying by 2}$$
$$h^2 - 8h = 112$$
$$h^2 - 8h - 112 = 0$$
We use the quadratic formula.
$$h = \frac{-b \pm \sqrt{b^2 - 4ac}}{2a}$$
$$= \frac{-(-8) \pm \sqrt{(-8)^2 - 4 \cdot 1 \cdot (-112)}}{2 \cdot 1}$$
$$= \frac{8 \pm \sqrt{64 + 448}}{2} = \frac{8 \pm \sqrt{512}}{2}$$
$$= \frac{8 \pm \sqrt{256 \cdot 2}}{2} = \frac{8 \pm 16\sqrt{2}}{2}$$
$$= \frac{8(1 \pm 2\sqrt{2})}{2} = 4(1 \pm 2\sqrt{2})$$
$$= 4 \pm 8\sqrt{2}$$
Check. The number $4 - 8\sqrt{2}$ is negative. We do not check it since the height cannot be negative. If the height is $4 + 8\sqrt{2}$ ft, then the base is $4 + 8\sqrt{2} - 8$, or $-4 + 8\sqrt{2}$ ft, and the area is $\dfrac{1}{2}(-4 + 8\sqrt{2})(4 + 8\sqrt{2})$, or $\dfrac{1}{2}(-16 + 128)$, or $\dfrac{1}{2}(112)$, or 56 ft^2. The answer checks.

State. The base of the sail is $-4 + 8\sqrt{2}$ ft, and the height is $4 + 8\sqrt{2}$ ft.

9. *Familiarize.* We make a drawing and label it. We let $x =$ the width of the frame.

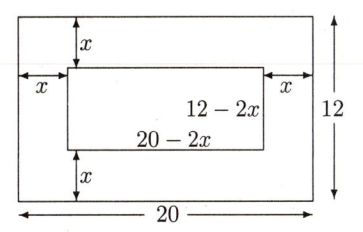

The length and width of the picture that shows are represented by $20 - 2x$ and $12 - 2x$. The area of the picture that shows is 84 cm^2.

Translate. Using the formula for the area of a rectangle, $A = l \cdot w$, we have
$$84 = (20 - 2x)(12 - 2x).$$

262 CHAPTER 11: Quadratic Equations and Functions

Solve. We solve the equation.
$$84 = (20 - 2x)(12 - 2x)$$
$$84 = 240 - 64x + 4x^2$$
$$0 = 156 - 64x + 4x^2$$
$$0 = 4x^2 - 64x + 156$$
$$0 = x^2 - 16x + 39 \quad \text{Dividing by 4}$$
$$0 = (x - 13)(x - 3)$$
$$x - 13 = 0 \quad \text{or} \quad x - 3 = 0$$
$$x = 13 \quad \text{or} \quad x = 3$$

Check. We see that 13 is not a solution, because when $x = 13$, then $20 - 2x = -6$ and $12 - 2x = -14$ and the dimensions of the frame cannot be negative. We check 3. When $x = 3$, then $20 - 2x = 14$ and $12 - 2x = 6$ and $14 \cdot 6 = 84$, the area of the picture that shows. The number 3 checks.

State. The width of the frame is 3 cm.

11. *Familiarize*. Using the labels on the drawing in the text, we let x and $x + 2$ represent the lengths of the legs of the right triangle.

Translate. We use the Pythagorean equation.
$$a^2 + b^2 = c^2$$
$$x^2 + (x + 2)^2 = 10^2 \quad \text{Substituting}$$

Solve. We solve the equation.
$$x^2 + x^2 + 4x + 4 = 100$$
$$2x^2 + 4x + 4 = 100$$
$$2x^2 + 4x - 96 = 0$$
$$x^2 + 2x - 48 = 0 \quad \text{Dividing by 2}$$
$$(x + 8)(x - 6) = 0$$
$$x + 8 = 0 \quad \text{or} \quad x - 6 = 0$$
$$x = -8 \quad \text{or} \quad x = 6$$

Check. We only check 6 since the length of a leg cannot be negative. When $x = 6$, then $x + 2 = 8$, and $6^2 + 8^2 = 100 = 10^2$. The number 6 checks.

State. The lengths of the legs are 6 ft and 8 ft.

13. *Familiarize*. The page numbers on facing pages are consecutive integers. Let $x =$ the number on the left-hand page. Then $x + 1 =$ the number on the right-hand page.

Translate.

$$\underbrace{\text{The product of the page numbers}}_{x(x+1)} \text{ is } 812.$$
$$x(x + 1) = 812$$

Solve. We solve the equation.
$$x^2 + x = 812$$
$$x^2 + x - 812 = 0$$
$$(x + 29)(x - 28) = 0$$
$$x + 29 = 0 \quad \text{or} \quad x - 28 = 0$$
$$x = -29 \quad \text{or} \quad x = 28$$

Check. We only check 28 since a page number cannot be negative. If $x = 28$, then $x + 1 = 29$ and $28 \cdot 29 = 812$. The number 28 checks.

State. The page numbers are 28 and 29.

15. *Familiarize*. We make a drawing and label it. We let $x =$ the length and $x - 4 =$ the width.

Translate. We use the formula for the area of a rectangle.
$$A = lw$$
$$10 = x(x - 4) \quad \text{Substituting}$$

Solve. We solve the equation.
$$10 = x^2 - 4x$$
$$0 = x^2 - 4x - 10$$
$$x = \frac{-b \pm \sqrt{b^2 - 4ac}}{2a} = \frac{-(-4) \pm \sqrt{(-4)^2 - 4 \cdot 1 \cdot (-10)}}{2 \cdot 1}$$
$$x = \frac{4 \pm \sqrt{16 + 40}}{2} = \frac{4 \pm \sqrt{56}}{2} = \frac{4 \pm \sqrt{4 \cdot 14}}{2}$$
$$x = \frac{4 \pm 2\sqrt{14}}{2} = 2 \pm \sqrt{14}$$

Check. We only need to check $2 + \sqrt{14}$ since $2 - \sqrt{14}$ is negative and the length cannot be negative. If $x = 2 + \sqrt{14}$, then $x - 4 = (2 + \sqrt{14}) - 4$, or $\sqrt{14} - 2$. Using a calculator we find that the length is $2 + \sqrt{14} \approx 5.742$ ft and the width is $\sqrt{14} - 2 \approx 1.742$ ft, and $(5.742)(1.742) = 10.003 \approx 10$. Our result checks.

State. The length is $2 + \sqrt{14}$ ft ≈ 5.742 ft; the width is $\sqrt{14} - 2$ ft ≈ 1.742 ft.

17. *Familiarize*. We make a drawing and label it. We let $x =$ the width of the margin.

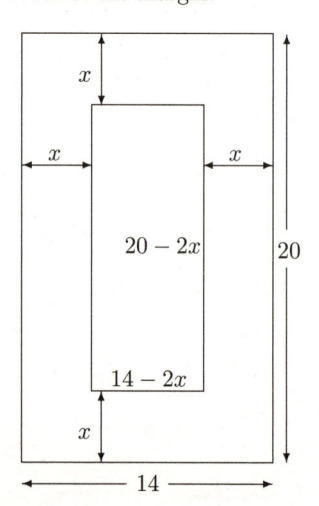

The length and width of the printed text are represented by $20 - 2x$ and $14 - 2x$. The area of the printed text is 100 in^2.

Translate. We use the formula for the area of a rectangle.
$$A = lw$$
$$100 = (20 - 2x)(14 - 2x)$$

Exercise Set 11.3

Solve. We solve the equation.
$$100 = 280 - 68x + 4x^2$$
$$0 = 4x^2 - 68x + 180$$
$$0 = x^2 - 17x + 45 \quad \text{Dividing by 4}$$
$$x = \frac{-b \pm \sqrt{b^2 - 4ac}}{2a} = \frac{-(-17) \pm \sqrt{(-17)^2 - 4 \cdot 1 \cdot 45}}{2 \cdot 1}$$
$$x = \frac{17 \pm \sqrt{289 - 180}}{2} = \frac{17 \pm \sqrt{109}}{2}$$
$$x \approx 13.720 \quad \text{or} \quad x \approx 3.280$$

Check. If $x \approx 13.720$, then $20 - 2x \approx -7.440$ and $14 - 2x \approx -13.440$. Since the width of the margin cannot be negative, 13.720 is not a solution. If $x \approx 3.280$, then $20 - 2x \approx 13.440$ and $14 - 2x \approx 7.440$ and $(13.440)(7.440) = 99.99 \approx 100$. The number $\frac{17 - \sqrt{109}}{2} \approx 3.280$ checks.

State. The width of the margin is $\frac{17 - \sqrt{109}}{2}$ in. ≈ 3.280 in.

19. Familiarize. We make a drawing. We let $x =$ the length of the shorter leg and $x + 14 =$ the length of the longer leg.

Translate. We use the Pythagorean equation.
$$a^2 + b^2 = c^2$$
$$x^2 + (x + 14)^2 = 24^2 \quad \text{Substituting}$$

Solve. We solve the equation.
$$x^2 + x^2 + 28x + 196 = 576$$
$$2x^2 + 28x - 380 = 0$$
$$x^2 + 14x - 190 = 0 \quad \text{Dividing by 2}$$
$$x = \frac{-b \pm \sqrt{b^2 - 4ac}}{2a} = \frac{-14 \pm \sqrt{14^2 - 4 \cdot 1 \cdot (-190)}}{2 \cdot 1}$$
$$x = \frac{-14 \pm \sqrt{196 + 760}}{2} = \frac{-14 \pm \sqrt{956}}{2} = \frac{-14 \pm \sqrt{4 \cdot 239}}{2}$$
$$x = \frac{-14 \pm 2\sqrt{239}}{2} = -7 \pm \sqrt{239}$$
$$x \approx 8.460 \quad \text{or} \quad x \approx -22.460$$

Check. Since the length of a leg cannot be negative, we only need to check 8.460. If $x = -7 + \sqrt{239} \approx 8.460$, then $x + 14 = -7 + \sqrt{239} + 14 = 7 + \sqrt{239} \approx 22.460$ and $(8.460)^2 + (22.460)^2 = 576.0232 \approx 576 = 24^2$. The number $-7 + \sqrt{239} \approx 8.460$ checks.

State. The lengths of the legs are $-7 + \sqrt{239}$ ft ≈ 8.460 ft and $7 + \sqrt{239}$ ft ≈ 22.460 ft.

21. Familiarize. We first make a drawing, labeling it with the known and unknown information. We can also organize the information in a table. We let r represent the speed and t the time for the first part of the trip.

Trip	Distance	Speed	Time
1st part	120	r	t
2nd part	100	$r - 10$	$4 - t$

Translate. Using $r = \frac{d}{t}$, we get two equations from the table, $r = \frac{120}{t}$ and $r - 10 = \frac{100}{4 - t}$.

Solve. We substitute $\frac{120}{t}$ for r in the second equation and solve for t.
$$\frac{120}{t} - 10 = \frac{100}{4 - t}, \quad \text{LCD is } t(4 - t)$$
$$t(4 - t)\left(\frac{120}{t} - 10\right) = t(4 - t) \cdot \frac{100}{4 - t}$$
$$120(4 - t) - 10t(4 - t) = 100t$$
$$480 - 120t - 40t + 10t^2 = 100t$$
$$10t^2 - 260t + 480 = 0 \quad \text{Standard form}$$
$$t^2 - 26t + 48 = 0 \quad \text{Multiplying by } \frac{1}{10}$$
$$(t - 2)(t - 24) = 0$$
$$t = 2 \quad \text{or} \quad t = 24$$

Check. Since the time cannot be negative (If $t = 24$, $4 - t = -20$.), we check only 2 hr. If $t = 2$, then $4 - t = 2$. The speed of the first part is $\frac{120}{2}$, or 60 mph. The speed of the second part is $\frac{100}{2}$, or 50 mph. The speed of the second part is 10 mph slower than the first part. The value checks.

State. The speed of the first part was 60 mph, and the speed of the second part was 50 mph.

23. Familiarize. We first make a drawing. We also organize the information in a table. We let $r =$ the speed and $t =$ the time of the slower trip.

Trip	Distance	Speed	Time
Slower	200	r	t
Faster	200	$r + 10$	$t - 1$

Translate. Using $t = d/r$, we get two equations from the table:
$$t = \frac{200}{r} \quad \text{and} \quad t - 1 = \frac{200}{r + 10}$$

Solve. We substitute $\frac{200}{r}$ for t in the second equation and solve for r.

$$\frac{200}{r} - 1 = \frac{200}{r+10}, \text{ LCD is } r(r+10)$$

$$r(r+10)\left(\frac{200}{r} - 1\right) = r(r+10) \cdot \frac{200}{r+10}$$

$$200(r+10) - r(r+10) = 200r$$

$$200r + 2000 - r^2 - 10r = 200r$$

$$0 = r^2 + 10r - 2000$$

$$0 = (r+50)(r-40)$$

$r = -50$ or $r = 40$

Check. Since negative speed has no meaning in this problem, we check only 40. If $r = 40$, then the time for the slower trip is $\frac{200}{40}$, or 5 hours. If $r = 40$, then $r + 10 = 50$ and the time for the faster trip is $\frac{200}{50}$, or 4 hours. This is 1 hour less time than the slower trip took, so we have an answer to the problem.

State. The speed is 40 mph.

25. **Familiarize.** We make a drawing and then organize the information in a table. We let r = the speed and t = the time of the Cessna.

    600 mi r mph t hr
    1000 mi $r + 50$ mph $t + 1$ hr

Plane	Distance	Speed	Time
Cessna	600	r	t
Beechcraft	1000	$r + 50$	$t + 1$

Translate. Using $t = d/r$, we get two equations from the table:

$$t = \frac{600}{r} \text{ and } t + 1 = \frac{1000}{r+50}$$

Solve. We substitute $\frac{600}{r}$ for t in the second equation and solve for r.

$$\frac{600}{r} + 1 = \frac{1000}{r+50},$$

LCD is $r(r+50)$

$$r(r+50)\left(\frac{600}{r} + 1\right) = r(r+50) \cdot \frac{1000}{r+50}$$

$$600(r+50) + r(r+50) = 1000r$$

$$600r + 30,000 + r^2 + 50r = 1000r$$

$$r^2 - 350r + 30,000 = 0$$

$$(r-150)(r-200) = 0$$

$r = 150$ or $r = 200$

Check. If $r = 150$, then the Cessna's time is $\frac{600}{150}$, or 4 hr and the Beechcraft's time is $\frac{1000}{150+50}$, or $\frac{1000}{200}$, or 5 hr. If $r = 200$, then the Cessna's time is $\frac{600}{200}$, or 3 hr and the Beechcraft's time is $\frac{1000}{200+50}$, or $\frac{1000}{250}$, or 4 hr. Since the Beechcraft's time is 1 hr longer in each case, both values check. There are two solutions.

State. The speed of the Cessna is 150 mph and the speed of the Beechcraft is 200 mph; or the speed of the Cessna is 200 mph and the speed of the Beechcraft is 250 mph.

27. **Familiarize.** We make a drawing and then organize the information in a table. We let r represent the speed and t the time of the trip to Hillsboro.

    40 mi r mph t hr   Hillsboro
    40 mi $r - 6$ mph $14 - t$ hr

Trip	Distance	Speed	Time
To Hillsboro	40	r	t
Return	40	$r - 6$	$14 - t$

Translate. Using $t = \frac{d}{r}$, we get two equations from the table,

$$t = \frac{40}{r} \text{ and } 14 - t = \frac{40}{r-6}.$$

Solve. We substitute $\frac{40}{r}$ for t in the second equation and solve for r.

$$14 - \frac{40}{r} = \frac{40}{r-6},$$

LCD is $r(r-6)$

$$r(r-6)\left(14 - \frac{40}{r}\right) = r(r-6) \cdot \frac{40}{r-6}$$

$$14r(r-6) - 40(r-6) = 40r$$

$$14r^2 - 84r - 40r + 240 = 40r$$

$$14r^2 - 164r + 240 = 0$$

$$7r^2 - 82r + 120 = 0$$

$$(7r - 12)(r - 10) = 0$$

$r = \frac{12}{7}$ or $r = 10$

Check. Since negative speed has no meaning in this problem (If $r = \frac{12}{7}$, then $r - 6 = -\frac{30}{7}$.), we check only 10 mph. If $r = 10$, then the time of the trip to Hillsboro is $\frac{40}{10}$, or 4 hr. The speed of the return trip is $10 - 6$, or 4 mph, and the time is $\frac{40}{4}$, or 10 hr. The total time for the round trip is 4 hr + 10 hr, or 14 hr. The value checks.

State. Naoki's speed on the trip to Hillsboro was 10 mph and it was 4 mph on the return trip.

29. **Familiarize.** We make a drawing and organize the information in a table. Let r represent the speed of the barge in still water, and let t represent the time of the trip upriver.

    24 mi $r - 4$ mph t hr
                                        Upriver
    Downriver   24 mi $r + 4$ mph $5 - t$ hr

Exercise Set 11.3

Trip	Distance	Speed	Time
Upriver	24	$r-4$	t
Downriver	24	$r+4$	$5-t$

Translate. Using $t = \dfrac{d}{r}$, we get two equations from the table,
$$t = \frac{24}{r-4} \text{ and } 5-t = \frac{24}{r+4}.$$

Solve. We substitute $\dfrac{24}{r-4}$ for t in the second equation and solve for r.
$$5 - \frac{24}{r-4} = \frac{24}{r+4},$$
LCD is $(r-4)(r+4)$
$$(r-4)(r+4)\left(5 - \frac{24}{r-4}\right) = (r-4)(r+4) \cdot \frac{24}{r+4}$$
$$5(r-4)(r+4) - 24(r+4) = 24(r-4)$$
$$5r^2 - 80 - 24r - 96 = 24r - 96$$
$$5r^2 - 48r - 80 = 0$$

We use the quadratic formula.
$$r = \frac{-(-48) \pm \sqrt{(-48)^2 - 4 \cdot 5 \cdot (-80)}}{2 \cdot 5}$$
$$r = \frac{48 \pm \sqrt{3904}}{10}$$
$$r \approx 11 \text{ or } r \approx -1.5$$

Check. Since negative speed has no meaning in this problem, we check only 11 mph. If $r \approx 11$, then the speed upriver is about $11 - 4$, or 7 mph, and the time is about $\dfrac{24}{7}$, or 3.4 hr. The speed downriver is about $11 + 4$, or 15 mph, and the time is about $\dfrac{24}{15}$, or 1.6 hr. The total time of the round trip is $3.4 + 1.6$, or 5 hr. The value checks.

State. The barge must be able to travel about 11 mph in still water.

31. $A = 6s^2$
$\dfrac{A}{6} = s^2$ Dividing by 6
$\sqrt{\dfrac{A}{6}} = s$ Taking the positive square root

33. $F = \dfrac{Gm_1m_2}{r^2}$
$Fr^2 = Gm_1m_2$ Multiplying by r^2
$r^2 = \dfrac{Gm_1m_2}{F}$ Dividing by F
$r = \sqrt{\dfrac{Gm_1m_2}{F}}$ Taking the positive square root

35. $E = mc^2$
$\dfrac{E}{m} = c^2$ Dividing by m
$\sqrt{\dfrac{E}{m}} = c$ Taking the square root

37. $a^2 + b^2 = c^2$
$b^2 = c^2 - a^2$ Subtracting a^2
$b = \sqrt{c^2 - a^2}$ Taking the square root

39. $N = \dfrac{k^2 - 3k}{2}$
$2N = k^2 - 3k$
$0 = k^2 - 3k - 2N$ Standard form
$a = 1, b = -3, c = -2N$
$k = \dfrac{-(-3) \pm \sqrt{(-3)^3 - 4 \cdot 1 \cdot (-2N)}}{2 \cdot 1}$ Using the quadratic formula
$k = \dfrac{3 \pm \sqrt{9 + 8N}}{2}$

Since taking the negative square root would result in a negative answer, we take the positive one.
$$k = \frac{3 + \sqrt{9 + 8N}}{2}$$

41. $A = 2\pi r^2 + 2\pi rh$
$0 = 2\pi r^2 + 2\pi rh - A$ Standard form
$a = 2\pi, b = 2\pi h, c = -A$
$r = \dfrac{-2\pi h \pm \sqrt{(2\pi h)^2 - 4 \cdot 2\pi \cdot (-A)}}{2 \cdot 2\pi}$ Using the quadratic formula
$r = \dfrac{-2\pi h \pm \sqrt{4\pi^2 h^2 + 8\pi A}}{4\pi}$
$r = \dfrac{-2\pi h \pm 2\sqrt{\pi^2 h^2 + 2\pi A}}{4\pi}$
$r = \dfrac{-\pi h \pm \sqrt{\pi^2 h^2 + 2\pi A}}{2\pi}$

Since taking the negative square root would result in a negative answer, we take the positive one.
$$r = \frac{-\pi h + \sqrt{\pi^2 h^2 + 2\pi A}}{2\pi}$$

43. $T = 2\pi \sqrt{\dfrac{L}{g}}$
$\dfrac{T}{2\pi} = \sqrt{\dfrac{L}{g}}$ Dividing by 2π
$\dfrac{T^2}{4\pi^2} = \dfrac{L}{g}$ Squaring
$gT^2 = 4\pi^2 L$ Multiplying by $4\pi^2 g$
$g = \dfrac{4\pi^2 L}{T^2}$ Dividing by T^2

45. $I = \dfrac{704.5W}{H^2}$
$H^2 I = 704.5W$ Multiplying by H^2
$H^2 = \dfrac{704.5W}{I}$ Dividing by I
$H = \sqrt{\dfrac{704.5W}{I}}$

47.
$$m = \frac{m_0}{\sqrt{1-\frac{v^2}{c^2}}}$$
$$m^2 = \frac{m_0^2}{1-\frac{v^2}{c^2}} \quad \text{Principle of powers}$$
$$m^2\left(1-\frac{v^2}{c^2}\right) = m_0^2$$
$$m^2 - \frac{m^2 v^2}{c^2} = m_0^2$$
$$m^2 - m_0^2 = \frac{m^2 v^2}{c^2}$$
$$c^2(m^2 - m_0^2) = m^2 v^2$$
$$\frac{c^2(m^2 - m_0^2)}{m^2} = v^2$$
$$\sqrt{\frac{c^2(m^2 - m_0^2)}{m^2}} = v$$
$$\frac{c\sqrt{m^2 - m_0^2}}{m} = v$$

49. Discussion and Writing Exercise

51.
$$\frac{1}{x-1} + \frac{1}{x^2 - 3x + 2}$$
$$= \frac{1}{x-1} + \frac{1}{(x-1)(x-2)}, \text{ LCD is } (x-1)(x-2)$$
$$= \frac{1}{x-1} \cdot \frac{x-2}{x-2} + \frac{1}{(x-1)(x-2)}$$
$$= \frac{x-2}{(x-1)(x-2)} + \frac{1}{(x-1)(x-2)}$$
$$= \frac{x-2+1}{(x-1)(x-2)}$$
$$= \frac{x-1}{(x-1)(x-2)}$$
$$= \frac{(x-1)\cdot 1}{(x-1)(x-2)}$$
$$= \frac{1}{x-2}$$

53.
$$\frac{2}{x+3} - \frac{x}{x-1} + \frac{x^2+2}{x^2 + 2x - 3}$$
$$= \frac{2}{x+3} - \frac{x}{x-1} + \frac{x^2+2}{(x+3)(x-1)},$$
$$\text{LCD is } (x+3)(x-1)$$
$$= \frac{2}{x+3} \cdot \frac{x-1}{x-1} - \frac{x}{x-1} \cdot \frac{x+3}{x+3} + \frac{x^2+2}{(x+3)(x-1)}$$
$$= \frac{2(x-1)}{(x+3)(x-1)} - \frac{x(x+3)}{(x-1)(x+3)} + \frac{x^2+2}{(x+3)(x-1)}$$
$$= \frac{2(x-1) - x(x+3) + x^2 + 2}{(x+3)(x-1)}$$
$$= \frac{2x - 2 - x^2 - 3x + x^2 + 2}{(x+3)(x-1)}$$
$$= \frac{-x}{(x+3)(x-1)}$$

55. $\sqrt{-20} = \sqrt{-1 \cdot 4 \cdot 5} = i \cdot 2 \cdot \sqrt{5} = 2\sqrt{5}i$, or $2i\sqrt{5}$

57.
$$\frac{\frac{4}{a^2 b}}{\frac{3}{a} - \frac{4}{b^2}} = \frac{\frac{4}{a^2 b}}{\frac{3}{a} - \frac{4}{b^2}} \cdot \frac{a^2 b^2}{a^2 b^2}$$
$$= \frac{\frac{4}{a^2 b} \cdot a^2 b^2}{\frac{3}{a} \cdot a^2 b^2 - \frac{4}{b^2} \cdot a^2 b^2}$$
$$= \frac{4b}{3ab^2 - 4a^2}, \text{ or } \frac{4b}{a(3b^2 - 4a)}$$

59.
$$\frac{1}{a-1} = a+1$$
$$\frac{1}{a-1} \cdot a - 1 = (a+1)(a-1)$$
$$1 = a^2 - 1$$
$$2 = a^2$$
$$\pm\sqrt{2} = a$$

61. Let s represent a length of a side of the cube, let S represent the surface area of the cube, and let A represent the surface area of the sphere. Then the diameter of the sphere is s, so the radius r is $s/2$. From Exercise 32, we know, $A = 4\pi r^2$, so when $r = s/2$ we have $A = 4\pi \left(\frac{s}{2}\right)^2 = 4\pi \cdot \frac{s^2}{4} = \pi s^2$. From the formula for the surface area of a cube (See Exercise 31.) we know that $S = 6s^2$, so $\frac{S}{6} = s^2$ and then $A = \pi \cdot \frac{S}{6}$, or $A(S) = \frac{\pi S}{6}$.

63.
$$\frac{w}{l} = \frac{l}{w+l}$$
$$l(w+l) \cdot \frac{w}{l} = l(w+l) \cdot \frac{l}{w+l}$$
$$w(w+l) = l^2$$
$$w^2 + lw = l^2$$
$$0 = l^2 - lw - w^2$$

Use the quadratic formula with $a = 1$, $b = -w$, and $c = -w^2$.
$$l = \frac{-(-w) \pm \sqrt{(-w)^2 - 4 \cdot 1 \cdot (-w^2)}}{2 \cdot 1}$$
$$l = \frac{w \pm \sqrt{w^2 + 4w^2}}{2} = \frac{w \pm \sqrt{5w^2}}{2}$$
$$l = \frac{w \pm w\sqrt{5}}{2}$$

Since $\frac{w - w\sqrt{5}}{2}$ is negative we use the positive square root:
$$l = \frac{w + w\sqrt{5}}{2}$$

Exercise Set 11.4

1. $x^2 - 8x + 16 = 0$

$a = 1$, $b = -8$, $c = 16$

We compute the discriminant.

Exercise Set 11.4

$b^2 - 4ac = (-8)^2 - 4 \cdot 1 \cdot 16$
$= 64 - 64$
$= 0$

Since $b^2 - 4ac = 0$, there is just one solution, and it is a real number.

3. $x^2 + 1 = 0$
$a = 1, b = 0, c = 1$
We compute the discriminant.
$b^2 - 4ac = 0^2 - 4 \cdot 1 \cdot 1$
$= -4$

Since $b^2 - 4ac < 0$, there are two nonreal solutions.

5. $x^2 - 6 = 0$
$a = 1, b = 0, c = -6$
We compute the discriminant.
$b^2 - 4ac = 0^2 - 4 \cdot 1 \cdot (-6)$
$= 24$

Since $b^2 - 4ac > 0$, there are two real solutions.

7. $4x^2 - 12x + 9 = 0$
$a = 4, b = -12, c = 9$
We compute the discriminant.
$b^2 - 4ac = (-12)^2 - 4 \cdot 4 \cdot 9$
$= 144 - 144$
$= 0$

Since $b^2 - 4ac = 0$, there is just one solution, and it is a real number.

9. $x^2 - 2x + 4 = 0$
$a = 1, b = -2, c = 4$
We compute the discriminant.
$b^2 - 4ac = (-2)^2 - 4 \cdot 1 \cdot 4$
$= 4 - 16$
$= -12$

Since $b^2 - 4ac < 0$, there are two nonreal solutions.

11. $9t^2 - 3t = 0$
$a = 9, b = -3, c = 0$
We compute the discriminant.
$b^2 - 4ac = (-3)^2 - 4 \cdot 9 \cdot 0$
$= 9 - 0$
$= 9$

Since $b^2 - 4ac > 0$, there are two real solutions.

13. $y^2 = \frac{1}{2}y + \frac{3}{5}$
$y^2 - \frac{1}{2}y - \frac{3}{5} = 0$ Standard form
$a = 1, b = -\frac{1}{2}, c = -\frac{3}{5}$
We compute the discriminant.

$b^2 - 4ac = \left(-\frac{1}{2}\right)^2 - 4 \cdot 1 \cdot \left(-\frac{3}{5}\right)$
$= \frac{1}{4} + \frac{12}{5}$
$= \frac{53}{20}$

Since $b^2 - 4ac > 0$, there are two real solutions.

15. $4x^2 - 4\sqrt{3}x + 3 = 0$
$a = 4, b = -4\sqrt{3}, c = 3$
We compute the discriminant.
$b^2 - 4ac = (-4\sqrt{3})^2 - 4 \cdot 4 \cdot 3$
$= 48 - 48$
$= 0$

Since $b^2 - 4ac = 0$, there is just one solution, and it is a real number.

17. The solutions are -4 and 4.
$x = -4 \quad \text{or} \quad x = 4$
$x + 4 = 0 \quad \text{or} \quad x - 4 = 0$
$(x + 4)(x - 4) = 0 \quad$ Principle of zero products
$x^2 - 16 = 0 \quad (A+B)(A-B) = A^2 - B^2$

19. The solutions are -2 and -7.
$x = -2 \quad \text{or} \quad x = -7$
$x + 2 = 0 \quad \text{or} \quad x + 7 = 0$
$(x + 2)(x + 7) = 0 \quad$ Principle of zero products
$x^2 + 9x + 14 = 0 \quad$ FOIL

21. The only solution is 8. It must be a double solution.
$x = 8 \quad \text{or} \quad x = 8$
$x - 8 = 0 \quad \text{or} \quad x - 8 = 0$
$(x - 8)(x - 8) = 0 \quad$ Principle of zero products
$x^2 - 16x + 64 = 0 \quad (A - B)^2 = A^2 - 2AB + B^2$

23. The solutions are $-\frac{2}{5}$ and $\frac{6}{5}$.
$x = -\frac{2}{5} \quad \text{or} \quad x = \frac{6}{5}$
$x + \frac{2}{5} = 0 \quad \text{or} \quad x - \frac{6}{5} = 0$
$5x + 2 = 0 \quad \text{or} \quad 5x - 6 = 0 \quad$ Clearing fractions
$(5x + 2)(5x - 6) = 0 \quad$ Principle of zero products
$25x^2 - 20x - 12 = 0 \quad$ FOIL

25. The solutions are $\frac{k}{3}$ and $\frac{m}{4}$.
$x = \frac{k}{3} \quad \text{or} \quad x = \frac{m}{4}$
$x - \frac{k}{3} = 0 \quad \text{or} \quad x - \frac{m}{4} = 0$
$3x - k = 0 \quad \text{or} \quad 4x - m = 0 \quad$ Clearing fractions

$$(3x-k)(4x-m) = 0 \quad \text{Principle of zero products}$$
$$12x^2 - 3mx - 4kx + km = 0 \quad \text{FOIL}$$
$$12x^2 - (3m+4k)x + km = 0 \quad \text{Collecting like terms}$$

27. The solutions are $-\sqrt{3}$ and $2\sqrt{3}$.
$$x = -\sqrt{3} \quad \text{or} \quad x = 2\sqrt{3}$$
$$x + \sqrt{3} = 0 \quad \text{or} \quad x - 2\sqrt{3} = 0$$
$$(x+\sqrt{3})(x-2\sqrt{3}) = 0 \quad \text{Principle of zero products}$$
$$x^2 - 2\sqrt{3}x + \sqrt{3}x - 2(\sqrt{3})^2 = 0 \quad \text{FOIL}$$
$$x^2 - \sqrt{3}x - 6 = 0$$

29. $x^4 - 6x^2 + 9 = 0$
Let $u = x^2$ and think of x^4 as $(x^2)^2$.
$$u^2 - 6u + 9 = 0 \quad \text{Substituting } u \text{ for } x^2$$
$$(u-3)(u-3) = 0$$
$$u - 3 = 0 \quad \text{or} \quad u - 3 = 0$$
$$u = 3 \quad \text{or} \quad u = 3$$
Now we substitute x^2 for u and solve the equation:
$$x^2 = 3$$
$$x = \pm\sqrt{3}$$
Both $\sqrt{3}$ and $-\sqrt{3}$ check. They are the solutions.

31. $x - 10\sqrt{x} + 9 = 0$
Let $u = \sqrt{x}$ and think of x as $(\sqrt{x})^2$.
$$u^2 - 10u + 9 = 0 \quad \text{Substituting } u \text{ for } \sqrt{x}$$
$$(u-9)(u-1) = 0$$
$$u - 9 = 0 \quad \text{or} \quad u - 1 = 0$$
$$u = 9 \quad \text{or} \quad u = 1$$
Now we substitute $\sqrt{x}$ for u and solve these equations:
$$\sqrt{x} = 9 \quad \text{or} \quad \sqrt{x} = 1$$
$$x = 81 \quad \text{or} \quad x = 1$$
The numbers 81 and 1 both check. They are the solutions.

33. $(x^2-6x)^2 - 2(x^2-6x) - 35 = 0$
Let $u = x^2 - 6x$.
$$u^2 - 2u - 35 = 0 \quad \text{Substituting } u \text{ for } x^2 - 6x$$
$$(u-7)(u+5) = 0$$
$$u - 7 = 0 \quad \text{or} \quad u + 5 = 0$$
$$u = 7 \quad \text{or} \quad u = -5$$
Now we substitute $x^2 - 6x$ for u and solve these equations:
$$x^2 - 6x = 7 \quad \text{or} \quad x^2 - 6x = -5$$
$$x^2 - 6x - 7 = 0 \quad \text{or} \quad x^2 - 6x + 5 = 0$$
$$(x-7)(x+1) = 0 \quad \text{or} \quad (x-5)(x-1) = 0$$
$$x = 7 \text{ or } x = -1 \text{ or } x = 5 \text{ or } x = 1$$
The numbers -1, 1, 5, and 7 check. They are the solutions.

35. $x^{-2} - 5^{-1} - 36 = 0$
Let $u = x^{-1}$.
$$u^2 - 5u - 36 = 0 \quad \text{Substituting } u \text{ for } x^{-1}$$
$$(u-9)(u+4) = 0$$
$$u - 9 = 0 \quad \text{or} \quad u + 4 = 0$$
$$u = 9 \quad \text{or} \quad u = -4$$
Now we substitute x^{-1} for u and solve these equations:
$$x^{-1} = 9 \quad \text{or} \quad x^{-1} = -4$$
$$\frac{1}{x} = 9 \quad \text{or} \quad \frac{1}{x} = -4$$
$$\frac{1}{9} = x \quad \text{or} \quad -\frac{1}{4} = x$$
Both $\frac{1}{9}$ and $-\frac{1}{4}$ check. They are the solutions.

37. $(1+\sqrt{x})^2 + (1+\sqrt{x}) - 6 = 0$
Let $u = 1 + \sqrt{x}$.
$$u^2 + u - 6 = 0 \quad \text{Substituting } u \text{ for } 1+\sqrt{x}$$
$$(u+3)(u-2) = 0$$
$$u + 3 = 0 \quad \text{or} \quad u - 2 = 0$$
$$u = -3 \quad \text{or} \quad u = 2$$
$$1 + \sqrt{x} = -3 \quad \text{or} \quad 1 + \sqrt{x} = 2 \quad \text{Substituting } 1+\sqrt{x} \text{ for } u$$
$$\sqrt{x} = -4 \quad \text{or} \quad \sqrt{x} = 1$$
$$\text{No solution} \quad \quad x = 1$$
The number 1 checks. It is the solution.

39. $(y^2-5y)^2 - 2(y^2-5y) - 24 = 0$
Let $u = y^2 - 5y$.
$$u^2 - 2u - 24 = 0 \quad \text{Substituting } u \text{ for } y^2-5y$$
$$(u-6)(u+4) = 0$$
$$u - 6 = 0 \quad \text{or} \quad u + 4 = 0$$
$$u = 6 \quad \text{or} \quad u = -4$$
$$y^2 - 5y = 6 \quad \text{or} \quad y^2 - 5y = -4$$
Substituting $y^2 - 5y$ for u
$$y^2 - 5y - 6 = 0 \quad \text{or} \quad y^2 - 5y + 4 = 0$$
$$(y-6)(y+1) = 0 \quad \text{or} \quad (y-4)(y-1) = 0$$
$$y = 6 \text{ or } y = -1 \text{ or } y = 4 \text{ or } y = 1$$
The numbers -1, 1, 4, and 6 check. They are the solutions.

41. $t^4 - 6t^2 - 4 = 0$

Let $u = t^2$.

$u^2 - 6u - 4 = 0$ Substituting u for t^2

$u = \dfrac{-(-6) \pm \sqrt{(-6)^2 - 4 \cdot 1 \cdot (-4)}}{2 \cdot 1}$

$u = \dfrac{6 \pm \sqrt{52}}{2} = \dfrac{6 \pm 2\sqrt{13}}{2}$

$u = 3 \pm \sqrt{13}$

Now we substitute t^2 for u and solve these equations:

$t^2 = 3 + \sqrt{13}$ or $t^2 = 3 - \sqrt{13}$

$t = \pm\sqrt{3 + \sqrt{13}}$ or $t = \pm\sqrt{3 - \sqrt{13}}$

All four numbers check. They are the solutions.

43. $2x^{-2} + x^{-1} - 1 = 0$

Let $u = x^{-1}$.

$2u^2 + u - 1 = 0$ Substituting u for x^{-1}

$(2u - 1)(u + 1) = 0$

$2u - 1 = 0$ or $u + 1 = 0$

$2u = 1$ or $u = -1$

$u = \dfrac{1}{2}$ or $u = -1$

$x^{-1} = \dfrac{1}{2}$ or $x^{-1} = -1$ Substituting x^{-1} for u

$\dfrac{1}{x} = \dfrac{1}{2}$ or $\dfrac{1}{x} = -1$

$x = 2$ or $x = -1$

Both 2 and -1 check. They are the solutions.

45. $6x^4 - 19x^2 + 15 = 0$

Let $u = x^2$.

$6u^2 - 19u + 15 = 0$ Substituting u for x^2

$(3u - 5)(2u - 3) = 0$

$3u - 5 = 0$ or $2u - 3 = 0$

$3u = 5$ or $2u = 3$

$u = \dfrac{5}{3}$ or $u = \dfrac{3}{2}$

$x^2 = \dfrac{5}{3}$ or $x^2 = \dfrac{3}{2}$ Substituting x^2 for u

$x = \pm\sqrt{\dfrac{5}{3}}$ or $x = \pm\sqrt{\dfrac{3}{2}}$

$x = \pm\dfrac{\sqrt{15}}{3}$ or $x = \pm\dfrac{\sqrt{6}}{2}$

Rationalizing denominators

All four numbers check. They are the solutions.

47. $x^{2/3} - 4x^{1/3} - 5 = 0$

Let $u = x^{1/3}$.

$u^2 - 4u - 5 = 0$ Substituting u for $x^{1/3}$

$(u - 5)(u + 1) = 0$

$u - 5 = 0$ or $u + 1 = 0$

$u = 5$ or $u = -1$

$x^{1/3} = 5$ or $x^{1/3} = -1$ Substituting $x^{1/3}$ for u

$(x^{1/3})^3 = 5^3$ or $(x^{1/3})^3 = (-1)^3$ Principle of powers

$x = 125$ or $x = -1$

Both 125 and -1 check. They are the solutions.

49. $\left(\dfrac{x-4}{x+1}\right)^2 - 2\left(\dfrac{x-4}{x+1}\right) - 35 = 0$

Let $u = \dfrac{x-4}{x+1}$.

$u^2 - 2u - 35 = 0$ Substituting u for $\dfrac{x-4}{x+1}$

$(u - 7)(u + 5) = 0$

$u - 7 = 0$ or $u + 5 = 0$

$u = 7$ or $u = -5$

$\dfrac{x-4}{x+1} = 7$ or $\dfrac{x-4}{x+1} = -5$ Substituting $\dfrac{x-4}{x+1}$ for u

$x - 4 = 7(x + 1)$ or $x - 4 = -5(x + 1)$

$x - 4 = 7x + 7$ or $x - 4 = -5x - 5$

$-6x = 11$ or $6x = -1$

$x = -\dfrac{11}{6}$ or $x = -\dfrac{1}{6}$

Both $-\dfrac{11}{6}$ and $-\dfrac{1}{6}$ check. They are the solutions.

51. $9\left(\dfrac{x+2}{x+3}\right)^2 - 6\left(\dfrac{x+2}{x+3}\right) + 1 = 0$

Let $u = \dfrac{x+2}{x+3}$.

$9u^2 - 6u + 1 = 0$ Substituting u for $\dfrac{x+2}{x+3}$

$(3u - 1)(3u - 1) = 0$

$3u - 1 = 0$ or $3u - 1 = 0$

$3u = 1$ or $3u = 1$

$u = \dfrac{1}{3}$ or $u = \dfrac{1}{3}$

Now we substitute $\dfrac{x+2}{x+3}$ for u and solve the equation:

$\dfrac{x+2}{x+3} = \dfrac{1}{3}$

$3(x + 2) = x + 3$ Multiplying by $3(x+3)$

$3x + 6 = x + 3$

$2x = -3$

$x = -\dfrac{3}{2}$

The number $-\dfrac{3}{2}$ checks. It is the solution.

53. $\left(\dfrac{x^2-2}{x}\right)^2 - 7\left(\dfrac{x^2-2}{x}\right) - 18 = 0$

Let $u = \dfrac{x^2-2}{x}$.

$u^2 - 7u - 18 = 0$ Substituting u for $\dfrac{x^2-2}{x}$

$(u-9)(u+2) = 0$

$u - 9 = 0$ or $u + 2 = 0$

$u = 9$ or $u = -2$

$\dfrac{x^2-2}{x} = 9$ or $\dfrac{x^2-2}{x} = -2$

Substituting $\dfrac{x^2-2}{x}$ for u

$x^2 - 2 = 9x$ or $x^2 - 2 = -2x$

$x^2 - 9x - 2 = 0$ or $x^2 + 2x - 2 = 0$

$x = \dfrac{-(-9) \pm \sqrt{(-9)^2 - 4 \cdot 1 \cdot (-2)}}{2 \cdot 1}$

$x = \dfrac{9 \pm \sqrt{89}}{2}$

or

$x = \dfrac{-2 \pm \sqrt{2^2 - 4 \cdot 1 \cdot (-2)}}{2 \cdot 1} = \dfrac{-2 \pm \sqrt{12}}{2}$

$x = \dfrac{-2 \pm 2\sqrt{3}}{2} = -1 \pm \sqrt{3}$

All four numbers check. They are the solutions.

55. The x-intercepts occur where $f(x) = 0$. Thus, we must have $5x + 13\sqrt{x} - 6 = 0$.

Let $u = \sqrt{x}$.

$5u^2 + 13u - 6 = 0$ Substituting

$(5u - 2)(u + 3) = 0$

$u = \dfrac{2}{5}$ or $u = -3$

Now replace u with $\sqrt{x}$ and solve these equations:

$\sqrt{x} = \dfrac{2}{5}$ or $\sqrt{x} = -3$

$x = \dfrac{4}{25}$ No solution

The number $\dfrac{4}{25}$ checks. Thus, the x-intercept is $\left(\dfrac{4}{25}, 0\right)$.

57. The x-intercepts occur where $f(x) = 0$. Thus, we must have $(x^2 - 3x)^2 - 10(x^2 - 3x) + 24 = 0$.

Let $u = x^2 - 3x$.

$u^2 - 10u + 24 = 0$ Substituting

$(u - 6)(u - 4) = 0$

$u = 6$ or $u = 4$

Now replace u with $x^2 - 3x$ and solve these equations:

$x^2 - 3x = 6$ or $x^2 - 3x = 4$

$x^2 - 3x - 6 = 0$ or $x^2 - 3x - 4 = 0$

$x = \dfrac{-(-3) \pm \sqrt{(-3)^2 - 4(1)(-6)}}{2 \cdot 1}$ or $(x - 4)(x + 1) = 0$

$x = \dfrac{3 \pm \sqrt{33}}{2}$ or $x = 4$ or $x = -1$

All four numbers check. Thus, the x-intercepts are $\left(\dfrac{3 + \sqrt{33}}{2}, 0\right)$, $\left(\dfrac{3 - \sqrt{33}}{2}, 0\right)$, $(4, 0)$, and $(-1, 0)$.

59. Discussion and Writing Exercise

61. *Familiarize.* Let $x =$ the number of pounds of Kenyan coffee and $y =$ the number of pounds of Peruvian coffee in the mixture. We organize the information in a table.

Type of Coffee	Kenyan	Peruvian	Mixture
Price per pound	$6.75	$11.25	$8.55
Number of pounds	x	y	50
Total cost	$6.75x$	$11.25y$	8.55×50, or $427.50

Translate. From the last two rows of the table we get a system of equations.

$x + y = 50$,

$6.75x + 11.25y = 427.50$

Solve. Solving the system of equations, we get $(30, 20)$.

Check. The total number of pounds in the mixture is $30 + 20$, or 50. The total cost of the mixture is $\$6.75(30) + \$11.25(20) = \$427.50$. The values check.

State. The mixture should consist of 30 lb of Kenyan coffee and 20 lb of Peruvian coffee.

63. $\sqrt{8x} \cdot \sqrt{2x} = \sqrt{8x \cdot 2x} = \sqrt{16x^2} = \sqrt{(4x)^2} = 4x$

65. $\sqrt[4]{9a^2} \cdot \sqrt[4]{18a^3} = \sqrt[4]{9a^2 \cdot 18a^3} =$
$\sqrt[4]{3 \cdot 3 \cdot a^2 \cdot 3 \cdot 3 \cdot 2 \cdot a^2 \cdot a} = \sqrt[4]{3^4 a^4 \cdot 2a} = \sqrt[4]{3^4} \sqrt[4]{a^4} \sqrt[4]{2a} = 3a\sqrt[4]{2a}$

67. Graph $f(x) = -\dfrac{3}{5}x + 4$.

Choose some values for x, find the corresponding values of $f(x)$, plot the points $(x, f(x))$, and draw the graph.

x	$f(x)$	$(x, f(x))$
-5	7	$(-5, 7)$
0	4	$(0, 4)$
5	1	$(5, 1)$

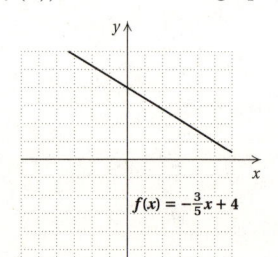

Exercise Set 11.4

69. Graph $y = 4$.

The graph of $y = 4$ is a horizontal line with y-intercept $(0, 4)$.

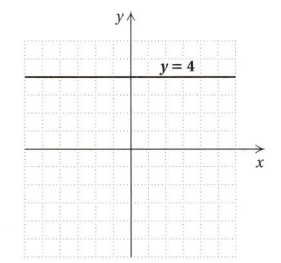

71. Left to the student

73. a) $kx^2 - 2x + k = 0$; one solution is -3

We first find k by substituting -3 for x.
$$k(-3)^2 - 2(-3) + k = 0$$
$$9k + 6 + k = 0$$
$$10k = -6$$
$$k = -\frac{6}{10}$$
$$k = -\frac{3}{5}$$

b) $-\frac{3}{5}x^2 - 2x + \left(-\frac{3}{5}\right) = 0$ Substituting $-\frac{3}{5}$ for k

$3x^2 + 10x + 3 = 0$ Multiplying by -5

$(3x + 1)(x + 3) = 0$

$3x + 1 = 0$ or $x + 3 = 0$

$3x = -1$ or $x = -3$

$x = -\frac{1}{3}$ or $x = -3$

The other solution is $-\frac{1}{3}$.

75. For $ax^2 + bx + c = 0$, $-\frac{b}{a}$ is the sum of the solutions and $\frac{c}{a}$ is the product of the solutions. Thus $-\frac{b}{a} = \sqrt{3}$ and $\frac{c}{a} = 8$.

$$ax^2 + bx + c = 0$$
$$x^2 + \frac{b}{a}x + \frac{c}{a} = 0 \quad \text{Multiplying by } \frac{1}{a}$$
$$x^2 - \left(-\frac{b}{a}\right)x + \frac{c}{a} = 0$$
$$x^2 - \sqrt{3}x + 8 = 0 \quad \text{Substituting } \sqrt{3} \text{ for } -\frac{b}{a} \text{ and } 8 \text{ for } \frac{c}{a}$$

77. The graph includes the points $(-3, 0)$, $(0, -3)$, and $(1, 0)$. Substituting in $y = ax^2 + bx + c$, we have three equations.
$$0 = 9a - 3b + c,$$
$$-3 = \phantom{9a - 3b + {}}c,$$
$$0 = a + b + c$$

The solution of this system of equations is $a = 1$, $b = 2$, $c = -3$.

79. $\dfrac{x}{x-1} - 6\sqrt{\dfrac{x}{x-1}} - 40 = 0$

Let $u = \sqrt{\dfrac{x}{x-1}}$.

$u^2 - 6u - 40 = 0$ Substituting for $\sqrt{\dfrac{x}{x-1}}$

$(u - 10)(u + 4) = 0$

$u = 10$ or $u = -4$

$\sqrt{\dfrac{x}{x-1}} = 10$ or $\sqrt{\dfrac{x}{x-1}} = -4$

Substituting for u

$\dfrac{x}{x-1} = 100$ or No solution

$x = 100x - 100$ Multiplying by $(x-1)$

$100 = 99x$

$\dfrac{100}{99} = x$

This number checks. It is the solution.

81. $\sqrt{x-3} - \sqrt[4]{x-3} = 12$

$(x-3)^{1/2} - (x-3)^{1/4} - 12 = 0$

Let $u = (x-3)^{1/4}$.

$u^2 - u - 12 = 0$ Substituting for $(x-3)^{1/4}$

$(u - 4)(u + 3) = 0$

$u = 4$ or $u = -3$

$(x-3)^{1/4} = 4$ or $(x-3)^{1/4} = -3$

Substituting for u

$x - 3 = 4^4$ or No solution

$x - 3 = 256$

$x = 259$

This number checks. It is the solution.

83. $x^6 - 28x^3 + 27 = 0$

Let $u = x^3$.

$u^2 - 28u + 27 = 0$ Substituting for x^3

$(u - 27)(u - 1) = 0$

$u = 27$ or $u = 1$

$x^3 = 27$ or $x^3 = 1$ Substituting for u

$x = 3$ or $x = 1$

Both 3 and 1 check. They are the solutions.

Exercise Set 11.5

1. $f(x) = 4x^2$

$f(x) = 4x^2$ is of the form $f(x) = ax^2$. Thus we know that the vertex is $(0,0)$ and $x = 0$ is the line of symmetry.

We know that $f(x) = 0$ when $x = 0$ since the vertex is $(0,0)$.

For $x = 1$, $f(x) = 4x^2 = 4 \cdot 1^2 = 4$.
For $x = -1$, $f(x) = 4x^2 = 4 \cdot (-1)^2 = 4$.
For $x = 2$, $f(x) = 4x^2 = 4 \cdot 2^2 = 16$.
For $x = -2$, $f(x) = 4x^2 = 4 \cdot (-2)^2 = 16$.

We complete the table.

x	$f(x)$	
0	0	← Vertex
1	4	
2	16	
-1	4	
-2	16	

We plot the ordered pairs $(x, f(x))$ from the table and connect them with a smooth curve.

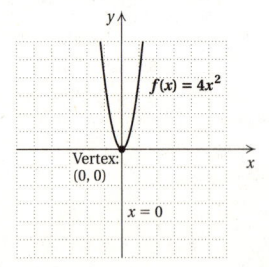

3. $f(x) = \frac{1}{3}x^2$ is of the form $f(x) = ax^2$. Thus we know that the vertex is $(0,0)$ and $x = 0$ is the line of symmetry. We choose some numbers for x and find the corresponding values for $f(x)$. Then we plot the ordered pairs $(x, f(x))$ and connect them with a smooth curve.

For $x = 3$, $f(x) = \frac{1}{3}x^2 = \frac{1}{3} \cdot 3^2 = 3$.

For $x = -3$, $f(x) = \frac{1}{3}x^2 = \frac{1}{3} \cdot (-3)^2 = 3$.

For $x = 6$, $f(x) = \frac{1}{3}x^2 = \frac{1}{3} \cdot 6^2 = 12$.

For $x = -6$, $f(x) = \frac{1}{3}x^2 = \frac{1}{3} \cdot (-6)^2 = 12$.

x	$f(x)$	
0	0	← Vertex
1	$\frac{1}{3}$	
2	$\frac{4}{3}$	
-1	$\frac{1}{3}$	
-2	$\frac{4}{3}$	

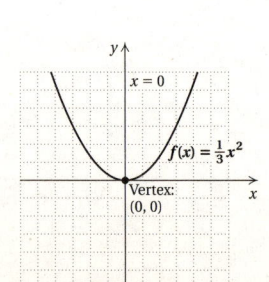

5. $f(x) = -\frac{1}{2}x^2$ is of the form $f(x) = ax^2$. Thus we know that the vertex is $(0,0)$ and $x = 0$ is the line of symmetry. We choose some numbers for x and find the corresponding values for $f(x)$. Then we plot the ordered pairs $(x, f(x))$ and connect them with a smooth curve.

For $x = 2$, $f(x) = -\frac{1}{2}x^2 = -\frac{1}{2} \cdot 2^2 = -2$.

For $x = -2$, $f(x) = -\frac{1}{2}x^2 = -\frac{1}{2} \cdot (-2)^2 = -2$.

For $x = 4$, $f(x) = -\frac{1}{2}x^2 = -\frac{1}{2} \cdot 4^2 = -8$.

For $x = -4$, $f(x) = -\frac{1}{2}x^2 = -\frac{1}{2} \cdot (-4)^2 = -8$.

x	$f(x)$	
0	0	← Vertex
2	-2	
-2	-2	
4	-8	
-4	-8	

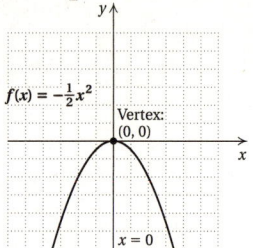

7. $f(x) = -4x^2$ is of the form $f(x) = ax^2$. Thus we know that the vertex is $(0,0)$ and $x = 0$ is the line of symmetry. We choose some numbers for x and find the corresponding values for $f(x)$. Then we plot the ordered pairs $(x, f(x))$ and connect them with a smooth curve.

For $x = 1$, $f(x) = -4x^2 = -4 \cdot 1^2 = -4$.
For $x = -1$, $f(x) = -4x^2 = -4 \cdot (-1)^2 = -4$.
For $x = 2$, $f(x) = -4x^2 = -4 \cdot 2^2 = -16$.
For $x = -2$, $f(x) = -4x^2 = -4 \cdot (-2)^2 = -16$.

x	$f(x)$	
0	0	← Vertex
1	-4	
-1	-4	
2	-16	
-2	-16	

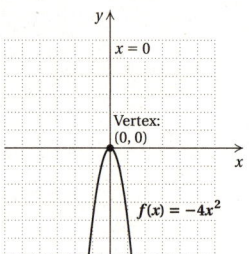

9. $f(x) = (x+3)^2 = [x - (-3)]^2$ is of the form $f(x) = a(x-h)^2$.

Thus we know that the vertex is $(-3, 0)$ and $x = -3$ is the line of symmetry. We choose some numbers for x and find the corresponding values for $f(x)$. Then we plot the ordered pairs $(x, f(x))$ and connect them with a smooth curve.

x	$f(x)$	
-3	0	← Vertex
-2	1	
-1	4	
-4	1	
-5	4	

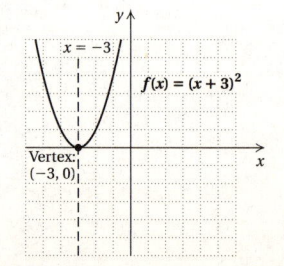

Exercise Set 11.5

11. Graph: $f(x) = 2(x-4)^2$

We choose some values of x and compute $f(x)$. Then we plot these ordered pairs and connect them with a smooth curve.

x	$f(x)$
4	0
5	2
3	2
6	8
2	8

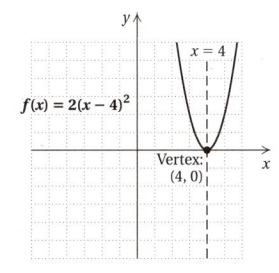

The graph of $f(x) = 2(x-4)^2$ looks like the graph of $f(x) = 2x^2$ except that it is translated 4 units to the right. The vertex is $(4, 0)$, and the line of symmetry is $x = 4$.

13. Graph: $f(x) = -2(x+2)^2$

We choose some values of x and compute $f(x)$. Then we plot these ordered pairs and connect them with a smooth curve.

x	$f(x)$
1	-18
0	-8
-1	-2
-2	0
-3	-2
-4	-8

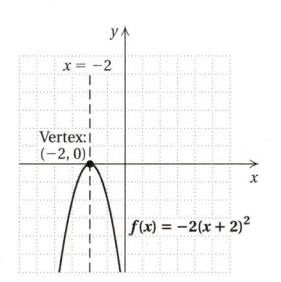

We can express the equation in the equivalent form $f(x) = -2[x-(-2)]^2$. Then we know that the graph looks like the graph of $f(x) = -2x^2$ translated 2 units to the left. The vertex is $(-2, 0)$, and the line of symmetry is $x = -2$.

15. Graph: $f(x) = 3(x-1)^2$

We choose some values of x and compute $f(x)$. Then we plot these ordered pairs and connect them with a smooth curve.

x	$f(x)$
1	0
2	3
0	3
3	12
-1	12

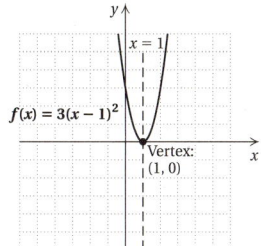

The graph of $f(x) = 3(x-1)^2$ looks like the graph of $f(x) = 3x^2$ except that it is translated 1 unit to the right. The vertex is $(1, 0)$, and the line of symmetry is $x = 1$.

17. Graph: $f(x) = -\dfrac{3}{2}(x+2)^2$

We choose some values of x and compute $f(x)$. Then we plot these ordered pairs and connect them with a smooth curve.

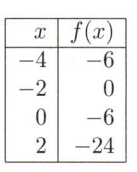

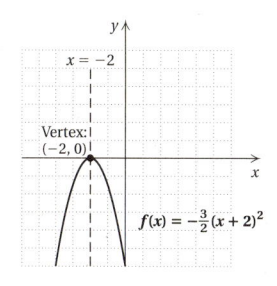

x	$f(x)$
-4	-6
-2	0
0	-6
2	-24

We can express the equation in the equivalent form $f(x) = -\dfrac{3}{2}[x-(-2)]^2$. Then we know that the graph looks like the graph of $f(x) = -\dfrac{3}{2}x^2$ translated 2 units to the left. The vertex is $(-2, 0)$, and the line of symmetry is $x = -2$.

19. Graph: $f(x) = (x-3)^2 + 1$

We choose some values of x and compute $f(x)$. Then we plot these ordered pairs and connect them with a smooth curve.

x	$f(x)$
3	1
4	2
2	2
5	5
1	5

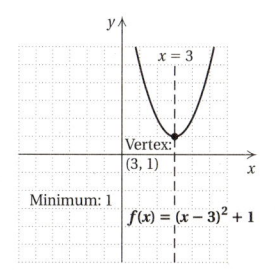

The graph of $f(x) = (x-3)^2 + 1$ looks like the graph of $f(x) = x^2$ except that it is translated 3 units right and 1 unit up. The vertex is $(3, 1)$, and the line of symmetry is $x = 3$. The equation is of the form $f(x) = a(x-h)^2 + k$ with $a = 1$. Since $1 > 0$, we know that 1 is the minimum value.

21. Graph: $f(x) = -3(x+4)^2 + 1$
$\qquad f(x) = -3[x-(-4)]^2 + 1$

We choose some values of x and compute $f(x)$. Then we plot these ordered pairs and connect them with a smooth curve.

x	$f(x)$
-4	1
$-3\dfrac{1}{2}$	$\dfrac{1}{4}$
$-4\dfrac{1}{2}$	$\dfrac{1}{4}$
-3	-2
-5	-2
-2	-11
-6	-11

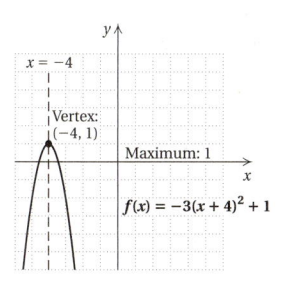

The graph of $f(x) = -3(x+4)^2 + 1$ looks like the graph of $f(x) = 3x^2$ except that it is translated 4 units left and 1 unit up and opens downward. The vertex is $(-4, 1)$, and the line of symmetry is $x = -4$. Since $-3 < 0$, we know that 1 is the maximum value.

23. Graph: $f(x) = \frac{1}{2}(x+1)^2 + 4$

$$f(x) = \frac{1}{2}[x-(-1)]^2 + 4$$

We choose some values of x and compute $f(x)$. Then we plot these ordered pairs and connect them with a smooth curve.

x	$f(x)$
1	6
2	$8\frac{1}{2}$
0	$4\frac{1}{2}$
-1	4
-2	$4\frac{1}{2}$
-3	6

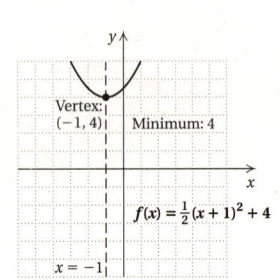

The graph of $f(x) = \frac{1}{2}(x+1)^2 + 4$ looks like the graph of $f(x) = \frac{1}{2}x^2$ except that it is translated 1 unit left and 4 units up. The vertex is $(-1, 4)$, and the line of symmetry is $x = -1$. Since $\frac{1}{2} > 0$, we know that 4 is the minimum value.

25. Graph: $f(x) = -(x+1)^2 - 2$
$f(x) = -[x-(-1)]^2 + (-2)$

We choose some values of x and compute $f(x)$. Then we plot these ordered pairs and connect them with a smooth curve.

x	$f(x)$
-1	-2
0	-3
-2	-3
1	-6
-3	-6

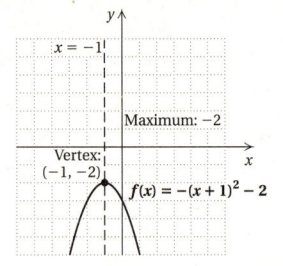

The graph of $f(x) = -(x+1)^2 - 2$ looks like the graph of $f(x) = x^2$ except that it is translated 1 unit left and 2 units down and opens downward. The vertex is $(-1, -2)$, and the line of symmetry is $x = -1$. Since $-1 < 0$, we know that -2 is the maximum value.

27. Discussion and Writing Exercise

29.
$$x - 5 = \sqrt{x+7}$$
$$(x-5)^2 = (\sqrt{x+7})^2 \quad \text{Principle of powers}$$
$$x^2 - 10x + 25 = x + 7$$
$$x^2 - 11x + 18 = 0$$
$$(x-9)(x-2) = 0$$
$$x - 9 = 0 \quad or \quad x - 2 = 0$$
$$x = 9 \quad or \quad x = 2$$

Check: For 9:
$$\begin{array}{c|c} x - 5 = \sqrt{x+7} \\ \hline 9 - 5 \;?\; \sqrt{9+7} \\ 4 \;|\; \sqrt{16} \\ \;|\; 4 \quad \text{TRUE} \end{array}$$

For 2:
$$\begin{array}{c|c} x - 5 = \sqrt{x+7} \\ \hline 2 - 5 \;?\; \sqrt{2+7} \\ -3 \;|\; \sqrt{9} \\ \;|\; 3 \quad \text{FALSE} \end{array}$$

Only 9 checks. It is the solution.

31. $\sqrt{x+4} = -11$

The equation has no solution, because the principal square root of a number is always nonnegative.

33. Left to the student

Exercise Set 11.6

1. $f(x) = x^2 - 2x - 3 = (x^2 - 2x) - 3$

We complete the square inside the parentheses. We take half the x-coefficient and square it.

$\frac{1}{2} \cdot (-2) = -1$ and $(-1)^2 = 1$

Then we add $1 - 1$ inside the parentheses.

$$f(x) = (x^2 - 2x + 1 - 1) - 3$$
$$= (x^2 - 2x + 1) - 1 - 3$$
$$= (x-1)^2 - 4$$
$$= (x-1)^2 + (-4)$$

Vertex: $(1, -4)$

Line of symmetry: $x = 1$

The coefficient of x^2 is 1, which is positive, so the graph opens up. This tells us that -4 is a minimum.

We plot a few points and draw the curve.

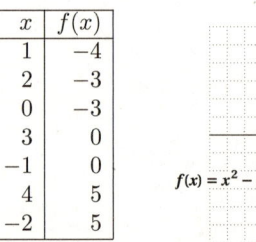

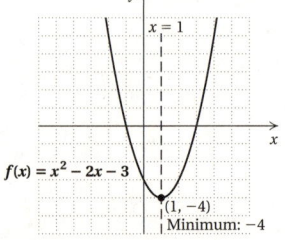

x	$f(x)$
1	-4
2	-3
0	-3
3	0
-1	0
4	5
-2	5

3. $f(x) = -x^2 - 4x - 2 = -(x^2 + 4x) - 2$

We complete the square inside the parentheses. We take half the x-coefficient and square it.

$\frac{1}{2} \cdot 4 = 2$ and $2^2 = 4$

Then we add $4 - 4$ inside the parentheses.

Exercise Set 11.6

$$f(x) = -(x^2 + 4x + 4 - 4) - 2$$
$$= -(x^2 + 4x + 4) + (-1)(-4) - 2$$
$$= -(x+2)^2 + 4 - 2$$
$$= -(x+2)^2 + 2$$
$$= -[x - (-2)]^2 + 2$$

Vertex: $(-2, 2)$

Line of symmetry: $x = -2$

The coefficient of x^2 is -1, which is negative, so the graph opens down. This tells us that 2 is a maximum.

We plot a few points and draw the curve.

x	$f(x)$
-2	2
-4	-2
-3	1
-1	1
0	-2

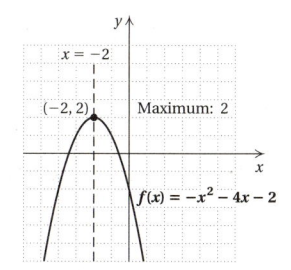

5. $f(x) = 3x^2 - 24x + 50 = 3(x^2 - 8x) + 50$

We complete the square inside the parentheses. We take half the x-coefficient and square it.

$$\frac{1}{2} \cdot (-8) = -4 \text{ and } (-4)^2 = 16$$

Then we add $16 - 16$ inside the parentheses.

$$f(x) = 3(x^2 - 8x + 16 - 16) + 50$$
$$= 3(x^2 - 8x + 16) - 48 + 50$$
$$= 3(x-4)^2 + 2$$

Vertex: $(4, 2)$

Line of symmetry: $x = 4$

The coefficient of x^2 is 3, which is positive, so the graph opens up. This tells us that 2 is a minimum.

We plot a few points and draw the curve.

x	$f(x)$
4	2
5	5
3	5
6	14
2	14

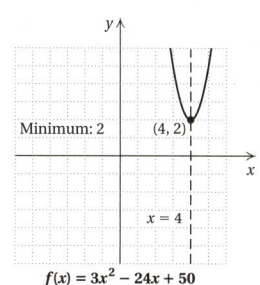

7. $f(x) = -2x^2 - 2x + 3 = -2(x^2 + x) + 3$

We complete the square inside the parentheses. We take half the x-coefficient and square it.

$$\frac{1}{2} \cdot 1 = \frac{1}{2} \text{ and } \left(\frac{1}{2}\right)^2 = \frac{1}{4}$$

Then we add $\frac{1}{4} - \frac{1}{4}$ inside the parentheses.

$$f(x) = -2\left(x^2 + x + \frac{1}{4} - \frac{1}{4}\right) + 3$$
$$= -2\left(x^2 + x + \frac{1}{4}\right) + (-2)\left(-\frac{1}{4}\right) + 3$$
$$= -2\left(x + \frac{1}{2}\right)^2 + \frac{1}{2} + 3$$
$$= -2\left(x + \frac{1}{2}\right)^2 + \frac{7}{2}$$
$$= -2\left[x - \left(-\frac{1}{2}\right)\right]^2 + \frac{7}{2}$$

Vertex: $\left(-\frac{1}{2}, \frac{7}{2}\right)$

Line of symmetry: $x = -\frac{1}{2}$

The coefficient of x^2 is -2, which is negative, so the graph opens down. This tells us that $\frac{7}{2}$ is a maximum.

We plot a few points and draw the curve.

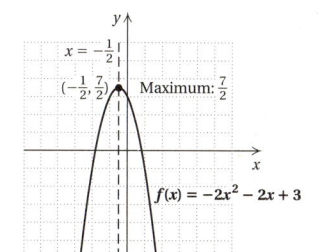

x	$f(x)$
$-\frac{1}{2}$	$\frac{7}{2}$
-2	-1
-1	3
0	3
1	-1

9. $f(x) = 5 - x^2 = -x^2 + 5 = -(x - 0)^2 + 5$

Vertex: $(0, 5)$

Line of symmetry: $x = 0$

The coefficient of x^2 is -1, which is negative, so the graph opens down. This tells us that 5 is a maximum.

We plot a few points and draw the curve.

x	$f(x)$
0	5
1	4
-1	4
2	1
-2	1
3	-4
-3	-4

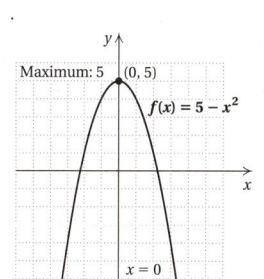

11. $f(x) = 2x^2 + 5x - 2 = 2\left(x^2 + \frac{5}{2}x\right) - 2$

We complete the square inside the parentheses. We take half the x-coefficient and square it.

$$\frac{1}{2} \cdot \frac{5}{2} = \frac{5}{4} \text{ and } \left(\frac{5}{4}\right)^2 = \frac{25}{16}$$

Then we add $\frac{25}{16} - \frac{25}{16}$ inside the parentheses.

276 Chapter 11: Quadratic Equations and Functions

$$f(x) = 2\left(x^2 + \frac{5}{2}x + \frac{25}{16} - \frac{25}{16}\right) - 2$$
$$= 2\left(x^2 + \frac{5}{2}x + \frac{25}{16}\right) + 2\left(-\frac{25}{16}\right) - 2$$
$$= 2\left(x + \frac{5}{4}\right)^2 - \frac{25}{8} - 2$$
$$= 2\left(x + \frac{5}{4}\right)^2 - \frac{41}{8}$$
$$= 2\left[x - \left(-\frac{5}{4}\right)\right]^2 + \left(-\frac{41}{8}\right)$$

Vertex: $\left(-\frac{5}{4}, -\frac{41}{8}\right)$

Line of symmetry: $x = -\frac{5}{4}$

The coefficient of x^2 is 2, which is positive, so the graph opens up. This tells us that $-\frac{41}{8}$ is a minimum.

We plot a few points and draw the curve.

x	$f(x)$
$-\frac{5}{4}$	$-\frac{41}{8}$
-3	1
-2	-4
-1	-5
0	-2
1	5

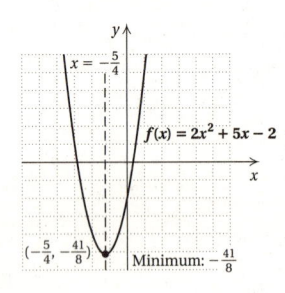

13. $f(x) = x^2 - 6x + 1$

The y-intercept is $(0, f(0))$. Since $f(0) = 0^2 - 6 \cdot 0 + 1 = 1$, the y-intercept is $(0, 1)$.

To find the x-intercepts, we solve $x^2 - 6x + 1 = 0$. Using the quadratic formula gives us $x = 3 \pm 2\sqrt{2}$.

Thus, the x-intercepts are $(3 - 2\sqrt{2}, 0)$ and $(3 + 2\sqrt{2}, 0)$, or approximately $(0.172, 0)$ and $(5.828, 0)$.

15. $f(x) = -x^2 + x + 20$

The y-intercept is $(0, f(0))$. Since $f(0) = -0^2 + 0 + 20 = 20$, the y-intercept is $(0, 20)$.

To find the x-intercepts, we solve $-x^2 + x + 20 = 0$. Factoring and using the principle of zero products gives us $x = -4$ or $x = 5$. Thus, the x-intercepts are $(-4, 0)$ and $(5, 0)$.

17. $f(x) = 4x^2 + 12x + 9$

The y-intercept is $(0, f(0))$. Since $f(0) = 4 \cdot 0^2 + 12 \cdot 0 + 9 = 9$, the y-intercept is $(0, 9)$.

To find the x-intercepts, we solve $4x^2 + 12x + 9 = 0$. Factoring and using the principle of zero products gives us $x = -\frac{3}{2}$. Thus, the x-intercept is $\left(-\frac{3}{2}, 0\right)$.

19. $f(x) = 4x^2 - x + 8$

The y-intercept is $(0, f(0))$. Since $f(0) = 4 \cdot 0^2 - 0 + 8 = 8$, the y-intercept is $(0, 8)$.

To find the x-intercepts, we solve $4x^2 - x + 8 = 0$. Using the quadratic formula gives us $x = \frac{1 \pm i\sqrt{127}}{8}$. Since there are no real-number solutions, there are no x-intercepts.

21. Discussion and Writing Exercise

23. a) $\quad D = kw$
$\quad 420 = k \cdot 28$
$\quad \dfrac{420}{28} = k$
$\quad 15 = k$

The equation of variation is $D = 15w$.

b) We substitute 42 for w and compute D.
$\quad D = 15w$
$\quad D = 15 \cdot 42$
$\quad D = 630$

630 mg would be recommended for a child who weighs 42 kg.

25. $\quad y = \dfrac{k}{x}$
$\quad 125 = \dfrac{k}{2}$
$\quad 250 = k \quad$ Variation constant
$\quad y = \dfrac{250}{x} \quad$ Equation of variation

27. $\quad y = kx$
$\quad 125 = k \cdot 2$
$\quad \dfrac{125}{2} = k \quad$ Variation constant
$\quad y = \dfrac{125}{2}x \quad$ Equation of variation

29. a) Minimum: -6.954

b) Maximum: 7.014

31. $f(x) = |x^2 - 1|$

We plot some points and draw the curve. Note that it will lie entirely on or above the x-axis since absolute value is never negative.

x	$f(x)$
-3	8
-2	3
-1	0
0	1
1	0
2	3
3	8

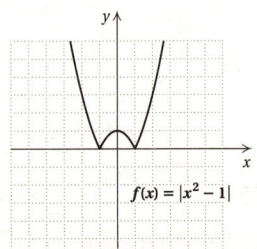

33. $f(x) = |x^2 - 3x - 4|$

We plot some points and draw the curve. Note that it will lie entirely on or above the x–axis since absolute value is never negative.

x	$f(x)$
-4	24
-3	14
-2	6
-1	0
0	4
1	6
2	6
3	4
4	0
5	6
6	14

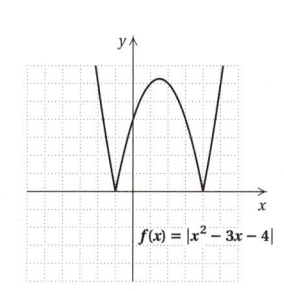

$f(x) = |x^2 - 3x - 4|$

35. The horizontal distance from $(-1, 0)$ to $(3, -5)$ is $|3-(-1)|$, or 4, so by symmetry the other x-intercept is $(3+4, 0)$, or $(7, 0)$. Substituting the three ordered pairs $(-1, 0)$, $(3, -5)$, and $(7, 0)$ in the equation $f(x) = ax^2 + bx + c$ yields a system of equations:

$$0 = a(-1)^2 + b(-1) + c,$$
$$-5 = a \cdot 3^2 + b \cdot 3 + c,$$
$$0 = a \cdot 7^2 + b \cdot 7 + c$$

or

$$0 = a - b + c,$$
$$-5 = 9a + 3b + c,$$
$$0 = 49a + 7b + c$$

The solution of this system of equations is $\left(\dfrac{5}{16}, -\dfrac{15}{8}, -\dfrac{35}{16}\right)$, so $f(x) = \dfrac{5}{16}x^2 - \dfrac{15}{8}x - \dfrac{35}{16}$.

37. $f(x) = \dfrac{x^2}{8} + \dfrac{x}{4} - \dfrac{3}{8}$

The x-coordinate of the vertex is $-b/2a$:

$$-\dfrac{b}{2a} = -\dfrac{\frac{1}{4}}{2 \cdot \frac{1}{8}} = -\dfrac{\frac{1}{4}}{\frac{1}{4}} = -1$$

The second coordinate is $f(-1)$:

$$f(-1) = \dfrac{(-1)^2}{8} + \dfrac{-1}{4} - \dfrac{3}{8}$$
$$= \dfrac{1}{8} - \dfrac{1}{4} - \dfrac{3}{8}$$
$$= -\dfrac{1}{2}$$

The vertex is $\left(-1, -\dfrac{1}{2}\right)$.

The line of symmetry is $x = -1$.

The coefficient of x^2 is $\dfrac{1}{8}$, which is positive, so the graph opens up. This tells us that $-\dfrac{1}{2}$ is a minimum.

We plot some points and draw the graph.

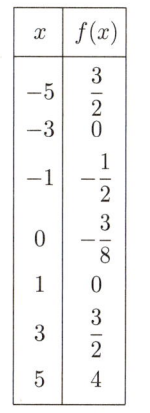

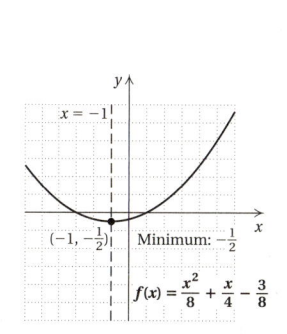

x	$f(x)$
-5	$\dfrac{3}{2}$
-3	0
-1	$-\dfrac{1}{2}$
0	$-\dfrac{3}{8}$
1	0
3	$\dfrac{3}{2}$
5	4

$f(x) = \dfrac{x^2}{8} + \dfrac{x}{4} - \dfrac{3}{8}$, Minimum: $-\dfrac{1}{2}$, $(-1, -\dfrac{1}{2})$, $x = -1$

39. Graph $y_1 = x^2 - 4x + 2$ and $y_2 = 2 + x$ and use INTERSECT to find the points of intersection. They are $(0, 2)$ and $(5, 7)$.

Exercise Set 11.7

1. Familiarize. Referring to the drawing in the text, we let l = the length of the atrium and w = the width. Then the perimeter of each floor is $2l + 2w$, and the area is $l \cdot w$.

Translate. Using the formula for perimeter we have:

$$2l + 2w = 720$$
$$2l = 720 - 2w$$
$$l = \dfrac{720 - 2w}{2}$$
$$l = 360 - w$$

Substituting $360 - w$ for l in the formula for area, we get a quadratic function.

$$A = lw = (360 - w)w = 360w - w^2 = -w^2 + 360w$$

Carry out. We complete the square in order to find the vertex of the quadratic function.

$$A = -w^2 + 360w$$
$$= -(w^2 - 360w)$$
$$= -(w^2 - 360w + 32,400 - 32,400)$$
$$= -(w^2 - 360w + 32,400) + (-1)(-32,400)$$
$$= -(w - 180)^2 + 32,400$$

The vertex is $(180, 32,400)$. The coefficient of w^2 is negative, so the graph of a function is a parabola that opens down. This tells us that the function has a maximum value and that value occurs when $w = 180$. When $w = 180$, $l = 360 - w = 360 - 180 = 180$.

Check. We could find the value of the function for some values of w less than 180 and for some values greater than 180, determining that the maximum value we found, 32,400, is larger than these function values. We could also use the graph of the function to check the maximum value. Our answer checks.

State. Floors with dimensions 180 ft by 180 ft will allow an atrium with maximum area.

3. **Familiarize**. Let x represent the height of the file and y represent the width. We make a drawing.

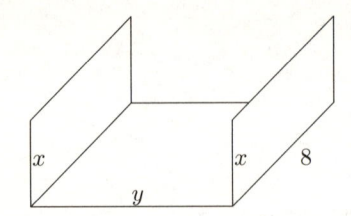

Translate. We have two equations.
$2x + y = 14$
$V = 8xy$

Solve the first equation for y.
$y = 14 - 2x$

Substitute for y in the second equation.
$V = 8x(14 - 2x)$
$V = -16x^2 + 112x$

Carry out. Completing the square, we get
$$V = -16\left(x - \frac{7}{2}\right)^2 + 196.$$
The maximum function value of 196 occurs when $x = \frac{7}{2}$. When $x = \frac{7}{2}$, $y = 14 - 2 \cdot \frac{7}{2} = 7$.

Check. Check a function value for x less than $\frac{7}{2}$ and for x greater than $\frac{7}{2}$.
$V(3) = -16 \cdot 3^2 + 112 \cdot 3 = 192$
$V(4) = -16 \cdot 4^2 + 112 \cdot 4 = 192$

Since 196 is greater than these numbers, it looks as though we have a maximum.

We could also use the graph of the function to check the maximum value.

State. The file should be $\frac{7}{2}$ in., or 3.5 in., tall.

5. **Familiarize and Translate**. We want to find the value of x for which $C(x) = 0.1x^2 - 0.7x + 2.425$ is a minimum.

Carry out. We complete the square.
$C(x) = 0.1(x^2 - 7x + 12.25) + 2.425 - 1.225$
$C(x) = 0.1(x - 3.5)^2 + 1.2$

The minimum function value of 1.2 occurs when $x = 3.5$.

Check. Check a function value for x less than 3.5 and for x greater than 3.5.
$C(3) = 0.1(3)^2 - 0.7(3) + 2.425 = 1.225$
$C(4) = 0.1(4)^2 - 0.7(4) + 2.425 = 1.225$

Since 1.2 is less than these numbers, it looks as though we have a minimum.

We could also use the graph of the function to check the minimum value.

State. The shop should build 3.5 hundred, or 350 bicycles.

7. **Familiarize**. We make a drawing and label it.

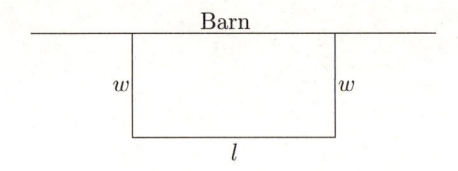

Translate. We have two equations.
$l + 2w = 40$
$A = lw$

Solve the first equation for l.
$l = 40 - 2w$

Substitute for l in the second equation.
$A = (40 - 2w)w = 40w - 2w^2$
$= -2w^2 + 40w$

Carry out. Completing the square, we get
$A = -2(w - 10)^2 + 200$

The maximum function value is 200. It occurs when $w = 10$. When $w = 10$, $l = 40 - 2 \cdot 10 = 20$.

Check. Check a function value for w less than 10 and for w greater than 10.
$A(9) = -2 \cdot 9^2 + 40 \cdot 9 = 198$
$A(11) = -2 \cdot 11^2 + 40 \cdot 11 = 198$

Since 200 is greater than these numbers, it looks as though we have a maximum. We could also use the graph of the function to check the maximum value.

State. The maximum area of 200 ft^2 will occur when the dimensions are 10 ft by 20 ft.

9. **Familiarize and Translate**. We are given the function $N(x) = -0.4x^2 + 9x + 11$.

Carry out. To find the value of x for which $N(x)$ is a maximum, we first find $-\frac{b}{2a}$:
$$-\frac{b}{2a} = -\frac{9}{2(-0.4)} = 11.25$$

Now we find the maximum value of the function $N(11.25)$:
$N(11.25) = -0.4(11.25)^2 + 9(11.25) + 11 = 61.625$

Check. We can go over the calculations again. We could also solve the problem again by completing the square. The answer checks.

State. Daily ticket sales will peak 11 days after the concert was announced. About 62 tickets will be sold that day.

11. Find the total profit:
$P(x) = R(x) - C(x)$
$P(x) = (1000x - x^2) - (3000 + 20x)$
$P(x) = -x^2 + 980x - 3000$

To find the maximum value of the total profit and the value of x at which it occurs we complete the square:

$$P(x) = -(x^2 - 980x) - 3000$$
$$= -(x^2 - 980x + 240{,}100 - 240{,}100) - 3000$$
$$= -(x^2 - 980x + 240{,}100) - (-240{,}100) - 3000$$
$$= -(x - 490)^2 + 237{,}100$$

The maximum profit of $237,100 occurs at $x = 490$.

13. **Familiarize**. Let x and y represent the numbers.

    Translate. The sum of the numbers is 22, so we have $x + y = 22$. Solving for y, we get $y = 22 - x$. The product of the numbers is xy. Substituting $22 - x$ for y in the product, we get a quadratic function:
    $$P = xy = x(22 - x) = 22x - x^2 = -x^2 + 22x$$

    Carry out. The coefficient of x^2 is negative, so the graph of the function is a parabola that opens down and a maximum exists. We complete the square in order to find the vertex of the quadratic function.
    $$P = -x^2 + 22x$$
    $$= -(x^2 - 22x)$$
    $$= -(x^2 - 22x + 121 - 121)$$
    $$= -(x^2 - 22x + 121) + (-1)(-121)$$
    $$= -(x - 11)^2 + 121$$

    The vertex is $(11, 121)$. This tells us that the maximum product is 121. The maximum occurs when $x = 11$. Note that when $x = 11$, $y = 22 - x = 22 - 11 = 11$, so the numbers that yield the maximum product are 11 and 11.

    Check. We could find the value of the function for some values of x less than 11 and for some greater than 11, determining that the maximum value we found is larger than these function values. We could also use the graph of the function to check the maximum value. Our answer checks.

    State. The maximum product is 121. The numbers 11 and 11 yield this product.

15. **Familiarize**. Let x and y represent the numbers.

    Translate. The difference of the numbers is 4, so we have $x - y = 4$. Solve for x, we get $x = y + 4$. The product of the numbers is xy. Substituting $y + 4$ for x in the product, we get a quadratic function:
    $$P = xy = (y + 4)y = y^2 + 4y$$

    Carry out. The coefficient of y^2 is positive, so the graph of the function opens up and a minimum exists. We complete the square in order to find the vertex of the quadratic function.
    $$P = y^2 + 4y$$
    $$= y^2 + 4y + 4 - 4$$
    $$= (y + 2)^2 - 4$$
    $$= [y - (-2)]^2 + (-4)$$

    The vertex is $(-2, 4)$. This tells us that the minimum product is -4. The minimum occurs when $y = -2$. Note that when $y = -2$, $x = y + 4 = -2 + 4 = 2$, so the numbers that yield the minimum product are 2 and -2.

    Check. We could find the value of the function for some values of y less than -2 and for some greater than -2, determining that the minimum value we found is smaller than these function values. We could also use the graph of the function to check the minimum value. Our answer checks.

    State. The minimum product is -4. The numbers 2 and -2 yield this product.

17. **Familiarize**. We let x and y represent the two numbers, and we let P represent their product.

    Translate. We have two equations.
    $$x + y = -12,$$
    $$P = xy$$

    Solve the first equation for y.
    $$y = -12 - x$$

    Substitute for y in the second equation.
    $$P = x(-12 - x) = -12x - x^2$$
    $$= -x^2 - 12x$$

    Carry out. Completing the square, we get
    $$P = -(x + 6)^2 + 36$$

    The maximum function value is 36. It occurs when $x = -6$. When $x = -6$, $y = -12 - (-6)$, or -6.

    Check. Check a function value for x less than -6 and for x greater than -6.
    $$P(-7) = -(-7)^2 - 12(-7) = 35$$
    $$P(-5) = -(-5)^2 - 12(-5) = 35$$

    Since 36 is greater than these numbers, it looks as though we have a maximum.

    We could also use the graph of the function to check the maximum value.

    State. The maximum product of 36 occurs for the numbers -6 and -6.

19. The data seem to fit a linear function $f(x) = mx + b$.

21. The data fall and then rise in a curved manner fitting a quadratic function $f(x) = ax^2 + bx + c$, $a > 0$.

23. The data fall, then rise, then fall again so they do not fit a linear or a quadratic function but might fit a polynomial function that is neither quadratic nor linear.

25. The data rise and then fall in a curved manner fitting a quadratic function $f(x) = ax^2 + bx + c$, $a < 0$.

27. We look for a function of the form $f(x) = ax^2 + bx + c$. Substituting the data points, we get
    $$4 = a(1)^2 + b(1) + c,$$
    $$-2 = a(-1)^2 + b(-1) + c,$$
    $$13 = a(2)^2 + b(2) + c,$$
    or
    $$4 = a + b + c,$$
    $$-2 = a - b + c,$$
    $$13 = 4a + 2b + c.$$

    Solving this system, we get
    $$a = 2, b = 3, \text{ and } c = -1.$$

Therefore the function we are looking for is
$$f(x) = 2x^2 + 3x - 1.$$

29. We look for a function of the form $f(x) = ax^2 + bx + c$. Substituting the data points, we get
$$0 = a(2)^2 + b(2) + c,$$
$$3 = a(4)^2 + b(4) + c,$$
$$-5 = a(12)^2 + b(12) + c,$$
or
$$0 = 4a + 2b + c,$$
$$3 = 16a + 4b + c,$$
$$-5 = 144a + 12b + c.$$
Solving this system, we get
$$a = -\frac{1}{4}, b = 3, c = -5.$$
Therefore the function we are looking for is
$$f(x) = -\frac{1}{4}x^2 + 3x - 5.$$

31. a) We look for a function of the form $A(s) = as^2 + bs + c$, where $A(s)$ represents the number of nighttime accidents (for every 200 million km) and s represents the travel speed (in km/h). We substitute the given values of s and $A(s)$.
$$400 = a(60)^2 + b(60) + c,$$
$$250 = a(80)^2 + b(80) + c,$$
$$250 = a(100)^2 + b(100) + c,$$
or
$$400 = 3600a + 60b + c,$$
$$250 = 6400a + 80b + c,$$
$$250 = 10,000a + 100b + c.$$
Solving the system of equations, we get
$$a = \frac{3}{16}, b = -\frac{135}{4}, c = 1750.$$
Thus, the function $A(s) = \frac{3}{16}s^2 - \frac{135}{4}s + 1750$ fits the data.

b) Find $A(50)$.
$$A(50) = \frac{3}{16}(50)^2 - \frac{135}{4}(50) + 1750 = 531.25$$
About 531 accidents per 200,000,000 km driven occur at 50 km/h.

33. *Familiarize*. Think of a coordinate system placed on the drawing in the text with the origin at the point where the arrow is released. Then three points on the arrow's parabolic path are $(0,0)$, $(63, 27)$, and $(126, 0)$. We look for a function of the form $h(d) = ad^2 + bd + c$, where $h(d)$ represents the arrow's height and d represents the distance the arrow has traveled horizontally.

Translate. We substitute the values given above for d and $h(d)$.

$$0 = a \cdot 0^2 + b \cdot 0 + c,$$
$$27 = a \cdot 63^2 + b \cdot 63 + c,$$
$$0 = a \cdot 126^2 + b \cdot 126 + c$$
or
$$0 = c,$$
$$27 = 3969a + 63b + c,$$
$$0 = 15,876a + 126b + c$$

Solve. Solving the system of equations, we get
$$a = -\frac{1}{147} \approx -0.0068, b = \frac{6}{7} \approx 0.8571, \text{ and } c = 0.$$

Check. Recheck the calculations.

State. The function $h(d) = -\frac{1}{147}d^2 + \frac{6}{7}d \approx -0.0068d^2 + 0.8571d$ expresses the arrow's height as a function of the distance it has traveled horizontally.

35. Discussion and Writing Exercise

37. $\sqrt[4]{5x^3y^5}\sqrt[4]{125x^2y^3} = \sqrt[4]{625x^5y^8} = \sqrt[4]{625x^4y^8 \cdot x} = \sqrt[4]{625x^4y^8} \cdot \sqrt[4]{x} = 5xy^2\sqrt[4]{x}$

39.
$$\sqrt{4x-4} = \sqrt{x+4} + 1$$
$$(\sqrt{4x-4})^2 = (\sqrt{x+4}+1)^2$$
$$4x - 4 = x + 4 + 2\sqrt{x+4} + 1$$
$$3x - 9 = 2\sqrt{x+4}$$
$$(3x-9)^2 = (2\sqrt{x+4})^2$$
$$9x^2 - 54x + 81 = 4(x+4)$$
$$9x^2 - 54x + 81 = 4x + 16$$
$$9x^2 - 58x + 65 = 0$$
$$(9x-13)(x-5) = 0$$
$$9x - 13 = 0 \quad \text{or} \quad x - 5 = 0$$
$$9x = 13 \quad \text{or} \quad x = 5$$
$$x = \frac{13}{9} \quad \text{or} \quad x = 5$$
Only 5 checks. It is the solution.

41. $-35 = \sqrt{2x+5}$

The equation has no solution, because the principle square root of a number is always nonnegative.

43. We will let x represent the number of years since 1997 and y represent the gross profit, in billions of dollars. Enter the data points (x, y) in STAT lists and use the quartic regression feature to find the equation $y = -0.290x^4 + 2.699x^3 - 8.306x^2 + 9.190x + 12.235$, where x is the number of years after 1997 and y is in billions of dollars.

Exercise Set 11.8

1. $(x-6)(x+2) > 0$

The solutions of $(x-6)(x+2) = 0$ are 6 and -2. They divide the real-number line into three intervals as shown:

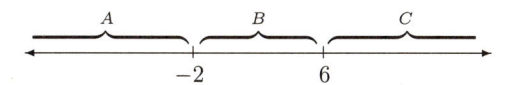

Exercise Set 11.8

```
        A           B           C
<---------|-----------|--------->
         -2           6
```

We try test numbers in each interval.

A: Test -3, $y = (-3-6)(-3+2) = 9 > 0$

B: Test 0, $y = (0-6)(0+2) = -12 < 0$

C: Test 7, $y = (7-6)(7+2) = 9 > 0$

The expression is positive for all values of x in intervals A and C. The solution set is $\{x|x < -2 \text{ or } x > 6\}$, or $(-\infty, -2) \cup (6, \infty)$.

From the graph in the text we see that the value of $(x-6)(x+2)$ is positive to the left of -2 and to the right of 6. This verifies the answer we found algebraically.

3. $4 - x^2 \geq 0$

$(2+x)(2-x) \geq 0$

The solutions of $(2+x)(2-x) = 0$ are -2 and 2. They divide the real-number line into three intervals as shown.

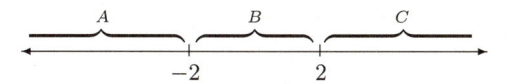

```
        A           B           C
<---------|-----------|--------->
         -2           2
```

We try test numbers in each interval.

A: Test -3, $y = 4 - (-3)^2 = -5 < 0$

B: Test 0, $y = 4 - 0^2 = 4 > 0$

C: Test 3, $y = 4 - 3^2 = -5 < 0$

The expression is positive for values of x in interval B. Since the inequality symbol is $\geq$ we also include the intercepts. The solution set is $\{x|-2 \leq x \leq 2\}$, or $[-2, 2]$.

From the graph we see that $4 - x^2 \geq 0$ at the intercepts and between them. This verifies the answer we found algebraically.

5. $3(x+1)(x-4) \leq 0$

The solutions of $3(x+1)(x-4) = 0$ are -1 and 4. They divide the real-number line into three intervals as shown:

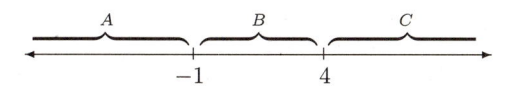

```
        A           B           C
<---------|-----------|--------->
         -1           4
```

We try test numbers in each interval.

A: Test -2, $y = 3(-2+1)(-2-4) = 18 > 0$

B: Test 0, $y = 3(0+1)(0-4) = -12 < 0$

C: Test 5, $y = 3(5+1)(5-4) = 18 > 0$

The expression is negative for all numbers in interval B. The inequality symbol is $\leq$, so we need to include the intercepts. The solution set is $\{x|-1 \leq x \leq 4\}$, or $[-1, 4]$.

7. $x^2 - x - 2 < 0$

$(x+1)(x-2) < 0$ Factoring

The solutions of $(x+1)(x-2) = 0$ are -1 and 2. They divide the real-number line into three intervals as shown:

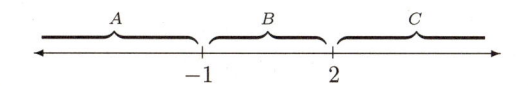

```
        A           B           C
<---------|-----------|--------->
         -1           2
```

We try test numbers in each interval.

A: Test -2, $y = (-2+1)(-2-2) = 4 > 0$

B: Test 0, $y = (0+1)(0-2) = -2 < 0$

C: Test 3, $y = (3+1)(3-2) = 4 > 0$

The expression is negative for all numbers in interval B. The solution set is $\{x|-1 < x < 2\}$, or $(-1, 2)$.

9. $x^2 - 2x + 1 \geq 0$

$(x-1)^2 \geq 0$

The solution of $(x-1)^2 = 0$ is 1. For all real-number values of x except 1, $(x-1)^2$ will be positive. Thus the solution set is $\{x|x \text{ is a real number}\}$, or $(-\infty, \infty)$.

11. $\quad\quad x^2 + 8 < 6x$

$x^2 - 6x + 8 < 0$

$(x-4)(x-2) < 0$

The solutions of $(x-4)(x-2) = 0$ are 4 and 2. They divide the real-number line into three intervals as shown:

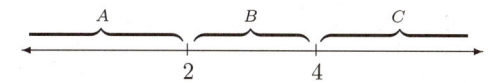

```
        A           B           C
<---------|-----------|--------->
          2           4
```

We try test numbers in each interval.

A: Test 0, $y = (0-4)(0-2) = 8 > 0$

B: Test 3, $y = (3-4)(3-2) = -1 < 0$

C: Test 5, $y = (5-4)(5-2) = 3 > 0$

The expression is negative for all numbers in interval B. The solution set is $\{x|2 < x < 4\}$, or $(2, 4)$.

13. $3x(x+2)(x-2) < 0$

The solutions of $3x(x+2)(x-2) = 0$ are 0, -2, and 2. They divide the real-number line into four intervals as shown:

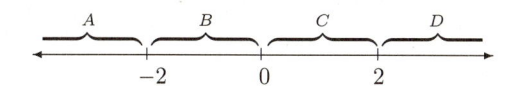

```
     A        B        C        D
<------|--------|--------|---------->
      -2        0        2
```

We try test numbers in each interval.

A: Test -3, $y = 3(-3)(-3+2)(-3-2) = -45 < 0$

B: Test -1, $y = 3(-1)(-1+2)(-1-2) = 9 > 0$

C: Test 1, $y = 3(1)(1+2)(1-2) = -9 < 0$

D: Test 3, $y = 3(3)(3+2)(3-2) = 45 > 0$

The expression is negative for all numbers in intervals A and C. The solution set is $\{x|x < -2 \text{ or } 0 < x < 2\}$, or $(-\infty, -2) \cup (0, 2)$.

15. $(x+9)(x-4)(x+1) > 0$

The solutions of $(x+9)(x-4)(x+1) = 0$ are -9, 4, and -1. They divide the real-number line into four intervals as shown:

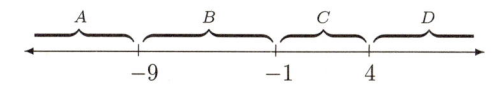

```
     A        B        C        D
<------|--------|--------|---------->
      -9       -1        4
```

We try test numbers in each interval.

281

A: Test -10, $y = (-10+9)(-10-4)(-10+1) = -126 < 0$
B: Test -2, $y = (-2+9)(-2-4)(-2+1) = 42 > 0$
C: Test 0, $y = (0+9)(0-4)(0+1) = -36 < 0$
D: Test 5, $y = (5+9)(5-4)(5+1) = 84 > 0$

The expression is positive for all values of x in intervals B and D. The solution set is $\{x | -9 < x < -1 \text{ or } x > 4\}$, or $(-9, -1) \cup (4, \infty)$.

17. $(x+3)(x+2)(x-1) < 0$

The solutions of $(x+3)(x+2)(x-1) = 0$ are -3, -2, and 1. They divide the real-number line into four intervals as shown:

```
      A    B      C        D
  <---|----|------|--------|--->
      -3   -2     1
```

We try test numbers in each interval.

A: Test -4, $y = (-4+3)(-4+2)(-4-1) = -10 < 0$
B: Test $-\dfrac{5}{2}$, $y = \left(-\dfrac{5}{2}+3\right)\left(-\dfrac{5}{2}+2\right)\left(-\dfrac{5}{2}-1\right) = \dfrac{7}{8} > 0$
C: Test 0, $y = (0+3)(0+2)(0-1) = -6 < 0$
D: Test 2, $y = (2+3)(2+2)(2-1) = 20 > 0$

The expression is negative for all numbers in intervals A and C. The solution set is $\{x | x < -3 \text{ or } -2 < x < 1\}$, or $(-\infty, -3) \cup (-2, 1)$.

19. $\dfrac{1}{x-6} < 0$

We write the related equation by changing the $<$ symbol to $=$:

$$\dfrac{1}{x-6} = 0$$

We solve the related equation.

$$(x-6) \cdot \dfrac{1}{x-6} = (x-6) \cdot 0$$
$$1 = 0$$

We get a false equation, so the related equation has no solution.

Next we find the numbers for which the rational expression is undefined by setting the denominator equal to 0 and solving:

$$x - 6 = 0$$
$$x = 6$$

We use 6 to divide the number line into two intervals as shown:

```
         A              B
  <------|--------------|------>
         6
```

We try test numbers in each interval.

A: Test 0, $\dfrac{1}{x-6} < 0$

$\dfrac{1}{0-6}$? 0

$-\dfrac{1}{6}$ | TRUE

The number 0 is a solution of the inequality, so the interval A is part of the solution set.

B: Test 7, $\dfrac{1}{x-6} < 0$

$\dfrac{1}{7-6}$? 0

1 | FALSE

The number 7 is not a solution of the inequality, so the interval B is not part of the solution set. The solution set is $\{x | x < 6\}$, or $(-\infty, 6)$.

21. $\dfrac{x+1}{x-3} > 0$

Solve the related equation.

$$\dfrac{x+1}{x-3} = 0$$
$$x + 1 = 0$$
$$x = -1$$

Find the numbers for which the rational expression is undefined.

$$x - 3 = 0$$
$$x = 3$$

Use the numbers -1 and 3 to divide the number line into intervals as shown:

```
        A         B          C
  <-----|---------|----------|---->
        -1        3
```

Try test numbers in each interval.

A: Test -2, $\dfrac{x+1}{x-3} > 0$

$\dfrac{-2+1}{-2-3}$? 0

$\dfrac{-1}{-5}$

$\dfrac{1}{5}$ | TRUE

The number -2 is a solution of the inequality, so the interval A is part of the solution set.

B: Test 0, $\dfrac{x+1}{x-3} > 0$

$\dfrac{0+1}{0-3}$? 0

$-\dfrac{1}{3}$ | FALSE

The number 0 is not a solution of the inequality, so the interval B is not part of the solution set.

C: Test 4, $\dfrac{x+1}{x-3} > 0$

$\dfrac{4+1}{4-3}$? 0

$\dfrac{5}{1}$

5 | TRUE

Exercise Set 11.8

The number 4 is a solution of the inequality, so the interval C is part of the solution set. The solution set is
$\{x | x < -1 \text{ or } x > 3\}$, or $(-\infty, -1) \cup (3, \infty)$.

23. $\dfrac{3x+2}{x-3} \le 0$

Solve the related equation.
$$\dfrac{3x+2}{x-3} = 0$$
$$3x + 2 = 0$$
$$3x = -2$$
$$x = -\dfrac{2}{3}$$

Find the numbers for which the rational expression is undefined.
$$x - 3 = 0$$
$$x = 3$$

Use the numbers $-\dfrac{2}{3}$ and 3 to divide the number line into intervals as shown:

```
      A         B         C
<----------|---------|---------->
          -2/3       3
```

Try test numbers in each interval.

A: Test -1,
$$\dfrac{3x+2}{x-3} \le 0$$
$$\dfrac{3(-1)+2}{-1-3} \;?\; 0$$
$$\dfrac{-1}{-4}$$
$$\dfrac{1}{4} \quad | \quad \text{FALSE}$$

The number -1 is not a solution of the inequality, so the interval A is not part of the solution set.

B: Test 0,
$$\dfrac{3x+2}{x-3} \le 0$$
$$\dfrac{3 \cdot 0 + 2}{0 - 3} \;?\; 0$$
$$\dfrac{2}{-3}$$
$$-\dfrac{2}{3} \quad | \quad \text{TRUE}$$

The number 0 is a solution of the inequality, so the interval B is part of the solution set.

C: Test 4,
$$\dfrac{3x+2}{x-3} \le 0$$
$$\dfrac{3 \cdot 4 + 2}{4 - 3} \;?\; 0$$
$$14 \quad | \quad \text{FALSE}$$

The number 4 is not a solution of the inequality, so the interval C is not part of the solution set. The solution set includes the interval B. The number $-\dfrac{2}{3}$ is also included since the inequality symbol is $\le$ and $-\dfrac{2}{3}$ is the solution of the related equation. The number 3 is not included because the rational expression is undefined for 3. The solution set is $\left\{x \middle| -\dfrac{2}{3} \le x < 3\right\}$, or $\left[-\dfrac{2}{3}, 3\right)$.

25. $\dfrac{x-1}{x-2} > 3$

Solve the related equation.
$$\dfrac{x-1}{x-2} = 3$$
$$x - 1 = 3(x - 2)$$
$$x - 1 = 3x - 6$$
$$5 = 2x$$
$$\dfrac{5}{2} = x$$

Find the numbers for which the rational expression is undefined.
$$x - 2 = 0$$
$$x = 2$$

Use the numbers $\dfrac{5}{2}$ and 2 to divide the number line into intervals as shown:

```
      A         B         C
<----------|---------|---------->
           2        5/2
```

Try test numbers in each interval.

A: Test 0,
$$\dfrac{x-1}{x-2} > 3$$
$$\dfrac{0-1}{0-2} \;?\; 3$$
$$\dfrac{1}{2} \quad | \quad \text{FALSE}$$

The number 0 is not a solution of the inequality, so the interval A is not part of the solution set.

B: Test $\dfrac{9}{4}$,
$$\dfrac{x-1}{x-2} > 3$$
$$\dfrac{\frac{9}{4} - 1}{\frac{9}{4} - 2} \;?\; 3$$
$$\dfrac{\frac{5}{4}}{\frac{1}{4}}$$
$$5 \quad | \quad \text{TRUE}$$

The number $\dfrac{9}{4}$ is a solution of the inequality, so the interval B is part of the solution set.

C: Test 3,
$$\dfrac{x-1}{x-2} > 3$$
$$\dfrac{3-1}{3-2} \;?\; 3$$
$$2 \quad | \quad \text{FALSE}$$

The number 3 is not a solution of the inequality, so the interval C is not part of the solution set. The solution set is $\left\{x \mid 2 < x < \dfrac{5}{2}\right\}$, or $\left(2, \dfrac{5}{2}\right)$.

27. $\dfrac{(x-2)(x+1)}{x-5} < 0$

Solve the related equation.
$$\dfrac{(x-2)(x+1)}{x-5} = 0$$
$$(x-2)(x+1) = 0$$
$$x = 2 \text{ or } x = -1$$

Find the numbers for which the rational expression is undefined.
$$x - 5 = 0$$
$$x = 5$$

Use the numbers 2, -1, and 5 to divide the number line into intervals as shown:

$$\begin{array}{c|c|c|c}
A & B & C & D \\
\hline
& -1 & 2 & 5
\end{array}$$

Try test numbers in each interval.

A: Test -2,
$$\dfrac{(x-2)(x+1)}{x-5} < 0$$
$$\dfrac{(-2-2)(-2+1)}{-2-5} \;?\; 0$$
$$\dfrac{-4(-1)}{-7}$$
$$-\dfrac{4}{7} \quad \text{TRUE}$$

Interval A is part of the solution set.

B: Test 0,
$$\dfrac{(x-2)(x+1)}{x-5} < 0$$
$$\dfrac{(0-2)(0+1)}{0-5} \;?\; 0$$
$$\dfrac{-2 \cdot 1}{-5}$$
$$\dfrac{2}{5} \quad \text{FALSE}$$

Interval B is not part of the solution set.

C: Test 3,
$$\dfrac{(x-2)(x+1)}{x-5} < 0$$
$$\dfrac{(3-2)(3+1)}{3-5} \;?\; 0$$
$$\dfrac{1 \cdot 4}{-2}$$
$$-2 \quad \text{TRUE}$$

Interval C is part of the solution set.

D: Test 6,
$$\dfrac{(x-2)(x+1)}{x-5} < 0$$
$$\dfrac{(6-2)(6+1)}{6-5} \;?\; 0$$
$$\dfrac{4 \cdot 7}{1}$$
$$28 \quad \text{FALSE}$$

Interval D is not part of the solution set.

The solution set is $\{x \mid x < -1 \text{ or } 2 < x < 5\}$, or $(-\infty, -1) \cup (2, 5)$.

29. $\dfrac{x+3}{x} \le 0$

Solve the related equation.
$$\dfrac{x+3}{x} = 0$$
$$x + 3 = 0$$
$$x = -3$$

Find the numbers for which the rational expression is undefined.
$$x = 0$$

Use the numbers -3 and 0 to divide the number line into intervals as shown:

$$\begin{array}{c|c|c}
A & B & C \\
\hline
& -3 & 0
\end{array}$$

Try test numbers in each interval.

A: Test -4,
$$\dfrac{x+3}{x} \le 0$$
$$\dfrac{-4+3}{-4} \;?\; 0$$
$$\dfrac{1}{4} \quad \text{FALSE}$$

Interval A is not part of the solution set.

B: Test -1,
$$\dfrac{x+3}{x} \le 0$$
$$\dfrac{-1+3}{-1} \;?\; 0$$
$$-2 \quad \text{TRUE}$$

Interval B is part of the solution set.

C: Test 1,
$$\dfrac{x+3}{x} \le 0$$
$$\dfrac{1+3}{1} \;?\; 0$$
$$4 \quad \text{FALSE}$$

Interval C is not part of the solution set.

The solution set includes the interval B. The number -3 is also included since the inequality symbol is $\le$ and -3 is a solution of the related equation. The number 0 is not included because the rational expression is undefined for 0. The solution set is $\{x \mid -3 \le x < 0\}$, or $[-3, 0)$.

31. $\dfrac{x}{x-1} > 2$

Solve the related equation.
$$\dfrac{x}{x-1} = 2$$
$$x = 2x - 2$$
$$2 = x$$

Find the numbers for which the rational expression is undefined.
$$x - 1 = 0$$
$$x = 1$$

Use the numbers 1 and 2 to divide the number line into intervals as shown:

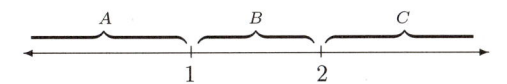

$$\xleftarrow{\qquad}\underbrace{\qquad}_{A}\underset{1}{|}\underbrace{\qquad}_{B}\underset{2}{|}\underbrace{\qquad}_{C}\xrightarrow{\qquad}$$

Try test numbers in each interval.

A: Test 0,
$$\dfrac{x}{x-1} > 2$$
$$\dfrac{0}{0-1} \; ? \; 2$$
$$0 \; \bigg| \; \text{FALSE}$$

Interval A is not part of the solution set.

B: Test $\dfrac{3}{2}$,
$$\dfrac{x}{x-1} > 2$$
$$\dfrac{\tfrac{3}{2}}{\tfrac{3}{2} - 1} \; ? \; 2$$
$$\dfrac{\tfrac{3}{2}}{\tfrac{1}{2}}$$
$$3 \; \bigg| \; \text{TRUE}$$

Interval B is part of the solution set.

C: Test 3,
$$\dfrac{x}{x-1} > 2$$
$$\dfrac{3}{3-1} \; ? \; 2$$
$$\dfrac{3}{2} \; \bigg| \; \text{FALSE}$$

Interval C is not part of the solution set.

The solution set is $\{x | 1 < x < 2\}$, or $(1, 2)$.

33. $\dfrac{x-1}{(x-3)(x+4)} < 0$

Solve the related equation.
$$\dfrac{x-1}{(x-3)(x+4)} = 0$$
$$x - 1 = 0$$
$$x = 1$$

Find the numbers for which the rational expression is undefined.
$$(x-3)(x+4) = 0$$
$$x = 3 \text{ or } x = -4$$

Use the numbers 1, 3, and -4 to divide the number line into intervals as shown:

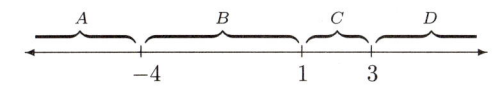

$$\xleftarrow{\qquad}\underbrace{\qquad}_{A}\underset{-4}{|}\underbrace{\qquad}_{B}\underset{1}{|}\underbrace{\qquad}_{C}\underset{3}{|}\underbrace{\qquad}_{D}\xrightarrow{\qquad}$$

Try test numbers in each interval.

A: Test -5,
$$\dfrac{x-1}{(x-3)(x+4)} < 0$$
$$\dfrac{-5-1}{(-5-3)(-5+4)} \; ? \; 0$$
$$\dfrac{-6}{-8(-1)}$$
$$-\dfrac{3}{4} \; \bigg| \; \text{TRUE}$$

Interval A is part of the solution set.

B: Test 0,
$$\dfrac{x-1}{(x-3)(x+4)} < 0$$
$$\dfrac{0-1}{(0-3)(0+4)} \; ? \; 0$$
$$\dfrac{-1}{-3 \cdot 4}$$
$$\dfrac{1}{12} \; \bigg| \; \text{FALSE}$$

Interval B is not part of the solution set.

C: Test 2,
$$\dfrac{x-1}{(x-3)(x+4)} < 0$$
$$\dfrac{2-1}{(2-3)(2+4)} \; ? \; 0$$
$$\dfrac{1}{-1 \cdot 6}$$
$$-\dfrac{1}{6} \; \bigg| \; \text{TRUE}$$

Interval C is part of the solution set.

D: Test 4,
$$\dfrac{x-1}{(x-3)(x+4)} < 0$$
$$\dfrac{4-1}{(4-3)(4+4)} \; ? \; 0$$
$$\dfrac{3}{1 \cdot 8}$$
$$\dfrac{3}{8} \; \bigg| \; \text{FALSE}$$

Interval D is not part of the solution set.

The solution set is $\{x | x < -4 \text{ or } 1 < x < 3\}$, or $(-\infty, -4) \cup (1, 3)$.

35. $3 < \dfrac{1}{x}$

Solve the related equation.
$$3 = \dfrac{1}{x}$$
$$x = \dfrac{1}{3}$$

Find the numbers for which the rational expression is undefined.
$$x = 0$$

Use the numbers 0 and $\dfrac{1}{3}$ to divide the number line into intervals as shown:

```
      A         B         C
<----------|---------|---------->
           0        1/3
```

Try test numbers in each interval.

A: Test -1,
$$3 < \dfrac{1}{x}$$
$$3 \;?\; \dfrac{1}{-1}$$
$$\mid \;-1\quad \text{FALSE}$$

Interval A is not part of the solution set.

B: Test $\dfrac{1}{6}$,
$$3 < \dfrac{1}{x}$$
$$3 \;?\; \dfrac{1}{\frac{1}{6}}$$
$$\mid \;6\quad \text{TRUE}$$

Interval B is part of the solution set.

C: Test 1,
$$3 < \dfrac{1}{x}$$
$$3 \;?\; \dfrac{1}{1}$$
$$\mid \;1\quad \text{FALSE}$$

Interval C is not part of the solution set.

The solution set is $\left\{x \,\middle|\, 0 < x < \dfrac{1}{3}\right\}$, or $\left(0, \dfrac{1}{3}\right)$.

37. $\dfrac{(x-1)(x+2)}{(x+3)(x-4)} > 0$

Solve the related equation.
$$\dfrac{(x-1)(x+2)}{(x+3)(x-4)} = 0$$
$$(x-1)(x+2) = 0$$
$$x = 1 \text{ or } x = -2$$

Find the numbers for which the rational expression is undefined.

$$(x+3)(x-4) = 0$$
$$x = -3 \text{ or } x = 4$$

Use the numbers 1, -2, -3, and 4 to divide the number line into intervals as shown:

```
    A       B       C       D       E
<------|-------|-------|-------|------>
      -3     -2       1       4
```

Try test numbers in each interval.

A: Test -4,
$$\dfrac{(x-1)(x+2)}{(x+3)(x-4)} > 0$$
$$\dfrac{(-4-1)(-4+2)}{(-4+3)(-4-4)} \;?\; 0$$
$$\dfrac{-5(-2)}{-1(-8)}$$
$$\dfrac{5}{4} \quad \text{TRUE}$$

Interval A is part of the solution set.

B: Test $-\dfrac{5}{2}$,
$$\dfrac{(x-1)(x+2)}{(x+3)(x-4)} > 0$$
$$\dfrac{\left(-\tfrac{5}{2}-1\right)\left(-\tfrac{5}{2}+2\right)}{\left(-\tfrac{5}{2}+3\right)\left(-\tfrac{5}{2}-4\right)} \;?\; 0$$
$$\dfrac{-\tfrac{7}{2}\left(-\tfrac{1}{2}\right)}{\tfrac{1}{2}\left(-\tfrac{13}{2}\right)}$$
$$-\dfrac{7}{13} \quad \text{FALSE}$$

Interval B is not part of the solution set.

C: Test 1,
$$\dfrac{(x-1)(x+2)}{(x+3)(x-4)} > 0$$
$$\dfrac{(0-1)(0+2)}{(0+3)(0-4)} \;?\; 0$$
$$\dfrac{-1 \cdot 2}{3(-4)}$$
$$\dfrac{1}{6} \quad \text{TRUE}$$

Interval C is part of the solution set.

D: Test 2,
$$\dfrac{(x-1)(x+2)}{(x+3)(x-4)} > 0$$
$$\dfrac{(2-1)(2+2)}{(2+3)(2-4)} \;?\; 0$$
$$\dfrac{1 \cdot 4}{5(-2)}$$
$$-\dfrac{2}{5} \quad \text{FALSE}$$

Interval D is not part of the solution set.

Exercise Set 11.8

E: Test 5,
$$\frac{(x-1)(x+2)}{(x+3)(x-4)} > 0$$
$$\begin{array}{c|c} \dfrac{(5-1)(5+2)}{(5+3)(5-4)} \;?\; 0 \\ \dfrac{4\cdot 7}{8\cdot 1} \\ \dfrac{7}{2} & \text{TRUE} \end{array}$$

Interval E is part of the solution set.

The solution set is $\{x | x < -3 \text{ or } -2 < x < 1 \text{ or } x > 4\}$, or $(-\infty, -3) \cup (-2, 1) \cup (4, \infty)$.

39. Discussion and Writing Exercise

41. $\sqrt[3]{\dfrac{125}{27}} = \dfrac{\sqrt[3]{125}}{\sqrt[3]{27}} = \dfrac{5}{3}$

43. $\sqrt{\dfrac{16a^3}{b^4}} = \dfrac{\sqrt{16a^3}}{\sqrt{b^4}} = \dfrac{\sqrt{16a^2 \cdot a}}{\sqrt{b^4}} = \dfrac{\sqrt{16a^2}\sqrt{a}}{\sqrt{b^4}} = \dfrac{4a}{b^2}\sqrt{a}$

45. $3\sqrt{8} - 5\sqrt{2} = 3\sqrt{4 \cdot 2} - 5\sqrt{2}$
$= 3\sqrt{4}\sqrt{2} - 5\sqrt{2}$
$= 3 \cdot 2\sqrt{2} - 5\sqrt{2}$
$= 6\sqrt{2} - 5\sqrt{2}$
$= \sqrt{2}$

47. $5\sqrt[3]{16a^4} + 7\sqrt[3]{2a} = 5\sqrt[3]{8a^3 \cdot 2a} + 7\sqrt[3]{2a}$
$= 5\sqrt[3]{8a^3}\sqrt[3]{2a} + 7\sqrt[3]{2a}$
$= 5 \cdot 2a\sqrt[3]{2a} + 7\sqrt[3]{2a}$
$= 10a\sqrt[3]{2a} + 7\sqrt[3]{2a}$
$= (10a + 7)\sqrt[3]{2a}$

49. For Exercise 11, graph $y_1 = x^2 + 8$ and $y_2 = 6x$. Then determine the values of x for which the graph of y_1 lies below the graph of y_2.

For Exercise 22, graph $y_1 = \dfrac{x-2}{x+5}$ and $y_2 = 0$. Then determine the values of x for which the graph of y_1 lies below the graph of y_2. Since the graph of $y_2 = 0$ is the x-axis, this could also be done by graphing $y_1 = \dfrac{x-2}{x+5}$ and determining the values of x for which the graph of y_1 lies below the x-axis.

For Exercise 25, graph $y_1 = \dfrac{x-1}{x-2}$ and $y_2 = 3$. Then determine the values of x for which the graph of y_1 lies above the graph of y_2.

51. $x^2 - 2x \leq 2$
$x^2 - 2x - 2 \leq 0$

The solutions of $x^2 - 2x - 2 = 0$ are found using the quadratic formula. They are $1 \pm \sqrt{3}$, or about 2.7 and -0.7. These numbers divide the number line into three intervals as shown:

$$\begin{array}{ccc} A & B & C \\ \hline & 1-\sqrt{3} & 1+\sqrt{3} \end{array}$$

We try test numbers in each interval.

A: Test -1, $y = (-1)^2 - 2(-1) - 2 = 1 > 0$
B: Test 0, $y = 0^2 - 2 \cdot 0 - 2 = -2 < 0$
C: Test 3, $y = 3^2 - 2 \cdot 3 - 2 = 1 > 0$

The expression is negative for all values of x in interval B. The inequality symbol is $\leq$, so we must also include the intercepts. The solution set is $\{x | 1 - \sqrt{3} \leq x \leq 1 + \sqrt{3}\}$, or $[1 - \sqrt{3}, 1 + \sqrt{3}]$.

53. $x^4 + 2x^2 > 0$
$x^2(x^2 + 2) > 0$

$x^2 > 0$ for all $x \neq 0$, and $x^2 + 2 > 0$ for all values of x. Then $x^2(x^2 + 2) > 0$ for all $x \neq 0$. The solution set is $\{x | x \neq 0\}$, or the set of all real numbers except 0, or $(-\infty, 0) \cup (0, \infty)$.

55. $\left|\dfrac{x+2}{x-1}\right| < 3$
$-3 < \dfrac{x+2}{x-1} < 3$

We rewrite the inequality using "and."

$-3 < \dfrac{x+2}{x-1}$ and $\dfrac{x+2}{x-1} < 3$

We will solve each inequality and then find the intersection of their solution sets.

Solve: $-3 < \dfrac{x+2}{x-1}$

Solve the related equation.
$-3 = \dfrac{x+2}{x-1}$
$-3x + 3 = x + 2$
$1 = 4x$
$\dfrac{1}{4} = x$

Find the numbers for which the rational expression is undefined.
$x - 1 = 0$
$x = 1$

Use the numbers $\dfrac{1}{4}$ and 1 to divide the number line into intervals as shown:

$$\begin{array}{ccc} A & B & C \\ \hline & \frac{1}{4} & 1 \end{array}$$

Try test numbers in each interval.

A: Test 0,
$$\begin{array}{c|c} -3 < \dfrac{x+2}{x-1} \\ -3 \;?\; \dfrac{0+2}{0-1} \\ -2 & \text{TRUE} \end{array}$$

Interval A is part of the solution set.

B: Test $\frac{1}{2}$,

$$\begin{array}{c|c} -3 < \dfrac{x+2}{x-1} \\ \hline -3 \;?\; \dfrac{\frac{1}{2}+2}{\frac{1}{2}-1} \\ \dfrac{\frac{5}{2}}{-\frac{1}{2}} \\ -5 \quad \text{FALSE} \end{array}$$

Interval B is not part of the solution set.

C: Test 2,

$$\begin{array}{c|c} -3 < \dfrac{x+2}{x-1} \\ \hline -3 \;?\; \dfrac{2+2}{2-1} \\ 4 \quad \text{TRUE} \end{array}$$

Interval C is part of the solution set.

The solution set of $-3 < \dfrac{x+2}{x-1}$ is $\left\{x \big| x < \dfrac{1}{4} \text{ or } x > 1\right\}$, or $\left(-\infty, \dfrac{1}{4}\right) \cup (1, \infty)$.

Solve: $\dfrac{x+2}{x-1} > 3$

Solve the related equation.

$$\dfrac{x+2}{x-1} = 3$$
$$x + 2 = 3x - 3$$
$$5 = 2x$$
$$\dfrac{5}{2} = x$$

From our work above we know that the rational expression is undefined for 1.

Use the numbers $\dfrac{5}{2}$ and 1 to divide the number line into intervals as shown:

```
         A       B       C
    ─────────┼───────┼─────────
             1      5/2
```

Try test numbers in each interval.

A: Test 0,

$$\begin{array}{c|c} \dfrac{x+2}{x-1} < 3 \\ \hline \dfrac{0+2}{0-1} \;?\; 3 \\ -2 \quad \text{TRUE} \end{array}$$

Interval A is part of the solution set.

B: Test 2,

$$\begin{array}{c|c} \dfrac{x+2}{x-1} < 3 \\ \hline \dfrac{2+2}{2-1} \;?\; 3 \\ 4 \quad \text{FALSE} \end{array}$$

Interval B is not part of the solution set.

C: Test 3,

$$\begin{array}{c|c} \dfrac{x+2}{x-1} < 3 \\ \hline \dfrac{3+2}{3-1} \;?\; 3 \\ \dfrac{5}{2} \quad \text{TRUE} \end{array}$$

Interval C is part of the solution set.

The solution set of $\dfrac{x+2}{x-1} < 3$ is $\left\{x \big| x < 1 \text{ or } x > \dfrac{5}{2}\right\}$, or $(-\infty, 1) \cup \left(\dfrac{5}{2}, \infty\right)$.

The solution set of the original inequality is

$\left\{x \big| x < \dfrac{1}{4} \text{ or } x > 1\right\} \cap \left\{x \big| x < 1 \text{ or } x > \dfrac{5}{2}\right\}$, or

$\left\{x \big| x < \dfrac{1}{4} \text{ or } x > \dfrac{5}{2}\right\}$, or $\left(-\infty, \dfrac{1}{4}\right) \cup \left(\dfrac{5}{2}, \infty\right)$.

57. a) Solve: $-16t^2 + 32t + 1920 > 1920$

$$-16t^2 + 32t > 0$$
$$t^2 - 2t < 0$$
$$t(t-2) < 0$$

The solutions of $t(t-2) = 0$ are 0 and 2. They divide the number line into three intervals as shown:

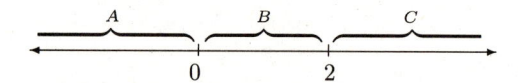

```
         A       B       C
    ─────────┼───────┼─────────
             0       2
```

Try test numbers in each interval.

A: Test -1, $y = -1(-1-2) = 3 > 0$

B: Test 1, $y = 1(1-2) = -1 < 0$

C: Test 3, $y = 3(3-2) = 3 > 0$

The expression is negative for all values of t in interval B. The solution set is $\{t | 0 < t < 2\}$, or $(0, 2)$.

b) Solve: $-16t^2 + 32t + 1920 < 640$

$$-16t^2 + 32t + 1280 < 0$$
$$t^2 - 2t - 80 > 0$$
$$(t - 10)(t + 8) > 0$$

The solutions of $(t-10)(t+8) = 0$ are 10 and -8. They divide the number line into three intervals as shown:

```
         A       B       C
    ─────────┼───────┼─────────
            -8      10
```

Try test numbers in each interval.

A: Test -10, $y = (-10 - 10)(-10 + 8) = 40 > 0$

B: Test 0, $y = (0 - 10)(0 + 8) = -80 < 0$

C: Test 20, $y = (20 - 10)(20 + 8) = 80 = 280 > 0$

The expression is positive for all values of t in intervals A and C. However, since negative values of t have no meaning in this problem, we disregard interval A. Thus, the solution set is $\{t | t > 10\}$, or $(10, \infty)$.

Chapter 12
Exponential and Logarithmic Functions

Exercise Set 12.1

1. Graph: $f(x) = 2^x$

We compute some function values and keep the results in a table.

$f(0) = 2^0 = 1$
$f(1) = 2^1 = 2$
$f(2) = 2^2 = 4$
$f(-1) = 2^{-1} = \frac{1}{2^1} = \frac{1}{2}$
$f(-2) = 2^{-2} = \frac{1}{2^2} = \frac{1}{4}$

x	$f(x)$
0	1
1	2
2	4
3	8
-1	$\frac{1}{2}$
-2	$\frac{1}{4}$
-3	$\frac{1}{8}$

Next we plot these points and connect them with a smooth curve.

f(x) = 2^x

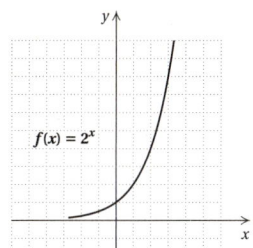

3. Graph: $f(x) = 5^x$

We compute some function values and keep the results in a table.

$f(0) = 5^0 = 1$
$f(1) = 5^1 = 5$
$f(2) = 5^2 = 25$
$f(-1) = 5^{-1} = \frac{1}{5^1} = \frac{1}{5}$
$f(-2) = 5^{-2} = \frac{1}{5^2} = \frac{1}{25}$

x	$f(x)$
0	1
1	5
2	25
-1	$\frac{1}{5}$
-2	$\frac{1}{25}$

Next we plot these points and connect them with a smooth curve.

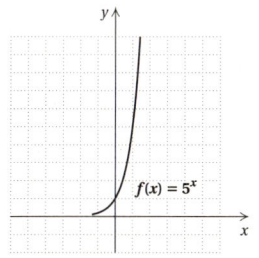

f(x) = 5^x

5. Graph: $f(x) = 2^{x+1}$

We compute some function values and keep the results in a table.

$f(0) = 2^{0+1} = 2^1 = 2$
$f(-1) = 2^{-1+1} = 2^0 = 1$
$f(-2) = 2^{-2+1} = 2^{-1} = \frac{1}{2^1} = \frac{1}{2}$
$f(-3) = 2^{-3+1} = 2^{-2} = \frac{1}{2^2} = \frac{1}{4}$
$f(1) = 2^{1+1} = 2^2 = 4$
$f(2) = 2^{2+1} = 2^3 = 8$

x	$f(x)$
0	2
-1	1
-2	$\frac{1}{2}$
-3	$\frac{1}{4}$
1	4
2	8

Next we plot these points and connect them with a smooth curve.

f(x) = 2^{x+1}

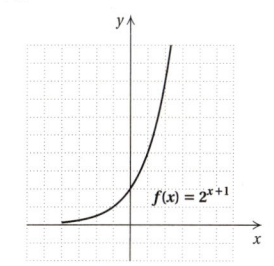

7. Graph: $f(x) = 3^{x-2}$

We compute some function values and keep the results in a table.

$f(0) = 3^{0-2} = 3^{-2} = \frac{1}{3^2} = \frac{1}{9}$
$f(1) = 3^{1-2} = 3^{-1} = \frac{1}{3^1} = \frac{1}{3}$
$f(2) = 3^{2-2} = 3^0 = 1$
$f(3) = 3^{3-2} = 3^1 = 3$
$f(4) = 3^{4-2} = 3^2 = 9$
$f(-1) = 3^{-1-2} = 3^{-3} = \frac{1}{3^3} = \frac{1}{27}$
$f(-2) = 3^{-2-2} = 3^{-4} = \frac{1}{3^4} = \frac{1}{81}$

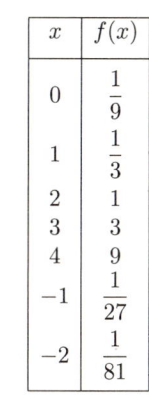

x	$f(x)$
0	$\frac{1}{9}$
1	$\frac{1}{3}$
2	1
3	3
4	9
-1	$\frac{1}{27}$
-2	$\frac{1}{81}$

Next we plot these points and connect them with a smooth curve.

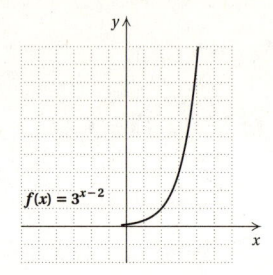

290   Chapter 12: Exponential and Logarithmic Functions

[Graph of $f(x) = 3^{x-2}$]

9. Graph: $f(x) = 2^x - 3$

We construct a table of values. Then we plot the points and connect them with a smooth curve.

$f(0) = 2^0 - 3 = 1 - 3 = -2$
$f(1) = 2^1 - 3 = 2 - 3 = -1$
$f(2) = 2^2 - 3 = 4 - 3 = 1$
$f(3) = 2^3 - 3 = 8 - 3 = 5$
$f(-1) = 2^{-1} - 3 = \dfrac{1}{2} - 3 = -\dfrac{5}{2}$
$f(-2) = 2^{-2} - 3 = \dfrac{1}{4} - 3 = -\dfrac{11}{4}$

x	$f(x)$
0	-2
1	-1
2	1
3	5
-1	$-\dfrac{5}{2}$
-2	$-\dfrac{11}{4}$

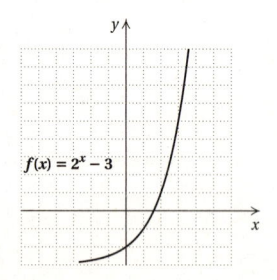

[Graph of $f(x) = 2^x - 3$]

11. Graph: $f(x) = 5^{x+3}$

We construct a table of values. Then we plot the points and connect them with a smooth curve.

$f(0) = 5^{0+3} = 5^3 = 125$
$f(-1) = 5^{-1+3} = 5^2 = 25$
$f(-2) = 5^{-2+3} = 5^1 = 5$
$f(-3) = 5^{-3+3} = 5^0 = 1$
$f(-4) = 5^{-4+3} = 5^{-1} = \dfrac{1}{5}$
$f(-5) = 5^{-5+3} = 5^{-2} = \dfrac{1}{25}$

x	$f(x)$
0	125
-1	25
-2	5
-3	1
-4	$\dfrac{1}{5}$
-5	$\dfrac{1}{25}$

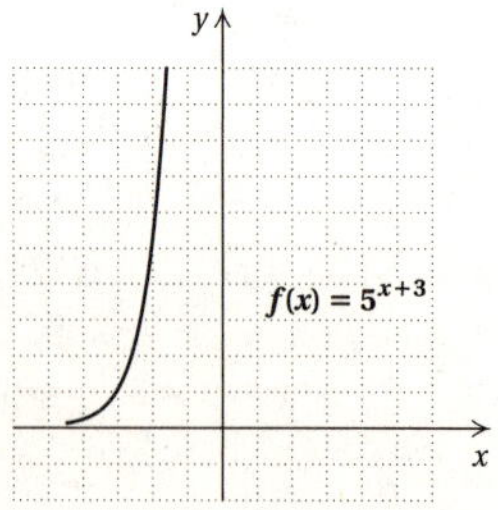

[Graph of $f(x) = 5^{x+3}$]

13. Graph: $f(x) = \left(\dfrac{1}{2}\right)^x$

We construct a table of values. Then we plot the points and connect them with a smooth curve.

$f(0) = \left(\dfrac{1}{2}\right)^0 = 1$
$f(1) = \left(\dfrac{1}{2}\right)^1 = \dfrac{1}{2}$
$f(2) = \left(\dfrac{1}{2}\right)^2 = \dfrac{1}{4}$
$f(3) = \left(\dfrac{1}{2}\right)^3 = \dfrac{1}{8}$
$f(-1) = \left(\dfrac{1}{2}\right)^{-1} = \dfrac{1}{\left(\dfrac{1}{2}\right)^1} = \dfrac{1}{\dfrac{1}{2}} = 2$
$f(-2) = \left(\dfrac{1}{2}\right)^{-2} = \dfrac{1}{\left(\dfrac{1}{2}\right)^2} = \dfrac{1}{\dfrac{1}{4}} = 4$
$f(-3) = \left(\dfrac{1}{2}\right)^{-3} = \dfrac{1}{\left(\dfrac{1}{2}\right)^3} = \dfrac{1}{\dfrac{1}{8}} = 8$

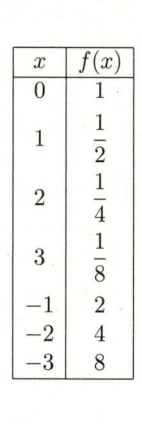

x	$f(x)$
0	1
1	$\dfrac{1}{2}$
2	$\dfrac{1}{4}$
3	$\dfrac{1}{8}$
-1	2
-2	4
-3	8

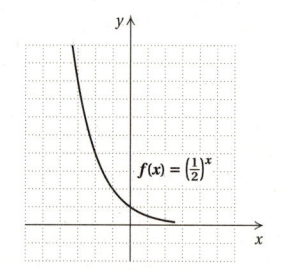

[Graph of $f(x) = \left(\dfrac{1}{2}\right)^x$]

15. Graph: $f(x) = \left(\dfrac{1}{5}\right)^x$

We construct a table of values. Then we plot the points and connect them with a smooth curve.

$f(0) = \left(\dfrac{1}{5}\right)^0 = 1$
$f(1) = \left(\dfrac{1}{5}\right)^1 = \dfrac{1}{5}$
$f(2) = \left(\dfrac{1}{5}\right)^2 = \dfrac{1}{25}$
$f(-1) = \left(\dfrac{1}{5}\right)^{-1} = \dfrac{1}{\dfrac{1}{5}} = 5$
$f(-2) = \left(\dfrac{1}{5}\right)^{-2} = \dfrac{1}{\dfrac{1}{25}} = 25$

x	$f(x)$
0	1
1	$\dfrac{1}{5}$
2	$\dfrac{1}{25}$
-1	5
-2	25

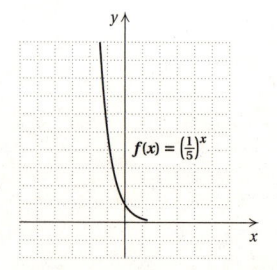

[Graph of $f(x) = \left(\dfrac{1}{5}\right)^x$]

Exercise Set 12.1

17. Graph: $f(x) = 2^{2x-1}$

We construct a table of values. Then we plot the points and connect them with a smooth curve.

$f(0) = 2^{2 \cdot 0 - 1} = 2^{-1} = \dfrac{1}{2}$

$f(1) = 2^{2 \cdot 1 - 1} = 2^1 = 2$

$f(2) = 2^{2 \cdot 2 - 1} = 2^3 = 8$

$f(-1) = 2^{2(-1)-1} = 2^{-3} = \dfrac{1}{8}$

$f(-2) = 2^{2(-2)-1} = 2^{-5} = \dfrac{1}{32}$

x	$f(x)$
0	$\dfrac{1}{2}$
1	2
2	8
-1	$\dfrac{1}{8}$
-2	$\dfrac{1}{32}$

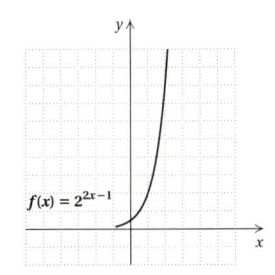

19. Graph: $x = 2^y$

We can find ordered pairs by choosing values for y and then computing values for x.

For $y = 0$, $x = 2^0 = 1$.

For $y = 1$, $x = 2^1 = 2$.

For $y = 2$, $x = 2^2 = 4$.

For $y = 3$, $x = 2^3 = 8$.

For $y = -1$, $x = 2^{-1} = \dfrac{1}{2^1} = \dfrac{1}{2}$.

For $y = -2$, $x = 2^{-2} = \dfrac{1}{2^2} = \dfrac{1}{4}$.

For $y = -3$, $x = 2^{-3} = \dfrac{1}{2^3} = \dfrac{1}{8}$.

x	y
1	0
2	1
4	2
8	3
$\dfrac{1}{2}$	-1
$\dfrac{1}{4}$	-2
$\dfrac{1}{8}$	-3

(1) Choose values for y.

(2) Compute values for x.

We plot these points and connect them with a smooth curve.

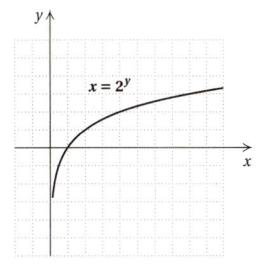

21. Graph: $x = \left(\dfrac{1}{2}\right)^y$

We can find ordered pairs by choosing values for y and then computing values for x. Then we plot these points and connect them with a smooth curve.

For $y = 0$, $x = \left(\dfrac{1}{2}\right)^0 = 1$.

For $y = 1$, $x = \left(\dfrac{1}{2}\right)^1 = \dfrac{1}{2}$.

For $y = 2$, $x = \left(\dfrac{1}{2}\right)^2 = \dfrac{1}{4}$.

For $y = 3$, $x = \left(\dfrac{1}{2}\right)^3 = \dfrac{1}{8}$.

For $y = -1$, $x = \left(\dfrac{1}{2}\right)^{-1} = \dfrac{1}{\dfrac{1}{2}} = 2$.

For $y = -2$, $x = \left(\dfrac{1}{2}\right)^{-2} = \dfrac{1}{\dfrac{1}{4}} = 4$.

For $y = -3$, $x = \left(\dfrac{1}{2}\right)^{-3} = \dfrac{1}{\dfrac{1}{8}} = 8$.

x	y
1	0
$\dfrac{1}{2}$	1
$\dfrac{1}{4}$	2
$\dfrac{1}{8}$	3
2	-1
4	-2
8	-3

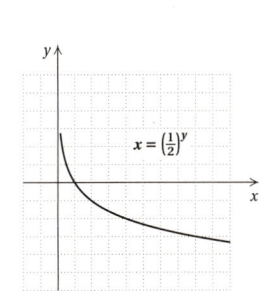

23. Graph: $x = 5^y$

We can find ordered pairs by choosing values for y and then computing values for x. Then we plot these points and connect them with a smooth curve.

For $y = 0$, $x = 5^0 = 1$.

For $y = 1$, $x = 5^1 = 5$.

For $y = 2$, $x = 5^2 = 25$.

For $y = -1$, $x = 5^{-1} = \dfrac{1}{5}$.

For $y = -2$, $x = 5^{-2} = \dfrac{1}{25}$.

x	y
1	0
5	1
25	2
$\dfrac{1}{5}$	-1
$\dfrac{1}{25}$	-2

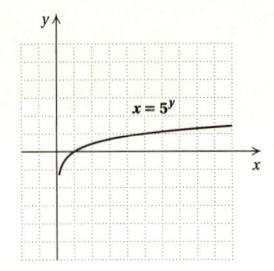

[Graph showing $x = 5^y$]

25. Graph $y = 2^x$ (see Exercise 1) and $x = 2^y$ (see Exercise 19) using the same set of axes.

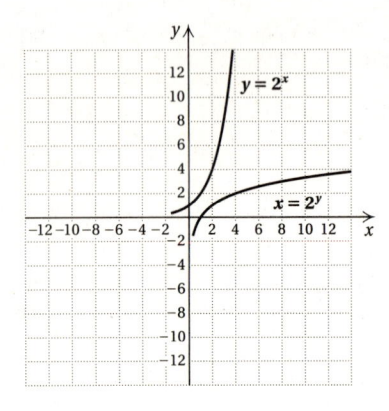

[Graph showing $y = 2^x$ and $x = 2^y$]

27. a) We substitute $50,000 for P and 6%, or 0.06, for r in the formula $A = P(1+r)^t$:
$$A(t) = \$50,000(1+0.06)^t = \$50,000(1.06)^t$$

b) $A(0) = \$50,000(1.06)^0 = \$50,000$
$A(1) = \$50,000(1.06)^1 = \$53,000$
$A(2) = \$50,000(1.06)^2 = \$56,180$
$A(4) = \$50,000(1.06)^4 \approx \$63,123.85$
$A(8) = \$50,000(1.06)^8 \approx \$76,692.40$
$A(10) = \$50,000(1.06)^{10} \approx \$89,542.38$
$A(20) = \$50,000(1.06)^{20} \approx \$160,356.77$

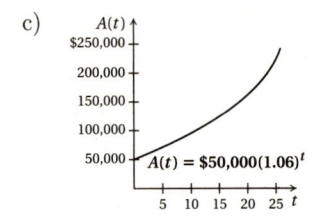

c) [Graph of $A(t) = \$50,000(1.06)^t$]

29. $V(t) = 4000(1.22)^t$

a) As the exercise states, $t = 0$ corresponds to 1958.
$V(0) = 4000(1.22)^0 = \$4000$

$1970 - 1958 = 12$, so $t = 12$ corresponds to 1970.
$V(12) = 4000(1.22)^{12} \approx \$43,489$

$1980 - 1958 = 22$, so $t = 22$ corresponds to 1980.
$V(22) = 4000(1.22)^{22} \approx \$317,670$

$1990 - 1958 = 32$, so $t = 32$ corresponds to 1990.
$V(32) = 4000(1.22)^{32} \approx \$2,320,463$

$1998 - 1958 = 40$, so $t = 40$ corresponds to 1998.
$V(40) = 4000(1.22)^{40} \approx \$11,388,151$

b) $1999 - 1958 = 41$, so $t = 41$ corresponds to 1999.
$V(41) = 4000(1.22)^{41} \approx \$13,893,544$

c) $2001 - 1958 = 43$, so $t = 43$ corresponds to 2001.
$V(43) = 4000(1.22)^{43} \approx \$20,679,151$

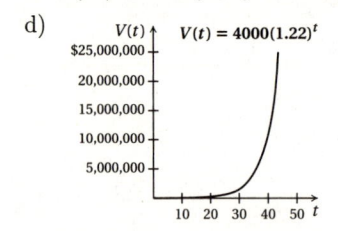

d) [Graph of $V(t) = 4000(1.22)^t$]

31. a) In 1930, $t = 1930 - 1900 = 30$.
$$P(t) = 150(0.960)^t$$
$$P(30) = 150(0.960)^{30}$$
$$\approx 44.079$$

In 1930, about 44.079 thousand, or 44,079, humpback whales were alive.
In 1960, $t = 1960 - 1900 = 60$.
$$P(t) = 150(0.960)^t$$
$$P(60) = 150(0.960)^{60}$$
$$\approx 12.953$$

In 1960, about 12.953 thousand, or 12,953, humpback whales were alive.

b) Plot the points found in part (a), (30, 44,079) and (60, 12,953) and additional points as needed and graph the function.

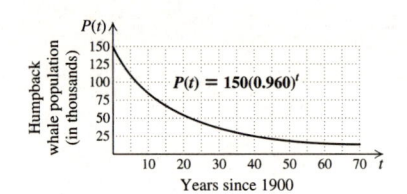

[Graph of $P(t) = 150(0.960)^t$]

33. $N(t) = 3000(2)^{t/20}$

a) $N(10) = 3000(2)^{10/20} \approx 4243$

There will be approximately 4243 bacteria after 10 min.

$N(20) = 3000(2)^{20/20} = 6000$

There will be 6000 bacteria after 20 min.

$N(30) = 3000(2)^{30/20} \approx 8485$

There will be approximately 8485 bacteria after 30 min.

$N(40) = 3000(2)^{40/20} = 12,000$

There will be 12,000 bacteria after 40 min.

$N(60) = 3000(2)^{60/20} = 24,000$

There will be 24,000 bacteria after 60 min.

b) We use the function values computed in part (a) to draw the graph. Other values can also be computed if needed.

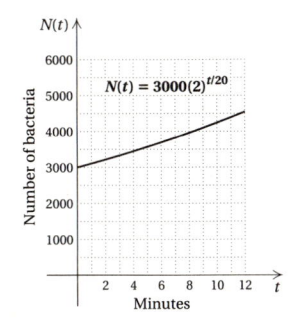

Exercise Set 12.1

35. Discussion and Writing Exercise

37. $x^{-5} \cdot x^3 = x^{-5+3} = x^{-2} = \dfrac{1}{x^2}$

39. $9^0 = 1$ (For any nonzero number a, $a^0 = 1$.)

41. $\left(\dfrac{2}{3}\right)^1 = \dfrac{2}{3}$ (For any number a, $a^1 = a$.)

43. $\dfrac{x^{-3}}{x^4} = x^{-3-4} = x^{-7} = \dfrac{1}{x^7}$

45. $\dfrac{x}{x^0} = x^{1-0} = x^1 = x$

(This exercise could also be done as follows:
$\dfrac{x}{x^0} = \dfrac{x}{1} = x$.)

47. $(5^{\sqrt{2}})^{2\sqrt{2}} = 5^{\sqrt{2} \cdot 2\sqrt{2}} = 5^4$, or 625

49. Graph: $y = 2^x + 2^{-x}$

Construct a table of values, thinking of y as $f(x)$. Then plot these points and connect them with a curve.

$f(0) = 2^0 + 2^{-0} = 1 + 1 = 2$

$f(1) = 2^1 + 2^{-1} = 2 + \dfrac{1}{2} = 2\dfrac{1}{2}$

$f(2) = 2^2 + 2^{-2} = 4 + \dfrac{1}{4} = 4\dfrac{1}{4}$

$f(3) = 2^3 + 2^{-3} = 8 + \dfrac{1}{8} = 8\dfrac{1}{8}$

$f(-1) = 2^{-1} + 2^{-(-1)} = \dfrac{1}{2} + 2 = 2\dfrac{1}{2}$

$f(-2) = 2^{-2} + 2^{-(-2)} = \dfrac{1}{4} + 4 = 4\dfrac{1}{4}$

$f(-3) = 2^{-3} + 2^{-(-3)} = \dfrac{1}{8} + 8 = 8\dfrac{1}{8}$

x	y, or $f(x)$
0	2
1	$2\dfrac{1}{2}$
2	$4\dfrac{1}{4}$
3	$8\dfrac{1}{8}$
-1	$2\dfrac{1}{2}$
-2	$4\dfrac{1}{4}$
-3	$8\dfrac{1}{8}$

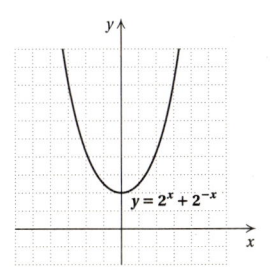

51. $y = \left|\left(\dfrac{1}{2}\right)^x - 1\right|$

Construct a table of values, thinking of y as $f(x)$. Then plot these points and connect them with a curve.

$f(-4) = \left|\left(\dfrac{1}{2}\right)^{-4} - 1\right| = |16 - 1| = |15| = 15$

$f(-2) = \left|\left(\dfrac{1}{2}\right)^{-2} - 1\right| = |4 - 1| = |3| = 3$

$f(-1) = \left|\left(\dfrac{1}{2}\right)^{-1} - 1\right| = |2 - 1| = |1| = 1$

$f(0) = \left|\left(\dfrac{1}{2}\right)^0 - 1\right| = |1 - 1| = |0| = 0$

$f(1) = \left|\left(\dfrac{1}{2}\right)^1 - 1\right| = \left|\dfrac{1}{2} - 1\right| = \left|-\dfrac{1}{2}\right| = \dfrac{1}{2}$

$f(2) = \left|\left(\dfrac{1}{2}\right)^2 - 1\right| = \left|\dfrac{1}{4} - 1\right| = \left|-\dfrac{3}{4}\right| = \dfrac{3}{4}$

$f(3) = \left|\left(\dfrac{1}{2}\right)^3 - 1\right| = \left|\dfrac{1}{8} - 1\right| = \left|-\dfrac{7}{8}\right| = \dfrac{7}{8}$

x	y, or $f(x)$
-4	15
-2	3
-1	1
0	0
1	$\dfrac{1}{2}$
2	$\dfrac{3}{4}$
3	$\dfrac{7}{8}$

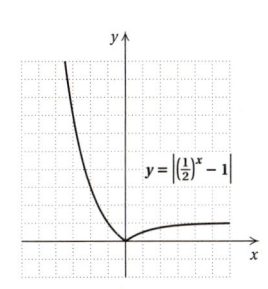

53. Construct a table of values for each equation and then draw the graphs on the same set of axes.

For $y = 3^{-(x-1)}$:

x	y
-3	81
-2	27
-1	9
0	3
1	1
2	$\dfrac{1}{3}$
3	$\dfrac{1}{9}$
4	$\dfrac{1}{27}$

For $x = 3^{-(y-1)}$:

x	y
81	-3
27	-2
9	-1
3	0
1	1
$\dfrac{1}{3}$	2
$\dfrac{1}{9}$	3
$\dfrac{1}{27}$	4

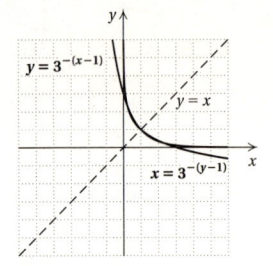

55. Left to the student

Exercise Set 12.2

1. To find the inverse of the given relation we interchange the first and second coordinates of each ordered pair. The inverse of the relation is $\{(2,1),(-3,6),(-5,-3)\}$.

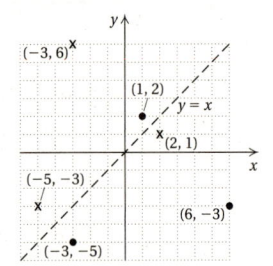

3. We interchange x and y to obtain an equation of the inverse of the relation. It is $x = 2y + 6$. The x-values in the first table become the y-values in the second table. We have

x	y
4	-1
6	0
8	1
10	2
12	3

We graph the original relation and its inverse. Since there is no horizontal line that crosses the graph more than once, the function is one-to-one.

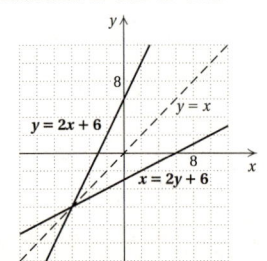

5. The graph of $f(x) = x - 5$ is shown below. Since no horizontal line crosses the graph more than once, the function is one-to-one.

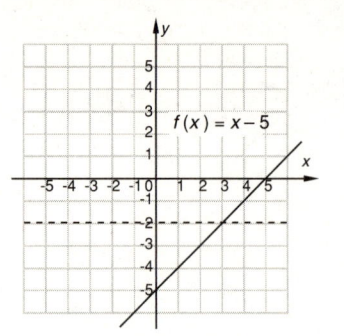

7. The graph of $f(x) = x^2 - 2$ is shown below. There are many horizontal lines that cross the graph more than once, so the function is not one-to-one.

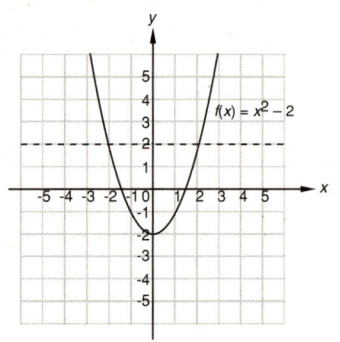

9. The graph of $g(x) = |x| - 3$ is shown below. There are many horizontal lines that cross the graph more than once, so the function is not one-to-one.

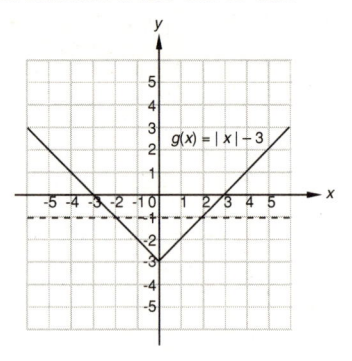

11. The graph of $g(x) = 3^x$ is shown below. Since no horizontal line crosses the graph more than once, the function is one-to-one.

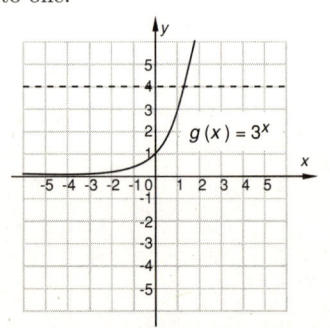

13. The graph of $f(x) = 5x - 2$ is shown below. It passes the horizontal-line test, so it is one-to-one.

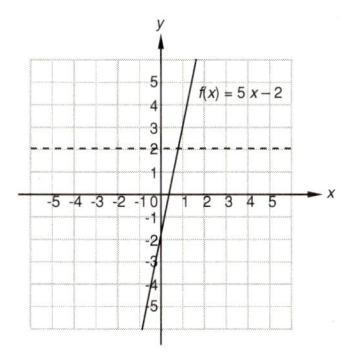

Exercise Set 12.2

[Graph of $f(x) = 5x - 2$]

We find a formula for the inverse.
1. Replace $f(x)$ by y: $y = 5x - 2$
2. Interchange x and y: $x = 5y - 2$
3. Solve for y: $x + 2 = 5y$
$$\frac{x+2}{5} = y$$
4. Replace y by $f^{-1}(x)$: $f^{-1}(x) = \frac{x+2}{5}$

15. The graph of $f(x) = \frac{-2}{x}$ is shown below. It passes the horizontal-line test, so it is one-to-one.

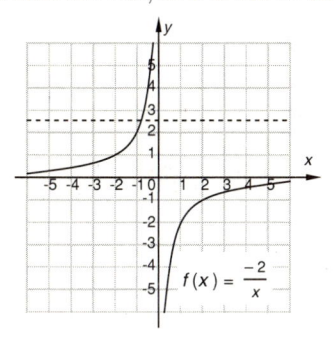

[Graph of $f(x) = \frac{-2}{x}$]

We find a formula for the inverse.
1. Replace $f(x)$ by y: $y = \frac{-2}{x}$
2. Interchange x and y: $x = \frac{-2}{y}$
3. Solve for y: $y = \frac{-2}{x}$
4. Replace y by $f^{-1}(x)$: $f^{-1}(x) = \frac{-2}{x}$

17. The graph of $f(x) = \frac{4}{3}x + 7$ is shown below. It passes the horizontal line test, so it is one-to-one.

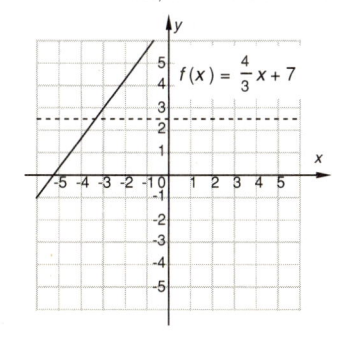

[Graph of $f(x) = \frac{4}{3}x + 7$]

We find a formula for the inverse.
1. Replace $f(x)$ by y: $y = \frac{4}{3}x + 7$
2. Interchange x and y: $x = \frac{4}{3}y + 7$
3. Solve for y: $x - 7 = \frac{4}{3}y$
$$\frac{3}{4}(x - 7) = y$$
4. Replace y by $f^{(-1)}(x)$: $f^{-1}(x) = \frac{3}{4}(x - 7)$

19. The graph of $f(x) = \frac{2}{x+5}$ is shown below. It passes the horizontal line test, so it is one-to-one.

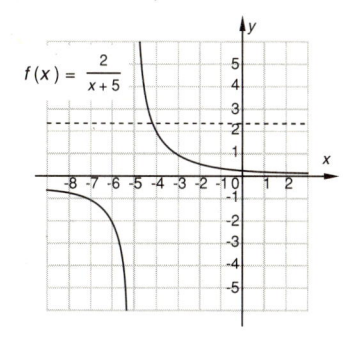

[Graph of $f(x) = \frac{2}{x+5}$]

We find a formula for the inverse.
1. Replace $f(x)$ by y: $y = \frac{2}{x+5}$
2. Interchange x and y: $x = \frac{2}{y+5}$
3. Solve for y: $x(y+5) = 2$
$$y + 5 = \frac{2}{x}$$
$$y = \frac{2}{x} - 5$$
4. Replace y by $f^{-1}(x)$: $f^{-1}(x) = \frac{2}{x} - 5$

21. The graph of $f(x) = 5$ is shown below. The horizontal line $y = 5$ crosses the graph more than once, so the function is not one-to-one.

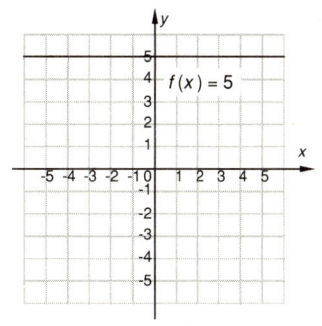

[Graph of $f(x) = 5$]

23. The graph of $f(x) = \dfrac{2x+1}{5x+3}$ is shown below. It passes the horizontal line test, so it is one-to-one.

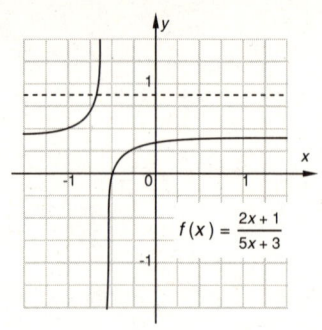

We find a formula for the inverse.

1. Replace $f(x)$ by y: $\quad y = \dfrac{2x+1}{5x+3}$
2. Interchange x and y: $\quad x = \dfrac{2y+1}{5y+3}$
3. Solve for y: $\quad 5xy + 3x = 2y + 1$
$$5xy - 2y = 1 - 3x$$
$$y(5x - 2) = 1 - 3x$$
$$y = \dfrac{1-3x}{5x-2}$$
4. Replace y by $f^{-1}(x)$: $\quad f^{-1}(x) = \dfrac{1-3x}{5x-2}$

25. The graph of $f(x) = x^3 - 1$ is shown below. It passes the horizontal line test, so it is one-to-one.

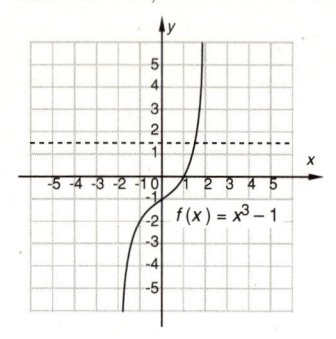

1. Replace $f(x)$ by y: $\quad y = x^3 - 1$
2. Interchange x and y: $\quad x = y^3 - 1$
3. Solve for y: $\quad x + 1 = y^3$
$$\sqrt[3]{x+1} = y$$
4. Replace y by $f^{-1}(x)$: $\quad f^{-1}(x) = \sqrt[3]{x+1}$

27. The graph of $f(x) = \sqrt[3]{x}$ is shown below. It passes the horizontal line test, so it is one-to-one.

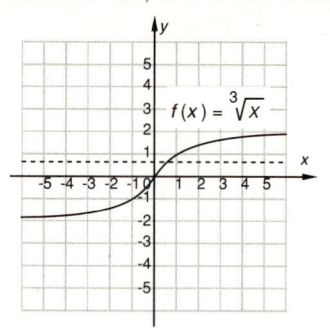

1. Replace $f(x)$ by y: $\quad y = \sqrt[3]{x}$
2. Interchange x and y: $\quad x = \sqrt[3]{y}$
3. Solve for y: $\quad x^3 = y$
4. Replace y by $f^{-1}(x)$: $\quad f^{-1}(x) = x^3$

29. We first graph $f(x) = \dfrac{1}{2}x - 3$. The graph of f^{-1} can be obtained by reflecting the graph of f across the line $y = x$.

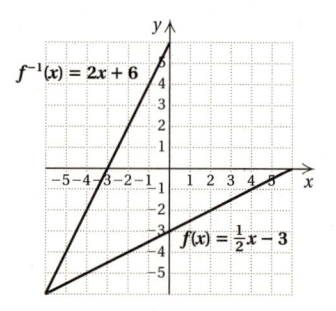

31. We first graph $f(x) = x^3$. The graph of f^{-1} can be obtained by reflecting the graph of f across the line $y = x$.

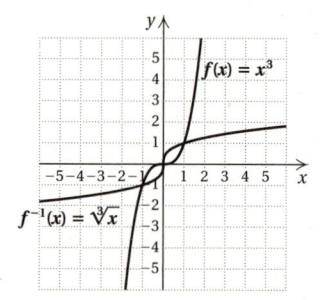

33. $f \circ g(x) = f(g(x)) = f(6 - 4x) = 2(6 - 4x) - 3 =$
$$12 - 8x - 3 = -8x + 9$$

$g \circ f(x) = g(f(x)) = g(2x - 3) = 6 - 4(2x - 3) =$
$$6 - 8x + 12 = -8x + 18$$

35. $f \circ g(x) = f(g(x)) = f(2x - 1) = 3(2x - 1)^2 + 2 =$
$$3(4x^2 - 4x + 1) + 2 = 12x^2 - 12x + 3 + 2 =$$
$$12x^2 - 12x + 5$$

$g \circ f(x) = g(f(x)) = g(3x^2 + 2) = 2(3x^2 + 2) - 1 =$
$$6x^2 + 4 - 1 = 6x^2 + 3$$

37. $f \circ g(x) = f(g(x)) = f\left(\frac{2}{x}\right) = 4\left(\frac{2}{x}\right)^2 - 1 =$
$$4\left(\frac{4}{x^2}\right) - 1 = \frac{16}{x^2} - 1$$
$g \circ f(x) = g(f(x)) = g(4x^2 - 1) = \dfrac{2}{4x^2 - 1}$

39. $f \circ g(x) = f(g(x)) = f(x^2 - 5) = (x^2 - 5)^2 + 5 =$
$\quad x^4 - 10x^2 + 25 + 5 = x^4 - 10x^2 + 30$
$g \circ f(x) = g(f(x)) = g(x^2 + 5) = (x^2 + 5)^2 - 5 =$
$\quad x^4 + 10x^2 + 25 - 5 = x^4 + 10x^2 + 20$

41. $h(x) = (5 - 3x)^2$
This is $5 - 3x$ raised to the second power, so the two most obvious functions are $f(x) = x^2$ and $g(x) = 5 - 3x$.

43. $h(x) = \sqrt{5x + 2}$
This is the square root of $5x + 2$, so the two most obvious functions are $f(x) = \sqrt{x}$ and $g(x) = 5x + 2$.

45. $h(x) = \dfrac{1}{x - 1}$
This is the reciprocal of $x - 1$, so the two most obvious functions are $f(x) = \dfrac{1}{x}$ and $g(x) = x - 1$.

47. $h(x) = \dfrac{1}{\sqrt{7x + 2}}$
This is the reciprocal of the square root of $7x + 2$. Two functions that can be used are $f(x) = \dfrac{1}{\sqrt{x}}$ and $g(x) = 7x + 2$.

49. $h(x) = (\sqrt{x} + 5)^4$
This is $\sqrt{x} + 5$ raised to the fourth power, so the two most obvious functions are $f(x) = x^4$ and $g(x) = \sqrt{x} + 5$.

51. We check to see that $f^{-1} \circ f(x) = x$ and $f \circ f^{-1}(x) = x$.
$f^{-1} \circ f(x) = f^{-1}(f(x)) = f^{-1}\left(\dfrac{4}{5}x\right) =$
$\dfrac{5}{4} \cdot \dfrac{4}{5}x = x$
$f \circ f^{-1}(x) = f(f^{-1}(x)) = f\left(\dfrac{5}{4}x\right) =$
$\dfrac{4}{5} \cdot \dfrac{5}{4}x = x$

53. We check to see that $f^{-1} \circ f(x) = x$ and $f \circ f^{-1}(x) = x$.
$f^{-1} \circ f(x) = f^{-1}(f(x)) = f^{-1}\left(\dfrac{1 - x}{x}\right) =$
$\dfrac{1}{\frac{1-x}{x} + 1} = \dfrac{1}{\frac{1-x}{x} + 1} \cdot \dfrac{x}{x} = \dfrac{x}{1 - x + x} =$
$\dfrac{x}{1} = x$
$f \circ f^{-1}(x) = f(f^{-1}(x)) = f\left(\dfrac{1}{x + 1}\right) =$

$\dfrac{1 - \frac{1}{x+1}}{\frac{1}{x+1}} = \dfrac{1 - \frac{1}{x+1}}{\frac{1}{x+1}} \cdot \dfrac{x+1}{x+1} =$
$\dfrac{x + 1 - 1}{1} = \dfrac{x}{1} = x$

55. The function $f(x) = 3x$ multiplies an input by 3, so the inverse would divide an input by 3. We have $f^{-1}(x) = \dfrac{x}{3}$.
Now we check to see that $f^{-1} \circ f(x) = x$ and $f \circ f^{-1}(x) = x$.
$f^{-1} \circ f(x) = f^{-1}(f(x)) = f^{-1}(3x) = \dfrac{3x}{3} = x$
$f \circ f^{-1}(x) = f(f^{-1}(x)) = f\left(\dfrac{x}{3}\right) = 3 \cdot \dfrac{x}{3} = x$
The inverse is correct.

57. The function $f(x) = -x$ takes the opposite of an input so the inverse would also take the opposite of an input. We have $f^{-1}(x) = -x$.
Now we check to see that $f^{-1} \circ f(x) = x$ and $f \circ f^{-1}(x) = x$.
$f^{-1} \circ f(x) = f^{-1}(f(x)) = f^{-1}(-x) = -(-x) = x$
$f \circ f^{-1}(x) = f(f^{-1}(x)) = f(-x) = -(-x) = x$
The inverse is correct.

59. The function $f(x) = \sqrt[3]{x - 5}$ subtracts 5 from an input and then takes the cube root of the difference, so the inverse would cube an input and then add 5. We have $f^{-1}(x) = x^3 + 5$.
Now we check to see that $f^{-1} \circ f(x) = x$ and $f \circ f^{-1}(x) = x$.
$f^{-1} \circ f(x) = f^{-1}(f(x)) = f^{-1}(\sqrt[3]{x - 5}) = (\sqrt[3]{x - 5})^3 + 5 = x - 5 + 5 = x$
$f \circ f^{-1}(x) = f(f^{-1}(x)) = f(x^3 + 5) = \sqrt[3]{x^3 + 5 - 5} = \sqrt[3]{x^3} = x$
The inverse is correct.

61. a) $f(8) = 8 + 32 = 40$
Size 40 in France corresponds to size 8 in the U.S.
$f(10) = 10 + 32 = 42$
Size 42 in France corresponds to size 10 in the U.S.
$f(14) = 14 + 32 = 46$
Size 46 in France corresponds to size 14 in the U.S.
$f(18) = 18 + 32 = 50$
Size 50 in France corresponds to size 18 in the U.S.

b) The graph of $f(x) = x + 32$ is shown below. It passes the horizontal-line test, so the function is one-to-one and, hence, has an inverse that is a function.

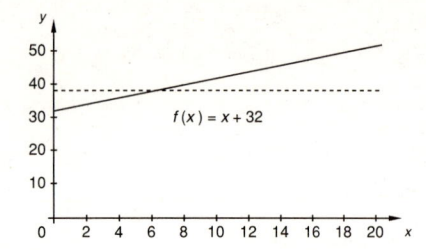

We now find a formula for the inverse.

1. Replace $f(x)$ by y: $y = x + 32$
2. Interchange x and y: $x = y + 32$
3. Solve for y: $x - 32 = y$
4. Replace y by $f^{-1}(x)$: $f^{-1}(x) = x - 32$

c) $f^{-1}(40) = 40 - 32 = 8$

Size 8 in the U.S. corresponds to size 40 in France.

$f^{-1}(42) = 42 - 32 = 10$

Size 10 in the U.S. corresponds to size 42 in France.

$f^{-1}(46) = 46 - 32 = 14$

Size 14 in the U.S. corresponds to size 46 in France.

$f^{-1}(50) = 50 - 32 = 18$

Size 18 in the U.S. corresponds to size 50 in France.

63. Discussion and Writing Exercise

65. $\sqrt[6]{a^2} = a^{2/6} = a^{1/3} = \sqrt[3]{a}$

67. $\sqrt{a^4 b^6} = (a^4 b^6)^{1/2} = a^2 b^3$

69. $\sqrt[8]{81} = (3^4)^{1/8} = 3^{1/2} = \sqrt{3}$

71. $\sqrt[12]{64x^6 y^6} = (2^6 x^6 y^6)^{1/12} = 2^{1/2} x^{1/2} y^{1/2} = (2xy)^{1/2} = \sqrt{2xy}$

73. $\sqrt[5]{32 a^{15} b^{40}} = (2^5 a^{15} b^{40})^{1/5} = 2 a^3 b^8$

75. $\sqrt[4]{81 a^8 b^8} = (3^4 a^8 b^8)^{1/4} = 3 a^2 b^2$

77. Graph the functions in a square window and determine whether one is a reflection of the other across the line $y = x$. The graphs show that these functions are not inverses of each other.

79. Graph the functions in a square window and determine whether one is a reflection of the other across the line $y = x$. The graphs show that these functions are inverses of each other.

81. (1) C; (2) A; (3) B; (4) D

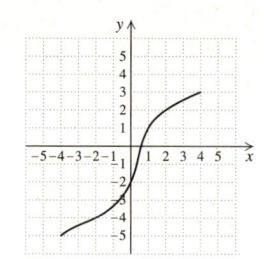

83. Reflect the graph of f across the line $y = x$.

85. $f(x) = \dfrac{1}{2} x + 3$, $g(x) = 2x - 6$

Since $(f \circ g)(x) = x$ and $(g \circ f)(x) = x$, the functions are inverse.

Exercise Set 12.3

1. Graph: $f(x) = \log_2 x$

The equation $f(x) = y = \log_2 x$ is equivalent to $2^y = x$. We can find ordered pairs by choosing values for y and computing the corresponding x-values.

For $y = 0$, $x = 2^0 = 1$.

For $y = 1$, $x = 2^1 = 2$.

For $y = 2$, $x = 2^2 = 4$.

For $y = 3$, $x = 2^3 = 8$.

For $y = -1$, $x = 2^{-1} = \dfrac{1}{2}$.

For $y = -2$, $x = 2^{-2} = \dfrac{1}{4}$.

x, or 2^y	y
1	0
2	1
4	2
8	3
$\dfrac{1}{2}$	-1
$\dfrac{1}{4}$	-2
$\dfrac{1}{8}$	-3

(1) Select y.

(2) Compute x.

We plot the set of ordered pairs and connect the points with a smooth curve.

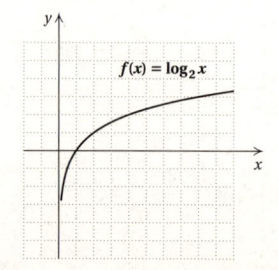

Exercise Set 12.3

3. Graph: $f(x) = \log_{1/3} x$

The equation $f(x) = y = \log_{1/3} x$ is equivalent to $\left(\frac{1}{3}\right)^y = x$. We can find ordered pairs by choosing values for y and computing the corresponding x-values.

For $y = 0$, $x = \left(\frac{1}{3}\right)^0 = 1$.

For $y = 1$, $x = \left(\frac{1}{3}\right)^1 = \frac{1}{3}$.

For $y = 2$, $x = \left(\frac{1}{3}\right)^2 = \frac{1}{9}$.

For $y = -1$, $x = \left(\frac{1}{3}\right)^{-1} = 3$.

For $y = -2$, $x = \left(\frac{1}{3}\right)^{-2} = 9$.

x, or $\left(\frac{1}{3}\right)^y$	y
1	0
$\frac{1}{3}$	1
$\frac{1}{9}$	2
3	-1
9	-2

We plot the set of ordered pairs and connect the points with a smooth curve.

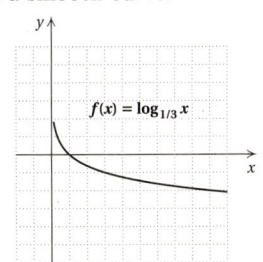

5. Graph $f(x) = 3^x$ (see Exercise Set 10.1, Exercise 2) and $f^{-1}(x) = \log_3 x$ on the same set of axes. We can obtain the graph of f^{-1} by reflecting the graph of f across the line $y = x$.

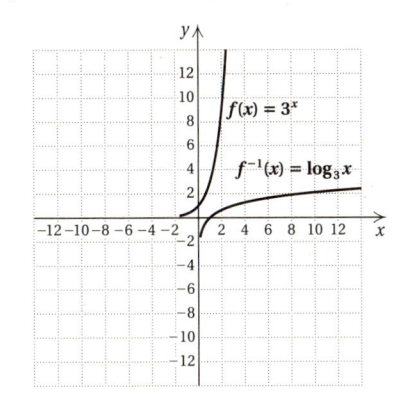

7. The exponent is the logarithm.
$10^3 = 1000 \Rightarrow 3 = \log_{10} 1000$
The base remains the same.

9. The exponent is the logarithm.
$5^{-3} = \frac{1}{125} \Rightarrow -3 = \log_5 \frac{1}{125}$
The base remains the same.

11. $8^{1/3} = 2 \Rightarrow \frac{1}{3} = \log_8 2$

13. $10^{0.3010} = 2 \Rightarrow 0.3010 = \log_{10} 2$

15. $e^2 = t \Rightarrow 2 = \log_e t$

17. $Q^t = x \Rightarrow t = \log_Q x$

19. $e^2 = 7.3891 \Rightarrow 2 = \log_e 7.3891$

21. $e^{-2} = 0.1353 \Rightarrow -2 = \log_e 0.1353$

23. The logarithm is the exponent.
$w = \log_4 10 \Rightarrow 4^w = 10$
The base remains the same.

25. The logarithm is the exponent.
$\log_6 36 = 2 \Rightarrow 6^2 = 36$
The base remains the same.

27. $\log_{10} 0.01 = -2 \Rightarrow 10^{-2} = 0.01$

29. $\log_{10} 8 = 0.9031 \Rightarrow 10^{0.9031} = 8$

31. $\log_e 100 = 4.6052 \Rightarrow e^{4.6052} = 100$

33. $\log_t Q = k \Rightarrow t^k = Q$

35. $\log_3 x = 2$
$3^2 = x$ Converting to an exponential equation
$9 = x$ Computing 3^2

37. $\log_x 16 = 2$
$x^2 = 16$ Converting to an exponential equation
$x = 4$ or $x = -4$ Principle of square roots
$\log_4 16 = 2$ because $4^2 = 16$. Thus, 4 is a solution. Since all logarithm bases must be positive, $\log_{-4} 16$ is not defined and -4 is not a solution.

39. $\log_2 16 = x$
$2^x = 16$ Converting to an exponential equation
$2^x = 2^4$
$x = 4$ The exponents are the same.

41. $\log_3 27 = x$
$3^x = 27$ Converting to an exponential equation
$3^x = 3^3$
$x = 3$ The exponents are the same.

43. $\log_x 25 = 1$
$x^1 = 25$ Converting to an exponential equation
$x = 25$

45. $\log_3 x = 0$

$3^0 = x$ Converting to an exponential equation

$1 = x$

47. $\log_2 x = -1$

$2^{-1} = x$ Converting to an exponential equation

$\dfrac{1}{2} = x$ Simplifying

49. $\log_8 x = \dfrac{1}{3}$

$8^{1/3} = x$

$2 = x$

51. Let $\log_{10} 100 = x$. Then

$10^x = 100$

$10^x = 10^2$

$x = 2$

Thus, $\log_{10} 100 = 2$.

53. Let $\log_{10} 0.1 = x$. Then

$10^x = 0.1 = \dfrac{1}{10}$

$10^x = 10^{-1}$

$x = -1$

Thus, $\log_{10} 0.1 = -1$.

55. Let $\log_{10} 1 = x$. Then

$10^x = 1$

$10^x = 10^0$ $(10^0 = 1)$

$x = 0$

Thus, $\log_{10} 1 = 0$.

57. Let $\log_5 625 = x$. Then

$5^x = 625$

$5^x = 5^4$

$x = 4$

Thus, $\log_5 625 = 4$.

59. Think of the meaning of $\log_7 49$. It is the exponent to which you raise 7 to get 49. That exponent is 2. Therefore, $\log_7 49 = 2$.

61. Think of the meaning of $\log_2 8$. It is the exponent to which you raise 2 to get 8. That exponent is 3. Therefore, $\log_2 8 = 3$.

63. Let $\log_9 \dfrac{1}{81} = x$. Then

$9^x = \dfrac{1}{81}$

$9^x = 9^{-2}$

$x = -2$

Thus, $\log_9 \dfrac{1}{81} = -2$.

65. Let $\log_8 1 = x$. Then

$8^x = 1$

$8^x = 8^0$ $(8^0 = 1)$

$x = 0$

Thus, $\log_8 1 = 0$.

67. Let $\log_e e = x$. Then

$e^x = e$

$e^x = e^1$

$x = 1$

Thus, $\log_e e = 1$.

69. Let $\log_{27} 9 = x$. Then

$27^x = 9$

$(3^3)^x = 3^2$

$3^{3x} = 3^2$

$3x = 2$

$x = \dfrac{2}{3}$

Thus, $\log_{27} 9 = \dfrac{2}{3}$.

71. 4.8970

73. -0.1739

75. Does not exist

77. 0.9464

79. $6 = 10^{0.7782}$; $84 = 10^{1.9243}$; $987,606 = 10^{5.9946}$; $0.00987606 = 10^{-2.0054}$; $98,760.6 = 10^{4.9946}$; $70,000,000 = 10^{7.8451}$; $7000 = 10^{3.8451}$

81. Discussion and Writing Exercise

83. $f(x) = 4 - x^2$

a) The x-coordinate of the vertex is $-\dfrac{b}{2a} = -\dfrac{0}{-2} = 0$.

The y-coordinate is $f(0) = 4 - 0^2 = 4$.

The vertex is $(0, 4)$.

b) The line of symmetry is $x = 0$.

c) Since the coefficient of x^2 is negative, the graph opens down and, hence, has a maximum. The maximum value is 4.

d) We find some points on each side of the vertex and use them to draw the graph.

x	y
0	4
-1	3
1	3
-2	0
2	0
-3	-5
3	-5

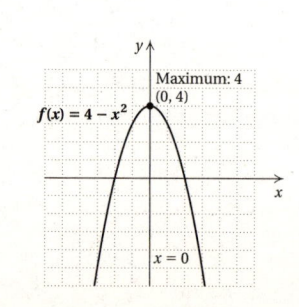

85. $f(x) = -2(x-1)^2 - 3$

  a) The equation is in the form $f(x) = a(x-h)^2 + k$ so we know that the vertex is (h, k), or $(1, -3)$.

  b) The line of symmetry is $x = 1$.

  c) Since the coefficient of x^2 is negative, the graph opens down and, hence, has a maximum. The maximum value is -3.

  d) We find some points on each side of the vertex and use them to draw the graph.

x	y
0	-5
2	-5
-1	-11
3	-11

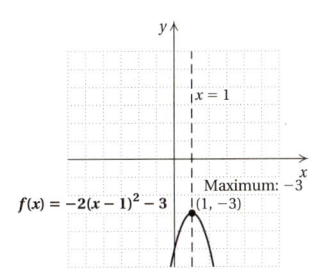

87. $E = mc^2$

$\dfrac{E}{m} = c^2$ Dividing by m

$\sqrt{\dfrac{E}{m}} = c$ Taking the positive square root

89. $A = \sqrt{3ab}$

$A^2 = (\sqrt{3ab})^2$ Squaring both sides

$A^2 = 3ab$

$\dfrac{A^2}{3a} = b$ Dividing by $3a$

91. a) We substitute in the equation $y = ax^2 + bx + c$ and get a system of equations.

$31 = a(20)^2 + b(20) + c$, or $31 = 400a + 20b + c$;

$34 = a(24)^2 + b(24) + c$, or $34 = 576a + 24b + c$;

$22 = a(34)^2 + b(34) + c$, or $22 = 1156a + 34b + c$

The solution of the system of equation is $\left(-\dfrac{39}{280}, \dfrac{963}{140}, -\dfrac{356}{7}\right)$, so the quadratic function that fits the data is

$f(x) = -\dfrac{39}{280}x^2 + \dfrac{963}{140}x - \dfrac{356}{7}$.

  b) $f(30) = -\dfrac{39}{280}(30)^2 + \dfrac{963}{140}(30) - \dfrac{356}{7} \approx 30$

  About 30% of drivers will be involved in accidents at age 30.

  $f(37) = -\dfrac{39}{280}(37)^2 + \dfrac{963}{140}(37) - \dfrac{356}{7} \approx 13$

  About 13% of drivers will be involved in accidents at age 37.

93. Graph: $f(x) = \log_3 |x+1|$

x	$f(x)$
0	0
2	1
8	2
-2	0
-4	1
-9	2

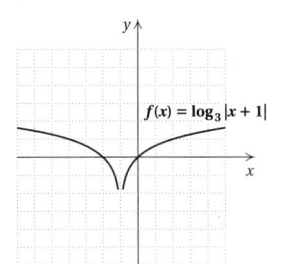

95. $\log_{125} x = \dfrac{2}{3}$

$125^{2/3} = x$

$(5^3)^{2/3} = x$

$5^2 = x$

$25 = x$

97. $\log_8(2x+1) = -1$

$8^{-1} = 2x + 1$

$\dfrac{1}{8} = 2x + 1$

$1 = 16x + 8$ Multiplying by 8

$-7 = 16x$

$-\dfrac{7}{16} = x$

99. Let $\log_{1/4} \dfrac{1}{64} = x$. Then

$\left(\dfrac{1}{4}\right)^x = \dfrac{1}{64}$

$\left(\dfrac{1}{4}\right)^x = \left(\dfrac{1}{4}\right)^3$

$x = 3$

Thus, $\log_{1/4} \dfrac{1}{64} = 3$.

101. $\log_{10}(\log_4(\log_3 81))$

$= \log_{10}(\log_4 4)$ $(\log_3 81 = 4)$

$= \log_{10} 1$ $(\log_4 4 = 1)$

$= 0$

103. Let $\log_{1/5} 25 = x$. Then

$\left(\dfrac{1}{5}\right)^x = 25$

$(5^{-1})^x = 25$

$5^{-x} = 5^2$

$-x = 2$

$x = -2$

Thus, $\log_{1/5} 25 = -2$.

Exercise Set 12.4

1. $\log_2 (32 \cdot 8) = \log_2 32 + \log_2 8$ Property 1
3. $\log_4 (64 \cdot 16) = \log_4 64 + \log_4 16$ Property 1
5. $\log_a Qx = \log_a Q + \log_a x$ Property 1
7. $\log_b 3 + \log_b 84 = \log_b (3 \cdot 84)$ Property 1
         $= \log_b 252$
9. $\log_c K + \log_c y = \log_c K \cdot y$ Property 1
         $= \log_c Ky$
11. $\log_c y^4 = 4 \log_c y$ Property 2
13. $\log_b t^6 = 6 \log_b t$ Property 2
15. $\log_b C^{-3} = -3 \log_b C$ Property 2
17. $\log_a \dfrac{67}{5} = \log_a 67 - \log_a 5$ Property 3
19. $\log_b \dfrac{2}{5} = \log_b 2 - \log_b 5$ Property 3
21. $\log_c 22 - \log_c 3 = \log_c \dfrac{22}{3}$ Property 3
23. $\log_a x^2 y^3 z$
     $= \log_a x^2 + \log_a y^3 + \log_b z$ Property 1
     $= 2 \log_a x + 3 \log_a y + \log_a z$ Property 2
25. $\log_b \dfrac{xy^2}{z^3}$
     $= \log_b xy^2 - \log_b z^3$ Property 3
     $= \log_b x + \log_b y^2 - \log_b z^3$ Property 1
     $= \log_b x + 2 \log_b y - 3 \log_b z$ Property 2
27. $\log_c \sqrt[3]{\dfrac{x^4}{y^3 z^2}}$
     $= \log_c \left(\dfrac{x^4}{y^3 z^2}\right)^{1/3}$
     $= \dfrac{1}{3} \log_c \dfrac{x^4}{y^3 z^2}$ Property 2
     $= \dfrac{1}{3}(\log_c x^4 - \log_c y^3 z^2)$ Property 3
     $= \dfrac{1}{3}[\log_c x^4 - (\log_c y^3 + \log_c z^2)]$ Property 1
     $= \dfrac{1}{3}(\log_c x^4 - \log_c y^3 - \log_c z^2)$ Removing parentheses
     $= \dfrac{1}{3}(4 \log_c x - 3 \log_c y - 2 \log_c z)$ Property 2
     $= \dfrac{4}{3} \log_c x - \log_c y - \dfrac{2}{3} \log_c z$

29. $\log_a \sqrt[4]{\dfrac{m^8 n^{12}}{a^3 b^5}}$
     $= \log_a \left(\dfrac{m^8 n^{12}}{a^3 b^5}\right)^{1/4}$
     $= \dfrac{1}{4} \log_a \dfrac{m^8 n^{12}}{a^3 b^5}$ Property 2
     $= \dfrac{1}{4}(\log_a m^8 n^{12} - \log_a a^3 b^5)$ Property 3
     $= \dfrac{1}{4}[\log_a m^8 + \log_a n^{12} - (\log_a a^3 + \log_a b^5)]$ Property 1
     $= \dfrac{1}{4}(\log_a m^8 + \log_a n^{12} - \log_a a^3 - \log_a b^5)$
                                           Removing parentheses
     $= \dfrac{1}{4}(\log_a m^8 + \log_a n^{12} - 3 - \log_a b^5)$ Property 4
     $= \dfrac{1}{4}(8 \log_a m + 12 \log_a n - 3 - 5 \log_a b)$ Property 2
     $= 2 \log_a m + 3 \log_a n - \dfrac{3}{4} - \dfrac{5}{4} \log_a b$

31. $\dfrac{2}{3} \log_a x - \dfrac{1}{2} \log_a y$
     $= \log_a x^{2/3} - \log_a y^{1/2}$ Property 2
     $= \log_a \dfrac{x^{2/3}}{y^{1/2}}$, or   Property 3
     $\log_a \dfrac{\sqrt[3]{x^2}}{\sqrt{y}}$

33. $\log_a 2x + 3(\log_a x - \log_a y)$
     $= \log_a 2x + 3 \log_a x - 3 \log_a y$
     $= \log_a 2x + \log_a x^3 - \log_a y^3$ Property 2
     $= \log_a 2x^4 - \log_a y^3$ Property 1
     $= \log_a \dfrac{2x^4}{y^3}$ Property 3

35. $\log_a \dfrac{a}{\sqrt{x}} - \log_a \sqrt{ax}$
     $= \log_a ax^{-1/2} - \log_a a^{1/2} x^{1/2}$
     $= \log_a \dfrac{ax^{-1/2}}{a^{1/2} x^{1/2}}$ Property 3
     $= \log_a \dfrac{a^{1/2}}{x}$, or
     $\log_a \dfrac{\sqrt{a}}{x}$

37. $\log_b 15 = \log_b (3 \cdot 5)$
            $= \log_b 3 + \log_b 5$ Property 1
            $= 1.099 + 1.609$
            $= 2.708$

39. $\log_b \dfrac{5}{3} = \log_b 5 - \log_b 3$ Property 3
       $= 1.609 - 1.099$
       $= 0.51$

Exercise Set 12.5

41. $\log_b \frac{1}{5} = \log_b 1 - \log_b 5$ Property 3
$= 0 - 1.609$ $(\log_b 1 = 0)$
$= -1.609$

43. $\log_b \sqrt{b^3} = \log_b b^{3/2} = \frac{3}{2}$ Property 4

45. $\log_b 5b = \log_b 5 + \log_b b$ Property 1
$= 1.609 + 1$ $(\log_b b = 1)$
$= 2.609$

47. $\log_e e^t = t$ Property 4

49. $\log_p p^5 = 5$ Property 4

51. $\log_2 2^7 = x$
$7 = x$ Property 4

53. $\log_e e^x = -7$
$x = -7$ Property 4

55. Discussion and Writing Exercise

57. $i^{29} = i^{28} \cdot i = (i^2)^{14} \cdot i = (-1)^{14} \cdot i = 1 \cdot i = i$

59. $(2+i)(2-i) = 4 - i^2 = 4 - (-1) = 4 + 1 = 5$

61. $(7 - 8i) - (-16 + 10i) = 7 - 8i + 16 - 10i = 23 - 18i$

63. $(8 + 3i)(-5 - 2i) = -40 - 16i - 15i - 6i^2 =$
$-40 - 16i - 15i + 6 = -34 - 31i$

65. Enter $y_1 = \log x^2$ and $y_2 = (\log x)(\log x)$ and show that the graphs are different and that the y-values in a table of values are not the same.

67. $\log_a (x^8 - y^8) - \log_a (x^2 + y^2)$
$= \log_a \frac{x^8 - y^8}{x^2 + y^2}$ Property 3
$= \log_a \frac{(x^4 + y^4)(x^2 + y^2)(x + y)(x - y)}{x^2 + y^2}$ Factoring
$= \log_a [(x^4 + y^4)(x + y)(x - y)]$ Simplifying
$= \log_a (x^6 - x^4 y^2 + x^2 y^4 - y^6)$ Multiplying

69. $\log_a \sqrt{1 - s^2}$
$= \log_a (1 - s^2)^{1/2}$
$= \frac{1}{2} \log_a (1 - s^2)$
$= \frac{1}{2} \log_a [(1 - s)(1 + s)]$
$= \frac{1}{2} \log_a (1 - s) + \frac{1}{2} \log_a (1 + s)$

71. False. For example, let $a = 10$, $P = 100$, and $Q = 10$.
$\frac{\log 100}{\log 10} = \frac{2}{1} = 2$, but
$\log \frac{100}{10} = \log 10 = 1$.

73. True, by Property 1

75. False. For example, let $a = 2$, $P = 1$, and $Q = 1$.
$\log_2(1 + 1) = \log_2 2 = 1$, but
$\log_2 1 + \log_2 1 = 0 + 0 = 0$.

Exercise Set 12.5

1. 0.6931

3. 4.1271

5. 8.3814

7. -5.0832

9. -1.6094

11. Does not exist

13. -1.7455

15. 1

17. 15.0293

19. 0.0305

21. 109.9472

23. 5

25. We will use common logarithms for the conversion. Let $a = 10$, $b = 6$, and $M = 100$ and substitute in the change-of-base formula.
$\log_b M = \frac{\log_a M}{\log_a b}$
$\log_6 100 = \frac{\log_{10} 100}{\log_{10} 6}$
$\approx \frac{2}{0.7782}$
≈ 2.5702

27. We will use common logarithms for the conversion. Let $a = 10$, $b = 2$, and $M = 100$ and substitute in the change-of-base formula.
$\log_2 100 = \frac{\log_{10} 100}{\log_{10} 2}$
$\approx \frac{2}{0.3010}$
≈ 6.6439

29. We will use natural logarithms for the conversion. Let $a = e$, $b = 7$, and $M = 65$ and substitute in the change-of-base formula.
$\log_7 65 = \frac{\ln 65}{\ln 7}$
$\approx \frac{4.1744}{1.9459}$
≈ 2.1452

31. We will use natural logarithms for the conversion. Let $a = e$, $b = 0.5$, and $M = 5$ and substitute in the change-of-base formula.

$$\log_{0.5} 5 = \frac{\ln 5}{\ln 0.5}$$
$$\approx \frac{1.6094}{-0.6931}$$
$$\approx -2.3219$$

33. We will use common logarithms for the conversion. Let $a = 10$, $b = 2$, and $M = 0.2$ and substitute in the change-of-base formula.

$$\log_2 0.2 = \frac{\log_{10} 0.2}{\log_{10} 2}$$
$$\approx \frac{-0.6990}{0.3010}$$
$$\approx -2.3219$$

35. We will use natural logarithms for the conversion. Let $a = e$, $b = \pi$, and $M = 200$.

$$\log_\pi 200 = \frac{\ln 200}{\ln \pi}$$
$$\approx \frac{5.2983}{1.1447}$$
$$\approx 4.6285$$

If $\ln 200$ and $\ln \pi$ are not rounded before the division is performed, the result is 4.6284.

37. Graph: $f(x) = e^x$

We find some function values with a calculator. We use these values to plot points and draw the graph.

x	e^x
0	1
1	2.7
2	7.4
3	20.1
-1	0.4
-2	0.1
-3	0.05

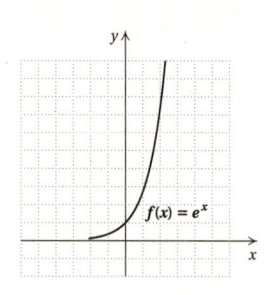

39. Graph: $f(x) = e^{-0.5x}$

We find some function values, plot points, and draw the graph.

x	$e^{-0.5x}$
0	1
1	0.61
2	0.37
-1	1.65
-2	2.72
-3	4.48
-4	7.39

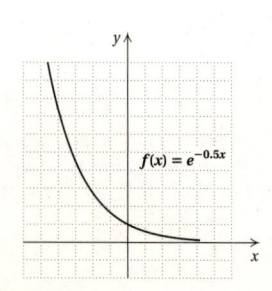

41. Graph: $f(x) = e^{x-1}$

We find some function values, plot points, and draw the graph.

x	e^{x-1}
0	0.4
1	1
2	2.7
3	7.4
4	20.1
-1	0.1
-2	0.05

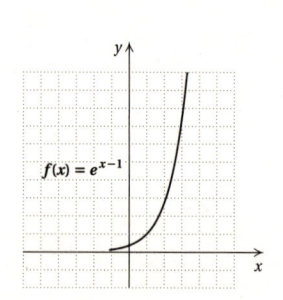

43. Graph: $f(x) = e^{x+2}$

We find some function values, plot points, and draw the graph.

x	e^{x+2}
1	20.1
0	7.4
-2	1
-3	0.4
-4	0.1

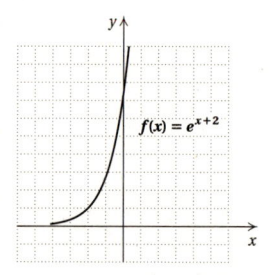

45. Graph: $f(x) = e^x - 1$

We find some function values, plot points, and draw the graph.

x	$e^x - 1$
0	0
1	1.72
2	6.39
3	19.09
-1	-0.63
-2	-0.86
-4	-0.98

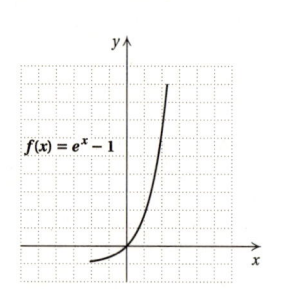

47. Graph: $f(x) = \ln(x+2)$

We find some function values, plot points, and draw the graph.

x	$\ln(x+2)$
0	0.69
1	1.10
2	1.39
3	1.61
-0.5	0.41
-1	0
-1.5	-0.69

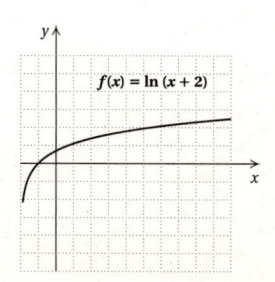

Exercise Set 12.6

49. Graph: $f(x) = \ln(x-3)$

We find some function values, plot points, and draw the graph.

x	$\ln(x-3)$
3	Undefined
4	0
5	0.69
6	1.10
8	1.61
10	1.95

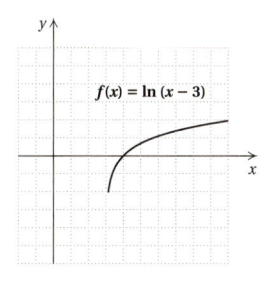

51. Graph: $f(x) = 2\ln x$

x	$2\ln x$
0.5	−1.4
1	0
2	1.4
3	2.2
4	2.8
5	3.2
6	3.6

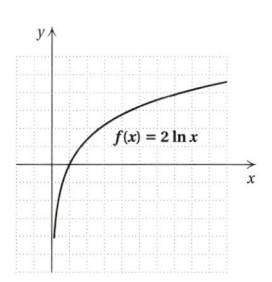

53. Graph: $f(x) = \dfrac{1}{2}\ln x + 1$

x	$\dfrac{1}{2}\ln x + 1$
1	1
2	1.35
3	1.55
4	1.69
6	1.90

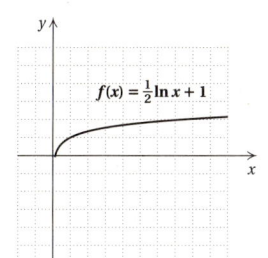

55. Graph: $f(x) = |\ln x|$

x	$\ln x$
$\frac{1}{4}$	1.4
$\frac{1}{2}$	0.7
1	0
3	1.1
5	1.6

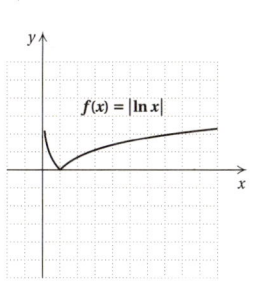

57. Discussion and Writing Exercise

59. $x^{1/2} - 6x^{1/4} + 8 = 0$

Let $u = x^{1/4}$.

$u^2 - 6u + 8 = 0$ Substituting

$(u-4)(u-2) = 0$

$u = 4$ or $u = 2$

$x^{1/4} = 4$ or $x^{1/4} = 2$

$x = 256$ or $x = 16$ Raising both sides to the fourth power

Both numbers check. The solutions are 256 and 16.

61. $x - 18\sqrt{x} + 77 = 0$

Let $u = \sqrt{x}$.

$u^2 - 18u + 77 = 0$ Substituting

$(u-7)(u-11) = 0$

$u = 7$ or $u = 11$

$\sqrt{x} = 7$ or $\sqrt{x} = 11$

$x = 49$ or $x = 121$ Squaring both sides

Both numbers check. The solutions are 49 and 121.

63. Domain: $(-\infty, \infty)$, range: $[0, \infty)$

65. Domain: $(-\infty, \infty)$, range: $(-\infty, 100)$

67. $f(x)$ can be calculated for positive values of $2x - 5$. We have:

$2x - 5 > 0$

$2x > 5$

$x > \dfrac{5}{2}$

The domain is $\left\{x \middle| x > \dfrac{5}{2}\right\}$, or $\left(\dfrac{5}{2}, \infty\right)$.

Exercise Set 12.6

1. $2^x = 8$

$2^x = 2^3$

$x = 3$ The exponents are the same.

3. $4^x = 256$

$4^x = 4^4$

$x = 4$ The exponents are the same.

5. $2^{2x} = 32$

$2^{2x} = 2^5$

$2x = 5$

$x = \dfrac{5}{2}$

7. $3^{5x} = 27$

$3^{5x} = 3^3$

$5x = 3$

$x = \dfrac{3}{5}$

9. $2^x = 11$

$\log 2^x = \log 11$ Taking the common logarithm on both sides

$x \log 2 = \log 11$ Property 2

$x = \dfrac{\log 11}{\log 2}$

$x \approx 3.4594$

11. $2^x = 43$

$\log 2^x = \log 43$ Taking the common logarithm on both sides

$x \log 2 = \log 43$ Property 2

$x = \dfrac{\log 43}{\log 2}$

$x \approx 5.4263$

13. $5^{4x-7} = 125$

$5^{4x-7} = 5^3$

$4x - 7 = 3$ The exponents are the same.

$4x = 10$

$x = \dfrac{10}{4}$, or $\dfrac{5}{2}$

15. $3^{x^2} \cdot 3^{4x} = \dfrac{1}{27}$

$3^{x^2+4x} = 3^{-3}$

$x^2 + 4x = -3$

$x^2 + 4x + 3 = 0$

$(x+3)(x+1) = 0$

$x = -3$ or $x = -1$

17. $4^x = 8$

$(2^2)^x = 2^3$

$2^{2x} = 2^3$

$2x = 3$ The exponents are the same.

$x = \dfrac{3}{2}$

19. $e^t = 100$

$\ln e^t = \ln 100$ Taking ln on both sides

$t = \ln 100$ Property 4

$t \approx 4.6052$ Using a calculator

21. $e^{-t} = 0.1$

$\ln e^{-t} = \ln 0.1$ Taking ln on both sides

$-t = \ln 0.1$ Property 4

$-t \approx -2.3026$

$t \approx 2.3026$

23. $e^{-0.02t} = 0.06$

$\ln e^{-0.02t} = \ln 0.06$ Taking ln on both sides

$-0.02t = \ln 0.06$ Property 4

$t = \dfrac{\ln 0.06}{-0.02}$

$t \approx \dfrac{-2.8134}{-0.02}$

$t \approx 140.6705$

25. $2^x = 3^{x-1}$

$\log 2^x = \log 3^{x-1}$

$x \log 2 = (x-1) \log 3$

$x \log 2 = x \log 3 - \log 3$

$\log 3 = x \log 3 - x \log 2$

$\log 3 = x(\log 3 - \log 2)$

$\dfrac{\log 3}{\log 3 - \log 2} = x$

$\dfrac{0.4771}{0.4771 - 0.3010} \approx x$

$2.7095 \approx x$

27. $(3.6)^x = 62$

$\log (3.6)^x = \log 62$

$x \log 3.6 = \log 62$

$x = \dfrac{\log 62}{\log 3.6}$

$x \approx 3.2220$

29. $\log_4 x = 4$

$x = 4^4$ Writing an equivalent exponential equation

$x = 256$

31. $\log_2 x = -5$

$x = 2^{-5}$ Writing an equivalent exponential equation

$x = \dfrac{1}{32}$

33. $\log x = 1$ The base is 10.

$x = 10^1$

$x = 10$

35. $\log x = -2$ The base is 10.

$x = 10^{-2}$

$x = \dfrac{1}{100}$

37. $\ln x = 2$

$x = e^2 \approx 7.3891$

39. $\ln x = -1$

$x = e^{-1}$

$x = \dfrac{1}{e} \approx 0.3679$

41. $\log_3 (2x+1) = 5$

$2x + 1 = 3^5$ Writing an equivalent exponential equation

$2x + 1 = 243$

$2x = 242$

$x = 121$

Exercise Set 12.6

43. $\log x + \log (x-9) = 1$ The base is 10.
$\log_{10}[x(x-9)] = 1$ Property 1
$x(x-9) = 10^1$
$x^2 - 9x = 10$
$x^2 - 9x - 10 = 0$
$(x-10)(x+1) = 0$
$x = 10 \text{ or } x = -1$

Check: For 10:

$$\begin{array}{c|c} \log x + \log (x-9) = 1 \\ \hline \log 10 + \log(10-9) \ ? \ 1 \\ \log 10 + \log 1 \\ 1 + 0 \\ 1 & \text{TRUE} \end{array}$$

For -1:

$$\begin{array}{c|c} \log x + \log (x-9) = 1 \\ \hline \log(-1) + \log(-1-9) \ ? \ 1 & \text{FALSE} \end{array}$$

The number -1 does not check, because negative numbers do not have logarithms. The solution is 10.

45. $\log x - \log (x+3) = -1$ The base is 10.
$\log_{10} \dfrac{x}{x+3} = -1$ Property 3
$\dfrac{x}{x+3} = 10^{-1}$
$\dfrac{x}{x+3} = \dfrac{1}{10}$
$10x = x + 3$
$9x = 3$
$x = \dfrac{1}{3}$

The answer checks. The solution is $\dfrac{1}{3}$.

47. $\log_2 (x+1) + \log_2 (x-1) = 3$
$\log_2[(x+1)(x-1)] = 3$ Property 1
$(x+1)(x-1) = 2^3$
$x^2 - 1 = 8$
$x^2 = 9$
$x = \pm 3$

The number 3 checks, but -3 does not. The solution is 3.

49. $\log_4 (x+6) - \log_4 x = 2$
$\log_4 \dfrac{x+6}{x} = 2$ Property 3
$\dfrac{x+6}{x} = 4^2$
$\dfrac{x+6}{x} = 16$
$x + 6 = 16x$
$6 = 15x$
$\dfrac{2}{5} = x$

The answer checks. The solution is $\dfrac{2}{5}$.

51. $\log_4 (x+3) + \log_4 (x-3) = 2$
$\log_4[(x+3)(x-3)] = 2$ Property 1
$(x+3)(x-3) = 4^2$
$x^2 - 9 = 16$
$x^2 = 25$
$x = \pm 5$

The number 5 checks, but -5 does not. The solution is 5.

53. $\log_3 (2x-6) - \log_3 (x+4) = 2$
$\log_3 \dfrac{2x-6}{x+4} = 2$ Property 3
$\dfrac{2x-6}{x+4} = 3^2$
$\dfrac{2x-6}{x+4} = 9$
$2x - 6 = 9x + 36$ Multiplying by $(x+4)$
$-42 = 7x$
$-6 = x$

Check:

$$\begin{array}{c|c} \log_3 (2x-6) - \log_3 (x+4) = 2 \\ \hline \log_3 [2(-6)-6] - \log_3 (-6+4) \ ? \ 2 \\ \log_3 (-18) - \log_3 (-2) & \text{FALSE} \end{array}$$

The number -6 does not check, because negative numbers do not have logarithms. There is no solution.

55. Discussion and Writing Exercise

57. $x^4 + 400 = 104x^2$
$x^4 - 104x^2 + 400 = 0$
Let $u = x^2$.
$u^2 - 104u + 400 = 0$
$(u-100)(u-4) = 0$
$u = 100 \text{ or } u = 4$
$x^2 = 100 \text{ or } x^2 = 4$ Replacing u with x^2
$x = \pm 10 \text{ or } x = \pm 2$

The solutions are ± 10 and ± 2.

59. $(x^2+5x)^2 + 2(x^2+5x) = 24$
$(x^2+5x)^2 + 2(x^2+5x) - 24 = 0$
Let $u = x^2 + 5x$.
$u^2 + 2u - 24 = 0$
$(u+6)(u-4) = 0$
$u = -6$ or $u = 4$
$x^2 + 5x = 6$ or $x^2 + 5x = 4$ Replacing u with $x^2 + 5x$
$x^2 + 5x + 6 = 0$ or $x^2 + 5x - 4 = 0$
$(x+3)(x+2) = 0$ or $x = \dfrac{-5 \pm \sqrt{5^2 - 4 \cdot 1 \cdot (-4)}}{2 \cdot 1}$
$x = -3$ or $x = -2$ or $x = \dfrac{-5 \pm \sqrt{41}}{2}$

The solutions are -3, -2, and $\dfrac{-5 \pm \sqrt{41}}{2}$.

61. $(125x^3 y^{-2} z^6)^{-2/3} =$
$(5^3)^{-2/3}(x^3)^{-2/3}(y^{-2})^{-2/3}(z^6)^{-2/3} =$
$5^{-2} x^{-2} y^{4/3} z^{-4} = \dfrac{1}{25} x^{-2} y^{4/3} z^{-4}$, or
$\dfrac{y^{4/3}}{25 x^2 z^4}$

63. Find the first coordinate of the point of intersection of $y_1 = \ln x$ and $y_2 = \log x$. The value of x for which the natural logarithm of x is the same as the common logarithm of x is 1.

65. a) 0.3770
b) -1.9617
c) 0.9036
d) -1.5318

67. $2^{2x} + 128 = 24 \cdot 2^x$
$2^{2x} - 24 \cdot 2^x + 128 = 0$
Let $u = 2^x$.
$u^2 - 24u + 128 = 0$
$(u-8)(u-16) = 0$
$u = 8$ or $u = 16$
$2^x = 8$ or $2^x = 16$ Replacing u with 2^x
$2^x = 2^3$ or $2^x = 2^4$
$x = 3$ or $x = 4$

The solutions are 3 and 4.

69. $8^x = 16^{3x+9}$
$(2^3)^x = (2^4)^{3x+9}$
$2^{3x} = 2^{12x+36}$
$3x = 12x + 36$
$-36 = 9x$
$-4 = x$

71. $\log_6 (\log_2 x) = 0$
$\log_2 x = 6^0$
$\log_2 x = 1$
$x = 2^1$
$x = 2$

73. $\log_5 \sqrt{x^2 - 9} = 1$
$\sqrt{x^2 - 9} = 5^1$
$x^2 - 9 = 25$ Squaring both sides
$x^2 = 34$
$x = \pm\sqrt{34}$

Both numbers check. The solutions are $\pm\sqrt{34}$.

75. $\log (\log x) = 5$ The base is 10.
$\log x = 10^5$
$\log x = 100{,}000$
$x = 10^{100{,}000}$

The number checks. The solution is $10^{100{,}000}$.

77. $\log x^2 = (\log x)^2$
$2 \log x = (\log x)^2$
$0 = (\log x)^2 - 2 \log x$
Let $u = \log x$.
$0 = u^2 - 2u$
$0 = u(u-2)$
$u = 0$ or $u = 2$
$\log x = 0$ or $\log x = 2$
$x = 10^0$ or $x = 10^2$
$x = 1$ or $x = 100$

Both numbers check. The solutions are 1 and 100.

79. $\log_a a^{x^2+4x} = 21$
$x^2 + 4x = 21$ Property 4
$x^2 + 4x - 21 = 0$
$(x+7)(x-3) = 0$
$x = -7$ or $x = 3$

Both numbers check. The solutions are $= -7$ and 3.

81. $3^{2x} - 8 \cdot 3^x + 15 = 0$
Let $u = 3^x$ and substitute.
$u^2 - 8u + 15 = 0$
$(u-5)(u-3) = 0$
$u = 5$ or $u = 3$
$3^x = 5$ or $3^x = 3$ Substituting 3^x for u
$\log 3^x = \log 5$ or $3^x = 3^1$
$x \log 3 = \log 5$ or $x = 1$
$x = \dfrac{\log 5}{\log 3}$ or $x = 1$, or
$x \approx 1.4650$ or $x = 1$

Both numbers check. Note that we can also express $\dfrac{\log 5}{\log 3}$ as $\log_3 5$ using the change-of-base formula.

Exercise Set 12.7

1. $L = 10 \cdot \log \dfrac{I}{I_0}$
$= 10 \cdot \log \dfrac{3.2 \times 10^{-3}}{10^{-12}}$ Substituting
$= 10 \cdot \log(3.2 \times 10^9)$
$\approx 10(9.5)$
≈ 95

The sound level is about 95 dB.

3. $L = 10 \cdot \log \dfrac{I}{I_0}$
$105 = 10 \cdot \log \dfrac{I}{10^{-12}}$
$10.5 = \log \dfrac{I}{10^{-12}}$
$10.5 = \log I - \log 10^{-12}$ $\quad (\log 10^a = a)$
$10.5 = \log I - (-12)$
$10.5 = \log I + 12$
$-1.5 = \log I$
$10^{-1.5} = I$ Converting to an exponential equation
$3.2 \times 10^{-2} \approx I$

The intensity of the sound is $10^{-1.5}$ W/m^2, or about 3.2×10^{-2} W/m^2.

5. $\text{pH} = -\log[H^+]$
$= -\log[1.6 \times 10^{-7}]$
$\approx -(-6.795880)$
≈ 6.8

The pH of milk is about 6.8.

7. $\text{pH} = -\log[H^+]$
$7.8 = -\log[H^+]$
$-7.8 = \log[H^+]$
$10^{-7.8} = [H^+]$
$1.58 \times 10^{-8} \approx [H^+]$

The hydrogen ion concentration is about 1.58×10^{-8} moles per liter.

9. $3{,}251{,}876 = 3251.876$ thousands
$w(P) = 0.37 \ln P + 0.05$
$w(3251.876) = 0.37 \ln 3251.876 + 0.05$
≈ 3.04 ft/sec

11. $311{,}121 = 311.121$ thousands
$w(P) = 0.37 \ln P + 0.05$
$w(311.121) = 0.37 \ln 311.121 + 0.05$
≈ 2.17 ft/sec

13. $N(t) = 3^t$

a) $N(5) = 3^5 = 243$ people

b) $6{,}200{,}000{,}000 = 3^t$
$\ln 6{,}200{,}000{,}000 = \ln 3^t$
$\ln 6{,}200{,}000{,}000 = t \ln 3$
$\dfrac{\ln 6{,}200{,}000{,}000}{\ln 3} = t$
$20.5 \approx t$

The acts of kindness will reach the entire world in about 20.5 months.

c) $2 = 3^t$
$\ln 2 = \ln 3^t$
$\ln 2 = t \ln 3$
$\dfrac{\ln 2}{\ln 3} = t$
$0.6 \approx t$

The doubling time is about 0.6 month.

15. a) In 2005, $t = 2005 - 2000$, or 5.
$C(5) = 11{,}054(1.06)^5 \approx \$14{,}793$

b) $21{,}000 = 11{,}054(1.06)^t$
$\dfrac{21{,}000}{11{,}054} = 1.06^t$
$\log \dfrac{21{,}000}{11{,}054} = \log 1.06^t$
$\log \dfrac{21{,}000}{11{,}054} = t \log 1.06$
$\dfrac{\log \dfrac{21{,}000}{11{,}054}}{\log 1.06} = t$
$11 \approx t$

The cost will be \$21,000 11 yr after 2000, or in 2011.

c) $22{,}108 = 11{,}054(1.06)^t$
$2 = (1.06)^t$
$\log 2 = \log(1.06)^t$
$\log 2 = t \log 1.06$
$\dfrac{\log 2}{\log 1.06} = t$
$11.9 \approx t$

The doubling time is about 11.9 years.

17. a) $P(t) = 6e^{0.015t}$, where $P(t)$ is in billions and t is the number of years after 1998.

b) In 2010, $t = 2010 - 1998$, or 12.
$P(12) = 6e^{0.015(12)} \approx 7.2$

In 2010, the population will be about 7.2 billion.

c)
$$10 = 6e^{0.015t}$$
$$\frac{5}{3} = e^{0.015t}$$
$$\ln\left(\frac{5}{3}\right) = \ln e^{0.015t}$$
$$\ln\left(\frac{5}{3}\right) = 0.015t$$
$$\frac{\ln\left(\frac{5}{3}\right)}{0.015} = t$$
$$34 \approx t$$

The population will be 10 billion 34 years after 1998, or in 2032.

d)
$$12 = 6e^{0.015t}$$
$$2 = e^{0.015t}$$
$$\ln 2 = \ln e^{0.015t}$$
$$\ln 2 = 0.015t$$
$$\frac{\ln 2}{0.015} = t$$
$$46.2 \approx t$$

The doubling time is about 46.2 yr.

19. a) $P(t) = P_0 e^{0.06t}$

b) To find the balance after one year, replace P_0 with 5000 and t with 1. We find $P(1)$:

$P(1) = 5000 e^{0.06(1)} \approx \$5309.18$

To find the balance after 2 years, replace P_0 with 5000 and t with 2. We find $P(2)$:

$P(2) = 5000 e^{0.06(2)} \approx \$5637.48$

To find the balance after 10 years, replace P_0 with 5000 and t with 10. We find $P(10)$: $P(10) = 5000 e^{0.06(10)} \approx \$9110.59$

c) To find the doubling time, replace P_0 with 5000 and $P(t)$ with 10,000 and solve for t.

$$10,000 = 5000 e^{0.06t}$$
$$2 = e^{0.06t}$$
$$\ln 2 = \ln e^{0.06t} \quad \text{Taking the natural logarithm on both sides}$$
$$\ln 2 = 0.06t \quad \text{Finding the logarithm of the base to a power}$$
$$\frac{\ln 2}{0.06} = t$$
$$11.6 \approx t$$

The investment will double in about 11.6 years.

21. a)
$$P(t) = P_0 e^{kt}$$
$$1,563,282 = 852,737 e^{k \cdot 10}$$
$$\frac{1,563,282}{852,737} = e^{10k}$$
$$\ln \frac{1,563,282}{852,737} = \ln e^{10k}$$
$$\ln \frac{1,563,282}{852,737} = 10k$$
$$\frac{\ln \frac{1,563,282}{852,737}}{10} = k$$
$$0.061 \approx k$$

The exponential growth rate is 0.061, or 6.1%.

The exponential growth function is $P(t) = 852,737 e^{0.061t}$, where t is the number of years after 1990.

b) In 2010, $t = 2010 - 1990$, or 20.
$$P(20) = 852,737 e^{0.061(20)}$$
$$= 852,737 e^{1.22}$$
$$\approx 2,888,380$$

In 2010, the population of Las Vegas will be about 2,888,380.

c)
$$8,000,000 = 852,737 e^{0.061t}$$
$$\frac{8,000,000}{852,737} = e^{0.061t}$$
$$\ln \frac{8,000,000}{852,737} = \ln e^{0.061t}$$
$$\ln \frac{8,000,000}{852,737} = 0.061t$$
$$\frac{\ln \frac{8,000,000}{852,737}}{0.061} = t$$
$$37 \approx t$$

The population will reach 8,000,000 about 37 yr after 1990, or in 2027.

23. If the scrolls had lost 22.3% of their carbon-14 from an initial amount P_0, then $77.7\%(P_0)$ is the amount present. To find the age t of the scrolls, we substitute $77.7\%(P_0)$, or $0.777P_0$, for $P(t)$ in the carbon-14 decay function and solve for t.

$$P(t) = P_0 e^{-0.00012t}$$
$$0.777 P_0 = P_0 e^{-0.00012t}$$
$$0.777 = e^{-0.00012t}$$
$$\ln 0.777 = \ln e^{-0.00012t}$$
$$-0.2523 \approx -0.00012t$$
$$t \approx \frac{-0.2523}{-0.00012} \approx 2103$$

The scrolls are about 2103 years old.

Exercise Set 12.7

25. The function $P(t) = P_0 e^{-kt}$, $k > 0$, can be used to model decay. For iodine-131, $k = 9.6\%$, or 0.096. To find the half-life we substitute 0.096 for k and $\frac{1}{2} P_0$ for $P(t)$, and solve for t.

$$\frac{1}{2} P_0 = P_0 e^{-0.096t}, \text{ or } \frac{1}{2} = e^{-0.096t}$$

$$\ln \frac{1}{2} = \ln e^{-0.096t} = -0.096t$$

$$t = \frac{\ln 0.5}{-0.096} \approx \frac{-0.6931}{-0.096} \approx 7.2 \text{ days}$$

27. The function $P(t) = P_0 e^{-kt}$, $k > 0$, can be used to model decay. We substitute $\frac{1}{2} P_0$ for $P(t)$ and 1 for t and solve for the decay rate k.

$$\frac{1}{2} P_0 = P_0 e^{-k \cdot 1}$$

$$\frac{1}{2} = e^{-k}$$

$$\ln \frac{1}{2} = \ln e^{-k}$$

$$-0.693 \approx -k$$

$$0.693 \approx k$$

The decay rate is 0.693, or 69.3% per year.

29. a) We use the exponential decay equation $W(t) = W_0 e^{-kt}$, where t is the number of years after 1996 and $W(t)$ is in millions of tons. In 1996, at $t = 0$, 17.5 million tons of yard waste were discarded. We substitute 17.5 for W_0.

$$W(t) = 17.5 e^{-kt}.$$

To find the exponential decay rate k, observe that 2 years after 1996, in 1998, 14.5 million tons of yard waste were discarded. We substitute 2 for t and 14.5 for $W(t)$.

$$14.5 = 17.5 e^{-k \cdot 2}$$

$$0.8286 \approx e^{-2k}$$

$$\ln 0.8286 \approx \ln e^{-2k}$$

$$\ln 0.8286 \approx -2k$$

$$\frac{\ln 0.8286}{-2} \approx k$$

$$0.094 \approx k$$

Then we have $W(t) = 17.5 e^{-0.094t}$, where t is the number of years after 1996 and $W(t)$ is in millions of tons.

b) In 2010, $t = 2010 - 1996 = 14$.

$$W(14) = 17.5 e^{-0.094(14)}$$

$$= 17.5 e^{-1.316}$$

$$\approx 4.7$$

In 2010, about 4.7 million tons of yard waste were discarded.

c) 1 ton is equivalent to 0.000001 million tons.

$$0.000001 = 17.5 e^{-0.094t}$$

$$5.71 \times 10^{-8} \approx e^{-0.094t}$$

$$\ln(5.71 \times 10^{-8}) \approx \ln e^{-0.094t}$$

$$\ln(5.71 \times 10^{-8}) \approx -0.094t$$

$$\frac{\ln(5.71 \times 10^{-8})}{-0.094} \approx t$$

$$177 \approx t$$

Only one ton of yard waste will be discarded about 177 years after 1996, or in 2173.

31. a) We start with the exponential growth equation

$$V(t) = V_0 e^{kt}, \text{ where } t \text{ is the number of years after 1991.}$$

Substituting 451,000 for V_0, we have

$$V(t) = 451,000 e^{kt}.$$

To find the exponential growth rate k, observe that the card sold for \$640,500 in 1996, or 5 years after 1991. We substitute and solve for k.

$$V(5) = 451,000 e^{k \cdot 5}$$

$$640,500 = 451,000 e^{5k}$$

$$1.42 = e^{5k}$$

$$\ln 1.42 = \ln e^{5k}$$

$$\ln 1.42 = 5k$$

$$\frac{\ln 1.42}{5} = k$$

$$0.07 \approx k$$

Thus the exponential growth function is $V(t) = 451,000 e^{0.07t}$, where t is the number of years after 1991.

b) In 2005, $t = 2005 - 1991$, or 14.

$$V(14) = 451,000 e^{0.07(14)} \approx 1,201,670$$

The card's value in 2005 will be about \$1,201,670.

c) Substitute \$902,000 for $V(t)$ and solve for t.

$$902,000 = 451,000 e^{0.07t}$$

$$2 = e^{0.07t}$$

$$\ln 2 = \ln e^{0.07t}$$

$$\ln 2 = 0.07t$$

$$\frac{\ln 2}{0.07} = t$$

$$9.9 \approx t$$

The doubling time is about 9.9 years.

d) Substitute $1,000,000 for $V(t)$ and solve for t.
$$1,000,000 = 451,000\, e^{0.07t}$$
$$2.217 \approx e^{0.07t}$$
$$\ln 2.217 \approx \ln e^{0.07t}$$
$$\ln 2.217 \approx 0.07t$$
$$\frac{\ln 2.217}{0.07} \approx t$$
$$11.4 \approx t$$

The value of the card will be $1,000,000 in 1991+11, or 2002.

e) In 2001, $t = 2001 - 1991$, or 10.
$$V(10) = 451,000 e^{0.07(10)}$$
$$= 451,000 e^{0.7}$$
$$\approx 908,202$$

The function estimates that the card's value in 2001 would be about $908,202. According to this, the card would not be a good buy at $1.1 million.

33. a)
$$P(t) = P_0 e^{-kt}$$
$$2,242,798 = 2,394,811 e^{-k \cdot 10}$$
$$\frac{2,242,798}{2,394,811} = e^{-10k}$$
$$\ln \frac{2,242,798}{2,394,811} = \ln e^{-10k}$$
$$\ln \frac{2,242,798}{2,394,811} = -10k$$
$$\frac{\ln \frac{2,242,798}{2,394,811}}{-10} = k$$
$$0.007 \approx k$$

The exponential decay rate is 0.007, or 0.7%.
The exponential decay function is $P(t) = 2,394,811 e^{-0.007t}$, where t is the number of years after 1990.

b) In 2010, $t = 2010 - 1990$, or 20.
$$P(20) = 2,394,811 e^{-0.007(20)}$$
$$= 2,394,811 e^{-0.14}$$
$$\approx 2,081,949$$

In 2010, the population of Pittsburgh will be about 2,081,949.

c)
$$1,000,000 = 2,394,811 e^{-0.007t}$$
$$\frac{1,000,000}{2,394,811} = e^{-0.007t}$$
$$\ln \frac{1,000,000}{2,394,811} = \ln e^{-0.007t}$$
$$\ln \frac{1,000,000}{2,394,811} = -0.007t$$
$$\frac{\ln \frac{1,000,000}{2,394,811}}{-0.007} = t$$
$$125 \approx t$$

The population will decline to 1 million about 125 yr after 1990, or in 2115.

35. Discussion and Writing Exercise

37. $i^{46} = (i^2)^{23} = (-1)^{23} = -1$

39. $i^{53} = (i^2)^{26} \cdot i = (-1)^{26} \cdot i = i$

41. $i^{14} + i^{15} = (i^2)^7 + (i^2)^7 \cdot i = (-1)^7 + (-1)^7 \cdot i = -1 - i$

43.
$$\frac{8-i}{8+i} = \frac{8-i}{8+i} \cdot \frac{8-i}{8-i}$$
$$= \frac{64 - 16i + i^2}{64 - i^2}$$
$$= \frac{64 - 16i - 1}{64 - (-1)}$$
$$= \frac{63 - 16i}{65}$$
$$= \frac{63}{65} - \frac{16}{65}i$$

45. $(5 - 4i)(5 + 4i) = 25 - 16i^2 = 25 + 16 = 41$

47. $-0.937, 1.078, 58.770$

49. $-0.767, 2, 4$

51. We will use the exponential growth equation $V(t) = V_0 e^{kt}$, where t is the number of years after 2001 and $V(t)$ is in millions of dollars. We substitute 21 for $V(t)$, 0.05 for k, and 9 for t and solve for V_0.
$$21 = V_0 e^{0.05(9)}$$
$$21 = V_0 e^{0.45}$$
$$\frac{21}{e^{0.45}} = V_0$$
$$13.4 \approx V_0$$

George Steinbrenner needs to invest $13.4 million at 5% interest compounded continuously in order to have $21 million to pay Derek Jeter in 2010.